Optimization and Learning via Stochastic Gradient Search

PRINCETON SERIES IN APPLIED MATHEMATICS

The Princeton Series in Applied Mathematics publishes high quality advanced texts and monographs in all areas of applied mathematics. Books include those of a theoretical and general nature as well as those dealing with the mathematics of specific applications areas and real-world situations.

Series Editors: Ingrid Daubechies (Princeton University); Weinan E (Princeton University); Jan Karel Lenstra (Eindhoven University); Endre Süli (University of Oxford)

For a full list of titles in the series, go to https://press.princeton.edu/series/princeton-series-in-applied-mathematics

Books in the series:

Optimization and Learning via Stochastic Gradient Search, *Felisa Vázquez-Abad and Bernd Heidergott*

Totally Nonnegative Matrices, *Shaun M. Fallat and Charles R. Johnson*

PDE Control of String-Actuated Motion, *Ji Wang and Miroslav Krstic*

Delay-Adaptive Linear Control, *Yang Zhu and Miroslav Krstic*

Statistical Inference via Convex Optimization, *Anatoli Juditsky and Arkadi Nemirovski*

A Dynamical Systems Theory of Thermodynamics, *Wassim M. Haddad*

Formal Verification of Control System Software, *Pierre-Loïc Garoche*

Rays, Waves, and Scattering: Topics in Classical Mathematical Physics, *John Adam*

Positive Definite Matrices, *Rajendra Bhatia*

Mathematical Methods in Elasticity Imaging, *Habib Ammari, Elie Bretin, Josselin Garnier, Hyeonbae Kang, Hyundae Lee, and Abdul Wahab*

Hidden Markov Processes: Theory and Applications to Biology, *M. Vidyasagar*

Topics in Quaternion Linear Algebra, *Leiba Rodman*

Mathematical Analysis of Deterministic and Stochastic Problems in Complex Media Electromagnetics, *G. F. Roach, I. G. Stratis, and A. N. Yannacopoulos*

Stability and Control of Large-Scale Dynamical Systems: A Vector Dissipative Systems Approach, *Wassim M. Haddad and Sergey G. Nersesov*

Matrix Completions, Moments, and Sums of Hermitian Squares, *Mihály Bakonyi and Hugo J. Woerdeman*

Modern Anti-windup Synthesis: Control Augmentation for Actuator Saturation, *Luca Zaccarian and Andrew R. Teel*

Graph Theoretic Methods in Multiagent Networks, *Mehran Mesbahi and Magnus Egerstedt*

Matrices, Moments and Quadrature with Applications, *Gene H. Golub and Gérard Meurant*

Optimization and Learning via Stochastic Gradient Search

Felisa Vázquez-Abad
Bernd Heidergott

PRINCETON UNIVERSITY PRESS
PRINCETON AND OXFORD

Copyright © 2025 by Felisa Vázquez-Abad and Bernd Heidergott

Princeton University Press is committed to the protection of copyright and the intellectual property our authors entrust to us. Copyright promotes the progress and integrity of knowledge created by humans. By engaging with an authorized copy of this work, you are supporting creators and the global exchange of ideas. As this work is protected by copyright, any reproduction or distribution of it in any form for any purpose requires permission; permission requests should be sent to permissions@press.princeton.edu. Ingestion of any IP for any AI purposes is strictly prohibited.

Published by Princeton University Press
41 William Street, Princeton, New Jersey 08540
99 Banbury Road, Oxford OX2 6JX

press.princeton.edu

GPSR Authorized Representative: Easy Access System Europe - Mustamäe tee 50, 10621 Tallinn, Estonia, gpsr.requests@easproject.com

All Rights Reserved

Library of Congress Control Number: 2025937538

ISBN 9780691245867
ISBN (e-book) 9780691245874

British Library Cataloging-in-Publication Data is available

Editorial: Diana Gillooly, Whitney Rauenhorst
Production Editorial: Elizabeth Byrd
Jacket: Heather Hansen
Production: Lauren Reese
Publicity: William Pagdatoon

Cover image: Adapted by Heather Hansen from the authors' schematic of Markovian dynamics

10 9 8 7 6 5 4 3 2 1

Contents

Preface   ix

I   Theory of Stochastic Optimization and Learning   1

1   Gradient-Based Methods for Deterministic Continuous Optimization   3
- 1.1   Unconstrained Optimization   3
- 1.2   Numerical Methods for Unconstrained Optimization   9
- 1.3   Constrained Optimization   18
- 1.4   Numerical Methods for Constrained Optimization   23
- 1.5   Practical Considerations   33
- 1.6   Exercises   36

2   The Iterative Method Seen as an Ordinary Differential Equation   39
- 2.1   Motivation   39
- 2.2   Stability of ODEs   42
- 2.3   Projected ODEs   48
- 2.4   On Boundedness of the Trajectories of an ODE   51
- 2.5   ODE Limit of Recursive Algorithms   53
- 2.6   The ODE Method for Optimization and Learning   61
- 2.7   Specific Algorithms for Constrained Optimization   66
- 2.8   Practical Considerations   70
- 2.9   Exercises   71

3   Stochastic Approximation: An Introduction   76
- 3.1   Motivation   76
- 3.2   Root Finding, Statistical Fitting, and Target Tracking   79
- 3.3   A Taxonomy for Stochastic Approximation   83
- 3.4   Overview on Stochastic Approximation   91
- 3.5   The Sample Average Approach   91
- 3.6   Practical Considerations   93
- 3.7   Exercises   96

4   Stochastic Approximation: The Static Model   99
- 4.1   Martingale Difference Noise Model   99
- 4.2   Analysis of Decreasing Stepsize SA   101
- 4.3   Analysis of Constant Stepsize SA   109
- 4.4   Practical Considerations   114
- 4.5   Exercises   115

5 Stochastic Approximation: Markovian Dynamics — 118
- 5.1 Long-Term Stationary Dynamics: Markovian Model — 118
- 5.2 Analysis of the Decreasing Stepsize SA — 123
- 5.3 Analysis of the Constant Stepsize SA — 129
- 5.4 Practical Considerations — 141
- 5.5 Exercises — 141

6 Asymptotic Efficiency — 143
- 6.1 Motivation — 143
- 6.2 Functional CLT — 144
- 6.3 Asymptotic Efficiency — 154
- 6.4 Practical Considerations — 157
- 6.5 Exercises — 158

II Gradient Estimation — 163

7 A Primer for Gradient Estimation — 165
- 7.1 Motivation — 165
- 7.2 One-Dimensional Distributions — 166
- 7.3 A Taxonomy of Gradient Estimation — 186
- 7.4 Practical Considerations — 194
- 7.5 Exercises — 195

8 Gradient Estimation, Finite Horizon — 197
- 8.1 Perturbation Analysis: IPA and SPA — 197
- 8.2 Distributional Approach: Basic Results and Techniques — 213
- 8.3 The Score Function Method — 215
- 8.4 Measure-Valued Differentiation — 227
- 8.5 Practical Considerations — 244
- 8.6 Exercises — 245

9 Gradient Estimation, Markovian Dynamics — 252
- 9.1 The Infinite Horizon Problem — 252
- 9.2 The Random Horizon Problem — 259
- 9.3 The Stationary Problem — 266
- 9.4 Practical Considerations — 272
- 9.5 Exercises — 272

III Selected Topics in Stochastic Approximation — 275

10 Applications of Stochastic Approximation to Inventory Problems — 277
- 10.1 Optimization Using MVD Gradient Estimation — 277
- 10.2 Model Fitting (An IPA Application) — 285
- 10.3 Variations of the Model — 287

11 Pseudo-Gradient Methods — 295
- 11.1 Simultaneous Perturbation Stochastic Approximation — 295
- 11.2 Gaussian Smoothed Functional Approximation — 300
- 11.3 Feasible Perturbed Parameter Values for SPSA and GSFA — 301

12 IPA for Discrete Event Systems — 303
- 12.1 Discrete Event Systems — 303
- 12.2 The Commuting Condition — 307
- 12.3 Unbiasedness of IPA — 309
- 12.4 Sufficient Conditions for the Event Condition — 314
- 12.5 Concluding Remarks — 315

13 A Markov Operator Approach — 316
- 13.1 The Finite Horizon Problem — 321
- 13.2 The Random Horizon Problem — 326
- 13.3 The Stationary Problem — 334
- 13.4 The Infinite Horizon Problem — 336

14 Stochastic Approximation in Statistics — 339
- 14.1 The Score Function in Statistics — 339
- 14.2 Generalized Method of Moments — 342

15 Stochastic Gradient Techniques in AI and Machine Learning — 344
- 15.1 Gradient-Based Approaches — 344
- 15.2 Q-Learning and Reinforcement Learning — 347

IV Appendixes — 353

A Analysis and Linear Algebra — 355
- A.1 Convexity — 355
- A.2 Multidimensional Derivatives — 355
- A.3 Geometric Interpretation of the Gradient — 357
- A.4 Weierstrass Theorem — 359
- A.5 Positive and Negative Definite Matrices — 360
- A.6 Normed Spaces and Equicontinuity — 360
- A.7 Lipschitz and Uniform Continuity — 361
- A.8 Taylor Series Expansions — 362
- A.9 L'Hôpital's Rule — 363
- A.10 Cesàro Limits — 364

B Probability Theory — 365
- B.1 Information Structure — 365
- B.2 (Probability) Measures — 366
- B.3 Expectations and Conditioning — 368
- B.4 Convergence of Random Sequences — 369
- B.5 v-Norm Convergence of Measures — 373
- B.6 Martingale Processes — 374
- B.7 Regenerative Processes — 377

C Markov Chains — 378
- C.1 Harris Recurrence — 378
- C.2 Normed Ergodicity and Central Limit Theorem — 379
- C.3 The Poisson Equation for Markov Chains in Discrete Time — 381

D	**Confidence Intervals**	**384**
	D.1 Independent and Identically Distributed Random Variables	384
	D.2 Stationary Processes	388
	D.3 Markov Chains: Long-Term and Stationary Estimation	390

Bibliography   **393**

Index   **415**

Preface

This monograph covers stochastic optimization and learning, mostly focusing on the methodology of stochastic approximations and gradient estimation techniques. We focus on continuous parameter problems and do not discuss discrete optimization. Our approach is theoretical, but emphasis is given to algorithms that implement the methods. Part I covers many important applications of stochastic approximations for optimization and learning that do not require the use of gradient estimation. Part II introduces gradient estimation for stochastic processes, which then significantly broadens the range of application of stochastic approximations, some of which are described in Part III.

Stochastic approximation has been intensively covered in the past decades. For general treatments of stochastic approximation, we refer to Wasan [323], Polyak and Tsypkin [250], Ljung [205], Kushner and Clark [193], Polyak [248], Bertsekas and Tsitsiklis [32, 33], Benaïm [24, 25], Pflug [245], Kushner and Yin [197], Meyn [216], Borkar [39], and Nemirovski et al. [226]. Gradient estimation has been included in Ho and Cao [159], Cao [58, 60], Fu [101, 102], Kroese et al. [191], Glasserman [114, 119], Pflug [245], and Rubinstein with various coauthors [267, 268, 270].

This monograph differs from the aforementioned standard references in structure, choice of topics, and pedagogical format. We address an audience of graduate students and researchers. We have carefully worked out full examples of applications and have included a set of exercises and problems that aim to help the instructors as well as the students as they go through the material. Mathematical theory, the art of modeling, and numerical algorithms complement each other when solving real problems. We follow this principle as the underlying philosophy of the book, aiming to strengthen both the readers' theoretical and practical skills, neither of which should dominate the other. The book has three main parts:

- Part I provides a thorough treatment of first-order methods in optimization with an emphasis on stochastic approximation (SA). The key results are presented and proven, and many illustrating examples are provided. Part I follows the standard approach in SA by studying recursive learning algorithms via ordinary differential equations (ODEs). We take great care in Part I to present the theory and also the numerical methods. Guidelines on how to apply the algorithms in practice are included in a final section of each chapter called "Practical Considerations," which addresses in detail typical trade-offs one faces in practice.
- Part II is a self-contained treatment of gradient estimation and summarizes the research done in this area in the last three decades. Many examples are provided, and the available theory is explained in depth. It gives a unique, exhaustive presentation of gradient estimation of expectations of functionals of stochastic processes. We build the models from the "static" case of one random variable to stochastic processes, including models with random stopping times and long-term stationary expectations (called sometimes the "infinite horizon" model in simulation methodology). At a first reading, sections indicated by an asterisk (∗) may be omitted.

- Part III blends the algorithmic results of Part I with the gradient estimators provided in Part II, and presents complete SA application stories (from problem formulation, to choice of algorithm and gradient estimator, to output analysis). Moreover, applications of SA in other field of research such as statistics and machine learning are discussed, and some advanced material on infinitesimal perturbation analysis and measure-valued differentiation is provided.

Historical remarks. This book has evolved over the years to its present form. Parts of the material were developed from 1996 to 2004 as part of a course in stochastic modeling and simulation that Felisa taught at the computer science department at the University of Montreal, with students in various joint programs in computer science, management, industrial engineering, finance, and mathematics. She taught a preliminary version of this subject to graduate and postdoctoral students in electrical engineering at the University of Melbourne in 1999, and then as an honours course for students in mathematics and statistics at the University of Melbourne from 2004 to 2008. The next version of the material was developed in coauthorship with Bernd, expanding the presentation of first-order optimization in Part I and of gradient estimation in Part II and Part III. Felisa lectured with this version in a doctoral course for engineering students at the University of Vienna in 2009. Since 2010, Felisa teaches the course every year for the Graduate Center and Hunter College of the City University of New York, for doctoral students in computer science, applied mathematics, physics, economics, chemistry, and data science. In 2019 and 2020 Felisa also taught the course at the postgraduate level for the department of computer and information systems at the University of Melbourne. Since 2012 Bernd, together with his colleague Ad Ridder, have been using the material each year for a course in the master's program in econometrics and operations research at the Vrije Universiteit in Amsterdam, mostly for economics and operations research students, and since 2021 for an advanced master's level course at the Dutch Network on the Mathematics of Operations Research, mostly for students in operations research, applied mathematics, and computer science. In addition, Bernd teaches the material of Part I together with Chapter 11 for an undergraduate level course at Peking University to students with an engineering background.

Throughout the years, students have expressed in their evaluations that they have appreciated the learning process just as much as the acquired knowledge, particularly through the research projects. We deeply thank the students for their feedback and their contribution to this work, which would not have been completed without their criticism and comments.

The field of optimization and learning for stochastic problems covers a wide range of methods, and an overview of even the most prominent methods is beyond this monograph, which focuses on stochastic optimization via stochastic approximations. Our focus includes optimization by simulation (synthetic data) as well as real-time optimization (streaming data). Methods such as the cross-entropy method, and evolutionary algorithms are not covered in this monograph. Finally, global optimization methods, like the response surface method or simulated annealing, are also not covered in this monograph. For a general overview in simulation-based optimization, we refer to [73], [102], [209], [245], [283], [284], and [295].

For the instructors. There are several possible "road maps" for one, two, or three courses depending on the background of the students and the motivation of the instructors.

We originally designed the book chapters to follow the order in which we actually lecture the material. Roughly, each chapter may be lectured in one week (three to four hours). In practice, we add some lectures for revision of concepts such as probability theory and Markov chains (included in Appendix B). For these topics we have developed teaching materials that students can watch on their own, in the form of posted YouTube videos that are available to the instructors.

Chapters 1 and 2 are included at the start of the monograph as preliminary material needed to develop the main methodology of stochastic gradient-based optimization. This decision was greatly motivated by our computer science and business courses, where most students don't have the necessary knowledge in these topics. Chapter 1 is a summary of the salient results for gradient-based deterministic optimization methods. When teaching a course with an optimization prerequisite the instructor may wish to omit lecturing on Chapter 1. Similarly, Chapter 2 (except for Section 2.6) presents basic concepts of dynamical systems that students may not have already studied. Otherwise the instructor may choose to omit lecturing on Chapter 2. The special section Section 2.6 in this chapter is an original view that blends the theory of deterministic optimization of Chapter 1 with the analysis of algorithms via ODEs, and it is a foundation of the work to follow. We recommend that this section be included in the lectures.

Chapter 3 (an overview of stochastic approximation) and Chapter 7 (an overview of gradient estimation) present the overall ideas of the topics without advanced proofs. We have included pseudocode and computer results whenever possible within examples that can be understood even at an undergraduate level. These chapters could be used in other courses such as simulation or algorithms courses. Depending on the level of the students, these chapters may be left as independent reading material before the following lectures. Chapter 4 presents the theory and methods of stochastic approximation for the static case, also called the finite horizon case. In parallel, Chapter 8 presents the theory and methods of gradient estimation for the finite horizon case. We have made a special effort to include in Chapter 8 examples where gradient estimation is implemented in stochastic approximations that follow the models of Chapter 4. Instructors may choose to lecture this material following this order (thus zig-zagging chapters from Part I and II). Similarly, Chapters 5 and 9 present the theory and methods of stochastic approximation and gradient estimation for dynamic systems, including infinite horizon problems and random horizon problems. These chapters require more advanced knowledge of mathematical analysis and may be omitted in lower level courses.

A course on first-order methods in optimization can be designed by combining Part I with Chapter 11. This combination has been proven to be successful for computer scientists, operations researchers, and applied mathematicians in providing a thorough understanding of stochastic approximation in combination with versatile numerical efficient pseudo-gradient methods. Part II is a welcome extension to any Monte Carlo simulation course as it provides advanced simulation analytics. The first chapter of Part III explains how the theory developed in the book is used for solving real-life optimization problems.

We will provide instructors with materials for teaching, including slides, videos, and complete solutions of the exercises. We also have a website for the course following the format that Felisa Vázquez-Abad has been following. Video clips with supportive material are available at the YouTube channel @felisavazquez-abad9893.

In particular, the playlist on basic probability introduces the concepts and notation used in this book: https://www.youtube.com/playlist?list=PL33ylu0JIUoTJKF2DmvLgq IZqUdVCn-ak. We find it useful for students to watch before Chapter 3. As well, the playlist on Markov chains introduces important concepts and convergence results that are useful for Chapter 5: https://www.youtube.com/playlist?list=PL33ylu0JIUoQg3aZCIZqu5jErmz _52XjL.

Notation. Throughout this monograph, we assume that random variables are defined on a common probability space $(\Omega, \mathfrak{F}, \mathbb{P})$. Furthermore, we assume that $\mathfrak{F}$ contains all null sets with respect to $\mathbb{P}$. By convention, we equip discrete spaces with the discrete topology and the real numbers with the usual topology. Product spaces are equipped with the product topology and, unless stated otherwise, measurable spaces are equipped with the corresponding Borel σ-fields. If not stated otherwise, random variables are real-valued and, in line with the aforementioned conventions, measurable mappings from $(\Omega, \mathfrak{F}, \mathbb{P})$ onto $(\mathbb{R}, \mathcal{B})$, with $\mathcal{B}$ denoting the Borel σ-field on $\mathbb{R}$. Expectation of a random variable X with respect to $\mathbb{P}$ is denoted by $\mathbb{E}$, i.e., we write $\mathbb{E}[X] = \int_\Omega X(\omega) \mathbb{P}(d\omega)$. To simplify notation, we suppress the explicit dependency on ω when this causes no confusion. We use the symbol "$\sim$" to relate random variables and their corresponding distribution, i.e., we write $X \sim F$ if X has cumulative distribution function F. We use this notation in a similar way for measures. Equality in distribution of two random variables, say, X and Y is denoted by $X \stackrel{d}{=} Y$. We use the following abbreviations throughout: cdf for cumulative distribution function, pdf for probability density function, and iid for "independently identically distributed." We also use w.p.1 for "with probability one" and a.s. for "almost surely." We denote the normal distribution with mean μ and variance σ^2 by $\mathcal{N}(\mu, \sigma^2)$ and occasionally by "normal(μ, σ^2)."

We denote the n-times continuously differentiable mappings from $\mathbb{R}^d$ to $\mathbb{R}$ by C^n. For $J \in C^1$, we denote the gradient of $J(\cdot)$ by $\nabla J(\cdot)$, and for $J \in C^2$, we denote the Hessian of $J(\cdot)$ by $HJ(\cdot) = \nabla^2 J(\cdot)$. Following standard notation, vectors in $\mathbb{R}^d$ are *column* vectors. For $x \in \mathbb{R}^d$, we denote the i-th element of x by x_i. In case of a sequence of vectors $\{x_n\}$, with $x_n \in \mathbb{R}^d$, we denote the i-th element of x by $x_{n,i}$. The gradient is a *row* vector with components $\partial/\partial \theta_k, k=1,\ldots d$. The Hessian is a $d \times d$ matrix with (i,j)-components $\partial^2/\partial \theta_i \partial \theta_j$. For a vector $v \in \mathbb{R}^d$ we write $v \geq 0$ if $v_i \geq 0$ for all components $i = 1, \ldots, d$. We will use the Landau symbol $O(\theta)$ for a mapping $J : \mathbb{R}^d \mapsto \mathbb{R}$ such that $|J(\theta)| \leq \alpha ||\theta||$, for some finite constant α, where $||\cdot||$ denotes Euclidean norm.

Acknowledgments. In addition to the invaluable contribution of the students who have followed the progress of this book, we wish to thank our colleagues for the many fruitful discussions on the topic of this monograph. Specials thanks go out to Guy Cohen, Chris Franssen, Michael C. Fu, Vikram Krishnamurthy, Harold J. Kushner, Haralambie Leahu, Ad Ridder, and Georg Pflug.

A last word before we begin. No book is perfect, and ours is no exception, when *Titivillus* lurks around scribes. We welcome comments that will help improve the material for future revisions, extensions, and updates.

Part I

Theory of Stochastic Optimization and Learning

Chapter One

Gradient-Based Methods for Deterministic Continuous Optimization

This chapter presents a summary of salient results in deterministic optimization, particularly focusing on numerical methods. For basic definitions, and results we refer to standard textbooks.

1.1 UNCONSTRAINED OPTIMIZATION

Consider a cost function $J(\theta)$, with $J: \Theta \subseteq \mathbb{R}^d \mapsto \mathbb{R}$, where θ is a decision vector. Throughout this monograph, we seek to the find the minimum of $J(\theta)$ for $\theta \in \Theta$. As is standard in the literature, we are not only interested in the value of the global minimum (if it exists) but also its location, i.e., we seek the solution θ^* to the problem

$$\arg\min_{\theta \in \mathbb{R}^d} J(\theta). \tag{1.1}$$

In the case that the global minimum is attained at several locations, θ^* is one of these locations. In the case that $J(\cdot)$ is an (affine) linear mapping, the above optimization problem is called a *linear problem* and it can be addressed with methods from the theory of linear optimization. See, for example, [84, 85, 224, 235] for details. In the case that $J(\cdot)$ is a general "smooth" continuous real-valued function, the above problem is called a *non-linear problem* and it is referred to as an NLP. The theory presented in this monograph is devoted to the study of NLPs. It is worth noting that while the results presented here can also be applied to linear problems, there are often more efficient methods available for linear problems exploiting the linear nature of the problem.

We assume that $\mathbb{R}^d$ is equipped with a norm denoted by $||\cdot||$. Most results presented in the following are independent of the choice of $||\cdot||$. Occasionally, we will work with the Euclidean norm on $\mathbb{R}^d$ given by

$$||x|| = \sqrt{x_1^2 + \ldots + x_d^2},$$

and when results only hold for this particular norm it will be stated in the text.

A particular class of applications arises when an input data vector x and corresponding output data vector $h(x)$ is available. Letting $f(\theta, x)$ denote some parametrized mapping proposed for replacing the unknown mapping $h(x)$, considering

$$J(\theta, x) = ||f(\theta, x) - h(x)||^2$$

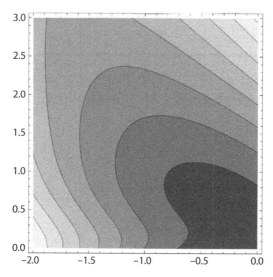

Figure 1.1. Plot showing various level curves for the function $x^2 + (x+2)y + 1/(y^2 + 0.5)$.

and solving (1.1) for given x, yields then the best fit to the output. This is called *supervised learning* in the literature. In this monograph, we will discuss classical optimization as well as learning applications.

Definition 1.1. The *level sets* of a function $J: \mathbb{R}^d \to \mathbb{R}$ are defined for every level $\alpha \in \mathbb{R}$ as:

$$\mathcal{L}_\alpha(J) = \{\theta \in \mathbb{R}^d : J(\theta) \le \alpha\}.$$

When no confusion arises, the notation will be simplified to $\mathcal{L}_\alpha$.

Notation. Denote the n-times continuously differentiable mappings from $\mathbb{R}^d$ to $\mathbb{R}$ by C^n. For $J \in C^1$, we denote the gradient of $J(\cdot)$ by $\nabla J(\cdot)$, and for $J \in C^2$, we denote the Hessian of $J(\cdot)$ by $HJ(\cdot) = \nabla^2 J(\cdot)$. Following standard notation, vectors in $\mathbb{R}^d$ are *column* vectors. For $x \in \mathbb{R}^d$, we denote the i-th element of x by x_i. In case of a sequence of vectors $\{x_n\}$, with $x_n \in \mathbb{R}^d$, we denote the i-th element of x by $x_{n,i}$. The gradient is a *row* vector with components $\partial/\partial\theta_k, k=1,\ldots d$. The Hessian is a $d \times d$ matrix with (i,j)-components $\partial^2/\partial\theta_i\partial\theta_j$. For a vector $v \in \mathbb{R}^d$ we write $v \ge 0$ if $v_i \ge 0$ for all components $i=1,\ldots,d$.

A matrix $B \in \mathbb{R}^{d\times d}$ is negative (positive) definite if $v^\top B v < (>) 0$ for all $v \in \mathbb{R}^d$ with $v \ne 0$, where $v^\top$ denotes the transpose of v. It is called "semi"-definite if the strict inequality equality "<" is replaced by inequality "$\le$." The notation $B < (>) 0$ is often used. A square matrix B is called symmetric if $B = B^\top$. For symmetric matrices the following characterization of positive definiteness exists: if B is symmetric, then $B > 0$ if and only if all its eigenvalues are strictly positive.

Remark 1.1. The visual interpretation of the gradient of a function will be very useful in the rest of this book. Refer to Figure 1.1. This is a "topographical" visualization of a two-dimensional function, where the shades of gray indicate height. Each of the level sets defines a boundary (in the example, they are ellipses). The gradient of the function (in this case, $x^2 + 2y^2$) records the rate of growth of the function along each of the axes. Now, take any point on a level set (refer to Figure 1.2). Because the function does not change *along* this curve, then necessarily the gradient ∇J must point *perpendicular* to the curve (i.e., the

GRADIENT-BASED METHODS

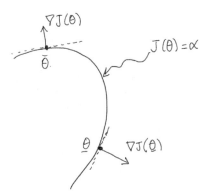

Figure 1.2. Illustration of the gradient of a convex function at two points $\bar{\theta}$ and $\underline{\theta}$.

projection of the gradient on the curve is zero). For the example, it points outward in the direction of growth.

Definition 1.2. A function $J: \mathbb{R}^d \to \mathbb{R}$ is called *concave (convex)* if for all $x, y \in \mathbb{R}^d$ and $\alpha \in (0, 1)$,
$$J(\alpha x + (1-\alpha)y) \geq (\leq) \alpha J(x) + (1-\alpha)J(y). \tag{1.2}$$

Strict concavity (convexity) is obtained when the above inequalities are strict. For $J \in C^2$, an equivalent condition is that the Hessian of the function be negative (positive) semi-definite: $\nabla^2 J(\cdot) \leq (\geq) 0$, and strict concavity (convexity) follow when the Hessian is negative (positive) definite throughout the domain of J.

Definition 1.3. A point $\theta^* \in \mathbb{R}^d$ can be characterized as follows:

- If $J(\theta^*) \leq J(\theta)$, for all $\theta \in \mathbb{R}^d$, then θ^* is a called *global* minimum. It is called a *local minimum* if there is a $\rho > 0$ such that $\|\theta - \theta^*\| \leq \rho$ implies $J(\theta) \geq J(\theta^*)$.
- If $J(\theta^*) \geq J(\theta)$, for all $\theta \in \mathbb{R}^d$, then θ^* is called a *global* maximum. It is called a *local maximum* if there is a $\rho > 0$ such that $\|\theta - \theta^*\| \leq \rho$ implies $J(\theta) \leq J(\theta^*)$.
- If the function may increase or decrease in a small neighborhood of the point, depending on the direction of motion, then $\theta^* \in \mathbb{R}^d$ is a *saddle point*.

If, in the definition of a maximum (respectively, minimum), the inequality $J(\theta^*) \leq J(\theta)$ (respectively, $J(\theta^*) \geq J(\theta)$) can be replaced by a strict inequality, then we say that the maximum (respectively, minimum) is *strict*.

Let $\alpha^* \stackrel{\text{def}}{=} \min J(\theta)$, with $\alpha^* = J(\theta^*)$. Then we say that $J(\theta)$ has a proper minimum, and we call α^* the *value of the minimum* and θ^* the *location of the minimum* (the concepts are defined for maxima analogously). Consider the mapping $J(\theta) = e^{-\theta}$ for $\theta \in \mathbb{R}$. For this function we have $0 = \alpha^* = \inf_\theta J(\theta)$ but there exists no value θ so that $J(\theta)$ attains 0. In this case, we will say that $J(\theta)$ has 0 as an improper minimum. Even at this early stage, it is conceivable that any gradient-based search algorithm will run into (numerical) difficulties in the presence of improper minima. In the following, we will only consider proper minima, and we call them minima for short. Whenever appropriate, we will also discuss improper minima, but this will be on an ad hoc basis. Note that the value of a (proper) minimum is unique (provided it exists) but there may be more than one location yielding the same minimal value of $J(\theta)$. In fact the level set of $J(\theta)$ for level α^*, denoted as $\mathcal{L}_{\alpha^*}$, yields the set of all locations of the global minima of $J(\theta)$. Figure 1.3 shows an example with various minima,

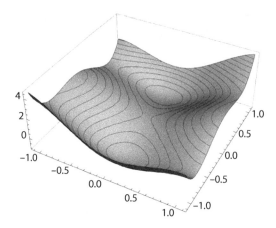

Figure 1.3. Example of a function with several maxima, minima, and saddle points.

maxima, and saddle points. The function is $(4 - 2.1\theta_1^2 + \frac{\theta_1^4}{3})\theta_1^2 + \theta_1\theta_2 + (-4 + 4\theta_2^2)^2\theta_2^2$, the so-called six-hump camel back function.

Definition 1.4. The points $\bar{\theta} \in \mathbb{R}^d$ that satisfy $\nabla J(\bar{\theta}) = 0$ are called *stationary* points of J.

The following theorem provides conditions for deciding the type of a stationary point.

Theorem 1.1. *Let $J \in C^2$.*

- *A local minimum (local maximum) θ^* of J is a stationary point, that is, it satisfies the first-order optimality condition:*

$$\nabla J(\theta^*) = 0. \tag{1.3}$$

- *A decision value θ^* is a local minimum (local maximum) of J if in addition to (1.3), the following is also satisfied*

$$\nabla^2 J(\theta^*) > 0 \qquad \left(\nabla^2 J(\theta^*) < 0\right). \tag{1.4}$$

Equation (1.4) is called second-order optimality condition.
- *If J is a* convex *(concave) function, then (1.3) is necessary and sufficient for θ^* being a global minimum (maximum).*

Proof. For a twice continuously differentiable function $J(\theta)$, i.e., $J \in C^2$, the Taylor series expansion with remainder yields the following expression of the value of J at a point $\theta + t\eta$, for $\eta \in \mathbb{R}^d$ and $t \in \mathbb{R}^+$:

$$J(\theta + t\eta) = J(\theta) + t \, \nabla J(\theta)\eta + \frac{t^2}{2}\eta^\top \nabla^2 J(\theta') \, \eta, \tag{1.5}$$

where $\theta' = \theta + t'\eta$, for some "intermediate" value $0 \le t' \le t$.

Use the above expression around θ^* to express $J(\theta^* + t\eta)$ for an arbitrary direction η. The definition of a positive definite matrix implies that $\eta^\top \nabla^2 J(\theta^*)\eta > 0$ for all $\eta \in \mathbb{R}^d$. In addition, $\nabla^2 J(\cdot)$ is continuous, so use a sufficiently small value of t to complete the arguments. The details are left as an exercise. □

Example 1.1. A well-known historical problem is that of explaining the phenomenon of refraction of light when traversing two different media. Since Ptolemy (circa 140 AD),

GRADIENT-BASED METHODS

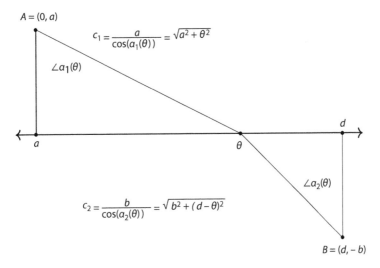

Figure 1.4. Problem is to find the optimal crossing point θ.

scientists were concerned with finding the relationship between the angles of refraction and the media's characteristics. The law of refraction was described by Ibn Sahl of Baghdad (978 AD), who used it to shape lenses, and in 1621 by the Dutch astronomer Snellius. It is now called Snell's Law in English. In 1637 Descartes found the same principle using conservation of moments, and in French it is called the Law of Snell-Descartes. We are mostly interested in Fermat, who in 1657 used variational calculus and his principle of "least time" to derive this law through an optimization problem.

The version of the problem that we give here is the pedagogical version of Richard Feynman. Imagine that you are walking on the beach when you see a person drowning and shouting for help. To get from where you are to the drowning person in the fastest way, you should not move along the straight line, because you run faster on the sand than you can swim. The distance from your position to the water is a, the distance from the water to the person is b, and the length of shoreline between the two points is d, as shown in Figure 1.4.

In Cartesian coordinates the drowning person is at $B = (d, -b)$ and your position is $A = (0, a)$. Here we assume that the waterfront is a straight line for simplicity. The speed on sand is v_1 and in the water v_2, with $v_2 < v_1$. We call θ the crossing point.

Fermat reasoned that light chooses not the shortest path but the one that saves more energy, which is the *fastest* path. It is easy to argue the existence of a solution for this problem: any value of θ to the left of the crossing point of the straight line between points A and B will give a slower path than the straight line because of $v_2 < v_1$. On the other hand, any point $(0, \theta)$, with $\theta > d$, will require unnecessary additional travel time compared to crossing at $(0, d)$, therefore there must be a minimum point between these two points. That the travel times are continuously differentiable follows from the linear relationships between distance and time.

At speed v, the distance traveled in time t is vt, so the total travel time can be expressed as

$$J(\theta) = \frac{1}{v_1} \frac{a}{\cos(\alpha_1(\theta))} + \frac{1}{v_2} \frac{b}{\cos(\alpha_2(\theta))}$$

$$= \frac{a}{v_1} \sec(\alpha_1(\theta)) + \frac{b}{v_2} \sec(\alpha_2(\theta)),$$

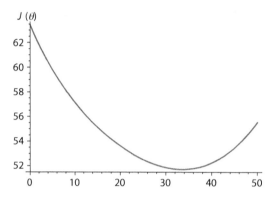

Figure 1.5. Plot of the function $J(\theta)$.

where the angles $\alpha_i(\theta)$ are as labeled in Figure 1.4. Figure 1.5 shows the plot of the time as a function of the crossing point θ.

According to Theorem 1.1, we now find the stationary points of $J(\theta)$. We use the following identities:

$$\tan \alpha_1(\theta) = \frac{\theta}{a}, \tag{1.6a}$$

$$\tan \alpha_2(\theta) = \frac{d-\theta}{b}, \tag{1.6b}$$

$$\frac{d}{d\alpha}\tan(\alpha) = \sec^2(\alpha) = \cos^{-2}(\alpha), \tag{1.6c}$$

$$\frac{d}{d\alpha}\sec(\alpha) = \sec(\alpha)\tan(\alpha). \tag{1.6d}$$

To obtain the first-order optimality condition, differentiate $J(\theta)$ and set it equal to zero:

$$J'(\theta) = \frac{a}{v_1} \sec(\alpha_1(\theta))\tan(\alpha_1(\theta))\frac{d\alpha_1(\theta)}{d\theta} + \frac{b}{v_2} \sec(\alpha_2(\theta))\tan(\alpha_2(\theta))\frac{d\alpha_2(\theta)}{d\theta} = 0.$$

By identity (1.6) it holds $\tan(\alpha_1(\theta))/\theta = 1/a =$ constant. Differentiating both sides of this equation with respect to θ yields

$$\frac{1}{\theta}\sec^2(\alpha_1(\theta))\left(\frac{d\alpha_1(\theta)}{d\theta}\right) - \frac{\tan \alpha_1(\theta)}{\theta^2} = 0 \Rightarrow \frac{d\alpha_1(\theta)}{d\theta} = \frac{1}{\theta}\sin(\alpha_1(\theta))\cos(\alpha_1(\theta)).$$

Similarly,

$$\frac{d\alpha_2(\theta)}{d\theta} = -\frac{1}{d-\theta}\sin(\alpha_2(\theta))\cos(\alpha_2(\theta)).$$

Replacing these values in $J'(\theta)$ and again using the identities in (1.6), one reaches the conclusion that $J'(\theta^*) = 0$ is achieved at the unique point that satisfies

$$\frac{\sin(\alpha_1(\theta^*))}{\sin(\alpha_2(\theta^*))} = \frac{v_1}{v_2}, \tag{1.7}$$

known as Snell's Law of refraction. Going back to the person at the beach, knowing Snell's Law is not very useful because he or she still has to determine the optimal crossing point $(0, \theta^*)$, however, (1.7) gives it as an implicit solution.

※※※

GRADIENT-BASED METHODS

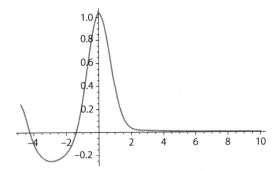

Figure 1.6. A function with "vanishing gradient."

When solving a problem of the form (1.1) analytically, one first looks for all points that satisfy (1.3). After the set of candidates is determined, one then evaluates the Hessian $\nabla^2 J(\theta)$ to verify which are local minima. If several local minima are found and if, in addition, it can be shown that $J(\theta)$ tends to ∞ as $\|\theta\|$ tends to infinity, then the location of the global minimum θ^* can be found by comparing the values of the local minima. It is worth noting that if there exists a unique stationary point, this analysis can be simplified. Indeed, if θ^* is the unique stationary point of $J(\theta)$ and if $J(\theta)$ tends to ∞ as $\|\theta\|$ tends to infinity, then θ^* is the unique location of the global minimum of $J(\theta)$.

Example 1.2. This example is provided to illustrate the terminology used in the field of optimization. Consider the function $J: \mathbb{R} \to \mathbb{R}$ plotted in Figure 1.6. This function has a unique global minimum that is attained at $\theta^* \approx -3$. Now consider the same function but limit its domain to $(0, \infty)$. Because $\lim_{\theta \to +\infty} J(\theta) = 0$ we find that $J(\theta)$ has 0 as an improper minimum. Indeed, 0 is not attained by any value for θ. In addition note that $J'(\theta)$ approaches 0 as θ tends to infinity. Loosely speaking we could express this by saying that "$\theta = +\infty$ is a stationary point of $J(\theta)$." The fact that $J'(\theta)$ tends to zero as θ tends to infinity is called the vanishing gradient problem, to which we shall return in later chapters. ✳✳✳

1.2 NUMERICAL METHODS FOR UNCONSTRAINED OPTIMIZATION

In most cases, as in Example 1.1, it is impossible to solve the inversion problem $\nabla J = 0$ analytically and numerical methods are used for finding a root θ^* of $\nabla_\theta J = 0$. A numerical iterative algorithm for approximating the solution θ^* of $\nabla_\theta J = 0$ is a recursion of the form

$$\theta_{n+1} = \theta_n + \epsilon_n d(\theta_n), \qquad (1.8)$$

where, for each n, ϵ_n is called the *stepsize* or *gain size*, $d(\theta_n)$ is called the *direction* of the algorithm, and $\{\epsilon_n\}$ is called the *stepsize sequence* or *gain sequence*. Occasionally, ϵ_n is also referred to as *learning rate*.

Methods for approximating θ^* can be classified according to the choice of the stepsize rule and the directions. Together with an initial value θ_0 and a *stopping rule*, (1.8) constitutes a numerical algorithm that terminates hopefully close to the true optimum, where "closeness" has to be defined appropriately. Analysis of such algorithms, however, are not based on a finite termination time, but are studied as the number of iterations grows to infinity. Stopping times for the algorithm are usually based on the convergence analysis.

Definition 1.5. A *descent direction of a differentiable function* $J(\theta)$ on Θ, or *descent direction* for short, is any vector $d(\theta)$ such that $\nabla J(\theta)\, d(\theta) < 0$ for all nonstationary points $\theta \in \Theta \subseteq \mathbb{R}^d$.

A descent direction $d(\theta)$ is pointing away from the direction $\nabla J(\theta)$. Indeed $-\nabla J(\theta)\, d(\theta) > 0$ implies that there is an angle of less than 90 degrees between $d(\theta)$ and $-\nabla J(\theta)$. The figure to the right depicts the situation. Recall that $\nabla J(\theta)$ points toward the direction of growth of the function, and it can be shown that $J(\theta)$ locally decreases along any descent direction $d(\theta)$, see Exercise 1.3. For a detailed geometric interpretation of the gradient, we refer to Section A.3 in Appendix A.

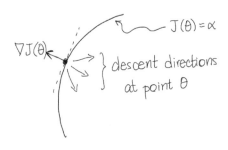

Algorithms that update along a decent direction are are called *descent algorithms*. Gradient-based methods for optimization, also called *gradient descent methods*, use $d(\theta_n) = -\nabla J(\theta_n)^\top$ as a direction in the algorithm, and are a subclass of descent algorithms. There are many methods available for gradient-based optimization. Typically these algorithms are tailored to specific classes of function such as the conjugate-gradient method and variations thereof, which are suitable for optimization of quadratic functions.

For later use we state here a result showing that the negative gradient rotated by a positive definite matrix remains a descent direction.

Lemma 1.1. *Let $J(\theta)$ be a differentiable function, and let $K(\theta) \in \mathbb{R}^{d \times d}$ be a positive definite matrix for all θ, then $-K(\theta)\nabla J(\theta)^\top$ is a descent direction on $\mathbb{R}^d$.*

Proof. Since $K(\theta)$ is positive definite, it holds that $x^\top K(\theta) x > 0$ for any non-zero vector x. Letting $x^\top = \nabla J(\theta)$ shows that $\nabla J(\theta) K \nabla J(\theta)^\top > 0$. Premultiplying by -1, then yields the result. □

Newton-Raphson Method

One of the most efficient methods for unconstrained optimization is the method developed by Newton (published in 1685) and Raphson (1690). It was originally designed to find the zeroes of a polynomial. In the context of finding stationary points of $J(\theta)$, $\theta \in \mathbb{R}^d$, let the vector $G(\theta)$ represent the gradient $\nabla J(\theta)^\top$. From a point θ_n, use a linear approximation of $G(\theta)$, that is, using Taylor's expansion

$$G(\theta_{n+1}) \approx G(\theta_n) + \nabla G(\theta)(\theta_{n+1} - \theta_n),$$

where now $\nabla G(\theta) = \nabla^2 J(\theta)$ is a $d \times d$ matrix for each θ.

To approximate the zero in one step, simply set $G(\theta_{n+1}) = 0$ in the approximation and solve the right hand side for θ_{n+1}. Assuming that the inverse matrix of $\nabla G(\theta)$ exists, this yields

$$\theta_{n+1} = \theta_n - [\nabla G(\theta_n)]^{-1} G(\theta_n). \tag{1.9}$$

Comparing (1.9) with (1.8), it follows that Newton's method is a gradient descent method with adaptive stepsize sequence $\epsilon_n := \epsilon(\theta_n) = [\nabla G(\theta_n)]^{-1}$.

Theorem 1.2. *Let $J: \mathbb{R}^d \to \mathbb{R} \in C^2$ be a convex function, and assume that the Hessian is invertible. Choose an initial point $\theta_0 \in \mathbb{R}^d$ and let $\{\theta_n\}$ be the sequence defined by (1.9),*

GRADIENT-BASED METHODS

with $G = \nabla J^\top$, that is,

$$\theta_{n+1} = \theta_n - [\nabla^2 J(\theta_n)]^{-1} \nabla J(\theta_n)^\top.$$

Suppose that $\bar{\theta}$ is an accumulation point of the sequence $\{\theta_n\}$ such that $\nabla^2 J(\bar{\theta}) > 0$, then $\bar{\theta}$ is a local minimum of $J(\theta)$ and the rate of convergence is superlinear, that is, there exists a sequence $\{c_n\}$ such that c_n tends to zero as n tends to ∞ and for some finite N itholds that

$$\|\theta_{n+1} - \bar{\theta}\| \leq c_n \|\theta_n - \bar{\theta}\|, \quad n \geq N.$$

Furthermore, if $J \in C^3$ then the rate of convergence is quadratic, that is, there exists a constant $c > 0$ such that, for large n

$$\|\theta_{n+1} - \bar{\theta}\| \leq c \|\theta_n - \bar{\theta}\|^2.$$

The proof of the result (omitted here) uses Taylor's approximation. Once the trajectory θ_n reaches a neighborhood of a local minimum θ^*, the Hessian $\nabla^2 J(\theta_n)$ becomes positive definite, which implies that it is invertible and that the Newton step moves along a descent direction, see Lemma 1.1. Although very efficient for convex functions, Newton's method has a number of practical problems when applied as a general-purpose optimization method:

- Newton's method finds zeros of the gradient, which may be locations of minima or inflection points for general functions. Consequently, it cannot be guaranteed that the Hessian is positive definite at every stationary point.
- The Hessian may not be invertible at every point.
- Finally, it needs calculation of gradients, Hessians, and Hessian inversion, all of which may be lengthy numerical operations, rendering the method slow. In some cases the Hessian can be approximately computed by repeated numerical function evaluation and we refer to Section 11.1.2 for details.

For deterministic problems, the *efficiency* of a method is defined in terms of CPU time to achieve a given precision δ. A number of algorithms have been proposed under the common name of "quasi-Newton" methods, which attempt to increase the efficiency of the method, overcoming the problems pointed out above.

★Cauchy's Method

The method known as *steepest descent* (or Cauchy's) for minimization of a cost function $J(\theta)$ chooses $d(\theta_n) = -\nabla_\theta J(\theta_n)$ at each iteration of (1.8). Originally proposed by Cauchy in 1847, instead of premultiplying by the matrix $[\nabla^2 J(\theta_n)]^{-1}$, the method chooses the stepsize to "move" along the direction $d(\theta_n)$ to reach the minimum on that line, that is,

$$\epsilon_n := \epsilon(\theta_n) = \arg\min_{\epsilon > 0}(J(\theta_n - \epsilon \nabla J(\theta_n)^\top)).$$

Gradient-Based Methods: Nonadaptive Stepsizes

As mentioned before, Newton's method has a good convergence rate, but every iteration may require too much computational time. Cauchy's method can have slow convergence due to possible zigzagging of the iterations, and several modifications have been proposed for adaptive stepsizes (where ϵ_n depends on $\theta_n, J(\theta_n), \nabla J(\theta_n)$, etc). Common methods use Wolfe's conditions [327, 328] and Armijo's rules [7] (and [29, 169] in combination with projection), which ensure that all accumulation points are local minima. For deterministic optimization adaptive stepsizes are undoubtedly superior to nonadaptive stepsizes. However, the focus of the present text is to extend the basic methodology for deterministic optimization to problems where the observations of the function $J(\theta)$ and its gradients (if available) are noisy, and the noise models may be very complex. For such scenarios, non adaptive stepsizes are simpler to analyze. The gradient-based methods use $d(\theta) = -\nabla J(\theta)$ as the direction of the algorithm, and the stepsizes can be of two kinds: either decreasing: $\epsilon_n \downarrow 0$, or constant: $\epsilon_n \equiv \epsilon$.

Without any detailed analysis, inspecting the mere structure of (1.8) allows us already to deduce properties of the stepsize sequence. To see this, insert the expression for θ_n on the right-hand side of (1.8), which yields $\theta_{n+1} = \theta_{n-1} + \epsilon_n d(\theta_n) + \epsilon_{n-1} d(\theta_{n-1})$ and continuing the recurrence

$$\theta_{n+1} = \theta_0 + \sum_{i=0}^{n} \epsilon_i d(\theta_i).$$

Suppose that $d(\cdot)$ is bounded. Then, for the algorithm to find θ^*, the stepsizes have to satisfy

$$\sum_{n=1}^{\infty} \epsilon_n = \infty, \tag{1.10}$$

so that the sequence $\{\theta_n\}$ is not confined to some bounded set (or, equivalently, will cover any bounded set as it can potentially reach any point in $\mathbb{R}^d$). Further conditions are required in order to ensure convergence of the algorithm to the optimal θ^*, as we will show in the upcoming theorem. Before we state and prove the main result in this section, we provide a useful technical result. The result and its proof is an adaptation of Lemma 1 in [34].

Lemma 1.2. *Consider the real-valued recursion:*

$$x_{n+1} = x_n - g_n + h_n, \quad x_0 \in \mathbb{R},$$

where $g_n \geq 0$ for all n, and the sequence h_n is summable, i.e., $\sum_n |h_n| < \infty$. Then either (i) $x_n \to -\infty$ or (ii) x_n converges to a finite value and $\sum_n g_n$ converges.

Proof. Note that

$$x_{n+2} = x_{n+1} - g_{n+1} + h_{n+1}$$
$$= x_n - (g_n + g_{n+1}) + (h_n + h_{n+1}).$$

Repeating this argument m times yields the telescopic sum

$$x_{m+n} = x_n - \sum_{i=n}^{m+n-1} g_i + \sum_{i=n}^{m+n-1} h_i. \tag{1.11}$$

By assumption $g_n \geq 0$, which implies

$$x_{m+n} \leq x_n + \sum_{i=n}^{m+n-1} |h_i| < \infty. \tag{1.12}$$

Use now $-\infty < \sum_{i=1}^{\infty} |h_i| < \infty$ to show that for all n

$$\limsup_{m \to \infty} \sum_{i=n}^{m+n-1} |h_i| = \lim_{m \to \infty} \sum_{i=n}^{m+n-1} |h_i| = \sum_{i=n}^{\infty} |h_i| < \infty \tag{1.13}$$

and
$$\liminf_{n\to\infty} \sum_{i=n}^{\infty} |h_i| = \lim_{n\to\infty} \sum_{i=n}^{\infty} |h_i| = 0. \tag{1.14}$$

Moreover, we have by (1.12) that
$$x_{m+n} \leq x_n + \sum_{i=n}^{\infty} |h_i| < \infty. \tag{1.15}$$

By (1.13), taking the limit superior on both sides of the inequality (1.15) as m tends to ∞ yields for all n
$$\limsup_{m\to\infty} x_{m+n} \leq x_n + \sum_{i=n}^{\infty} |h_i|,$$

and, since $\limsup_{m\to\infty} x_m = \limsup_{m\to\infty} x_{m+n}$, we arrive at
$$\limsup_{m} x_m \leq x_n + \sum_{i=n}^{\infty} |h_i|.$$

By (1.14) together with (1.15), taking the limit inferior on both sides of the above inequality gives
$$\limsup_{m\to\infty} x_m \leq \liminf_{n\to\infty} x_n < \infty,$$

and, as $\liminf_{m\to\infty} x_m \leq \limsup_{m\to\infty} x_m$ by definition, we arrive at
$$\limsup_{m\to\infty} x_m = \liminf_{m\to\infty} x_m,$$

which implies that either x_n converges to some finite $\bar{x} \in \mathbb{R}$, or $x_n \to -\infty$.

In the case that $\lim_n x_n = \bar{x} \in \mathbb{R}$, letting $n = 0$ in (1.11) yields
$$\sum_{i=0}^{m-1} g_i = \sum_{i=0}^{m-1} h_i - x_m + x_0,$$

and as the right-hand side of the above equation converges as $m \to \infty$ to a finite value so does the left-hand side, which proves the claim. $\square$

Next, we introduce two important concepts.

Definition 1.6. Let $\Theta \subset \mathbb{R}^d$ be an open connected set. A mapping $f : \Theta \to \mathbb{R}$ is called *Lipschitz continuous* if $L \in \mathbb{R}$ exists such that for any $x, x + \Delta \in \Theta$ is holds that
$$\|f(x) - f(x+\Delta)\| \leq L \|\Delta\|.$$

The constant L is called *Lipschitz constant*.

Definition 1.7. We say that a sequence $\{x_n\}$ with limit $\bar{x}$ achieves the limit in finite time if there exist a finite index $m < \infty$ such that $x_n = \bar{x}$ for $n \geq m$.

We are now ready to state the gradient-descent theorem for decreasing stepsize.

Theorem 1.3. *Let $J \in C^2$ and assume that ∇J is Lipschitz continuous on $\mathbb{R}^d$. For given initial value θ_0, let $\{\theta_n\}$ be given through the algorithm*
$$\theta_{n+1} = \theta_n - \epsilon_n \nabla J(\theta_n)^\top, \tag{1.16}$$

where the gain sequence $\{\epsilon_n\}$, with $\epsilon_n > 0$ for all n, satisfies

$$\sum_{n=1}^{\infty} \epsilon_n = +\infty, \quad \sum_{n=1}^{\infty} \epsilon_n^2 < \infty. \tag{1.17}$$

If $\{\|\nabla J(\theta_n)\| : n \geq 0\}$ is bounded, then any (finite) limit θ^* of $\{\theta_n\}$ is a stationary point of $J(\theta)$. If, in addition, θ^* is not attained in finite time, then, for n sufficiently large, $\{J(\theta_k), k \geq n\}$ is a strictly monotone decreasing sequence.

Proof. Approximating $J(\theta_{n+1})$ via a Taylor series expansion developed at θ_n (e.g., let $\eta = \theta_{n+1} - \theta_n$ and $t = 1$ in (1.5)), yields

$$J(\theta_{n+1}) = J(\theta_n) + \nabla J(\theta_n)(\theta_{n+1} - \theta_n) + \frac{1}{2}(\theta_{n+1} - \theta_n)^\top \nabla^2 J(\xi)(\theta_{n+1} - \theta_n), \tag{1.18}$$

where $\xi = \alpha \theta_n + (1 - \alpha)\theta_{n+1}$ for some $\alpha \in [0, 1]$. Inserting (1.16) into the above representation of $J(\theta_{n+1})$ yields

$$J(\theta_{n+1}) = J(\theta_n) - \epsilon_n \|\nabla J(\theta_n)\|^2 + \frac{\epsilon_n^2}{2} \nabla J(\theta_n) \nabla^2 J(\xi) \nabla J(\theta_n)^\top. \tag{1.19}$$

Recall that $\|\cdot\|$ denotes the Euclidean norm. Call $g_n = \epsilon_n \|\nabla J(\theta_n)\|^2$ and $h_n = \epsilon_n^2 \nabla J(\theta_n) \nabla^2 J(\xi) \nabla J(\theta_n)^\top / 2$, then

$$J(\theta_{n+1}) = J(\theta_n) - g_n + h_n.$$

From Lipschitz continuity of $\nabla J(\theta)$ it follows (see Exercise 1.7 below) that

$$|h_n| \leq \frac{\epsilon_n^2}{2} L \|\nabla J(\theta_n)\|^2,$$

for some finite constant L. Boundedness of the gradient along the trajectory together with $\sum \epsilon_n^2 < \infty$, shows that h_n is absolutely summable, so we can apply Lemma 1.2 to conclude that $J(\theta_n)$ either tends to $-\infty$, or it converges and

$$\sum_{n=0}^{\infty} \epsilon_n \|\nabla J(\theta_n)\|^2 < \infty. \tag{1.20}$$

Suppose that $\bar{\theta} \in \mathbb{R}^d$ is the limit of θ_n and achieved in finite time. Then $\theta_n = \bar{\theta}$ for all n larger than some k, which can only happen if the update $\epsilon_n \nabla J(\theta_n = \bar{\theta}) = 0$ for $n > k$. This shows that $\bar{\theta}$ is a stationary point. In case $\bar{\theta}$ is not achieved in finite time, we have from (1.20) together with continuity of ∇J that $\|\nabla J(\bar{\theta})\| = \lim_{i \to \infty} \|\nabla J(\theta_{m_i})\| = 0$. This shows that $\bar{\theta}$ is a *stationary* point.

We turn to the proof of the second part of the statement. As before, we denote the limit of θ_n by $\bar{\theta}$. We apply the bound

$$\|\nabla J(\theta_n) \nabla^2 J(\xi) \nabla J(\theta_n)^\top\| \leq L \|\nabla J(\theta_n)\|^2$$

(see Exercise 1.7) to (1.19) and thereby establish that

$$J(\theta_{i+1}) \leq J(\theta_i) - \left(\epsilon_i - \frac{1}{2} L \epsilon_i^2\right) \|\nabla J(\theta_i)\|^2.$$

Since ϵ_i tends to zero as i tends to infinity, we have for sufficiently large i that $L \epsilon_i < 2$, and thus $(\epsilon_i - L \epsilon_i^2/2) \|\nabla J(\theta_i)\|^2 > 0$, for $\|\nabla J(\theta_i)\| \neq 0$. Hence, if $\bar{\theta}$ is not attained in finite time, so that $\|\nabla J(\theta_i)\| \neq 0$ for all i, then there exists i_0, such that $\{J(\theta_i) : i \geq i_0\}$ is strictly monotone decreasing toward $J(\bar{\theta})$. □

GRADIENT-BASED METHODS

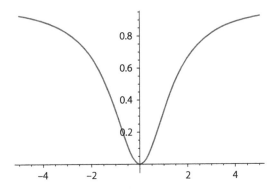

Figure 1.7. An example of a function with unique minimum and uniformly bounded gradient.

As the theorem shows, a gradient descent algorithm will find a stationary point of $J(\theta)$, but the nature of that point cannot be deduced from the algorithm alone and it requires some knowledge on the curvature of $J(\theta)$ in a neighborhood of θ^*; see Theorem 1.1. A sufficient condition for the algorithm to converge to a minimum, which then is also the unique global minimum, is convexity of $J(\theta)$. Exercise 1.12 asks to show this result.[1]

Theorem 1.3 provides sufficient conditions under which the sequence obtained via a gradient descent algorithm finds a stationary point of $J(\theta)$. Next to more generic conditions such as the choice of the stepsize and sufficient smoothness of $J(\theta)$, the key condition is that of boundedness of the gradient *along the trajectory* $\{\theta_n\}$. A nontrivial example of a mapping with bounded gradient is $J(\theta) = 1 - 2/(2 + \theta^2)$, see Figure 1.7. Note that $\lim_{|\theta| \to \infty} J'(\theta) = 0$, shown in Figure 1.7.

However, typically the assumption of boundedness of the gradient along $\{\theta_n\}$ is not straightforward to check except for simple cases as the following example shows.

Example 1.3. When the gradient is not bounded for all θ, it is sometimes useful to apply the argument that the algorithm will not persistently move "away from the minimizer." For illustration, consider $J(\theta) = \theta^2 + c$, for some constant c. The minimization problem has unique solution $\theta^* = 0$ and, by computation,

$$|\theta_{n+1}| = |\theta_n - \epsilon_n J'(\theta_n)| = |\theta_n - 2\epsilon_n \theta_n| = |\theta_n(1 - 2\epsilon_n)| = |\theta_n| \, |1 - 2\epsilon_n|.$$

So, as soon as $\epsilon_n < 1/2$ for some n, we see that $|\theta_{m+1}| < |\theta_m|$ for all $m \geq n$, and the trajectory stays inside a bounded set, which implies finiteness of the gradient along the trajectory.

※※※

Next we discuss a more challenging example.

Example 1.4. Let us consider again the function in Example 1.2 illustrated in Figure 1.6. Following the same argument as for the function $\theta^2 + c$ in the previous example, one can show that the gradient of the function in Example 1.2 is bounded along the trajectories that start at initial values $\theta_0 < 0$. Moreover, if $\theta_0 > 0$, then the negative gradient will be

---

[1] While it cannot be guaranteed that the gradient descent finds a local minimum, it is worth noting that the algorithm "goes in the right direction" and the likelihood that the algorithm gets trapped at a saddle point is small in practice. See Exercise 1.15 for an example of a case where the algorithm provably gets trapped in a saddle point. The problem of convergence to a saddle point can be avoided by using specific adaptive stepsizes. However, as explained in our discussion at the beginning of this section, non adaptive stepsizes are preferable when the observations of $J(\theta)$ are noisy, which is the main focus of this monograph; deterministic optimization is introduced here, but it is not the topic of our work.

positive and θ_n becomes an increasing sequence. While it is true that the limit here satisfies $\lim_{n \to \infty} J'(\theta_n) = 0$, the corresponding limit point $\lim_{n \to \infty} \theta_n = \infty$ is not finite, and 0 is an improper minimum on $(0, \infty)$. Even worse, for initial value $\theta_0 > 0$, the descent direction moves away from the actual solution, creating a numerical instability, and the algorithm should be properly modified.

This is an important situation that arises in practical applications, because the algorithm could diverge if applied directly. Interestingly, the function in Figure 1.7 also has vanishing gradients as $\theta \to \pm\infty$, but in that case this does not pose a problem because the negative gradient is a descent direction and thus it "pulls" the sequence θ_n toward the unique minimum. ✲✲✲

On occasion, it is possible to measure the outcome $J(\theta)$ of the performance of a system but the gradient $\nabla J(\cdot)$ is analytically unavailable. Instead of a gradient, some methods use a finite difference approximation. More generally, suppose that the algorithm is driven by a biased approximation of the gradient:

$$\theta_{n+1} = \theta_n - \epsilon_n \big(\nabla J(\theta_n)^\top + \beta_n(\theta_n) \big), \quad (1.21)$$

where the decreasing bias terms satisfy $\beta_n(\theta_n) \to 0$. Lemma 1.3 provides an important extension of Theorem 1.3 to biased algorithms. In the presence of bias, we have to exclude the case that for some n, the update $\nabla J(\theta_n)^\top + \beta_n(\theta_n)$ becomes zero for θ_n, so that the algorithm freezes at θ_n due to the bias.

Lemma 1.3. *Let $J \in C^2$ be such that the gradient is a Lipschitz continuous function on $\mathbb{R}^d$ and consider the biased algorithm (1.21), where the bias and the stepsize sequence $\{\epsilon_n\}$, with $\epsilon_n > 0$ for all n, satisfy*

$$\sum_{n=1}^{\infty} \epsilon_n = +\infty, \quad \sum_{n=1}^{\infty} \epsilon_n \|\beta_n(\theta_n)\| < \infty, \quad \sum_{n=1}^{\infty} \epsilon_n^2 < \infty. \quad (1.22)$$

If $\{\|\nabla J(\theta_n)\| : n \geq 0\}$ is bounded, then any (finite) limit θ^* of $\{\theta_n\}$ is a stationary point of $J(\theta)$. If, in addition, θ^* is not attained in finite time, then, for n sufficiently large, $\{J(\theta_k), k \geq n\}$ is a strictly monotone decreasing sequence.

Proof. The method of proof for this lemma is the same as for Theorem 1.3. The Taylor series for $J(\theta)$ now involves the bias terms, and under the assumptions, the corresponding terms g_n, h_n can be defined to apply Lemma 1.2. The details are left as an exercise, see Exercise 1.9. □

Lemma 1.3 can be adapted to a setting with non-vanishing bias. This is explained in the following example.

Example 1.5. Consider the biased algorithm (1.21) and assume that $\beta_n(\theta_n) = \hat{\beta}_n(\theta_n) + \beta$, for $\lim_n \|\hat{\beta}_n(\theta_n)\| = 0$ and some non-zero vector β. Hence, the bias does not asymptotically vanish. Let $\hat{J}(\theta) = J(\theta) + \beta^\top \theta$. Provided that the conditions in Lemma 1.3 are met for $J(\theta)$ and $\hat{\beta}_n(\theta_n)$, they straightforwardly extend to $\hat{J}(\theta)$ and $\hat{\beta}_n(\theta_n)$. The algorithm then finds a stationary point θ^* of $\hat{J}(\theta)$ that satisfies $\nabla J(\theta^*) + \beta^\top = 0$. Hence, the algorithm traces the stationary point of the shifted performance function $J(\theta)$. ✲✲✲

Example 1.6. Suppose that we do not know the function $J(\cdot)$ analytically, but for any point θ it is possible to obtain the numerical value of $J(\theta)$. In this situation, $\nabla J(\theta)$ is not available in closed-form either. A commonly used approximation to the derivative is given by finite differences (FD), which require that $J \in C^3$. In this example we will use a "centered" version of the approximation as follows. For simplicity, let $\theta \in \mathbb{R}$ and use a Taylor expansion around θ to obtain

$$\frac{J(\theta_n + c_n) - J(\theta_n)}{2c_n} = \frac{J'(\theta_n)}{2} + \frac{1}{4}J''(\theta_n)\, c_n + \beta_+(\theta_n, c_n)$$

$$\frac{J(\theta_n) - J(\theta_n - c_n)}{2c_n} = \frac{J'(\theta_n)}{2} - \frac{1}{4}J''(\theta_n)\, c_n + \beta_-(\theta_n, c_n)$$

so that the centered, or two-sided FD satisfies

$$\frac{J(\theta_n + c_n) - J(\theta_n - c_n)}{2c_n} = J'(\theta_n) + \beta_n(\theta_n, c_n),$$

where $\beta_n(\theta, x) = \beta_+(\theta, x) + \beta_-(\theta, x) = O(x^2)$, for fixed θ. Note that the terms containing $J''(\theta_n)$ cancel out.

When implementing FD in the descent algorithm, it is necessary to show that $\lim_{n \to \infty} \beta_n(\theta_n, c_n) = 0$ to conclude that the algorithm converges to the optimal value. Note that the main problem in showing convergence lies in the fact that we do not know beforehand the sequence $\{\theta_n\}$ visited by the algorithm. To establish convergence in (1.21), we need to verify either (a) that the third derivative $J'''(\cdot)$ is uniformly bounded in θ, or (b) that θ_n remains within a compact set along the sequence, which would imply that $J'''(\theta_n)$ is uniformly bounded (as $n \to \infty$). When either (a) or (b) hold, we know that $\beta_n(\theta_n, x) \to 0$ for any sequence $\{\theta_n\}$ visited by (1.21) as long as $x \to 0$. Hence, we can choose $c_n = O(n^{-c})$ for some constant $c > 0$, which implies $\beta_n(\theta_n, c_n) = O(n^{-2c})$. In general the choice of c_n will depend on how fast $\epsilon_n \to 0$. Assume that $\epsilon_n = O(n^{-\gamma})$, so that (1.22) holds for $\gamma \in (0, 1]$. From Lemma 1.3 it follows that Theorem 1.3 can be extended to finite difference algorithms provided that

$$\sum_{n \geq 1} \epsilon_n \beta_n < \infty \implies \sum_{n \geq 1} n^{-(\gamma + 2c)} < \infty,$$

so that we need $\gamma + 2c > 1$ for the algorithm to converge. When $\gamma = 1$, positive c is sufficient.

※※※

While gradient-based methods of the type (1.21) ensure convergence for functions with only one stationary point giving the location of the global minimum (called "unimodal") and which are continuously differentiable, the rate of convergence may be much slower than Newton's method. In particular, the steepest descent method has linear convergence, i.e., there is a constant $c \in (0, 1)$ such that $\|\theta_{n+1} - \theta^*\| \leq c\|\theta_n - \theta^*\|$, whereas Newton's method in general has quadratic convergence, see Theorem 1.2. On the other hand, the gradient descent algorithm shows remarkable resilience even for distorted gradient measurements as long as the size of the distortion decreases as $n \to \infty$, as shown in Lemma 1.3.

We complete this discussion by providing the equivalent statement to Theorem 1.3 for constant stepsize. It is worth noting that when the bias does not vanish asymptotically, the algorithm will find a stationary point of a modified objective function. Moreover, the effect a bias has on the fixed stepsize algorithm is different from the effect a bias has on the decreasing stepsize algorithm; compare Example 1.6 with the theorem below.

Theorem 1.4. *Let $J \in C^2$. Assume that $\nabla J(\theta)$ is Lipschitz continuous on $\mathbb{R}^d$ with Lipschitz constant L. Consider the constant stepsize algorithm*

$$\theta_{n+1} = \theta_n - \epsilon \nabla J(\theta_n)^\top,$$

for $\epsilon > 0$. Then any (finite) limit θ^* of $\{\theta_n\}$ is a stationary point of $J(\theta)$. Moreover, if (i) $\epsilon < 2/L$ and (ii) θ^* is not attained in finite time, then $\{J(\theta_n)\}$ is a strictly monotone decreasing sequence.

Let β_n denote the bias at the n-th iteration, and assume that $\lim_n \beta_n = \beta \in \mathbb{R}^d$. Then, any (finite) limit θ_β^* of $\{\theta_n^\beta\}$ given by

$$\theta_{n+1}^\beta = \theta_n^\beta - \epsilon \, (\nabla J(\theta_n^\beta)^\top + \beta_n))$$

solves $\nabla J(\theta_\beta^*) + \beta = 0$, i.e., in the asymptomatically unbiased case (given by $\beta = 0$), θ_β^* is a stationary point of $J(\theta)$, and in the asymptotically biased case (given by $\beta \neq 0$), θ_β^* is a stationary point of the adjusted objective $J(\theta) + \beta^\top \theta$.

Proof. If θ_n converges to some $\theta^* \in \mathbb{R}^d$, then this is only possible if $\lim_n \epsilon \nabla J(\theta_n) = 0$. Since ϵ is constant, this implies $\nabla J(\theta_n) = 0$, and by continuity of ∇J, is holds that $\nabla J(\theta^*) = 0$ and θ^* is thus a stationary point.

For the next part of the proof, we note that we have already shown in the proof of Theorem 1.3 that for any $i \geq 0$,

$$J(\theta_{i+1}) \leq J(\theta_i) - \epsilon \left(1 - \frac{1}{2} L \epsilon \right) \|\nabla J(\theta_i)\|^2.$$

Hence, for $\epsilon < 2/L$ we have that $(1 - \frac{1}{2} L \epsilon) > 0$ so that $J(\theta_{i+1}) < J(\theta_i)$, which shows that $J(\theta_i)$ is strictly monotone decreasing toward $J(\theta^*)$, with θ^* a stationary point.

For the biased case, we argue like before for showing that convergence of θ_n^β toward $\theta_\beta^* \in \mathbb{R}^d$ together with continuity of $\nabla J(\theta)$ implies $\nabla J(\theta_\beta^*) + \beta = 0$. This shows that θ_β^* is a stationary point of $J(\theta)$ for $\beta = 0$. For $\beta \neq 0$, note that $\nabla (J(\theta) + \beta^\top \theta) = \nabla J(\theta) + \beta^\top$, so that θ_β^* is a stationary point of $J(\theta) + \beta^\top \theta$. □

Typically, the Lipschitz constant for the gradient is hard to bound, and one applies the algorithm for ϵ "small." If in addition to the assumptions in Theorem 1.4, the function $J(\cdot)$ is convex, then the unbiased algorithm converges to the location of the minimum of J.

1.3 CONSTRAINED OPTIMIZATION

In this section we turn to optimization problems involving constraints. For ease of reference we introduce the general setting in the following definition.

Definition 1.8. For $J(\theta) \in C^1$

- the unconstrained optimization problem

$$\min J(\theta),$$

or

- for $g_i(\theta), i = 1, \ldots, p$, and $h_j(\theta), j = 1, \ldots, q$, all in C^1, the constrained optimization problem

$$\min_{\theta \in \Theta} J(\theta), \qquad (1.23)$$

$$\Theta = \{\theta \in \mathbb{R}^d : g(\theta) \leq 0, h(\theta) = 0\},$$

is called a non-linear problem (NLP). The function $J\colon \mathbb{R}^d \to \mathbb{R}$ is called the objective function, the set Θ is called the feasible region (including the case $\Theta = \mathbb{R}^d$), and a point $\theta \in \Theta$ is called a feasible point.

An NLP is called a (strictly) convex non-linear problem, or (strictly) convex problem for short, if $J(\theta)$ and—in case the problem has constraints—each $g_i(\theta), i = 1, \ldots, p$, are (strictly) convex, and each $h_j(\theta), j = 1, \ldots, q$, is an affine function (linear plus a constant).

An NLP is characterized by functions $g\colon \mathbb{R}^d \to \mathbb{R}^p, h\colon \mathbb{R}^d \to \mathbb{R}^q$, that represent p inequality and q equality constraints that must be satisfied. Note that since the constraints are convex by assumption, the feasible region Θ of an NLP is a convex set.

When we want to stress that a gradient or a Hessian is taken with respect to θ of a mapping with more arguments, we write ∇_θ and ∇_θ^2, respectively.

Definition 1.9. For an NLP the associated *Lagrangian* $\mathcal{L}\colon \mathbb{R}^d \times \mathbb{R}^p \times \mathbb{R}^q \to \mathbb{R}$ is defined as

$$\mathcal{L}(\theta, \lambda, \eta) = J(\theta) + \lambda^\top g(\theta) + \eta^\top h(\theta). \tag{1.24}$$

The vectors λ and η are called *Lagrange multipliers*.

Definition 1.10. A constraint g_i of an NLP is said to be *active at a feasible point* $\theta \in \Theta$ if $g_i(\theta) = 0$. Otherwise it is said to be *inactive*. The set $A(\theta)$ of active constraints at θ contains all indices i for which $g_i(\theta) = 0$. The *constraint qualification* condition at a feasible point θ requires that the set of vectors $\{\nabla_\theta g_i(\theta), i \in A(\theta); \nabla_\theta h_j(\theta), j = 1, \ldots, q\}$ be linearly independent, and that there exist a vector $v \in \mathbb{R}^d, v \neq 0$, such that:

(a) $\nabla h_j(\theta) v = 0, \quad 1 \leq j \leq q$,
(b) for all $i \in A(\theta)$ it holds that $\nabla g_i(\theta) v < 0$.

Definition 1.11. A *stationary point* $(\theta^*, \lambda^*, \eta^*)$ of an NLP is a point that satisfies the *Karush Kuhn-Tucker (KKT) conditions* if

$$\nabla_\theta \mathcal{L}(\theta^*, \lambda^*, \eta^*) = 0 \tag{1.25a}$$

$$\nabla_\lambda \mathcal{L}(\theta^*, \lambda^*, \eta^*) = g(\theta^*)^\top \leq 0, \lambda^* \geq 0, \text{ and } \forall i : \lambda_i^* g_i(\theta^*) = 0 \tag{1.25b}$$

$$\nabla_\eta \mathcal{L}(\theta^*, \lambda^*, \eta^*) = h(\theta^*)^\top = 0; \tag{1.25c}$$

where $\nabla_\lambda \mathcal{L}(\theta, \lambda, \eta)$ denotes the gradient of $\mathcal{L}(\theta, \lambda, \eta)$ with respect to λ and $\nabla_\eta \mathcal{L}(\theta, \lambda, \eta)$ the gradient with respect to η. A stationary point that satisfies the KKT conditions is called a *KKT point*.

Condition (1.25b) is called the *complementary slackness* property, from this property it follows that $i \notin A(\theta)$ implies $\lambda_i = 0$. The following theorem shows that that the KKT conditions are necessary conditions for a local minimum, i.e., local minima are KKT points. Moreover, if the problem is strictly convex, then the KKT conditions are also sufficient for a global minimum. The proof is standard and a proof is omitted.

Theorem 1.5. *Assume that for a given NLP the constraint qualification holds at a local minimum θ^* of $J(\theta)$ in (1.23). Then there exist $\lambda^* \in \mathbb{R}^p, \eta^* \in \mathbb{R}^q$, such that $(\theta^*, \lambda^*, \eta^*)$ is a KKT point of the NLP. The vectors λ^* and η^* are called* Lagrange multipliers.

If, in addition, if the problem is a convex NLP, then the KKT conditions hold at θ^* if and only if θ^* is the global minimum.

Example 1.7. Many canned products in the supermarket come in cans of similar shape, where the height is the same as the diameter of the container. What is the reason for this?

Allegedly, a similar question haunted Galileo about the leather bags used by traders. Here is the answer: if a fixed volume of a given good has to be canned, the containers should be produced at minimal cost (in particular using minimal amount of material) in order to maximize your profit.

This problem can be formulated as a surface minimization problem under the fixed volume constraint. Call $\theta = (r, y)^T$, where r is the radius and y is the height of the (cylindrical) can. Then we want to find

$$\min_{r,y} J(\theta) \stackrel{\text{def}}{=} 2(\pi r^2) + 2\pi r\, y$$

$$\text{subject to:} \quad \pi r^2 y = V,$$

where we have expressed the total surface as the rectangular surface for the side of the can, plus the two covers. The volume V is fixed. Call $h(\theta) = \pi r^2 y - V$.

We will show how to apply Theorem 1.5 in practice. The problem fails to be convex, as neither is $J(\theta)$ convex nor is h affine, and the second part of the theorem cannot be used. Instead, we proceed as follows. First, we find the KKT points that satisfy (1.25), and then we determine which one (if several) is the global optimizer. The Lagrangian is

$$\mathcal{L}(\theta; \eta) = \mathcal{L}(r, y; \eta) = 2(\pi r^2) + 2\pi r\, y + \eta(\pi r^2 y - V).$$

Condition (1.25a) for a KKT points reads

$$\frac{\partial}{\partial r}\mathcal{L}(r, y; \eta) = 4\pi r + 2\pi y + \eta\, 2\pi r y = 0 \qquad (1.26)$$

$$\frac{\partial}{\partial y}\mathcal{L}(r, y; \eta) = 2\pi r + \eta\, \pi r^2 = 0. \qquad (1.27)$$

From the second equality we get $\eta^* = -2/r^*$, replacing this value in the first we get: $2r + y - 2y = 2r - y = 0$, so that $y^* = 2r^*$, which is the actual proportion found in many commercial cans.

To illustrate the mathematical method, we will finish the example. Using (1.25c), i.e., $h(\theta) = 0$, we replace $y = V/\pi r^2$ to obtain the actual solution to (1.25), namely $(r^*)^3 = V/2\pi$ and $y^* = 2r^*$. The constraint qualification holds at this (unique) KKT point. Indeed there is only one constraint, and it satisfies

$$\nabla h(\theta) = (2\pi r y,\ \pi r^2),$$

which is non-zero at r^*, y^*, as required. Observe that taking $v^T = (r/2, -y)$ yields $\nabla h(\theta)\, v = 0$.

Because this is the only KKT point, it is the only candidate for the solution. To see that the KKT point is indeed a local minimum, note that $r \in (0, \infty)$ and as r either tends to 0 or to ∞, the value of $J(r, y)$ tends to ∞, so that we can conclude that J has to have a minimum for some value of $r \in (0, \infty)$. Since the only candidates for the location of a minimum are the KKT points, it follows from the uniqueness of the solution, that the KKT point is the location of the global minimum.

It is worth noting that this example is academic and placed here for illustrating the use of the theory. A more direct solution is readily obtained by direct substitution $y = V/\pi r$ into J to obtain a function of only one variable $f(r) = 2\pi(r^2 + V/\pi r^2)$. That this is convex follows from $f'(r) = 2\pi(2r - V/\pi r^2)$, and $f''(r) = 2\pi(2 + 2V/\pi r^3) > 0$ for all $r > 0$. The unique zero of $f'(r), r \geq 0$ is exactly at r^*. ✼✼✼

The theorem below provides the second-order conditions that help in determining if a KKT point is indeed a local minimum along the feasible set under no convexity.

Definition 1.12. Let $(\theta^*, \lambda^*, \eta^*)$ be a stationary point of an NLP. The *critical cone* $\mathbf{C}(\theta^*, \lambda^*)$ is

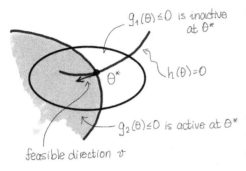

$$\mathbf{C}(\theta^*, \lambda^*) = \left\{ v \in \mathbb{R}^d : \nabla g_i(\theta^*) v \leq 0, \text{ if } i \in A(\theta^*), \right.$$
$$\lambda_i^* = 0, \nabla g_i(\theta^*) v = 0, \text{ if } \lambda_i^* > 0,$$
$$\left. \nabla h(\theta^*) v = 0 \right\}.$$

This cone defines the set of directions v that move along the active and equality constraints, as well as those that move "inside" the feasible set if the active constraint has a null multiplier.

Theorem 1.6. *Consider an NLP such that $J(\theta), g(\theta), h(\theta) \in C^2$ and that the constraint qualifications hold for $g(\theta), h(\theta)$ at θ^*. If $(\theta^*, \lambda^*, \eta^*)$ satisfies the first-order condition of being a stationary point (i.e., a KKT point), and the following second-order condition holds:*

$$v^\top \nabla_\theta^2 \mathcal{L}(\theta^*, \lambda^*, \eta^*) v > 0, \quad \text{for } 0 \neq v \in \mathbf{C}(\theta^*, \lambda^*), \tag{1.28}$$

then θ^* is a local minimum of (1.23), where $\nabla_\theta^2 \mathcal{L}$ denotes the Hessian of $\mathcal{L}$ with respect to θ.

Note that if the domain $\{\theta \in \mathbb{R}^d : g_i(\theta) \leq 0, 1 \leq i \leq p; h_j(\theta) = 0, 1 \leq j \leq q\}$ is compact, then the use of the second-order condition can be avoided as continuity of $J(\theta)$ already implies existence of a global maximum and minimum on a compact set. Evaluating all stationary points then solves the optimization problem. See, for example, [49]. Lagrange multipliers frequently have an interpretation in practical contexts. In economics, they can often be interpreted in terms of prices for constraints, so-called "shadow prices" while in physics, they can represent concrete physical quantities. Mathematically, Lagrange multipliers can be viewed as *rates of change* of the optimal cost as the level of constraint changes. These types of results are called *envelop theorems* in the literature. Next, we state, without proof, the fundamental envelop theorem.

Theorem 1.7. *Consider an NLP with no inequality constraints ($p = 0$) and $J(\theta) \in C^2$ and convex. Let (θ^*, η^*) be a local minimum and Lagrange multiplier, respectively, satisfying the KKT conditions and condition (1.28). Moreover, consider the family of continuous non-linear problems*

$$\min J(\theta), \theta \in \mathbb{R}^d \tag{1.29a}$$

$$\text{s.t.} \quad h(\theta) = u \tag{1.29b}$$

parameterized by $u \in \mathbb{R}^q$. Then there exists an open sphere S centred at $u = 0$ such that for every $u \in S$, there exist $\theta(u) \in \mathbb{R}^d$ and $\eta(u) \in \mathbb{R}^q$ such that $\theta(u)$ is the location of a local minimum of the above NLP and $\eta(u)$ the corresponding Lagrange multiplier.

Furthermore, $\theta(u), \eta(u)$ are continuously differentiable functions within S and we have $\theta(0) = \theta^*, \eta(0) = \eta^*$. In addition, for all $u \in S$,

$$\nabla_u F(u) = -\eta(u),$$

where $F(u) = J(\theta(u))$ is the optimal cost of the problem at value u.

In the case of inequality constraints, evidently $\{\theta: g(\theta) \leq 0\} \subset \{\theta: g(\theta) \leq u\}$ for $u > 0$. Thus, the optimal cost value of the modified problem must satisfy $F(u) \leq F(0)$, for $F(u)$ defined as in Theorem 1.7. For all inactive inequality constraints, $\lambda_i = 0$, and for all active constraints, $\lambda_i > 0$, indicating a potential marginal *decrease* in the cost function as a result of increased resources.

Example 1.8. A company has a budget of $10,000 for advertising, all of which must be spent. It costs $3,000 per minute to advertise on television and $1,000 per minute to advertise on radio. If the company buys x minutes of television advertising and y minutes of radio advertising, its revenue in thousands of dollars is determined by the company's data-mining oracle/statistician to be reasonably approximated by the function

$$f(x,y) = -2x^2 - y^2 + xy + 8x + 3y.$$

We can find the best solution to maximize profit solving the minimization problem:

$$\begin{aligned} \min_{x,y \in \mathbb{R}} \quad & f(x,y) = 2x^2 + y^2 - xy - 8x - 3y \\ \text{s.t.} \quad & h(x,y) = 3x + y - 10 = 0 \\ & g_1(x,y) = -x \leq 0 \\ & g_2(x,y) = -y \leq 0, \end{aligned}$$

where f and h are expressed in units of thousands of dollars. The Lagrangian is

$$\mathcal{L}(x, y, \lambda, \eta) = 2x^2 + y^2 - xy - 8x - 3y + \lambda_1(-x) + \lambda_2(-y) + \eta(3x + y - 10),$$

and $\nabla_{(x,y)} \mathcal{L}(x, y, \lambda, \eta) = (4x - y - 8 - \lambda_1 + 3\eta, 2y - x - 3 - \lambda_2 + \eta)^T$. By the first KKT condition, a local minimum (x^*, y^*) satisfies $\nabla_{(x,y)} \mathcal{L}(x^*, y^*, \lambda^*, \eta^*) = 0$, which gives the following simultaneous equations:

$$\begin{aligned} 4x^* - y^* - 8 - \lambda_1^* + 3\eta^* &= 0 \\ 2y^* - x^* - 3 - \lambda_2^* + \eta^* &= 0. \end{aligned}$$

There are four combinations of $g_1(x)$ and $g_2(x)$ being active/inactive.

Suppose both inequality constraints are inactive, so that complementary slackness gives $\lambda_1^* = \lambda_2^* = 0$. Together with the equality constraint, this gives three equations in three unknowns x^*, y^*, η^*. Their solutions yields the KKT point $(x^*, y^*, \lambda_1^*, \lambda_2^*, \eta^*)^\top = (\frac{69}{28}, \frac{73}{28}, 0, 0, \frac{1}{4})^\top$. This point satisfies a constraint qualification since the function $h(x, y)$ is linear, so it is a KKT point, and is thus a candidate for a local minimum. Furthermore, we have

$$\nabla^2 f(x,y) = \begin{pmatrix} 4 & -1 \\ -1 & 2 \end{pmatrix},$$

which is positive definite, and therefore is positive semi-definite (a sufficient condition for a function to be convex), thus f is convex and the KKT point is the unique global minimum of f, and is therefore the unique global maximum of the original maximization problem. The company can therefore maximize its revenue by purchasing $\frac{69}{28}$ minutes of television time and $\frac{73}{28}$ minutes of radio time. Since we have found the unique global maximum of the optimization problem, we do not need to search for any other KKT points.

Now suppose you have in front of you this solution and the company boss puts you "on the spot" during a meeting and asks for an estimate of the extra revenue which would be generated if she spent an extra $1,000 on advertising, what would be a reasonable answer?

Instead of solving again the problem with the budget changed to $11,000, you can use Theorem 1.7: $-\eta$ is the instantaneous rate of change of the minimum cost function value $F(u)$ as a function of the change in the level of constraint. Here $-\eta^* = -\eta(0) = -0.25$. In terms of the original maximization problem, this translates to an *increase* of $250 to the maximum revenue that can be generated if the the advertising budget is increased by $1,000. Thus, knowing $\eta^* = .25$ will be enough for you to answer promptly "Madam, an extra expense of $1,000 can only provide an extra revenue around $250. Actually, we would be better off *decreasing* the advertising budget." ✳✳✳

1.4 NUMERICAL METHODS FOR CONSTRAINED OPTIMIZATION

It should be apparent that even for seemingly small dimensions, finding all KKT points of an NLP may be an infeasible task. As in the case of unconstrained optimization, one often uses numerical iterative procedures to approximate the solution. We will now mention some of the methods that extend the simple recursive procedure (1.21). The main idea of the methods is to either approximate or reformulate the problem in terms of unconstrained optimization and then use an appropriate numerical algorithm.

Penalty Methods. These methods modify the original performance function to penalize the extent to which the constraints are not satisfied. Let $\|\cdot\|$ denote the Euclidean norm, then the penalized function is defined:

$$J_\alpha(\theta) = J(\theta) + \frac{\alpha}{2}\left(\|g(\theta)_+\|^2 + \|h(\theta)\|^2\right), \tag{1.30}$$

where $g(\theta)_+ = (g_1(\theta)_+, \ldots, g_j(\theta)_+)^\top$, and $g_i(\theta)_+ = \max(0, g_i(\theta))$.

Theorem 1.8. *Consider an NLP and let $\{\alpha_n\}$ be an increasing sequence such that $\lim_n \alpha_n = \infty$. For α_n given, let θ_n be the location of the minimum of $J_{\alpha_n}(\theta)$, i.e., let $\theta_n = \arg\min J_{\alpha_n}(\theta)$. If $\{\theta_n\}$ has an accumulation point θ^* and a constraint qualification holds at θ^*, then θ^* is (feasible and) stationary for the NLP. Moreover, if λ^*, η^* are the Lagrange multipliers for θ^*, then*

$$\lambda^* = \lim_{n\to\infty} \alpha_n (g(\theta_n))_+, \quad \eta^* = \lim_{n\to\infty} \alpha_n h(\theta_n),$$

and for each $i = 1, \ldots, j$, $\lambda_i^* \geq 0$ and $\lambda_i^* = 0$ if $g_i(\theta^*) < 0$.

Figure 1.4 illustrates the idea of penalizing the unsatisfaction of the constraint. Here $J(\theta) = \theta^2$, and the constraint is $\theta \geq 3$. Naturally for this example direct inspection yields that (a) the constraint must be active at the optimal value (because the unconstrained optimum is infeasible), and (b) thus $\theta^* = 3$. The function $J_\alpha(\theta)$ looks like $J(\theta)$, except that the segment of the curve to the left of $\theta = 3$ (in the infeasible region) is "lifted" more dramatically as α increases.

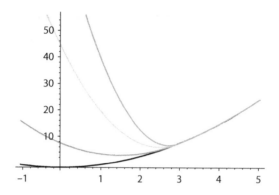

Figure 1.8. Functions $J_\alpha(\theta)$, $\alpha = 0, 1, 5, 10$.

The actual optimal values for the consecutive penalties are $0.0, 1.4, 2.5$, and 2.8. It follows from Theorem 1.8 that as $\alpha \to \infty$ this sequence converges to the optimal value. Theorem 1.8 requires solving each unconstrained problem $\min J_{\alpha_n}(\theta)$ exactly. However, this is often not possible, so one may use a gradient-based iterative method, for example, in order to *approximate* the solution θ_n. This is often referred to as "inexact optimization" for each step.

Numerical methods with inexact optimization typically use $\theta_{k+1}^n = \theta_k^n - \epsilon_k \nabla J_{\alpha_n}(\theta_k^n)^\top$, with $k = 1, 2, \ldots, T_n$, for minimizing $J_{\alpha_n}(\theta)$ with respect to θ. The terminal time T_n is either chosen to satisfy a stopping criterion, or sometimes an increasing sequence $T_n \to \infty$ is used. The idea is to approach the true solutions for the subsidiary problems as n increases, while using fewer iterations at first. The algorithm is

$$\theta_{k+1}^n = \theta_k^n - \epsilon_k \nabla J_{\alpha_n}(\theta_k^n)^\top, k = 0, \ldots T_n - 1 \qquad (1.31a)$$

$$\theta_0^{n+1} = \theta_{T_n}^n \qquad (1.31b)$$

$$\alpha_{n+1} = \alpha_n + \delta_n, \qquad (1.31c)$$

where $\sum \delta_n = +\infty$, and

$$\nabla J_{\alpha_n}(\theta_n) = \nabla_\theta J(\theta_n) + \alpha_n \left(g(\theta_n)^\top \nabla g(\theta_n) \mathbf{1}_{\{\|g(\theta_n)\|>0\}} + h(\theta_n)^\top \nabla h(\theta_n) \right). \qquad (1.32)$$

Under appropriate conditions, the sequence $\theta_{T_n}^n$ will converge to the constrained optimum. Different stopping schemes yield different overall rates of convergence. Notice that setting the initial value for step $n+1$ as the final value for the previous step is more convenient than re-initializing, provided that the final estimate $\theta_{T_n}^n$ is indeed close to the exact optimal value for J_{α_n}. Algorithm 1.1 corresponds to the updating scheme in (1.31).

When T_n is increasing, convergence of the auxiliary optimization problem for α_n ensures that the end point $\theta_{T_n}^n$ gets closer to the minimum of J_{α_n}; however, it also implies longer running times for Algorithm 1.1 than using a constant value. Indeed, the running time for this algorithm is proportional to $\sum_{n=1}^\tau T_n$, considering that each iteration inside the **for** loop has constant running time. Here, τ represents the stopping time, which is usually dependent on the current values of the gradients and consecutive end points $\theta_{T_n}^n$. Amortized analysis yields that the running time per (outer) iteration is the average of the consecutive lengths T_n, which grows as $n \to \infty$ unless T_n are constant. This may produce slow algorithms.

GRADIENT-BASED METHODS

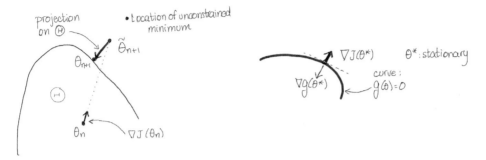

Figure 1.9. Visualization of the projection algorithm.

Algorithm 1.1 Penalty method

Read cost and constraint functions J, g, h.
Pre-define the increasing function $\text{ALPHA}(n)$.
Pre-define the non-decreasing sequence T_n.
Pre-define the function $\text{GRAD}(\alpha, \theta)$ that returns $\nabla J_\alpha^\top(\theta)$ in (1.32)
Initialize $\theta[0,0], \alpha_0 = \text{ALPHA}(0), n = 0$
while (**not** stopping-condition) **do**
    for $(k = 0, \ldots, T_n - 1)$ **do**
        $\theta[n, k+1] = \theta[n, k] - \epsilon_k \, \text{GRAD}(\alpha_n, \theta[n, k])$
    $\theta[n+1, 0] = \theta[n, T_n]$
    $\alpha_{n+1} = \text{ALPHA}(n+1)$
    $n \leftarrow n + 1$

Alternatively, one can introduce a two-timescale method. Let $T_n = 1$ and suppose that α_n grows ever so slowly that it "looks" constant for the iteration in θ_n when using Taylor expansions. The corresponding algorithm is of the form

$$\theta_{n+1} = \theta_n - \epsilon_n \nabla J_{\alpha_n}(\theta_n)^\top \tag{1.33a}$$

$$\alpha_{n+1} = \alpha_n + \delta_n, \tag{1.33b}$$

with $\delta_n \epsilon_n \to 0$, $\sum \delta_n = +\infty$, and $\nabla J_{\alpha_n}(\theta_n)$ as in (1.32). A convergence proof for this scheme with fixed ϵ_n, δ_n is provided in Theorem 2.12. For a treatment of general convergence results for two-timescale algorithms we refer to [222]. Algorithm 1.2 shows the pseudocode for the two-timescale implementation of the penalty method. In terms of the running time, it is now linear in the number of iterations performed in the **while** loop. However convergence of the two-timescale in terms of the stopping time τ may be slower than Algorithm 1.1 because α_n grows now very slowly. Inspecting Figure 1.4 it becomes apparent why the growth-rate of α_n has to be chosen with care: if α_n is too large, the penalty may push the algorithm far to the right (and away from the solution) and thus renders the method numerically inefficient.

Projection Methods. Gradient projection methods iterate successive solutions in the direction of improvement of the cost function (descent directions) where the value at each iteration *remains always feasible*. The projection method was introduced in [130, 131] and independently thereof in [202]. In the literature, the projection method also goes under the name *Goldstein-Levitin-Polyak projection method*, see [29, 141]. The algorithm is in

Algorithm 1.2 Penalty method: Two-timescale

Read cost and constraint functions J, g, h.
Pre-define the decreasing function $\text{DELTA}(n)$.
Pre-define the non-decreasing sequence T_n.
Pre-define the function $\text{GRAD}(\alpha, \theta)$ that returns $\nabla J_\alpha^\top(\theta)$ in (1.32)
Initialize $\theta_0, \alpha_0 = \text{ALPHA}(0), n = 0$
while (**not** stopping-condition) **do**
$\quad \theta_{n+1} = \theta_n - \epsilon_n \, \text{GRAD}(\alpha_n, \theta[n, k])$
$\quad \alpha_n = \alpha_n + \text{DELTA}(n)$
$\quad n \leftarrow n + 1$

general form:

$$\tilde{\theta}_{n+1} = \theta_n - \epsilon_n \nabla J(\theta_n)^\top \tag{1.34a}$$

$$\theta_{n+1} = \Pi_\Theta\left(\tilde{\theta}_{n+1}\right), \tag{1.34b}$$

where $\Pi_\Theta(v)$ is the projection of the vector $v \in \mathbb{R}^d$ onto the set Θ; and for $\Theta \subset \mathbb{R}^d$ a closed convex set, the projection Π_Θ on Θ is defined as

$$\Pi_\Theta(x) = \arg\min_{z \in \Theta} \|x - z\|. \tag{1.35}$$

In words, $\Pi_\Theta(v)$ is the point closest to v in Θ in Euclidean distance. Figure 1.9 (left) shows the geometric interpretation of the algorithm.

For mathematical analysis of the projection algorithm the following representation of the projection version of the gradient descent algorithm will be used in later chapters

$$\theta_{n+1} = \theta_n - \epsilon_n \bigl(\nabla J(\theta_n) + Z(\epsilon_n, \theta_n, -\nabla J(\theta_n))\bigr), \tag{1.36}$$

where where $Z(\epsilon_n, \theta_n, -\nabla J(\theta_n))$ is the "projection force" that keeps the algorithm on Θ. Specifically, if $\tilde{\theta}_{n+1} \in \Theta$, then

$$Z(\epsilon_n, \theta_n, -\nabla J(\theta_n)) = 0,$$

and otherwise

$$Z(\epsilon_n, \theta_n, -\nabla J(\theta_n)) = \frac{1}{\epsilon_n}\left(\theta_n - \epsilon_n \nabla J(\theta_n) - \Pi_\Theta\bigl(\theta_n - \epsilon_n \nabla J(\theta_n)\bigr)\right). \tag{1.37}$$

Note that by (1.35) the projection force on a convex set at some point $\theta \in \Theta$ is by construction no larger than the unconstrained increment given by the gradient at θ times the gain size, that is,

$$\|Z(\eta, \theta, -\nabla J(\theta))\| \leq \eta \|\nabla J(\theta)\|, \tag{1.38}$$

for $\eta > 0$, and that the projection force is monotone decreasing in the gain size

$$\|Z(\eta, \theta, -\nabla J(\theta))\| \leq \|Z(\hat{\eta}, \theta, -\nabla J(\theta))\|, \tag{1.39}$$

for $\eta \leq \hat{\eta}$, which stems from the fact that the force pushing the update outside of Θ is the negative gradient scaled by the gain size, and is therefore monotone decreasing in the gain size.

The actual evaluation of the projection operation is usually the main computational burden for each step in the algorithm. See, for example, [77], where the projection onto a

GRADIENT-BASED METHODS

simplex is provided. The simplest case is that of projection on a hypercube or a hyperball, which are detailed in the following examples.

Example 1.9. In case Θ is a d-dimensional hypercube, i.e., $\Theta = [-M, M]^d$ for some finite M, the projection is easily obtained through

$$\Pi_M(\theta) := \Pi_{[-M,M]^d}(\theta) = \Big(\max(\theta_i, -M)\mathbf{1}_{\{\theta_i \leq 0\}} + \min(\theta_i, M)\mathbf{1}_{\{\theta_i \geq 0\}} : 1 \leq i \leq d\Big)^\top.$$

In the special case of the projection on a hypercube we call the projection a *truncation* (on each coordinate). We call the constraint set Θ *box constraints*. Note that Θ can be encoded in the KKT setting, see (1.23), through $g_i(\theta) = \theta_i - M$ and $g_{i+d}(\theta) = -\theta_i - M$, for $1 \leq i \leq d$. ✻✻✻

Example 1.10. In case Θ is a d-dimensional ball around the origin of radius $r > 0$, i.e.,

$$\Theta = B_r := \{\theta \in \mathbb{R}^d : ||\theta|| \leq r\}, \tag{1.40}$$

the projection is obtained by rescaling vectors ourside of B_r:

$$\Pi_r(\theta) := \Pi_{B_r}(\theta) = \begin{cases} r\theta/||\theta|| & \text{if } \theta \notin B_r, \\ \theta & \text{if } \theta \in B_r. \end{cases} \tag{1.41}$$

Note that B_r can be encoded in KKT setting, see (1.23), through $g(\theta) = ||\theta|| - r$. ✻✻✻

In the following we consider NLP's without equality constraints, in which case the NLP becomes

$$\min_{\theta \in \Theta} J(\theta), \qquad \Theta = \{\theta \in \mathbb{R}^d : g(\theta) \leq 0\} \tag{1.42}$$

and solutions are characterized by the KKT conditions.

Before stating (a version) of the convergence result, we will motivate the result by the following consideration. Suppose that $\{\theta_n\}$ is obtained via (1.34), then the following cases can occur: (i) the minimizer θ^* is an inner point of Θ and the algorithm will (after possibly finitely many projections) stay inside Θ, and will behave just like the unconstrained version; (ii) the *unconstrained* minimizer $\tilde{\theta}^*$ lies outside of Θ (or on the boundary of Θ) and the algorithm will eventually converge to a point θ^* on the boundary of Θ; and finally (iii) the problem may be ill-posed so that θ_n has no accumulation points at all (e.g., minimizing $J(\theta) = -\theta^2$). Note that case (iii) is ruled out if we assume Θ to be compact which is a consequence of the Weierstrass theorem, see Theorem A.2 in Appendix A. Before turning to the study of the behavior of the algorithm, we provide some details on case (ii). For ease of argument we consider the fixed ϵ version of the algorithm. Suppose that θ^* lies outside Θ, and suppose that the algorithm converges to a point θ', then $\nabla J(\theta_n) + Z(\epsilon, \theta_n, -\nabla J(\theta_n))$ tends to zero as n tends to infinity. Assume, for simplicity, that only one constraint g_i is active at θ', i.e., $g(\theta') := g_i(\theta') = 0$. Then, the descent direction in θ' is $-\nabla J(\theta')^\top$. This implies that $-\nabla J(\theta')^\top$ is pointing outward of Θ. For the algorithm to have θ' as fixed point, it must hold that the projection of $\tilde{\theta}' = \theta' - \epsilon \nabla J(\theta')^\top$ on Θ is θ' itself. This means that $\tilde{\theta}' - \theta'$ is perpendicular to the tangent plane (an object in $\mathbb{R}^{d+1}$) to Θ at θ'. Since Θ is given as $\{\theta \in \mathbb{R}^d : g(\theta) \leq 0\}$, we know that $\nabla g(\theta)$ is a the projection of the normal vector to the tangent plane onto the parameter space $\mathbb{R}^d$, and due to the inequality we have that $\nabla g(\theta)$ is pointing outward of Θ. This shows that $-\nabla J(\theta')^\top$ and $\nabla g(\theta')$ are co-linear and pointing in the same direction. Hence, $-\nabla J(\theta')^\top = \lambda \nabla g(\theta')$, for some $\lambda > 0$, as illustrated in Figure 1.9. We conclude that θ' is a KKT point. Note that it thus holds that $Z(\epsilon, \theta', -\nabla J(\theta'))$

and $\nabla g(\theta')$ are co-linear for all ϵ; however, they point in opposite directions. To summarize, for $\nabla J(\theta') \neq 0$ and $g(\theta') = 0$, we have $\lambda \nabla g(\theta') = -\nabla J(\theta')$, which shows that θ' is a KKT point for (1.42) and under appropriate smoothness conditions a local minimizer for (1.42).

In the presence of bias, it may happen that the biased version is co-linear with the projection force at some point θ_n so that the algorithm does not advance any more (i.e., $\theta_{n+m} = \theta_n$ for $m \geq 1$) while the gradient is not co-linear with the projection force and θ_n is thus not a KKT point. As illustrating example for this phenomena consider the coordinate descent gradient

$$G(\theta) = \mathbf{e}_j (\partial J(\theta)/\partial \theta_j),$$

where $\mathbf{e}_j$ is the jth unit vector and

$$j = \arg\max_i |\partial J(\theta)/\partial \theta_i|.$$

It is easily seen that $-G(\theta)$ is a descent direction and, in general, a biased version of $-\nabla J(\theta)$. Let Θ be a hypercube. Suppose that θ_n is the first time that the algorithm steps outside hyercube Θ, so that $\tilde{\theta}_n$ is on the surface of the hypercube. Since $G(\theta)$ is by construction perpendicular to the surface of the hypercube, this implies that algorithm gets stuck at $\tilde{\theta}_n$, i.e., $\tilde{\theta}_n = \tilde{\theta}_{n+k}$ for $k \geq 0$. Letting k now tend to ∞, neither the value of $\tilde{\theta}_{n+k}$ nor that of the bias will change. Hence, the bias cannot tend to zero and we can rule this out by imposing the condition that β_n tends to zero as n tends to ∞. If, on the other hand, the algorithm comes to a halt at $\tilde{\theta}_n$ with $\beta_n = 0$, we have found a KKT point.

We now present the theorem.

Theorem 1.9. *Consider the NLP*

$$\min_{\theta \in \Theta} J(\theta), \qquad \Theta = \{\theta \in \mathbb{R}^d : g(\theta) \leq 0\} \tag{1.43}$$

with Θ being a compact and convex set, and let $J(\theta) \in C^2$ with L denoting the uniform Lipschtiz constant of ∇J on Θ. Consider the algorithm

$$\tilde{\theta}_{n+1} = \theta_n - \epsilon_n (\nabla J(\theta_n)^\top + \beta_n)$$
$$\theta_{n+1} = \Pi_\Theta(\tilde{\theta}_{n+1}),$$

with either

$$\sum_n \epsilon_n = \infty, \ \sum_{n=1}^{\infty} \epsilon_n \|\beta_n(\theta_n)\| < \infty \quad \text{and} \quad \sum_n \epsilon_n^2 < \infty,$$

where $\epsilon_n > 0$ for all n, or

$$0 < \epsilon_n = \epsilon < 2/L, \ \text{for } n \geq 0, \quad \text{and} \quad \lim_{n \to \infty} \|\beta_n\| = 0.$$

Then every accumulation point of $\{\theta_n\}$ of this algorithm is a KKT-point of the NLP in (1.43).

Proof. We proof the theorem in case of no bias. The extension to the biased case follows the line of argument provided in the discussion prior to the theorem.

As Θ is compact, then by the Bolzano-Weierstrass theorem, $\{\theta_n\}$ has accumulation points. Let θ^* be an accumulation point of $\{\theta_n\}$ and assume that θ^* is an inner point of Θ. Let $\theta_m := \theta_{n_m}$ denote the subsequence converging toward θ^*. Then, for N sufficiently large, $\theta_m \in \hat{\Theta}$, for $m \geq N$, for some compact proper subset $\hat{\Theta}$ of Θ (i.e., $\hat{\Theta}$ contains no boundary points of Θ). Continuity of the gradient and the Hessian implies that the gradient as well as the Hessian are bounded on $\hat{\Theta}$. We now apply the arguments put forward in the proof of Theorem 1.3 for the decreasing ϵ case and Theorem 1.4 for the fixed ϵ case, to show that

$$\lim_{m\to\infty} \nabla J(\theta_m) = 0 = \nabla J(\theta^*),$$

which shows that θ^* is a stationary point of $J(\theta)$. We have assumed that θ^* is an inner point of Θ, so that $g_i(\theta^*) < 0$ for all i, and it follows that θ^* is a KKT point for (1.42).

Now consider the case that θ^* lies on the boundary of Θ. Convergence of θ_m implies that $\|\nabla J(\theta_m) + Z(\epsilon, \theta_m, -\nabla J(\theta_m))\|$ converges toward zero. Note that projection on a convex set is continuous; see Exercise 1.5. By continuity of both gradient and projection it holds that

$$\lim_{m\to\infty} \|\nabla J(\theta_m) + Z(\epsilon, \theta_m, -\nabla J(\theta_m))\| = \|\nabla J(\theta^*) + Z(\epsilon, \theta^*, -\nabla J(\theta^*))\| = 0. \quad (1.44)$$

For ϵ sufficiently small, we apply Theorem 1.4 to conclude from the above that either $\nabla J(\theta^*) = 0$ (and therefore $Z(\epsilon, \theta^*, -\nabla J(\theta^*)) = 0$) and $g_i(\theta^*) = 0$, or $\nabla J(\theta^*) \neq 0$ in which case the negative gradient points outward from Θ. For the projection force to counter balance $-\nabla J(\theta^*)$, the negative gradient has to be perpendicular to the projection of hyperplane spanned by any g_i at θ^* onto $\mathbb{R}^d$. As, moreover the $\nabla g_i(\theta^*)$'s are pointing outward of Θ, we have that $-\nabla J(\theta^*) = \sum \lambda_i \nabla g_i(\theta^*)$ for some constants $\lambda_i > 0$ where the sum runs through the indices of the active constraints. This shows that θ^* is a KKT point for (1.42).

For the decreasing ϵ we take N such that $\epsilon_n \leq 2/L$ for $n \geq N$, and we use $\|Z(\epsilon_n, \theta_n, -\nabla J(\theta_n))\| \leq \|Z(\epsilon, \theta_n, -\nabla J(\theta_n))\|$ for $n \geq N$, which stems from the fact that the projection force is monotone in the gain size; see (1.39). The proof then follows from (1.44).

To conclude the proof, we evoke Theorem 1.5, to show that any KKT point for (1.42) is the location of a local minimum for the NLP in (1.42). □

Remark 1.2. In case that Θ represents hard constraints so that $J(\theta)$ is not defined outside of Θ, the gradient of $J(\theta)$ is only defined on interior points of Θ. As on the boundary of Θ only the directional derivatives along directions pointing inward of Θ are defined (and not the gradient as such), the projection method as presented here cannot be straightforwardly applied.

Remark 1.3. Unless the constraint set Θ is of a particular nice and simple form (e.g., a "box" or a "ball"), the projection step may require numerical approximation methods (see Example 1.11 below). In general, the projection method provides an analytically attractive tool that we will use extensively in the rest of this book. Indeed, many technical assumptions become less restrictive if the algorithm is projected onto a bounded set. For example, the condition that the gradient is bounded along trajectories in Theorem 1.3 rules out even a quadratic form of $J(\theta)$, but there is nothing wrong with a quadratic function as long as the trajectories remain inside a bounded set. In many cases, we study the projected version of the algorithm restricting the solutions $\{\theta_n\}$ to a hypothetical large hyperball; see Example 1.10. If the projected algorithm has accumulation points that are independent of the hyperball, then theoretical arguments can be used to establish that the original (unprojected) version has the same limiting behavior. It is worth mentioning that the projection method is well-studied in the area of deterministic optimization. A method for finding a projection on a general convex set through iterative projection on simpler convex sets is Dykstra's method [43], and for an exhaustive overview of these kind of methods we refer to [67]. It is worth noting that projection can be avoided by moving only along an update direction that stays inside the feasible set. An example of an algorithm that elaborates on this idea and that is popular in machine learning is the Frank-Wolfe algorithm, which uses a linear approximation of the objective for finding a descent direction that stays inside the feasible set; see [99].

We conclude this section on the projection method with a discussion on finding (approximate) projections when the feasible set is not of a simple form and the projection operation cannot be expressed analytically in closed form.

Example 1.11. In the case that the constraint function g is affine, Example 4.3 in Chapter 3 of [31] shows that the dual of this problem leads to a much simpler optimization problem with only positivity constraints. We now refer to [190] where a method is proposed for general $g \in C^2$ using the fact that $\theta_n \in \Theta$, and $\tilde{\theta}_{n+1} - \theta_n$ is of order ϵ_n so a Taylor approximation can be used to linearize the constraints around θ_n:

$$g(x) \approx g(\theta_n) + \nabla g(\theta_n)(x - \theta_n).$$

Let $v = \tilde{\theta}_{n+1}$. The constraint $x \in \Theta$ is approximated by the constraint

$$g(\theta_n) + \nabla g(\theta_n)x \leq \nabla g(\theta_n)\theta_n.$$

Call $\mathbb{A} = \nabla g(\theta_n)^\top$, and $b = \mathbb{A}\theta_n - g(\theta_n)$. The approximated, or "surrogate" subsidiary problem becomes

$$\min_x \left(\frac{1}{2}x^\top x - v^\top x\right) \tag{1.45}$$

$$\text{s.t.} \quad \mathbb{A}x \leq b, \tag{1.46}$$

The Lagrangian for this subsidiary problem is

$$\mathcal{L}(x, \mu) = \frac{1}{2}x^\top x - v^\top x + \mu^\top \mathbb{A}x - \mu^\top b.$$

Using Lagrange duality (Theorem 1.11), we seek $\max_{\mu \geq 0} (\min_{x \in \mathbb{R}^d} \mathcal{L}(x, \mu))$. Because of the quadratic form, we can solve the minimization step analytically by setting the gradient to zero, which readily yields $x^*(\mu) = v - \mathbb{A}^\top \mu$. Then the subsidiary problem is

$$\max_{\mu \geq 0} \left(\frac{1}{2}(v - \mathbb{A}^\top \mu)^\top (v - \mathbb{A}^\top \mu) - b^\top \mu\right),$$

which, after replacing the appropriate values, gives another quadratic maximization problem with a projection to the positive real numbers. Finding the zero of the derivative of this function, however, now requires an inversion of matrices depending on $g(\theta_n)$ and $\nabla g(\theta_n)$, which is generally computationally expensive. Instead, [190] propose a recursive gradient method to solve for μ. With this, one sets $\theta_{n+1} = v - \mathbb{A}^\top \mu^* = \tilde{\theta}_{n+1} - \nabla g(\theta_n)^\top \mu^*$.

The above analysis requires exact optimization for the subsidiary problem (i.e., solving μ^*), but this is often not possible, so inexact optimization can be used. Let T_n be the number of iterations used to approximate μ^* at step n of the procedure. Then the algorithm is

$$\tilde{\theta}_{n+1} = \theta_n - \epsilon_n \nabla J(\theta_n)^\top \tag{1.47a}$$

$$\mu^n_{k+1} = \max\left(0, \mu^n_k + \delta_k \left(\nabla g(\theta_n)\nabla g(\theta_n)^\top \mu^n_k + \nabla g(\theta_n)(\tilde{\theta}_{n+1} - \theta_n) - g(\theta_n)\right)\right);$$
$$k = 0, \ldots, T_n - 1 \tag{1.47b}$$

$$\theta_{n+1} = \tilde{\theta}_{n+1} - \nabla g(\theta_n)^\top \mu^n_{T_n}. \tag{1.47c}$$

Indeed, recalling our definitions $\mathbb{A} = \nabla g(\theta_n)^\top$, $v = \tilde{\theta}_{n+1}$, and $b = \mathbb{A}\theta_n - g(\theta_n)$ one notes that the second iteration above refers to the approximate solution of the subsidiary problem.

Because of the approximation of the non-linear constraint, this new point may be infeasible, but under appropriate conditions on $T_n, \epsilon_n, \delta_n$, this algorithm will converge. It is also possible to adapt this algorithm to a two-timescale version. ✳✳✳

Multiplier Methods. These methods are based on the following result for equality constraint problems; for a proof, we refer to [31].

Theorem 1.10. *Consider an equality constrained problem. Let*

$$\theta_n^* = \arg\min_\theta \mathcal{L}(\theta, \eta_n)$$

$$\eta_{n+1} = \eta_n + \rho_n h(\theta_n^*),$$

for a sequence $\rho_n \to \infty$, then $(\theta_n^*, \eta_n) \to (\theta^*, \eta^*)$ a local minimum and a KKT point of (1.23).

The *inexact* multipliers methods use an approximation to θ_n^* via Theorem 1.3. Let T_n be a stopping time for the approximation at step n. Then the algorithm is

$$\theta_{k+1}^n = \theta_k^n - \epsilon_k \nabla_\theta \mathcal{L}(\theta_k^n, \eta_n)^\top = \theta_k^n - \epsilon_k \left(\nabla_\theta J(\theta_k^n)^\top + \nabla_\theta h(\theta_k^n) \eta_n \right);$$
$$k = 0, \ldots, T_n \tag{1.48a}$$

$$\eta_{n+1} = \eta_n + \rho_n h(\theta_{T_n}^n). \tag{1.48b}$$

This method can be applied to inequality constraints as well via a transformation; see [31]. The choice of T_n will determine the convergence properties. It is common to either use a stopping criterion in terms of $\nabla \mathcal{L} \approx 0$, or to use an increasing sequence T_n. Compared with the analysis of the penalty method, one can also deduce that the amortized running time will be an average of the batch lengths T_n, which is an increasing function of n.

Algorithm 1.3 Multiplier method

Read cost and constraint functions J, h.
Pre-define the increasing function RHO(n).
Pre-define the non-decreasing sequence T_n.
Initialize $\theta[0, 0], \eta_n, \rho_0 = RHO(0), n = 0$
while (**not** stopping-condition) **do**
  for $(k = 0, \ldots, T_n - 1)$ **do**
    $\theta[n, k+1] = \theta[n, k] - \epsilon_k \left(\nabla J(\theta)^\top (\theta[n, k] + \nabla h(\theta[n, k]) \eta_n \right)$
  $\theta[n+1, 0] = \theta[n, T_n]$
  $\eta_{n+1} = \eta_n + RHO(n) h(\theta[n+1, 0])$
  $n \leftarrow n + 1$

A two-timescale algorithm can be implemented here, like for the penalty method, using $T_n = 1$ but making ρ_n grow "slower" than ϵ_n decreases so that the primal variable behaves locally in bounded intervals as if it was driven with a constant dual variable. This algorithm will have a constant amortized running time.

Lagrange Duality Methods. In both penalty and multiplier methods, the theory establishes convergence only when an exact minimization takes place for given multiplier values. The

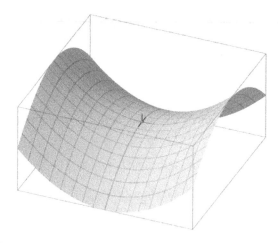

Figure 1.10. Saddle point illustration for θ, λ on the axes, no equality constraints.

numerical approximations often use inexact minimization by updating the decision variable θ_n for T_n iterations and then updating the multipliers. However, there is no guarantee that the algorithm will converge, and it is not clear how to tune the parameter T_n for better convergence.

An important class of methods is based on Lagrange Duality Theory. It is straightforward to note that the solution to (1.23) is the same as the solution of the minmax problem:

$$\min_{\theta \in \mathbb{R}^d} \max_{\lambda \geq 0, \eta} \mathcal{L}(\theta; \lambda, \eta) = \min_{\theta \in \mathbb{R}^d} \begin{cases} J(\theta) & \text{if } g(\theta) \leq 0, h(\theta) = 0, \\ +\infty & \text{otherwise.} \end{cases}$$

However, the above minmax problem is clearly not useful for an iterative algorithm. Instead, we use the following strong result.

Theorem 1.11 (Saddle Point Theorem). *For a given convex NLP, the triplet $(\theta^*, \lambda^*, \eta^*)$ is a KKT point if and only if it is a saddle point of the Lagrangian, that is,*

$$\mathcal{L}(\theta^*, \lambda, \eta) \leq \mathcal{L}(\theta^*, \lambda^*, \eta^*) \leq \mathcal{L}(\theta, \lambda^*, \eta^*)$$

for every $\theta \in \mathbb{R}^d, \lambda(\geq 0) \in \mathbb{R}^p, \eta \in \mathbb{R}^q$. Furthermore,

$$\min_{\theta \in \mathbb{R}^d} \max_{\lambda \geq 0, \eta} \mathcal{L}(\theta; \lambda, \eta) = \max_{\lambda \geq 0, \eta} \min_{\theta \in \mathbb{R}^d} \mathcal{L}(\theta; \lambda, \eta).$$

The saddle point theorem can be used to maximize first over the multipliers, and then perform a minimization over the decision variables. This is the motivation for the *Uzawa* algorithm [304]:

$$\theta_{n+1} = \arg\min_{\theta} \mathcal{L}(\theta, \lambda_n, \eta_n) \tag{1.49a}$$

$$\lambda_{n+1} = \lambda_n + \max(0, \lambda_n + \epsilon_n \nabla_\lambda \mathcal{L}(\theta_{n+1}, \lambda_n, \eta_n)^\top) \tag{1.49b}$$

$$\eta_{n+1} = \eta_n + \epsilon_n \nabla_\eta \mathcal{L}(\theta_{n+1}, \lambda_n, \eta_n)^\top, \tag{1.49c}$$

where $\nabla_\lambda \mathcal{L}(\theta, \lambda, \eta)^\top = g(\theta)$ and $\nabla_\lambda \mathcal{L}(\theta, \lambda, \eta)^\top = h(\theta)$. The $\max(0, \cdot)$ is a component-wise max operation on the vector.

Instead of exact minimization, the so-called *Arrow-Hurwicz iterative algorithm* [9] can be used for convex NLPs:

$$\theta_{n+1} = \theta_n - \epsilon_n \nabla_\theta \mathcal{L}(\theta_n, \lambda_n, \eta_n)^\top \quad (1.50a)$$

$$\lambda_{n+1} = \max\left(0, \lambda_n + \epsilon_n \nabla_\lambda \mathcal{L}(\theta_n, \lambda_n, \eta_n)^\top\right) \quad (1.50b)$$

$$\eta_{n+1} = \eta_n + \epsilon_n \nabla_\eta \mathcal{L}(\theta_n, \lambda_n, \eta_n)^\top, \quad (1.50c)$$

with

$$\nabla_\theta \mathcal{L}(\theta, \lambda, \eta) = \nabla J(\theta) + \lambda^\top \nabla g(\theta) + \eta^\top \nabla h(\theta),$$

and where the $\max(0, \cdot)$ is again the component-wise max operation on the vector. Algorithm 1.4 shows the corresponding pseudocode, where the function GRAD-LAGRANGE(θ, λ, η) returns the value $\nabla_\theta \mathcal{L}(\theta, \lambda, \eta)$ and the function Pos() returns the positive part of the argument, component-wise. It is possible to use different stepsize sequences for the multipliers, which often requires some parameter tuning. The algorithm has a running time that is linear in the number of iterations of the **while** loop, which is the reason why it is very convenient when Theorem 1.11 is applicable.

Algorithm 1.4 Arrow-Hurwicz

Read cost and constraint functions J, g, h.
Pre-define the decreasing functions ϵ_n.
Initialize $\theta_0, \lambda_0, \eta_0$
while (**not** stopping-condition) **do**
 $\theta_{n+1} = \theta_n - \epsilon_n$ GRAD-LAGRANGE$(\theta_n, \lambda_n, \eta_n)$
 $\lambda_{n+1} =$ Pos$(\lambda_n + \epsilon_n\, g(\theta_n))$
 $\eta_{n+1} = \eta_n + \epsilon_n h(\theta_n)$
 $n \leftarrow n + 1$

We will summarize the convergence properties of this algorithm in the following section.

Remark 1.4. As with the unconstrained methods, the above numerical methods can be (and are often) implemented with constant stepsizes ($\epsilon_n \equiv \epsilon$, $\delta_n \equiv \delta$ and $\rho_n \equiv \rho$). The following chapter presents general iterative algorithms and provides the analysis of the behavior of such algorithms both for constant and decreasing stepsizes.

1.5 PRACTICAL CONSIDERATIONS

All the assumptions in the theorems are there for a reason. However, some conditions are hard to verify analytically. In this section, we illustrate how modeling/meta-arguments can be used to justify that the conditions are verified.

Any optimization algorithm can only be successful if the optimization problem is well-posed, and part of the conditions in the theorems are there to ensure precisely this: there actually is a solution. For example, if $J(\theta)$ is convex, then the (unconstrained) minimization problem is well-posed and possesses a unique global solution, and in the strongly convex case the minimizer is also unique. For more general mappings, the theory identifies analytical conditions that characterize candidates for the global solution (stationary points, KKT conditions). Additional arguments are then typically required to determine the nature of a candidate. There is no one sizes fits all theory/algorithm, and building meaningful models that are well-posed for optimization is what constitutes the *art of modeling*, which although critical for application is not the primary focus of this monograph.

Generally speaking, it is recommendable first to get a good impression of the behavior of the algorithm by comparing several trajectories. For this, start the algorithm for several randomized initial values and let the algorithm run for some time. With today's computer technology, computing many trajectories in parallel is feasible. The following cases can be distinguished:

- If the trajectories tend to the same point, it is a good guess that the trajectories converged to the solution of the problem.
- If the limit point seems to be dependent on the initial value, then the problem may have several local minima (constrained or unconstrained), in which case finding the overall solution requires methods for exploration (which are outside the scope of this book). In practice, one explores the best solution (e.g., by evaluating the cost function at neighboring points).
- If some of the trajectories seem to wander off to infinity, this indicates that the algorithm is numerically unstable, which suggests that the problem might be ill-posed.

The golden rule in optimization can be phrased like this: *When something goes wrong in the numerical tests, your main suspects for explaining the wrongful behavior are those conditions in the theorems that you were not able to establish thoroughly.* In solving real problems, mathematical theory and the art of modeling complement each other; neither one dominates the other. Know your theory very well, but do not be restrained by all the technical and often hard to prove conditions. Beware of minimizing $-\theta^2$ or $e^{-\theta^2}$.

In what follows, we address aspects of the algorithms that deserve some further elaboration.

Choice of the Gain Sequence

The key issue is here that the stepsize should be such that it allows the algorithm to cover the relevant search space but should not lead the algorithm astray. Fixed stepsizes are very useful in understanding the way the algorithm works for a given problem. Visualizing $\{(\theta_n, J(\theta_n))\}$ and comparing outcomes for various initial values and various choices for ϵ, provides valuable insight into the algorithm. If your control variable θ_n is in high dimensions, θ_n may be replaced by $\|\theta_n\|$. For a fixed ϵ algorithm, such a pre-analysis is done to see whether ϵ is small enough to yield convergence and large enough to avoid unnecessary long trajectories. In case of decreasing ϵ_n the key question is whether the decrease is too slow, yielding rather erratic behavior of the algorithm, or too small, with the consequence that $\epsilon_n \nabla J(\theta_n)$ approaches zero only due to ϵ_n becoming very small. We call this the problem of the *vanishing update*. Although $\epsilon \to 0$ is mathematically not a problem, when running a program, due to floating-point accuracy, very small values of ϵ can be interpreted as zero, so it will give rise to numerical problems.

Algorithmic parameters such as the stepsize of the gain sequence or the initial value are called *hyperparameters* and choosing the right setting of the hyperparameters is a problem in its own right. We refer to [16] for results on hyperparameter optimization using gradient techniques.

Boundedness along Trajectories

Suppose that we can argue that the solution set of the original problem is unaltered when θ is confined to some hypercube $[-M, M]^d$, for $M > 0$ (or, equivalently, to some ball B_r of radius r around the origin). Then, we can apply the projection method. As the constant M can be chosen arbitrarily large, we may, in practice, simply neglect the projection to $[-M, M]^d$. An indication that this line of thought is not applicable to the problem under consideration is when a trajectory is found that seems to wander off to infinity. In this situation, one has to go back to the modeling table and rethink the problem formulation. If $J(\theta)$ tends to infinity as $\|\theta\|$ tends to infinity, then provided that $J(\theta)$ is continuous, it can be argued by the Weierstrass theorem that M exists satisfying the above condition.

Soft versus Hard Constraints

In Section 1.4 we discussed numerical methods for solving constrained optimization problems. Except for the projection method, the penalty methods, multiplier methods, and the Lagrange duality methods require that the objective together with the constraint mappings are well-defined on $\mathbb{R}^m$ for appropriate m. Indeed, the penalty method, for example, allows the algorithm to leave Θ and penalizes such a missstep so that the trajectory will eventually move back to Θ. Such an approach requires that the objective together with the constraint mappings are well-defined outside of Θ. Put differently, these methods can deal with constraints that are not essential for the well-definedness of the problem. These type of constraints are called *soft constraints*. When a constrain cannot be violated and must be satisfied under all circumstances, then we call it a *hard constraint*. If, e.g., we want to solve $\min_\theta (\theta + 1/\sqrt{\theta + 2})$, then $\theta > -2$ is a hard constraint. The only numerical method that can directly deal with such hard constraints is the projection method, and in the previous example we would choose $\Theta = [-2 + \eta, \infty)$, for $\eta > 0$ small, and project on Θ, to find the solution numerically.

Fewer Constraints Are Better

Sometimes it can be argued that a constraint is not active for the given problem, and therefore can be discarded. Alternatively, if the solution to the problem can be found with a constraint disregarded, and the solution remains feasible under the constraint, then the constraint has no influence on the solution (it is not active at the solution). Consider, for example, a simple economic model where θ denotes a production volume, the profit is increasing in θ, and cost of resources, denoted by $\hat{g}(\theta)$, is also increasing in θ, so that one would like to choose θ as large as possible, while $\hat{g}(\theta) \leq b$, for some budget b. From the model it is clear that one will use the maximal budget, i.e., $\{g(\theta) := \hat{g}(\theta) - b = 0\}$. In words, the constraint can be argued to be an equality constraint.

If a component of the trajectory $\{\theta_n\}$ tends toward the boundary of the feasible set, it is sometimes possible to argue that the particular component of θ_n should be set to the corresponding boundary value. Then, this component of the parameter vector can be excluded from the optimization, which makes the optimization algorithm more efficient.

It is sometimes possible to remove hard constraints such as non-negativity via a transformation of the model. For example, $\min_{\theta \geq 0} f(\theta)$ can be rephrased as the unconstrained problem $\min_\theta f(\theta^2)$. It is worth noting that any such transformation has to be done with care as the objective and therefore the gradient is effected. Indeed, by the chain rule,

$df(\theta^2)/d\theta = 2\theta f'(\theta^2)$ and next to the desired solution $f'(r) = 0$ so that if $\theta = \sqrt{r}$ we get the new and non-informative stationary point $\theta = 0$. In higher dimensions, this effect is known as hairy-ball theorem from algebraic topology.

We conclude by mentioning that a transformation of $\mathbb{R}$ to a bounded interval, say, $[a,b]$ can be achieved by means of the sigmoid function, i.e., we may rewrite $f(\theta)$ as $f(a + (b-a)e^\theta/(1+e^\theta))$. Unfortunately, when the solution of $\min_{\theta \in [a,b]} f(\theta)$ is at one of the boundary points, i.e., a or b, this transformation runs into the problem of the vanishing gradient as the solution has become an improper minimum. In the machine learning literature a wealth of similar transformation mappings can be found for dealing with hard constraints, which all come at the price of performing poorly in case the solution is close to or at the boundary.

Convexity

Most of the algorithms are designed to work well for convex functions, but in practice and particularly when the dimension d is very large, it is often impossible to verify this condition. Rather than a rigorous proof, it is common to argue that there is a unique minimum, as has been illustrated in Example 1.1. An alternative can be achieved by designing an experiment to test statistically if for any two (random) points x, y, (1.2) holds by choosing a random number $\alpha \in (0, 1)$ and testing if $J(\alpha x + (1-\alpha)y) \leq \alpha J(x) + (1-\alpha)J(y)$.

Many important problems are not convex, but as mentioned in Example 1.3, non-convexity does not prevent gradient search from working well. In other cases the cost function has several local minima, and global optimization is required, either restarting the algorithms at random initial points or adding a random perturbation in order to combine gradient search with random exploration methods.

1.6 EXERCISES

Exercise 1.1. Show that a convex function has *convex* level sets, that is, if $x, y \in \mathcal{L}_\alpha(J)$ then any convex combination of x and y is also in the set $\mathcal{L}_\alpha(J)$.

Exercise 1.2. Let $\bar{\theta}$ be a stationary point for which $\nabla^2 J(\bar{\theta})$ has at least one negative and one positive eigenvalue. Using the relationship $\nabla^2 J(\bar{\theta}) v_i = \lambda_i v_i, i = 1, \ldots, d$, for eigenvalues λ_i and eigenvectors $v_i \in \mathbb{R}^d$, and Taylor's expansion around $\bar{\theta}$ in the direction of appropriate eigenvectors, show that $\bar{\theta}$ is neither a local maximum nor a local minimum.

Exercise 1.3. Use a Taylor series expansion to show that for any descent direction $d(\theta)$ at a non-stationary point θ of a twice continuously differentiable function $J(\theta)$, there exists $\epsilon_0 > 0$ such that
$$J(\theta + \epsilon d(\theta)) \leq J(\theta), \text{ for all } 0 \leq \epsilon \leq \epsilon_0.$$

Exercise 1.4. Let $\Theta \subset \mathbb{R}^d$ be a convex set. Show that the Euclidean distance between projection $\Pi_\Theta(x)$ and $\Pi_\Theta(y)$ is bounded by $||x - y||$, for $x, y \in \mathbb{R}^d$.

Exercise 1.5. Give an example of a non-convex set such that the projection on the set is discontinuous. Hint: Consider $\mathbb{R}^2$ with an ellipsoid-shaped area removed.

Exercise 1.6. Consider $f(x): \mathbb{R} \to \mathbb{R}$. Show that if f is Lipschitz with Lipschitz constant L and differentiable, then $|f'(x)|$ is bounded by L on $\mathbb{R}$.

Exercise 1.7. For $J \in C^2$, prove that if $\nabla J(\cdot)$ is Lipschitz continuous on $\mathbb{R}^d$, that is, there is a constant $0 \leq L < \infty$ such that for every $x, y \in \mathbb{R}^d$ $\|\nabla J(x) - \nabla J(y)\| \leq L\|x - y\|$, then

$$-L\|x\|^2 \leq x^\top \nabla^2 J(\theta) x \leq L\|x\|^2,$$

for any $x \in \mathbb{R}^d$, where $\|A\| = \max_{\|x\|=1} \|Ax\|$.

Exercise 1.8. Let $J \in C^2$ and assume that ∇J is Lipschitz continuous. Consider the gradient descent algorithm $\theta_{n+1} = \theta_n - \epsilon_n \nabla J(\theta_n)$ and suppose that θ^* is the unique global minimum, so that $J(\theta) \geq J(\theta^*) > -\infty$ for all $\theta \in \mathbb{R}^d$. Assume furthermore that θ^* is the only stationary point of J. Show that for each initial point $\theta_0 \in \mathbb{R}^d$, θ_n converges to θ^*. (Hint: Inspect the proof of Theorem 1.3 and the use of Lemma 1.1 therein to show that θ_n will converge for any initial value to some finite limit point.)

Exercise 1.9. Prove the following result. Let $J \in C^2$ be such that the gradient is a bounded and Lipschitz continuous function and consider the biased algorithm

$$\theta_{n+1} = \theta_n - \epsilon_n (\nabla_\theta J(\theta_n)^\top + \beta_n(\theta_n)), \qquad (1.51)$$

with $\|\beta_n(\theta_n)\| > 0$ for all n, where

$$\sum_{n=1}^\infty \epsilon_n = +\infty, \quad \sum_{n=1}^\infty \epsilon_n \|\beta_n(\theta_n)\| < \infty, \quad \sum_{n=1}^\infty \epsilon_n^2 < \infty. \qquad (1.52)$$

If $\{\|\nabla J(\theta_n)\| : n \geq 0\}$ is bounded, then any limit θ^* of $\{\theta_n\}$ is a stationary point of $J(\theta)$.

Exercise 1.10. Let $F(u)$ be defined as in Theorem 1.7. Prove that $\nabla_u F(0) = -\eta^*$ in the special case of affine constraints $h(\theta) = a^\top \theta - b$.

Exercise 1.11. Consider a single machine that can operate one piece at a time, and let the service times of the machine constitute a sequence of independent and identically distributed (iid) exponentially distributed random variables. Parts arrive to the machine according to a Poisson process with unit rate. In other words, the time between arrivals of parts is exponentially distributed with mean 1 and that interarrival times are mutually independent. In order for the system to be stable, assume that the expected service time is strictly less than 1. Let $C(\theta) = 1/\theta^2$ be the cost of operating the system at service, θ. Let $P(\theta)$ denote the stationary probability that the queue length is larger than or equal to a threshold b.

(a) Find the solution to the constrained problem:

$$\min C(\theta), \quad \text{s.t.} \ P(\theta) \leq \alpha.$$

Interpret the constraint qualifications, the second-order condition, and the Lagrange multiplier. Hint: Use the fact that the probability that the stationary queue length equals n is given by $(1-\theta)\theta^n$, for $n \in \mathbb{N}$.

(b) Program two different numerical methods to solve this problem for $\alpha = .01$ and $b = 10$. Plot the consecutive values of θ_n and discuss your results, comparing with the theoretical answer in part (a).

Exercise 1.12. Let $J \in C^2$ be convex and assume that ∇J is Lipschitz continuous. For given initial value θ_0, let gradient descent algorithm $\{\theta_n\}$ given through

$$\theta_{n+1} = \theta_n - \epsilon_n \nabla J(\theta_n)^\top,$$

where the gain sequence satisfies

$$\sum_{n=1}^{\infty} \epsilon_n = +\infty, \quad \sum_{n=1}^{\infty} \epsilon_n^2 < \infty.$$

If $\{\|\nabla J(\theta_n)\| : n \geq 0\}$ is bounded, then any (finite) limit θ^* of $\{\theta_n\}$ is a location of the global minimum of $J(\theta)$.

Exercise 1.13. Let $J: \mathbb{R}^d \to \mathbb{R} \in C^2$ be a convex function. Assume that the Hessian of $J(\theta)$ is positive definite at θ. Show that the update of the Newton-Raphson method, given by

$$-[\nabla^2 J(\theta)]^{-1} \nabla J(\theta)^\top$$

is a descent direction at θ.

Exercise 1.14. Consider the function $f(x,y) = x^4 - x^3 + y^4 - 2y^3$. Show that $f(x,y)$ has global minimum at $(3/4, 3/2)$. Now consider the constrained problem

$$\min_{x^2+y^2=1} f(x,y)$$

and show that the solution to the constraint problem is $(0,1)$. Argue that the solution of the constraint problem is not the point on the constraint closest to the unconstrained problem.

Exercise 1.15. Consider $J(\theta) = \theta^3$, and the fixed gain-size gradient descent algorithm

$$\theta_{n+1} = \theta_n - \epsilon 3(\theta_n)^2, n \geq 0.$$

Show that for $0 \leq \theta_0 \leq 1/3\epsilon$, $\{\theta_n\}$ has limit point 0, whereas for $\theta_0 > 1/3\epsilon$ the sequence $\{\theta_n\}$ passes the saddle point and tends to $-\infty$.

Exercise 1.16. Let $J: \mathbb{R}^d \to \mathbb{R} \in C$ be given. Show that the gradient of $J(\theta)$ is the direction of the steepest ascent of $J(\theta)$.

Exercise 1.17. Let $J: \mathbb{R}^d \to \mathbb{R} \in C^1$ be given and let $\Theta \subset \mathbb{R}^d$ be a closed, convex set. Assume that J has no stationary points in Θ. Show that $\theta^* \in \Theta$ is a strict local minimum of J over Θ if there is no $\theta \in \Theta \setminus \{\theta^*\}$ such that

$$\nabla J(\theta^*)(\theta - \theta^*) < 0.$$

Exercise 1.18. Let $\Pi_M(x)$ be the projection onto the hypercube $[-M, M]^d$ for some finite $M > 0$. For $J: \mathbb{R}^d \to \mathbb{R} \in C^1$ show that the projected gradient, in formular, $\Pi_M(\nabla J(\theta))$ is a descent direction.

Chapter Two

The Iterative Method Seen as an Ordinary Differential Equation

Chapter 1 introduced the gradient descent method for unconstrained optimization (1.21), as well as a number of methods for constrained optimization, such as the penalty method of (1.30), the multiplier method in (1.48), the Arrow-Hurwicz method in (1.50), and the projection method of (1.34). All of these algorithms are, in general, of the form

$$\theta_{n+1} = \theta_n + \epsilon_n \, d(\theta_n). \tag{2.1}$$

The algorithms in the previous chapter that use a sequence of subproblems do not quite conform to this general recursive equation, although the subproblems do. In contrast, the two-time scale implementations for those algorithms does conform to this general form (considering a two-dimensional stepsize sequence as appropriate). While the theory provided in the following is easily extended to time homogeneous algorithms, in the following we will, for the sake of simplicity, consider time-homogeneous algorithms like (2.1) only.

This type of recursive algorithm is beneficial in numerical analysis, computer science, and adaptive learning algorithms. In the rest of this book we will be studying algorithms that have a stochastic direction $d(\theta_n)$, and to analyze them we will use some of the concepts introduced in this chapter. We start with a constant stepsize sequence, for ease of presentation.

2.1 MOTIVATION

To set the stage, we illustrate the issues with a simple example. Suppose that we wish to solve

$$\min_{\theta \in \mathbb{R}^2} J(\theta) = \min_{\theta^\top = (\theta_1, \theta_2)} \{2\theta_1^2 + \theta_2^2\}. \tag{2.2}$$

According to the results in Chapter 1, the gradient-based algorithm

$$\theta_{n+1} = \theta_n - \epsilon \nabla J(\theta_n)^\top = \theta_n - \epsilon \begin{pmatrix} 4\theta_{n,1} \\ 2\theta_{n,2} \end{pmatrix}, \tag{2.3}$$

with $\theta_n^\top = (\theta_{n,1}, \theta_{n,2})$, will approximate the solution.

Although our presentation is motivated by an optimization problem, the present analysis methodology is more general for iterative algorithms of the form (2.1). We now change the notation of the variables to x and y, respectively, when studying the convergence properties. We now let $x_n = \theta_{n,1}$ and $y_n = \theta_{n,2}$. Plotting consecutive values of the sequence (2.3):

$$\begin{aligned} x_{n+1} &= x_n - \epsilon \, 4x_n \\ y_{n+1} &= y_n - \epsilon \, 2y_n \end{aligned} \tag{2.4}$$

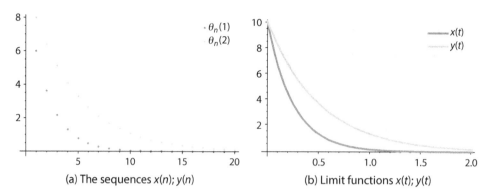

Figure 2.1. Visualizing the algorithm.

we obtain Figure (2.1)(a), using $\epsilon = 0.1$ for initial values $x(0) = y(0) = 10$. The count number on the x-axis is the iteration number. If we plot the same figure using closer points (i.e., smaller ϵ), our mind will interpolate the plots and interpret the graph as a *function* rather than a sequence of points.

Figure (2.2)(c) shows the plots of the continuous functions

$$x(t) = x(0)e^{-4t}$$
$$y(t) = y(0)e^{-2t}$$

and they are "very close" to the interpolation of the dots. How are these two processes related?

Differentiating w.r.t. t, the functions $x(t), y(t)$ satisfy the ordinary differential equations (ODEs):

$$\frac{dx(t)}{dt} = -4x(t)$$
$$\frac{dy(t)}{dt} = -2y(t).$$

To numerically "solve" an ODE over the interval $(0, T)$, Euler proposed the following method. First choose a grid of equally spaced points with spacing $\epsilon > 0$. Call the points $t_n = \epsilon n$, for $0 \leq n \leq T/\epsilon$. The total number of points in the interval is inversely proportional to the subinterval size ϵ. Next, use the finite difference approximation of a derivative at each point t_n to express

$$\frac{x(t_{n+1}) - x(t_n)}{\epsilon} \approx -4x(t_n), \quad \text{and} \quad \frac{y(t_{n+1}) - y(t_n)}{\epsilon} \approx -2y(t_n).$$

The resulting sequences approximate the continuous function and satisfy exactly the recursions (2.4). Figure 2.2 compares the plots of (2.4) using $\epsilon = 0.1$ and a smaller $\epsilon = .01$. As ϵ decreases the plots look more and more similar to Figure 2.2(c).

The iteration count is related to the computing time required to execute the sequential algorithm (2.4), and therefore, it is essential in analyzing the performance of algorithms. If we are only interested in the limit as $n \to \infty$, then we can use the results of Chapter 1. In this chapter, we explore this further, and we will provide a framework to identify not just the limit, but the behavior of the algorithm as a dynamic process. The corresponding

THE ITERATIVE METHOD

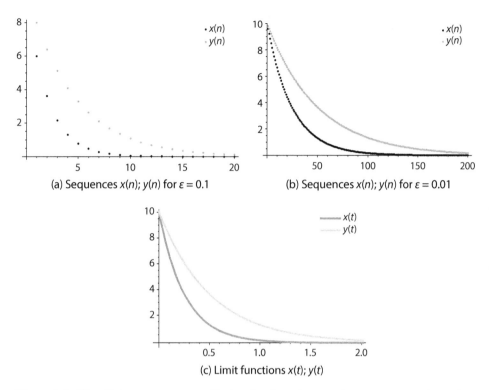

Figure 2.2. Visualizing convergence. Notice that iteration number n corresponds to time $t_n = \epsilon n$, so both plots cover the same interval $t \in [0, 2]$.

ODE is called the *limit process* of the algorithm, where the limit is taken with respect to ϵ.

Remark 2.1. Any function $d(\cdot)$ such that the recursion (2.1) converges to the solution in the original optimization problem in (2.2) is a possible choice, with the negative of the gradient being a natural choice for $d(\cdot)$. However, inspecting the behavior of the algorithm in the above example, it is apparent that the first component approaches zero (which is the solution to (2.2)) faster than the second (actually precisely twice as fast). On occasion, it is possible to first analyze the ODE and tweak the drift $d(\cdot)$ to achieve faster convergence. In this simple illustration, if instead of $-\nabla J^\top$ we use the scaled vector $d(x, y) = (-4x, -4y)^\top$ then both equations will follow the same trajectory, and it is in principle possible to achieve accuracy in both components faster. Observe that such scaling preserves the descent direction of the algorithm, but sometimes it is more convenient to use a surrogate gradient than the actual one. Naturally, the behavior of the ODE (and consequently of the algorithm) will also depend on the initial condition. When we study the problem in this manner, we call the desired trajectory the *target* ODE and then build the algorithm from it. These are concepts that will help us establish the *relative efficiency* of algorithms, a topic that we will discuss in Chapter 6 for the more general stochastic approximation algorithms.

This chapter contains a summary of the theoretical tools required to correctly define "close to" for sequences and functions, and the notion of "convergence." In addition, we present the full proofs of convergence and the methodology for analyzing the behavior of the (ϵ-) limit ODEs.

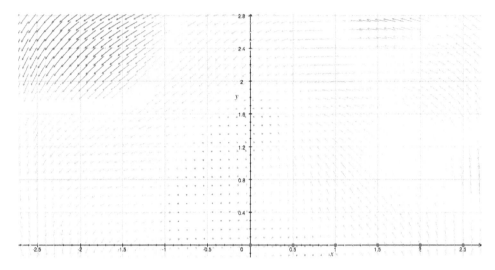

Figure 2.3. Example of a vector field.

2.2 STABILITY OF ODEs

We are concerned with the characterization of the *limit points* of dynamical systems governed by time homogeneous ODEs, because these will often describe the convergence of recursive numerical algorithms of interest. We say that $\mathbf{x} := \{x(t) : t \geq 0\}$ solves the (autonomous) ODE given by $G : E \mapsto E$, for $E \subseteq \mathbb{R}^d$ connected, if

$$\frac{dx(t)}{dt} = G(x(t)), \quad t \geq 0, \tag{2.5}$$

for $x(0) = x_0 \in E$, and we call $x_0 \in \mathbb{R}^d$ the *initial condition*, and $\mathbf{x}$ a *trajectory* of the ODE. For background on ODEs we refer to [78]. Throughout this monograph we assume that the domain E of G in (2.5) is a connected set.

The notion of a *vector field* can be used to understand the dynamic behavior of the successive iterations by visualizing the trajectories as well as their speed. A *vector field* on $E \subseteq \mathbb{R}^d$ is a mapping that assigns a vector $v(x) \in E$ to each point $x \in E$. Figure 2.3 shows an example of a vector field: the vector $v(\theta)$ is shown at each point $\theta \in \mathbb{R}^2$. The size of the arrow is proportional to the magnitude of $v(\theta)$. Such representations are useful to visualize the "drift" of recursive equations. The vector field associated with recursion (2.1) is defined by the "drift" mapping $d(\theta)$. For example, in the recursive description (2.3) we have $d(\theta) = (4\theta_1, \theta_2)^\top$. Successive values of numerical algorithms can be studied as a *dynamical system* that evolves in "time" (iteration number).

The function $G(\cdot)$ in (2.5) is a *vector field*, called the "drift" of the dynamical system. When possible, it can be plotted to help visualize the dynamics of trajectories, as illustrated in Figure 2.3. Note that the vector field displayed in Figure 2.3 is not the one given in (2.3). In the following we will discuss ODEs given via vector fields on $E \subseteq \mathbb{R}^d$.

Definition 2.1. A vector field $G : E \mapsto E$, for $E \subseteq \mathbb{R}^d$, is called *locally Lipschitz* on E if for each $\theta \in E$ an ϵ-neighborhood $N_\epsilon(\theta) = \{x \in E : ||x - \theta|| < \epsilon\} \subset E$ exists such that G is Lipschitz continuous on $N_\epsilon(\theta)$, i.e., there exists a constant $K_\epsilon(\theta)$ such that for all $r, s \in N_\epsilon(\theta)$ it holds that $||G(r) - G(s)|| \leq K_\epsilon(\theta) ||r - s||$.

THE ITERATIVE METHOD

The following theorem is a standard result from the theory of dynamical systems; for example, see Section 2.2 in [240].

Theorem 2.1. *If $G: E \mapsto E$, for $E \subseteq \mathbb{R}^d$, is locally Lipschitz continuous on E, then the initial value problem $dx(t)/dt = G(x(t))$, with $x(0) = x_0 \in E$, has a unique solution $x(t)$ on the time interval $(-T, T)$ for some finite T. Moreover, the solution $x(\cdot)$ is Lipschitz continuous on $(-T, T)$.*

If at time δ_0 the conditions of Theorem 2.1 are fulfilled for $x_0 = x(\delta_0)$, we can consider δ_0 as new starting point in time and $x(\delta_0)$ as new initial value condition and apply Theorem 2.1 again, and thus continue the solution further to larger values of time in a unique way. The question then arises whether the maximal interval onto which the solution can be extended is finite or infinite. It can be shown that if the maximal interval is finite with some endpoint a, then the only possibility for this is that

$$\lim_{t \to a} \|x(t)\| = \infty. \tag{2.6}$$

This leads to the following definition.

Definition 2.2. We say that a solution $x(t)$ of the initial value problem in (2.5) *blows up in finite time* if (2.6) holds for $x(t)$ some finite a.

Hence, if $x(t)$ does not blow up in finite time, then this solution to the initial value problem can be continued in a unique way on $[0, \infty)$. For a proof we refer to Theorem 1.10 in [168]. Later in the text, we will provide in Theorem 2.3 a sufficient condition for an ODE to have trajectories that do not blow up in finite time. We arrive at the following result.

Theorem 2.2. *If $G: E \mapsto E$, for $E \subseteq \mathbb{R}^d$, is locally Lipschitz, then the solution of the ODE $dx(t)/dt = G(x(t))$, with $x(0) = x_0 \in E$, can be uniquely continued (in time t) as long as it does not blow up at some finite time point.*

We illustrate the theorem with the following example.

Example 2.1. Consider $dx(t)/dt = (x(t))^2$, $t \geq 0$, for $x(0) = x_0$, $E = \mathbb{R}$. Then, the solution is given by $x(t) = x_0/(1 - x_0 t)$, and it exists for $t < 1/x_0$. The solution thus blows up in finite time and we cannot extend $x(t)$ beyond $1/x_0$ in a unique way. Alternatively, consider $dx(t)/dt = cx(t)$, for $c > 0$, then $x(t) = e^{ct}$ is the unique solution to the initial value problem with $x(0) = x_0 = c$ and is defined on $[0, \infty)$. Note that $\lim_{t \to \infty} x(t) = \infty$ though $x(t)$ does not blow up in finite time. ✳✳✳

Definition 2.3. Let $G: E \mapsto E$, for $E \subseteq \mathbb{R}^d$, be continuous. A point $\bar{x} \in E$ is called an *equilibrium (or stationary) point* of the ODE:

$$\frac{dx(t)}{dt} = G(x(t)), \quad t \geq 0, \tag{2.7}$$

if $G(\bar{x}) = 0$ for some $x(0) \in E$.

The interpretation is that when $\bar{x}$ is an equilibrium point, then the dynamical system does not move once it reaches $\bar{x}$: it remains at the equilibrium. When (2.7) is considered on the extended reals, i.e., $\mathbb{R} \cup \{-\infty, \infty\}$, then $\bar{x} = +\infty$ (resp. $-\infty$) is called an extended equilibrium point if $\lim_{t \to \infty} G(x(t)) = 0$ for $\lim_{t \to \infty} x(t) = +\infty$ (resp. $-\infty$).

Example 2.2. If the vector field G is parametrized by, say, μ, then the nature of an equilibrium point depends on the choice of μ. For example, $G(x) = \mu - x^2$ has a unique equilibrium point for $\mu = 0$ at $\bar{x} = 0$, and for $\mu > 0$ the equilibrium points are $\bar{x} = \pm\sqrt{\mu}$. If we plot the equilibrium points of G as a mapping of μ, we see that there are no equilibrium points for $\mu < 0$, a unique equilibrium point $\mu = 0$, and a split of the graph with an upper curve $\sqrt{\mu}$ and a lower curve $-\sqrt{\mu}$. Such a graph is called *bifurcation diagram* and the split at $\mu =$ is called a (saddle point) *bifurcation*. ✳✳✳

Definition 2.4. A finite equilibrium point $\bar{x}$ of (2.7) is said to be:

- *stable* if $\forall \epsilon > 0 \, \exists \delta > 0$ such that $\|x(0) - \bar{x}\| < \delta \Rightarrow \|x(t) - \bar{x}\| < \epsilon$ for all $t \geq 0$,
- *asymptotically stable* if it is stable and, in addition, $\exists \delta > 0$ such that if $\|x(0) - \bar{x}\| < \delta$ then $\lim_{t \to \infty} x(t) = \bar{x}$,
- *globally asymptotically stable* if it is stable and, in addition, for all $x(0)$ it holds that $\lim_{t \to \infty} x(t) = \bar{x}$,
- *unstable* if it is not stable.

Moreover, for a stable point $\bar{x}$ the largest set M such that for all initial points $x(0) \in M$, it holds that $\lim_{t \to \infty} x(t) = \bar{x}$, is called the domain of attraction of x^*.

Example 2.3. Consider, $G_\mu(x) = \mu - x^2$, $x \in \mathbb{R}$. For $\mu = 0$, the trajectories can be solved explicitly:

$$x(t) = \frac{1}{\frac{1}{x_0} + t}, \quad t \geq 0,$$

which implies that $x(t)$ tends to 0 as $t \to \infty$, and because $G_0(0) = 0$, then $\bar{x} = 0$ is a stable point. Moreover, it is globally asymptotically stable as $x(t)$ is monotone decreasing toward 0, i.e., the domain of attraction of 0 is $\mathbb{R}$.

For $\mu = -1 < 0$, it can be shown that

$$x(t) = -\tan(\arctan(x_0) + t), \quad t \geq 0,$$

which shows that $x(t)$ tends to $-\infty$ as $t \to \infty$ and G_{-1} has no stationary points. All points are thus unstable. This is of course expected as $G_{-1}(x) = -1 - x^2 = 0$ has no real-valued solution. In the same way, one can see that for $\mu > 0$ the solution $x(t)$ tends to the equilibrium point $\sqrt{\mu}$ as $t \to \infty$, provided that $x_0 > 0$, and as $|x(t) - \sqrt{\mu}|$ is strictly monotone decreasing in t, the equilibrium point is asymptotically stable, with domain of attraction given by $(0, \infty)$. All other points are unstable. ✳✳✳

Asymptotic stability is a desired property as it implies that once the ODE comes close to $\bar{x}$, it will eventually reach $\bar{x}$. The domain of attraction of $\bar{x}$ is then of the form $\{x : \|x - \bar{x}\| \leq \delta\}$, for some $\delta > 0$.

Example 2.4. Consider a linear drift on $\mathbb{R}^d$ with a symmetric $(d \times d)$ matrix $\mathbb{A}$ of full rank, that is,

$$\frac{dx(t)}{dt} = \mathbb{A}x(t), \qquad (2.8)$$

for $x(0) \in \mathbb{R}^d$. Because the matrix is full rank, the only solution to $\mathbb{A}x = 0$ is the zero vector, which shows that the origin is the unique equilibrium point of the above ODE. Recall that all eigenvalues, ρ_k, $1 \leq k \leq d$, of a symmetric matrix are real-valued. If, in addition, the matrix is full rank, then eigenvectors v_k corresponding the eigenvalues ρ_k are orthogonal and span

THE ITERATIVE METHOD

$\mathbb{R}^d$. Specifically, $\mathbb{A} v_k = \rho_k v_k$, and for each point $\bar{x} \in \mathbb{R}^d$ there are unique coefficients α_i such that $\bar{x} = \sum_{k=1}^{d} \alpha_k v_k$. The coefficients $(\alpha_k, k = 1, \ldots, d)$ are also called the *coordinates* of x in the orthogonal basis $\{v_k\}$.

To express $x(t)$ in this coordinate system we set $x(t) = \sum_{k=1}^{d} \alpha_k(t) v_k$, so that

$$\frac{dx(t)}{dt} = \frac{d}{dt} \sum_{k=1}^{d} \alpha_k(t) v_k = \sum_{k=1}^{d} \frac{d}{dt} \alpha_k(t) v_k$$

and

$$\mathbb{A} x(t) = \mathbb{A} \left(\sum_{k=1}^{d} \alpha_k(t) v_k \right) = \sum_{k=1}^{d} \alpha_k(t) \rho_k v_k.$$

By (2.8) this yields

$$\sum_{k=1}^{d} \left(\frac{d\alpha_k(t)}{dt} - \alpha_k(t) \rho_k \right) v_k = 0.$$

Because the eigenvectors are orthonormal they are linearly independent, which implies that for each component k it holds

$$\frac{d}{dt} \alpha_k(t) = \alpha_k(t) \rho_k,$$

which is a one-dimensional ODE with solution $\alpha_k(t) = \alpha_k(0) e^{\rho_k t}$. Therefore, the (unique) equilibrium point $\bar{x} = 0$ is

- stable $\iff$ $\max_k \rho_k \leq 0$. Indeed, if we assume $\max \rho_k = 0$, then at least one eigenvalue has real part null, in which case for each k either $\alpha_k(t) = \alpha_k(0)$, or $\alpha_k(t) = \alpha_k(0) e^{-|\rho_k| t}$. Thus if $\|x(0) - \bar{x}\| \leq \epsilon$, then $\|x(t) - \bar{x}\| \leq \epsilon$;
- globally asymptotically stable $\iff$ $\max_k \rho_k < 0$, so all coordinate processes satisfy $\alpha_k(t) = \alpha_k(0) e^{-|\rho_k| t}$. Therefore $x(t) \to \bar{x} = 0$;
- unstable $\iff$ $\max_k \rho_k > 0$. In this final case, all coordinate processes satisfy $\alpha_k(t) = \alpha_k(0) e^{\rho_k t}$, which means that they diverge to infinity.

✷✷✷

As detailed in the above example for a linear problem, the eigenvalues of the matrix defining the drift characterize the type of stability. This motivates the following definition.

Definition 2.5. A matrix $\mathbb{A}$ is called a *Hurwitz matrix* if the maximum of the real parts of its eigenvalues is strictly negative.

In the following, we will study the stability of an ODE around an equilibrium point via a linearization of the vector field around the stationary point. Let w^* be a stationary point of the vector field G, so that $G(w^*) = 0$, and consider the ODE

$$\frac{dw(t)}{dt} = G(w(t)).$$

Provided that G is twice continuously differentiable, using a Taylor expansion of G at w^* yields

$$G(w) = G(w^*) + \nabla G(w^*)(w - w^*) + O(\|w - w^*\|^2).$$

Close to the stationary point, we approximate the behavior of $w(t)$ by the behavior of the linearized ODE:

$$\frac{dw(t)}{dt} = \mathbb{A}(w(t) - w^*), \qquad (2.9)$$

where $\mathbb{A} = \nabla G(w^*)^T$ is the gradient of the vector field, evaluated at the stationary point, for which $G(w^*) = 0$. The linear system has the solution $w(t) = w^* + e^{\mathbb{A}t}(w(0) - w^*)$, so that the solution $w(t)$ converges to w^* if all the eigenvalues of $\mathbb{A}$ have strictly negative real part; that is, if $\mathbb{A}$ is Hurwitz then w^* is asymptotically stable. For ease of reference, we note this result in the following lemma.

Lemma 2.1. *Let θ^* be a stationary point of vector field G. If $\nabla G(\theta^*)$ is Hurwitz, then θ^* is an asymptotically stable point of the ODE $dx(t)/dt = G(x(t))$.*

Remark 2.2. Lemma 2.1 is the ODE counterpart of Theorem 1.3 for the very specific case of unconstrained optimization, using $G(\theta) = -\nabla J(\theta)$. Indeed, in Theorem 1.3 we have shown that $\{\theta_n\}$, obtained from a gradient descent algorithm started close to some stationary point θ^*, will converge to θ^* with $J(\theta^*)$ strictly monotone decreasing provided that the Hessian of J is positive definite at θ^*, i.e., $-\nabla^2 J(\theta^*)$ is Hurwitz.

Example 2.5. Consider the following problem:

$$\min_{\theta=(\theta_1,\theta_2)^\top \in \mathbb{R}^2} J(\theta) \stackrel{\text{def}}{=} 2\theta_1^2 + \theta_2^2,$$

with solution at the origin. Let us analyze the behavior of the ODE driven by the negative gradient:

$$\frac{dx(t)}{dt} = -(\nabla J(x(t)))^\top.$$

The gradient and Hessian at a point $\theta = (\theta_1, \theta_2)^\top \in \mathbb{R}^2$ are given by

$$\nabla J(\theta) = (4\theta_1, 2\theta_2)$$

and

$$\nabla^2 J(\theta) = \begin{pmatrix} 4 & 0 \\ 0 & 2 \end{pmatrix}.$$

We can verify that the vector field $G = -\nabla J$ here has a unique equilibrium point at the origin (the same as the stationary point for the optimization problem, so it satisfies the first-order condition for optimality). The eigenvalues ρ of $\mathbb{A}$, with $\mathbb{A} = -\nabla^2 J(\theta)$, satisfy $(-4-\rho)(-2-\rho) = 0$, so that $\rho_1 = -2$ and $\rho_2 = -4$ are negative, which implies that the origin is asymptotically stable (it also implies that the cost function is convex so it satisfies the second-order condition for optimality). Moreover, as the ODE is linear, all the trajectories $x(t)$ move exponentially fast to the origin, regardless of the initial value $x(0)$. The vector field for the gradient search algorithm is shown in Figure 2.4, with some trajectories streamlined for easy visualization. The corresponding trajectories are those shown in Figure 2.2(c).   ✳✳✳

Example 2.6. Consider the following problem:

$$\min_{\theta=(\theta_1,\theta_2)^\top \in \mathbb{R}^2} J(\theta) \stackrel{\text{def}}{=} 2\theta_1^2 + \theta_1^2 \theta_2^2,$$

THE ITERATIVE METHOD

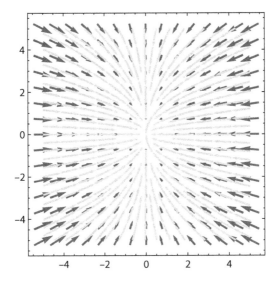

Figure 2.4. Stability for unconstrained optimization, $\mathbb{A}$ Hurwitz, Example 2.5.

with solution at the origin. Let us analyze the behavior of $dx(t)/dt = -(\nabla J(x(t)))^\top$. In this case the gradient and Hessian at $\theta \in \mathbb{R}^2$ are given by

$$\nabla J(\theta) = \begin{pmatrix} 4\theta_1 + 2\theta_1\theta_2^2 \\ 2\theta_1^2\theta_2 \end{pmatrix}, \quad \mathbb{A} = -\nabla^2 J(0) = \begin{pmatrix} -4 & 0 \\ 0 & 0 \end{pmatrix}.$$

The eigenvalues ρ of $\mathbb{A}$ satisfy $(-4-\rho)(-\rho) = 0$, so that $\rho_1 = -4, \rho_2 = 0$ and $\mathbb{A}$ fails to be Hurwitz. The vector field is shown in Figure 2.5(a) with some trajectories streamlined for easy visualization. Every point with $\theta_1 = 0$ is a stationary point of the ODE. Therefore, the limit point of a trajectory $x(t)$ may depend on the initial condition. Notice that although there is a unique value for the minimum $J(\theta^*) = 0$, the set of optimal values $\{\theta \in \mathbb{R}^2 : J(\theta) = 0\}$ is not unique: all stationary points are minimizers. Thus the ODE will drive the trajectories toward optimality, but the limit points depend on the initial condition $x(0)$. Put differently, different initial conditions will yield different limit points. Besides, the vector field is relatively much weaker as it approaches the stability region. Figure 2.5(b) plots the trajectory of the corresponding approximation $\theta_{n+1} = \theta_n - \epsilon \nabla J(\theta_n)$ starting at the initial point $(10, 10)$. It shows that many more iterations would be required to get reasonably close to the limit, as expected. ✳✳✳

The two examples above illustrate cases of the limit behavior associated with a gradient-driven ODE, and they demonstrate how the ODE trajectories can provide insight into the iterative gradient descent algorithm. We will present an example later on where the trajectories exhibit more complex behavior around the stable point. Specifically, we will show an example where the stable point is reached via spiral trajectories.

Remark 2.3. To link this material with Chapter 1, consider $J \in C^2$ as a function that we wish to minimize (without constraints), and let x^* be such that the Hessian of J at x^* is positive definite. Then, the function J behaves in a neighborhood of x^* like a convex function and it is conceivable that any descent direction algorithm, once entering the neighborhood of x^*, will be attracted to the stationary point x^*. Observe that in the special case that the descent direction is given by the negative gradient we have $\mathbb{A} = -\nabla^2 J(x^*)$ in (2.9), so that

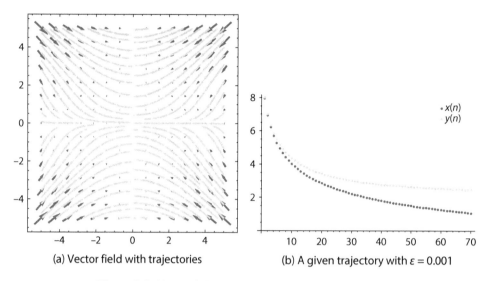

Figure 2.5. Vector field and approximation, Example 2.6.

the positive definiteness condition on the Hessian $\nabla^2 J(x^*)$ (introduced in Chapter 1 as the second-order condition for optimality) is equivalent to $\mathbb{A}$ being Hurwitz.

2.3 PROJECTED ODEs

For completeness and rigorous definition, we provide in this section the formal framework for working with ODEs in context of projections. The reader may skip this section at a first reading. Recall that for $\Theta \subset \mathbb{R}^d$, the projection Π_Θ on Θ is defined as

$$\Pi_\Theta(x) = \arg\min_{z \in \Theta} ||x - z||.$$

Projections are characterized by the normals to the surface of Θ to be defined presently.

Definition 2.6. Let Θ be a closed set, and denote by $\delta\Theta$ the boundary of Θ. For $x \in \delta\Theta$, we define the set of inward normals to Θ at x by

$$n(x) = \{\gamma : ||\gamma|| = 1, \gamma^\top (y - x) \geq 0, \forall y \in \Theta\},$$

where the condition $||\gamma|| = 1$ is for notational convenience. The construction of $n(x)$ is illustrated in Figure 2.6.

For $x \notin \Theta$, Π_Θ projects x onto Θ as $\Pi_\Theta(x) = x + \alpha y$ for some $\alpha > 0$ and $y \in n(x)$. Note that if Θ in Definition 2.6 is convex with smooth boundaries (so that the boudary has no "kinks"; see Figure 2.6), then $n(x)$ contains as a single element the (inward) normal vector to the tangent space to Θ at x.

ODEs can be numerically approximately solved by using the Euler scheme, or Euler method, which is a discrete approximation of the derivative dynamics. In the following, we work with the Euler scheme for solving approximately an ODE using a discretization grid of size Δt. It is given by

$$x(t + \Delta t) = \Pi_\Theta(x(t) + \Delta t G(x(t))) \quad (2.10)$$

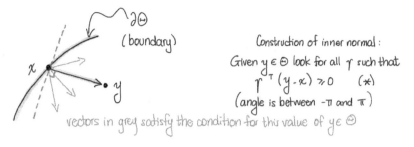
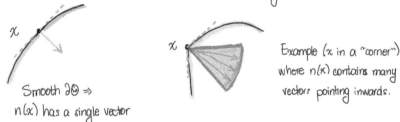

Figure 2.6. Set $n(x)$ of the inward normals at a point x on the boundary.

and writing this using the projection force (introduced in (1.36) for the special case of the vector field $G(\theta) = -\nabla J^\top(\theta)$), we obtain

$$x(t+\Delta t) = x(t) + \Delta t \left(G(x(t)) + \frac{1}{\Delta t}\Big(\Pi_\Theta\big(x(t)+\Delta t\, G(x(t))\big) - (x(t)+\Delta t\, G(x(t)))\Big)\right)$$

$$= x(t) + \Delta t\, (G(x(t)) + Z(\Delta t, x(t), G(x(t)))),$$

where

$$Z(\Delta t, x(t), G(x(t))) = \frac{1}{\Delta t}\Big(\Pi_\Theta\big(x(t)+\Delta t\, G(x(t))\big) - (x(t)+\Delta t\, G(x(t)))\Big),$$

which gives

$$\frac{x(t+\Delta t)-x(t)}{\Delta t} = \frac{1}{\Delta t}\Big(\Pi_\Theta\big(x(t)+\Delta t\, G(x(t))\big) - x(t)\Big)$$
$$= G(x(t)) + Z(\Delta t, x(t), G(x(t))).$$

For $x \in \Theta$ and $v \in \mathbb{R}^d$, we introduce the *directional derivative of the projection at x in direction of v* as

$$\pi(x,v) = \lim_{\delta \downarrow 0} \frac{1}{\delta}(\Pi_\Theta(x+\delta v) - x). \tag{2.11}$$

If x is an inner point of Θ, then, for δ sufficiently small, $x+\delta v$ lies in Θ, and $\Pi_\Theta(x+\delta v) - x = \delta v$, so that $\pi(x,v) = v$. It is shown in [94, 336] that if, on the other hand, x lies on the boundary of Θ, then

$$\pi(x,v) = v + \max(-v^\top \gamma(x,v), 0)\gamma(x,v), \tag{2.12}$$

where $\gamma(x,v)$ is an element $\rho \in n(x)$ that maximizes $-v^\top \rho$. In words, moving from x in direction of v along the boundary of Θ one achieves infinitesimal change $\pi(x,v)$. Note that

$\gamma(x, v)$ is uniquely defined for Θ convex with smooth boundaries, such as the d-dimensional ball $\Theta = B_r = \{\theta : ||\theta|| \leq r\}$. In case that the boundary has kinks, such as for the hypercube $\Theta = [-M, M]^d$, $\gamma(x, v)$ may not be uniquely defined.

From the above we see that

$$\lim_{\Delta t \to 0} \frac{1}{\Delta t} \left(\Pi_\Theta(x(t) + \Delta t\, G(x(t))) - x(t) \right) = \lim_{\Delta t \to 0} \left(G(x(t)) + Z(\Delta t, x(t), G(x(t))) \right)$$
$$= \pi(x(t), G(x(t))), \qquad (2.13)$$

which gives

$$\lim_{\Delta t \to 0} Z(\Delta t, x(t), G(x(t))) =: \mathbf{Z}(x(t), G(x(t))) \qquad (2.14)$$
$$= \pi(x(t), G(x(t))) - G(x(t)).$$

We summarize our discussion in the following definition.

Definition 2.7. Let $\Theta \subseteq E$ be a closed set and G a vector field on $E \subseteq \mathbb{R}^d$ such that the ODE

$$\frac{d}{dt} x(t) = G(x(t))$$

is well-defined for $x(0) \in \Theta$. Using the notation for the directional derivative of the projection in (2.12), we define the version of the above ODE projected on Θ, or the *projected ODE*, by

$$\frac{d}{dt} x(t) = G(x(t)) + \mathbf{Z}(x(t), G(x(t))) = \pi(x(t), G(x(t))),$$

which we will denote in shorthand notation as

$$\frac{d}{dt} x(t) = (\Pi_\Theta G)(x(t)).$$

Remark 2.4. Note that $\|G(x(t))\|$ bounds $\|\mathbf{Z}(x(t), G(x(t)))\|$ by construction, that is, $\|\mathbf{Z}(x(t), G(x(t)))\| \leq \|G(x(t))\|$ for all $x(t) \in \Theta$. Moreover, note that for continuous G on E the corresponding vector field of the projected ODE, namely $(\Pi_\Theta G)(\cdot)$, is discontinuous as a vector field on Θ as well as E. However, it can be shown that for continuous G the projected ODE has unique, continuous trajectories, and we refer to [94] and [336] for details.

Example 2.7. Let $x(t)$ be the solution of the ODE

$$\frac{d}{dt} x(t) = G(x(t)), t \geq 0,$$

with $x(0) = x_0$. Suppose that we wish to restrict the solution to the unit ball $B = \{x \in \mathbb{R}^d : \|x\| \leq 1\}$. In order to construct the projected ODE, first note that for x on the surface of B the inward normal is uniquely given by $-x/\|x\|$, i.e.,

$$n(x) = \left\{ -\frac{x}{\|x\|} \right\},$$

which implies that

$$\gamma(x, G(x)) = -\frac{x}{\|x\|}.$$

THE ITERATIVE METHOD   51

The projected ODE then becomes

$$\frac{d}{dt}x(t) = G(x(t)) - \max\left(G^\top(x(t))\frac{x(t)}{||x(t)||}, 0\right)\frac{x(t)}{||x(t)||}.$$

In words, the correction force is given by stretching the inward normal by the scalar projection of $G^\top(x(t))$ onto the outward normal at $x(t)$, given by $\frac{x(t)}{||x(t)||}$. If $G(x(t))$ is perpendicular to the tangent-space to B_r at $x(t)$ and pointing outward, then it holds that

$$G(x(t)) = \max\left(G^\top(x(t))\frac{x(t)}{||x(t)||}, 0\right)\frac{x(t)}{||x(t)||},$$

as it should be so that $dx(t)/dt = 0$.   ✺✺✺

As the above example shows, computing the projected ODE $(\Pi_\Theta G)(\cdot)$ for given Θ and G is a non trivial task. However, inspecting (2.10), it is easy to see that $(\Pi_\Theta G)(\cdot)$ can be approximated via the Euler method since for Δt sufficiently small

$$(\Pi_\Theta G)(x(t)) \approx \Pi_\Theta(x(t) + \Delta t G(x(t))) \tag{2.15}$$

and the above right-hand side can be used for numerical approximating of the ODE trajectory.

2.4 ON BOUNDEDNESS OF THE TRAJECTORIES OF AN ODE

Definition 2.8. Let $G : E \mapsto E$, for $E \subseteq \mathbb{R}^d$, be a vector field. We say that the ODE $dx(t)/dt = G(x(t))$ has bounded trajectories if for all $x_0 \in E$ there exits a bounded set $H_{x_0} \subset E$ such that

$$x(t) \in H_{x_0}, \quad t \geq 0,$$

with initial value $x(0) = x_0$.

An ODE fails to have bounded trajectories if there exists at least one initial value x_0 such that for the corresponding trajectory $||x(t)|| \to \infty$ for $t \to a$, with $a \leq \infty$. More specifically, if an ODE has bounded trajectories, then the trajectories do not blow up in finite time and, provided G is locally Lipschitz, this implies uniqueness of $x(t)$ on $[0, \infty)$ for any initial condition.

A sufficient condition for an ODE with drift vector field G to have bounded trajectories is the following *center of mass condition*: there exists $r > 0$ such that

$$G^\top(\theta)\theta < 0 \quad \text{for } ||\theta|| \geq r. \tag{2.16}$$

In words, outside a ball $B_r = \{\theta : ||\theta|| \leq r\}$ around the origin with radius r, the vector field $G(\theta)$ "points toward B_r." This implies that any trajectory started inside of B_r will stay inside B_r forever, and that for any initial point outside of B_r the trajectories move toward B_r and are therefore bounded.

In the following example, we illustrate how boundedness of the trajectories of an ODE can be established by *Lyapunov functions*.

Example 2.8. Let $J(\theta) \in C^1$ and denote the set of stationary points of $J(\theta)$ by S_J. Consider a continuous descent direction vector field $G(\theta)$ for $J(\theta)$ on $\mathbb{R}^d$, i.e., $\nabla J(\theta)G(\theta) < 0$ for all $\theta \notin S_J$. Assume that θ^* is the unique location of the minimum of $J(\theta)$, that is, $S_J = \{\theta^*\}$ and let

$$J^* \stackrel{\text{def}}{=} J(\theta^*) = \min_\theta J(\theta).$$

Introduce the *Lyapunov function* $V(t)$ by

$$V(t) = J(x(t)) - J^*, \qquad (2.17)$$

for $t \geq 0$, with initial value $V(0) = V(x(0))$ for $x(0) = x_0$.

Since $G(\theta)$ is a descent direction, then for $x(t) \neq \theta^*$ it follows that

$$\frac{dV(t)}{dt} = \nabla J(x(t)) \frac{dx(t)}{dt} = \nabla J(x(t)) G(x(t)) < 0. \qquad (2.18)$$

Because $0 \leq V(t) \leq V(0)$ and $V(t)$ is strictly monotone decreasing by (2.18), then

$$V(0) \geq V(0) - V(t) = -\int_0^t V'(u) du = \int_0^t (-\nabla J(x(u)) G(x(u))) \, du \geq 0.$$

This shows that $\int_0^\infty (-\nabla J(x(u)) G(x(u))) \, du$ is finite, and as $-\nabla J(x(u)) G(x(u)) \geq 0$, this gives

$$\lim_{t \to \infty} \nabla J(x(t)) G(x(t)) = 0.$$

Since we assumed that $J(\theta)$ has unique stationary point θ^*, continuity of $\nabla J(x(t)) G(x(t))$ implies that $x(t)$ approaches θ^* as $t \to \infty$ for any initial point. Hence, the ODE driven by a descent direction G traces the stationary point of $J(\theta)$ and thus the solution of the problem $\min_\theta J(\theta)$. Moreover, θ^* is the globally asymptotically stable point of G and, since $\{x(t)\}$ converges to a finite limit point, then the ODE driven by G has bounded trajectories. ✼✼✼

The following theorem provides an analytical condition for boundedness of a vector field along its trajectories, which is a weaker condition than that of boundedness of trajectories. The statement is a special case of the statement in Lemma 2.1. The condition of full rank of ∇G is used for ease of proof here.

Theorem 2.3. *For any $x \in \mathbb{R}^d$ assume that $\nabla G(x)$ exists and is of full rank. Let $\rho_k(x)$ denote the real part of the k-th eigenvalue of $\nabla G(x)$. If the following Hurwitz condition*

$$\forall x \in \mathbb{R}^d \quad \max_k (\rho_k(x)) =: -\rho_{\max} \leq 0$$

holds, then G is bounded along its trajectories and the solutions the of ODE do not blow up in finite time.

Proof. Differentiation yields

$$\frac{d}{dt} G(x(t)) = \nabla G(x(t)) \frac{dx(t)}{dt} = \nabla G(x(t)) G(x(t)).$$

At $x = x(t)$, the rate of growth of each of the coordinates of $G(x)$ is bounded by $-\rho_{\max}$, so that if $y(t) = |G(x(t))|$, where $|\cdot|$ denotes the Euclidean norm, then

$$\frac{dy(t)}{dt} \leq -\rho_{\max} y(t),$$

and using Grönwall's lemma, $y(t) \leq y(0) e^{-\rho_{\max} t} \leq y(0)$. This implies that $G(x(t))$ is bounded for all $t \geq 0$ (even if $G(x)$ itself may not be bounded for all x), and the trajectory $x(t)$ therefore can not blow up in finite time. □

2.5 ODE LIMIT OF RECURSIVE ALGORITHMS

Consider a recursive equation of the form

$$\theta_{n+1} = \theta_n + \epsilon_n G(\theta_n), \quad n \geq 0 \text{ and } \theta_0 \in E, \qquad (2.19)$$

for some appropriate vector field $G: E \mapsto E$, for $E \subseteq \mathbb{R}^d$. This formulation covers all the gradient-based optimization methods for unconstrained as well as constrained problems that we mentioned in Chapter 1, notably the Arrow-Hurwicz method (1.50). Although our notation is suggestive of a "gradient G," we will see that there is no need to restrict the analysis to the case of the gradient, and general descent directions are covered by our analysis as well. Importantly, we will give particular attention to the *constant gain* case, where $\epsilon_n \equiv \epsilon$ is a constant.

Remark 2.5. Interestingly, in his book *Institutionum Calculi Integralis* (published between 1768 and 1770) Euler proposed to use (2.19) with constant stepsize ϵ to solve the differential equation (2.7). We proceed in the inverse way here, starting with a difference equation and establishing conditions under which a "limit" ODE can be used to assess the behavior of successive iterations.

We wish to study the limit of the sequence of points θ_n given by (2.19) and compare it to the solution of an ODE. In order to do so, we will first interpret the sequence θ_n as mapping of t and thus obtaining an interpolation process to be defined formally in the following.

Definition 2.9. For any sequence $\epsilon = \{\epsilon_n\}$, with $\epsilon_n > 0$, the interpolation process $\vartheta^\epsilon(\cdot)$ of the recursion put forward in (2.19) is defined for $t \in \mathbb{R}^+$, $\vartheta^\varepsilon(0) = \theta_0$ by

$$\vartheta^\epsilon(t) = \theta_{m(t)},$$

where

$$m(t) \stackrel{\text{def}}{=} \max\{n : t_n \leq t\}, \text{ for } t > 0, m(0) = 0; \quad t_n \stackrel{\text{def}}{=} \sum_{i=0}^{n-1} \epsilon_i, \text{ for } n \geq 1, t_0 = 0. \qquad (2.20)$$

By convention, $m(t) = 0$ if the set on the above left-hand side is empty, and $t_0 = 0$. In case $\epsilon_n \equiv \epsilon$ we simply write $\vartheta^\varepsilon(\cdot)$ for the interpolation process. A typical interpolation process for a decreasing gain size algorithm is depicted in Figure 2.7. Moreover, for $n \geq 0$, we define the *shifted interpolation process* as $\vartheta^\varepsilon(t_n + t), t \geq 0$.

Example 2.9. Recall the recursion put forward in (2.3) in the motivation Section 2.1. The interpolation process with the fixed stepsize $\epsilon = 0.1$ corresponding to the recursion shown in Figure 2.1(a) is displayed in Figure 2.8. ✲✲✲

The following example illustrates that the limit of a point-wise convergent sequence of continuous functions may fail to be continuous.

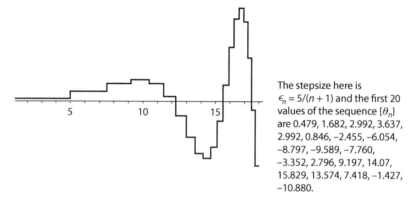

The stepsize here is
$\epsilon_n = 5/(n+1)$ and the first 20
values of the sequence $\{\theta_n\}$
are 0.479, 1.682, 2.992, 3.637,
2.992, 0.846, −2.455, −6.054,
−8.797, −9.589, −7.760,
−3.352, 2.796, 9.197, 14.07,
15.829, 13.574, 7.418, −1.427,
−10.880.

Figure 2.7. The interpolated process $\vartheta^\epsilon(t) = \theta_{m(t)}$.

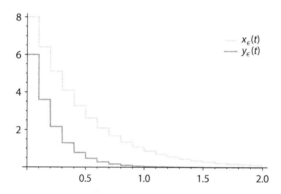

Figure 2.8. The interpolation process $x_\epsilon(t)$.

Example 2.10. This example illustrates that the limit of a point-wise convergent sequence of continuous functions may not converge in sup-norm and may also fail to be continuous. Consider the sequence of functions $f_n : [0, T] \mapsto \mathbb{R}$:

$$f_n(x) = \begin{cases} 0 & 0 \leq x < 1 - \frac{1}{n}, \\ nx + (1-n) & 1 - \frac{1}{n} \leq x < 1, \\ 1 & 1 \leq x \leq T, \end{cases}$$

so each f_n is a continuous and piecewise linear function. For each $x < 1$, and all $n > 1/(1-x)$, $f_n(x) = 0$, so that $\lim_{n \to \infty} f_n(x) = 0$. As well, if $x \geq 1$ then $f_n(x) = 1$ for all n. This implies that $f_n(x)$ converges point-wise to the *function*

$$f(x) = \begin{cases} 0 & \text{if } 0 \leq x < 1, \\ 1 & \text{if } 1 \leq x \leq T, \end{cases}$$

which is not a continuous function. That is, f_n does not converge in the sup-norm. ✻✻✻

To study the convergence of the interpolation process, we need to establish an appropriate norm on the functional space where the interpolations are defined.

THE ITERATIVE METHOD

Definition 2.10. For any $N \in \mathbb{N}$ and $a, b \in \mathbb{R}$, with a, b, call $F_N(a, b)$ the space of functions $f: [a, b] \mapsto \mathbb{R}^N$. The *sup-norm* of a mapping $f \in F_N(a, b)$ is given by

$$\|f\|_{\infty, [a,b]} = \sup_{x \in [a,b]} \|f(x)\|.$$

In the above example, the "limiting" function contains a jump, so it's discontinuous. For our project, we are faced with an even harder problem as we are studying the limit of interpolation processes, which are piecewise constant functions, and therefore not even continuous. The concept of equicontinuity in the extended sense, as introduced presently, is a generalization of uniform continuity and will provide sufficient conditions for convergence of our interpolation processes to a continuous limit.

Definition 2.11. A family $\{f_n\}$ of mappings from $(-\infty, \infty)$ to $\mathbb{R}^N$, for some $N \geq 1$, is called *equicontinuous in the extended sense* if (i) $\|f_n(c)\| \leq M < \infty$ for all n and some fixed point c, and (ii) for each $T > 0$ and $\eta > 0$ there exists $\delta_{\eta, T} > 0$ such that

$$\limsup_{n \to \infty} \sup_{0 \leq t-s \leq \delta_{\eta,T}, |t| \leq T} \|f_n(t) - f_n(s)\| < \eta.$$

For a proof of the following theorem we refer to [93, 265].

Theorem 2.4 (Ascoli-Arzelà). *Let the family $\{f_n\}$ be equicontinuous in the extended sense. Then every sequence has a convergent subsequence, and all accumulation points are continuous functions on $(-\infty, \infty)$.*

The equicontinuity condition puts a bound on the modulus of continuity of the whole family of functions, effectively bounding the slopes or growth rate across the whole family. In our example above, the slope of the functions is n, which is not bounded, thus creating a discontinuity for the limit function, even when each of the functions in the set fails to be continuous. For more details on equicontinuity in the extended sense, we refer to [198].

We now establish the main result of this chapter, which proves the existence of a unique solution of the ODE. In the light of the above discussion, this is a surprising result as, under the conditions put forward in the following theorem, the limit of a sequence of discontinuous functions (in this case, the interpolation processes) is a continuous function.

Theorem 2.5. *Let $G: \mathbb{R}^d \mapsto \mathbb{R}^d$ be a vector field and consider the recursion*

$$\theta_{n+1}^\epsilon = \theta_n^\epsilon + \epsilon \left(G(\theta_n^\epsilon) + \beta_\epsilon(\theta_n^\epsilon) \right),$$

with constant stepsize $\epsilon > 0$. Let $x_\epsilon(t) = \vartheta^\epsilon(t), 0 \leq t$, denote the interpolation process of $\{\theta_n^\epsilon\}$, with $x_\epsilon(0) = \theta_0^\epsilon = \theta_0$. Assume that

(C1) G is either Lipschitz continuous or bounded,
(C2) the solution $\{x(t) : t \geq 0\}$ to the initial value problem

$$\frac{dx(t)}{dt} = G(x(t)), \quad x(0) = \theta_0 \in \mathbb{R}^d, \quad (2.21)$$

is uniquely defined for $t \geq 0$.
(C3) $\sup_\theta \|\beta_\epsilon(\theta)\| = O(\epsilon)$.

Let $\{x(t): t \geq 0\}$ denote the limit of a convergent ϵ-subsequence of $\{x_\epsilon(t): t \geq 0\}$ and assume that $\sup_\theta \|\beta_\epsilon(\theta)\| = O(\epsilon)$. Then $x(t)$ is the unique solution of the ODE in (2.21). Moreover, the convergence of $x_\epsilon(t)$ happens uniformly over finite time-intervals, i.e., for any $0 \leq a, b < \infty$ it holds that $\sup_{a \leq t \leq b} |x_\epsilon(t) - x(t)| \to 0$ as ϵ tends to zero.

Remark 2.6. For Theorem 2.5, condition (C1) will ensure equicontinuity of the family $\{x_\epsilon(\cdot)\}$, as we will show in the proof. Condition (C2) ensures uniqueness of the target trajectory in (2.21). Condition (C3) assumes that the bias is asymptotically negligible. Note that we assume that G is vector field on $\mathbb{R}^d$ (i.e., $E = \mathbb{R}^d$), so that θ_n^ϵ is feasible for all n and ϵ.

The following proof is neither the simplest nor the most elegant one. The method of proof, however, will become an important tool when we analyze the stochastic version of the recursions (2.19). In the case of constant stepsize ϵ we have $t_n = n\epsilon$ and $m(t) = \lfloor t/\epsilon \rfloor$ is the integer part of t/ϵ.

Proof. The proof has the following main parts: (i) the integral representation, (ii) the "equicontinuity in the extended sense" argument, and (iii) the proof that the limiting process of the interpolation process is given as the solution of the ODE in (2.21), which is called "compactness argument" in [198, 305]. We first prove the statement for the unbiased case, i.e., $\beta_n(\theta_n^\epsilon) = 0$ for all n.

We now turn to part (i) of the proof. First, from (2.19), we obtain

$$\theta_{n+m}^\epsilon - \theta_n^\epsilon = \sum_{i=n}^{n+m-1} (\theta_{i+1}^\epsilon - \theta_i^\epsilon) = \sum_{i=n}^{n+m-1} \epsilon G(\theta_i^\epsilon), \tag{2.22}$$

which, using $x_\epsilon(t) = \theta_{m(t)}^\epsilon$ (and letting $m(t+s) = m+n$, $m(t) = n$) is equivalent to

$$x_\epsilon(t+s) - x_\epsilon(t) = \sum_{i=m(t)}^{m(t+s)-1} \epsilon G(\theta_i^\epsilon). \tag{2.23}$$

Because $x_\epsilon(\cdot)$ is piecewise constant, $G(x_\epsilon(\cdot))$ is also piecewise constant and its jump times are given by $\{t_n = n\epsilon, n \in \mathbb{N}\}$. Thus the definite integral on $[t, t+s]$ of $G(x_\epsilon(\cdot))$ is a sum that can be expressed via considering all jumps t_n that fall into the interval $[t, t+s]$:

$$\int_t^{t+s} G(x_\epsilon(u))\, du = \sum_{i=m(t)}^{m(t+s)-1} \epsilon G(\theta_i^\epsilon) + \rho(\epsilon),$$

where $\rho(\epsilon)$ is the error in the approximation, due to the discretization at the end points. See Figure 2.9 for an illustration of the integral approximation.

When both t and $t+s$ are multiples of ϵ, this approximation error is null; see Figure 2.9. Therefore,

$$x_\epsilon(t+s) - x_\epsilon(t) = \int_t^{t+s} G(x_\epsilon(u))\, du - \rho(\epsilon). \tag{2.24}$$

We now turn to part (ii). The proof of establishing equicontinuity of x_ϵ in the extended sense. Consider the interval $[r, q]$, for $0 \leq r < q < \infty$. Notice that the sum in (2.23) contains $m(q) - m(r)$ terms. We now treat the case of boundedness of G and Lipschtiz continuity seperately.

THE ITERATIVE METHOD

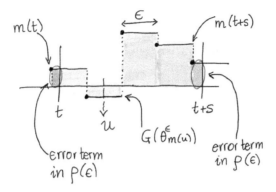

Figure 2.9. The sum and its approximation by an integral.

- If G is bounded by some finite constant $\bar{G}$, then this yields, for small ϵ,

$$\|x_\epsilon(q) - x_\epsilon(r)\| = \left\|\sum_{i=m(r)}^{m(q)-1} \epsilon G(\theta_i^\epsilon)\right\| \leq \epsilon \bar{G} \frac{(q-r)}{\epsilon} = \bar{G}(q-r). \quad (2.25)$$

- If G is Lipschtiz continuous we argue as follows:

$$\|G(\theta_{n+1}^\epsilon) - G(\theta_n^\epsilon)\| \leq L\|\theta_{n+1}^\epsilon - \theta_n^\epsilon\| = \epsilon L\|G(\theta_n^\epsilon)\|,$$

where L is the Lipschitz constant for G. Thus, by the triangle inequality

$$\|G(\theta_{n+1}^\epsilon)\| \leq \|G(\theta_n^\epsilon)\| + (G(\theta_{n+1}^\epsilon) - G(\theta_n^\epsilon)) \leq \|G(\theta_n^\epsilon)\|(1 + \epsilon L)$$

and by induction, $\|G(\theta_{n+m}^\epsilon)\| \leq \|G(\theta_n^\epsilon)\|(1+\epsilon L)^m$ for all m. We now apply this inequality for $n = 0$ and $m = m(q)$ (recall that $m(t) = \lfloor t/\epsilon \rfloor$), and obtain

$$\|G(\theta_m^\epsilon)\| \leq \|G(\theta_0)\|(1+\epsilon L)^{m(q)} \leq \|G(\theta_0)\|(1+\epsilon L)^{q/\epsilon}, \quad m \leq m(q).$$

Use now the fact that for ϵ tending to zero, $(1 + \epsilon L)^{q/\epsilon}$ is monotone increasing toward $\lim_{\epsilon \to 0}(1 + \epsilon L)^{q/\epsilon} = e^{Lq}$. Hence, for ϵ smaller than some sufficiently small ϵ_0 we obtain the uniform bound $\hat{G} = \|G(\theta_0)\| e^{Lq}$, for all $G(\theta_n^\epsilon)$ such that $n \leq m(q)$ and sufficiently small ϵ. Note that θ_0 is independent of ϵ as we keep the initial value fixed. This shows that $G(\theta_n^\epsilon)$ for $n < m(q)$ and $\epsilon \leq \epsilon_0$ is bounded by $\bar{G} = \|G(\theta_0)\| e^{LT}$, for $q \leq T$ and any $T \geq 0$. Therefore

$$|x_\epsilon(q) - x_\epsilon(r)| \leq \bar{G}(q-r), \quad \epsilon \leq \epsilon_0.$$

To summarize, for ϵ sufficiently small, we have shown that for any $\eta > 0$ and any $T < \infty$, it follows that $\|x_\epsilon(q) - x_\epsilon(r)\| \leq \eta$ whenever $|q - r| \leq \delta := \eta/\bar{G}$ and $q \leq T$, i.e.,

$$\sup_{0 \leq q, r, |q-r| \leq \delta, q \leq T} \|x_\epsilon(q) - x_\epsilon(r)\| \leq \eta,$$

for ϵ sufficiently small. This, together with the fact that $x_\epsilon(0) = \theta_0$ for all ϵ, shows equicontinuity in the extended sense of $\{x_\epsilon\}$. Moreover, the above arguments show that for any $T > 0$, $\bar{G}$ exists such that $\|\rho(\epsilon)\|$ is uniformly bounded for $t, t+s \leq T$ by $2\epsilon\bar{G}$.

The remainder of the proof is devoted to part (iii): characterize the limit of the interpolation process as the solution of the ODE in (2.21). The sequence $\{x_{\epsilon_k}(\cdot)\}$ is equicontinuous in the extended sense. By Theorem 2.4 (Ascoli-Arzelà) any infinite subsequence of $\{x_\epsilon(\cdot)\}$ has a convergent subsequence with a continuous limit on $[0, \infty)$. Consider a convergent subsequence along $\epsilon_r \to 0$, so that the limit $\hat{x}(\cdot) = \lim_{r\to\infty} x_{\epsilon_r}(\cdot)$ (in the sup norm) is continuous. Then

$$\lim_{r\to\infty}(x_{\epsilon_r}(t+s) - x_{\epsilon_r}(t)) \stackrel{(a)}{=} \lim_{r\to\infty}\int_t^{t+s} G(x_{\epsilon_r}(u))du$$

$$\stackrel{(b)}{=} \int_t^{t+s} \lim_{r\to\infty} G(x_{\epsilon_r}(u))du$$

$$\stackrel{(c)}{=} \int_t^{t+s} G(\hat{x}(u))du,$$

where (a) follows from the fact that $\rho(\epsilon)$ in (2.24) is bounded by $\|\rho(\epsilon)\|_\infty \le 2\epsilon\bar{G}$ and thus of order $O(\epsilon)$; (b) follows from Lebesgue Dominated Convergence Theorem (see, e.g., [18, 265]); and (c) is a consequence of the continuity of $G(\hat{x}(\cdot))$ on $[0, \infty)$. Therefore, for any $s > 0$ the limit $\hat{x}(\cdot)$ satisfies

$$\frac{\hat{x}(t+s) - \hat{x}(t)}{s} = \frac{1}{s}\int_t^{t+s} G(\hat{x}(u))du.$$

By continuity of $G(\hat{x}(\cdot))$ on $[t, t+s]$, taking the limit as s goes to zero, the above right-hand side converges to $G(\hat{x}(t))$, which establishes (2.21) for $\hat{x}(\cdot)$. This shows that the limit of any convergent subsequence of $\{x_\epsilon(\cdot)\}$ solves the ODE in (2.21), and uniqueness of the ODE follows from (C2) of theorem 2.5.

We now turn to the proof in case the bias term is present. The proof follows by noticing that the perturbations will appear in the part (i) of the above proof via

$$\int_t^{t+s} G(x_\epsilon(u))\,du = \sum_{i=m(t)}^{m(t+s)-1} \epsilon G(\theta_i^\epsilon) + \sum_{i=m(t)}^{m(t+s)-1} \epsilon \beta_\epsilon(\theta_i^\epsilon) + \rho(\epsilon),$$

and using the bound on the perturbations, for any $r < q$,

$$\sum_{i=m(r)}^{m(q)-1} \epsilon \beta_\epsilon(\theta_i^\epsilon) = (q-r)O(\epsilon),$$

so this term can be added to the approximation error $\rho(\epsilon)$ in (2.24) and the proof follows directly. $\square$

Remark 2.7. Suppose that the bias in Theorem 2.5 is not vanishing, i.e., $\lim_{\epsilon \to 0} \beta_\epsilon(\theta) = \beta(\theta) \ne 0$. Under appropriate smoothness assumptions on the bias function $\beta(\theta)$, it then follows that the limit of any convergent ϵ-subsequence of $\{x_\epsilon(t) : t \ge 0\}$, given by $\{x(t) : t \ge 0\}$, is the solution of the ODE $dx(t)/dt = G(x(t) + \beta(x(t)))$. In words, the bias shifts the target ODE.

The result put forward in Theorem 2.5 can be extended to the case of decreasing stepsizes. We leave the proof as exercise (see Exercise 2.4).

THE ITERATIVE METHOD

Theorem 2.6. *Assume that conditions (C1) and (C2) put forward in Theorem 2.5 hold*

$$\theta_{n+1} = \theta_n + \epsilon_n \left(G(\theta_n) + \beta_n(\theta_n) \right),$$

for $\theta_0 \in \mathbb{R}^d$, where $\sum_n \epsilon_n = \infty$, $\epsilon_n > 0$ for all n, $\epsilon_n \to 0$, and $\sum_n \epsilon_n \|\beta_n(\theta_n)\| < \infty$. Then the shifted interpolation process $x_n(t) = \vartheta^\varepsilon(t_n + t), t \geq 0$ and $\vartheta^\varepsilon(0) = \theta_0$, converge as $n \to \infty$ to the solution of the ODE

$$\frac{dx(t)}{dt} = G(x(t)), t \geq 0, \quad x(0) = \theta_0.$$

Theorems 2.5 and 2.6 establish the limit of the algorithm as an ODE for the different gain sizes. The following theorem extends this result to the case where a projected ODE is applied. It is worth noting that when an ODE is projected onto some compact set Θ, then the solutions to the initial value problem given by the ODE with initial value in Θ cannot blow up in finite time by construction.

Theorem 2.7. *Let Θ be closed and convex, and $G : E \mapsto E$, with $\Theta \subseteq E \subset \mathbb{R}^d$. Consider the projected ODE*

$$\frac{dx(t)}{dt} = (\Pi_\Theta G)(x(t)), \quad x(0) \in \Theta, \qquad (2.26)$$

and (i) let G be either Lipschitz continuous on Θ and assume that for any initial value in Θ the solution $\{x(t) : t \geq 0\}$ to the initial value problem in (2.26) is uniquely defined, or (ii) assume that G is continuous and Θ is also compact. Then the following holds.

- *Consider the recursion*

$$\theta^\epsilon_{n+1} = \Pi_\Theta \left(\theta^\epsilon_n + \epsilon(G(\theta^\epsilon_n) + \beta_\epsilon(\theta^\epsilon)) \right), \quad \theta^\epsilon_0 \in \Theta,$$

with constant stepsize $\epsilon > 0$. Let $x_\epsilon(t) = \vartheta^\varepsilon(t), 0 \leq t$, denote the interpolation process of $\{\theta^\epsilon_n\}$, with $x_\epsilon(0) = \theta^\epsilon_0 = \theta_0 \in \Theta$. Let $\{x(t) : t \geq 0\}$ denote the limit of a convergent ϵ-subsequence of $\{x_\epsilon(t) : t \geq 0\}$ and assume that $\sup_\theta \|\beta_\epsilon(\theta)\| = O(\epsilon)$. Then $x(t)$ is the unique solution of the ODE in (2.26). Moreover, the convergence of $x_\epsilon(t)$ happens uniformly over finite time-intervals, i.e., for any $0 \leq a, b < \infty$ it holds that $\sup_{a \leq t \leq b} |x_\epsilon(t) - x(t)| \to 0$ as ϵ tends to zero.

- *Alternatively, consider the recursion*

$$\theta_{n+1} = \Pi_\Theta \left(\theta_n + \epsilon_n (G(\theta_n) + \beta_n(\theta_n)) \right), \quad \theta_0 \in \Theta,$$

where $\sum_n \epsilon_n = \infty$, $\epsilon_n > 0$ for all n, $\epsilon_n \to 0$, and $\sum_n \epsilon_n \|\beta_n(\theta_n)\| < \infty$. Then the shifted interpolation process $x_n(t) = \vartheta^\varepsilon(t_n + t), t \geq 0$ and $\vartheta^\varepsilon(0) = \theta_0$, converge as $n \to \infty$ to the solution of the ODE in (2.26).

Proof. Note that condition (C1) and (C2) in Theorem 2.5 are satisfied on Θ. For the second set of conditions, we remark that in case that Θ is compact, continuity of G on Θ implies that G is bounded on Θ, which establishes (C1) on Θ. Compactness of Θ furthermore implies that the trajectories of the ODE started in Θ cannot blow up in finite time, since Θ is bounded, which yields uniqueness of the trajectory of the projected ODE.

The main line of the proof follows that of Theorem 2.5 and we only address the parts that deviate due to the projection. Recall that the projection operator can be modeled by means of the projection force; see Section 2.3. With this notation, the algorithm reads

$$\theta_{n+1}^\epsilon = \theta_n^\epsilon + \epsilon \left(G(\theta_n^\epsilon) + Z(\epsilon, \theta_n^\epsilon, G(\theta_n^\epsilon)) \right),$$

Following the line of argument put forward in the proof of Theorem 2.5 we obtain

$$x_\epsilon(t+s) - x_\epsilon(t) = \int_t^{t+s} \left(G(x_\epsilon(u)) + Z(\epsilon, x_\epsilon(u), G(x_\epsilon(u))) \right) du + \rho(\epsilon)$$

$$= \sum_{i=m(t)}^{m(t+s)-1} \epsilon \left(G(\theta_i^\epsilon) + Z(\epsilon, \theta_i^\epsilon, G(\theta_i^\epsilon)) \right) + \rho(\epsilon),$$

where $\rho(\epsilon)$ is the error in the approximation, due to the discretization at the end points.

By assumption $G(\theta)$ is locally Lipschitz continuous on Θ and therefore uniformly bounded by some finite constant $\bar{G}$ on Θ. Moreover, for $\theta \in \Theta$, the projection force $Z(\epsilon, \theta, G(\theta))$ is bounded by $G(\theta)$ for $\epsilon < 1$; see (1.38). This yields, for small ϵ,

$$\|x_\epsilon(q) - x_\epsilon(r)\| = \left\| \sum_{i=m(r)}^{m(q)-1} \epsilon \left(G(\theta_i^\epsilon) + Z(\epsilon, \theta_i^\epsilon, G(\theta_i^\epsilon)) \right) \right\|_\infty \leq 2\epsilon \bar{G} \frac{(q-r)}{\epsilon} = 2\bar{G}(q-r). \tag{2.27}$$

This establishes equicontinuity of $x_\epsilon(t)$ in the extended sense.

By Theorem 2.4 (Ascoli-Arzelà), we select a convergent subsequence with a continuous limit on $[0, \infty)$. Then, using (2.14) we obtain

$$\lim_{r \to \infty} (x_{\epsilon_r}(t+s) - x_{\epsilon_r}(t)) = \lim_{r \to \infty} \int_t^{t+s} \left(G(x_{\epsilon_r}(u)) + Z(\epsilon_r, x_{\epsilon_r}(u), G(x_{\epsilon_r}(u))) \right) du$$

$$= \int_t^{t+s} \lim_{r \to \infty} \left(G(x_{\epsilon_r}(u)) + Z(\epsilon_r, x_{\epsilon_r}(u), G(x_{\epsilon_r}(u))) \right) du$$

$$= \int_t^{t+s} \left(G(\hat{x}(u)) + \mathbf{Z}(\hat{x}(u), G(\hat{x}(u))) \right) du$$

$$= \int_t^{t+s} \pi(\hat{x}(u), G(\hat{x}(u))) du,$$

where we use next to the standard arguments put forward in the proof of Theorem 2.5 the continuity of directional derivative of the projection. Therefore, for any $s > 0$ the limit $\hat{x}(\cdot)$ satisfies

$$\frac{\hat{x}(t+s) - \hat{x}(t)}{s} = \frac{1}{s} \int_t^{t+s} \pi(\hat{x}(u), G(\hat{x}(u))) du.$$

By continuity of $\pi(\hat{x}(u), G(\hat{x}(u)))$ as a mapping of u on $[t, t+s]$, taking the limit as s goes to zero, the above right-hand side converges to $(\Pi_\Theta G)(\hat{x}(t))$, which establishes (2.26) for $\hat{x}(\cdot)$. This establishes the first part of the theorem. The second part follows from the same line of argument elaborating on Theorem 2.6. □

Theorem 2.7 provides sufficient conditions for the interpolation process obtained from projecting the updates onto Θ to converge toward the trajectory of the projected ODE in (2.26). As expected, when using the projected ODE for the updates, projection is not required, as the following lemma shows (the extension to the biased case is left to the reader).

Lemma 2.2. *Let $G : E \mapsto E$ be a continuous vector field, and $\Theta \subseteq E \subseteq \mathbb{R}^d$ compact and convex. Then, Theorem 2.7 applies to the constant stepsize algorithm $\theta_{n+1}^\epsilon = \theta_n^\epsilon + \epsilon (\Pi_\Theta G)(\theta_n^\epsilon)$ with initial value in Θ and $\epsilon > 0$; and to the decreasing stepsize algorithm $\theta_{n+1} = \theta_n + \epsilon_n (\Pi_\Theta G)(\theta_n)$ with initial value in Θ, $\sum_n \epsilon_n = \infty$, $\sum_n \epsilon_n^2 < \infty$, and $\epsilon_n > 0$ for all n.*

Proof. By construction, any trajectory $dx(t)/dt = (\Pi_\Theta G)(x(t))$, with $x(0) \in \Theta$, is bounded, and therefore does not blow up in finite time. Moreover, because the trajectories are continuous (for a proof we refer to [336] and [94]) and do not blow up in finite time, then the trajectories are unique. Hence, the arguments in the proof of Theorems 2.5 and 2.7 apply to $(\Pi_\Theta G)(\cdot)$. □

Remark 2.8. Due to the Euler approximation in the algorithm put forward in Lemma 2.2, it cannot be ruled out that θ_n wanders outside of Θ. Hence, in the case that Θ represents hard constraints in the sense that the model is not well-defined outside of Θ, an outer projection is needed, leading, for example for the decreasing gain size version, to the algorithm $\theta_{n+1} = \Pi_\Theta(\theta_n + \epsilon_n (G(\theta_n) + \beta_n(\theta_n)))$.

2.6 THE ODE METHOD FOR OPTIMIZATION AND LEARNING

So far, our limiting results establish conditions under which an interpolated difference equation converges in the sup-norm to the solution of an ODE. Independent of whether projection, penalty or multiplier methods are used in an optimization setting, the ODE is given through a field that depends on the gradient of the cost function and of the constraints. Establishing properties of the stable points of this ODE (which are stationary points) requires knowledge of the matrix $\mathbb{A}(\theta) = \nabla G(\theta)$ evaluated at the stable points. The limiting behavior of the algorithm for θ_n as $\epsilon \to 0$ is described by the stable points of the corresponding ODE $dx(t)/dt = G(x(t))$.

When dealing with an optimization problem, it is essential to find an appropriate model for the problem. Building a model allows for some flexibility as one may use alternative formulations of the problem as long as they have the same solution (i.e., the same KKT points; see Definition 1.11). A first step is to ensure that the optimzation problem itself is well-posed. For definition and standard setting of an NLP, we refer to Definition 1.8.

Definition 2.12. A NLP is said to be *well-posed* if the set of solutions is not empty, contains only KKT points, and the set of KKT points is confined to a bounded set. In addition, we require that there is no direction $v \in \mathbb{R}^d$ such that $J(tv)$ for some sequence $t \to \infty$ is strictly decreasing. It is called *ill-posed* otherwise.

We illustrate the concept of well-posedness in the following example.

Example 2.11. Let $J(\theta) = \sin(\theta)$, then the problem $\min_\mathbb{R} J(\theta)$ is not well-posed. To see this, note that while $J(\theta)$ has unique global minimum -1, the set of locations of the global minimum (KKT points), given by $\{2\pi k : k \in \mathbb{Z}\}$, is denumerable and not confined to a compact set. Obviously, $J(\theta) = \sin(\theta)$ as a cost function becomes feasible by considering the constrained optimization problem $\min_{\theta \in [a,b]} J(\theta)$, for $-\infty < a < b < \infty$. This shows that well-posedness is not a mathematical property per se but is related to "asking the right question." Indeed, in applications, one is typically searching for solutions within a reasonable range.

As a second example consider $J(\theta) = -\theta^2$ and $\min_{\mathbb{R}} J(\theta)$. This problem has no solutions and is thus not well-posed. In the appropriately constrained case, i.e., $\min_{\theta \in [a,b]} J(\theta)$, for $-\infty < a < b < \infty$, the problem becomes well-posed.

Finally, consider $J(\theta) = e^{-\theta^2}$ and $\min_{\mathbb{R}} J(\theta)$. In this case $J(\theta)$ has no proper minimum and any descent algorithm will tend to $\pm\infty$ following the decreasing tails of $J(\theta)$. This problem is not well-posed as $J(\theta)$ is strictly decreasing on $(-\infty, 0]$ and $[0, \infty)$ and thus has two improper minima at $\pm\infty$. Again, constrain the problem to any finite interval, and the problem becomes well-posed.
※※※

After having made sure that the actual optimization problem is well-posed, the next step is to define a function $G(\theta)$ such that the asymptotically stable points of the ODE with vector field G are KKT points of the optimization problem. This vector field G will then naturally define the recursive algorithm to find candiates for the solution of the problem, and we call $G(\theta)$ the *target vector field*, and the ODE that goes with $G(\theta)$ the *limiting ODE*. In other words, an iterative numerical method will work only if it serves as a numerical approximation of the limiting ODE that tracks the KKT points of the optimization problem.

It is worth noting that the way in which the sequence $\{\theta_n\}$, and thereby the corresponding interpolation process $\vartheta^\epsilon(t)$, tracks the limiting ODE is different for fixed and decreasing stepsize algorithm. Indeed, in case of a decreasing stepsize, the essential part of the proof is in showing that the shifted interpolated process $x_n(t)$ approaches the stable points of the limiting ODE. Hence, the embedded sequence $\{\theta_n\}$ of consecutive values of $\{\vartheta^\epsilon(t) : t \geq 0\}$ tracks the stable points of the limiting ODE and thus finds a KKT point.

This is different from the situation for fixed ϵ. As $\epsilon \to 0$ the process $\vartheta^\epsilon(t)$ approaches the solution of the ODE on the whole trajectory (which is a strong result). To understand how "close" we are to the optimal value we can use the analogy of approximating integrals with a finite grid of size ϵ. For any time T, given a precision, we may find a sufficiently small value of ϵ for which $\theta^\epsilon_{m(T)}$ is close to the solution $x(T)$ of the ODE (within this precision). Because $x(t)$ converges to a KKT point of the optimization problem, we will, for "large enough T," be close to the KKT point. Specifically,

$$\theta^* = \lim_{t \to \infty} \lim_{\epsilon \to 0} \theta^\epsilon_{m(t)}$$

This leads to the following definition.

Definition 2.13. Consider a well-posed NLP $\min_{\theta \in \Theta} J(\theta)$ (including the case $\Theta = \mathbb{R}^d$). The target vector field $G: E \mapsto E$, with $\Theta \subseteq E$, is said to be *coercive* for this well-posed NLP with respect to $H \subseteq \Theta$ if

- The vector field G is either Lipschitz continuous or bounded on Θ,
- The solution $\{x(t) : t \geq 0\}$ to the initial value problem

$$\frac{dx(t)}{dt} = G(x(t)), \quad t \geq 0, \quad \text{with } x(0) \in H, \qquad (2.28)$$

is uniquely defined for $t \geq 0$,

and, in addition,

- the set of asymptotically stable points of the ODE (2.28) is non-empty and a subset of the KKT points of the NLP.

THE ITERATIVE METHOD 63

It is often the case in applications that $H = \Theta = E$ in Definition 2.13. Notable exceptions are settings like the one depicted in Figure 1.6. Here, the negative gradient $G(\theta) = -J'(\theta)$ is defined on $\mathbb{R}$ so that $E = \mathbb{R}$. Now consider the NLP $\min_{\theta \in [-4,-2]} J(\theta)$, i.e., $\Theta = [-4, -2]$. Then, for any initial point in, say, $H = [-4.1, 0)$ the ODE trajectory of the ODE driven by $G(\theta)$ will converge to the solution of the NLP. Another instance where $E = \mathbb{R}^d \neq \Theta$ occurs is when G is the vector field corresponding to the penalty and multiplier method, or the Arrow-Hurwicz algorithm; see Section 1.4.

It follows from the development in the previous sections that, if G is coercive for a well-posed NLP, then the deterministic algorithm $\theta_{n+1} = \theta_n + \epsilon\, G(\theta_n)$, provided that $\theta_{n+1} \in E$ for $\theta_n \in E$, will approximate candidates for the solution of the NLP. We arrive at the following result that summarizes the results of the previous discussion into the context of optimization.

Theorem 2.8. *Let G be a vector field on $\mathbb{R}^d$ and assume that G is coercive for a well-posed NLP on H.*

(i) Assume that $\sup_{\theta \in \Theta} \|\beta_\epsilon(\theta)\| = O(\epsilon)$. Consider the constant gain size algorithm

$$\theta_{n+1}^\epsilon = \theta_n^\epsilon + \epsilon\, (G(\theta_n^\epsilon) + \beta_\epsilon(\theta_n^\epsilon)), \quad \theta_0 \in H,$$

for all n. Define the limit point, if it exits, by $\theta^* = \lim_{t \to \infty} \lim_{\epsilon \to 0} \theta_{\lfloor t/\epsilon \rfloor}^\epsilon$.

(ii) Assume that $\sum_n \epsilon_n = \infty$, $\epsilon_n > 0$ for all n, $\epsilon_n \to 0$, $\sum_n \epsilon_n \|\beta_n(\theta_n)\| < \infty$. Consider the decreasing stepsize algorithm

$$\theta_{n+1} = \theta_n + \epsilon_n\, (G(\theta_n) + \beta_n(\theta_n)), \quad \theta_0 \in H,$$

for all n. Let the limit point, if it exits, be

$$\theta^* = \lim_{n \to \infty} \theta_n.$$

Then θ^* is a KKT point of the NLP.

In case G is defined only on a subset of $\mathbb{R}^d$, the corresponding statements hold for the projected versions of the algorithms.

The following lemma provides the insight that any descent direction vector field—and not just the negative gradient—is coercive for standard unconstrained minimization problems. Recall Definition 1.5 for a descent direction.

Lemma 2.3. *Given is a well-posed optimization problem*

$$\min_{\theta \in \mathbb{R}^d} J(\theta). \tag{2.29}$$

Let G be a continuous descent-direction vector field G for $J(\theta)$ on $\mathbb{R}^d$. Then, any asymptotically stable point of G is a stationary point of $J(\theta)$. If G is, in addition, either bounded and locally Lipschitz continuous, or Lipschitz continuous, then G is coercive for (2.29) on $\mathbb{R}^d$. In the special case that $J(\theta)$ has a unique stationary point θ^* as location of the minimum, θ^* is the globally asymptotically stable point of G.

Proof. The first part of the lemma follows from a Lyapunov function argument as detailed in Example 2.8. The second part is a direct consequence of the definition of coercivity, and the third part again follows directly from the fact that $J(\theta)$ has a unique KKT point. □

Lemma 2.3 shows that we have flexibility in choosing the descent direction. This is a welcome property as it allows us to circumvent the restrictive condition that the negative gradient is either bounded or globally Lipschitz by considering a descent direction that is, for example, bounded and locally Lipschitz. The following example presents such a descent direction obtained from modifying the negative gradient vector field that is popular in machine learning.

Example 2.12. A popular descent direction used for stabilizing gradient updates is *clipping* [217, 218, 337], where clipping scales the norm of large outcomes of the gradient to a predefined value c; specifically, for $J \in C^1$ the clipped gradient is given by

$$\nabla J_c(\theta) = \begin{cases} \nabla J(\theta) & \text{if } ||\nabla J(\theta)|| < c, \\ c \frac{\nabla J(\theta)}{||\nabla J(\theta)||} & \text{otherwise,} \end{cases}$$

for some finite $c > 0$. By construction, ∇J_c is bounded and has the same zeros as ∇J. Hence, by Lemma 2.3, provided that ∇J is locally Lipschitz (so that ∇J_c is locally Lipschitz), it follows that ∇J_c is coercive for a well-posed optimization problem of the type $\min_\theta J(\theta)$. For polynomial objective functions and appropriate choice of c, the clipped gradient algorithm $\theta_{n+1} = \theta_n + \epsilon_n \nabla J_c(\theta_n)$ is known to show faster convergence than the standard algorithm $\theta_{n+1} = \theta_n + \epsilon_n \nabla J(\theta_n)$, and we refer to [337] for details. ✼✼✼

Clipping as introduced in Example 2.12 above has the welcome property that a clipped coercive vector field stays coercive. The precise statement is given in the following theorem (the extension to the biased case is left to the reader).

Theorem 2.9 (Clipping). *Let G be coercive for the NLP $\min_{\theta \in \Theta} J(\theta)$ on H, then the clipped vector field G_c, given by*

$$G_c(\theta) = \begin{cases} G(\theta) & \text{if } ||G(\theta)|| < c, \\ c \frac{\nabla G(\theta)}{||G(\theta)||} & \text{otherwise,} \end{cases}$$

for any finite $c > 0$, is coercive for the NLP on H as well.

- *Consider the recursion*

$$\theta_{n+1}^\epsilon = \theta_n^\epsilon + \epsilon G(\theta_n^\epsilon) \left(\mathbf{1}_{\{||G(\theta_n^\epsilon)|| \leq c\}} + \frac{c}{||G(\theta_n^\epsilon)||} \mathbf{1}_{\{||G(\theta_n^\epsilon)|| > c\}} \right),$$

  with constant stepsize $\epsilon > 0$ and let $x_\epsilon(t) = \vartheta^\epsilon(t), 0 \leq t$, denote the interpolation process of $\{\theta_n^\epsilon\}$, with $x_\epsilon(0) = \theta_0^\epsilon = \theta_0 \in H$. Let $\{x(t) : t \geq 0\}$ denote the limit of some convergent ϵ-subsequence of $\{x_\epsilon(t) : t \geq 0\}$. Then $x(t)$ is the unique solution of the corresponding ODE G_c. Moreover, the convergence of $x_\epsilon(t)$ happens uniformly over finite time-intervals, i.e., for any $0 \leq a, b < \infty$ it holds that $\sup_{a \leq t \leq b} |x_\epsilon(t) - x(t)| \to 0$ as ϵ tends to zero.

- *Alternatively, consider the recursion*

$$\theta_{n+1} = \theta_n + \epsilon_n G(\theta_n) \left(\mathbf{1}_{\{||G(\theta_n)|| \leq c\}} + \frac{c}{||G(\theta_n)||} \mathbf{1}_{\{||G(\theta_n)|| > c\}} \right),$$

  for $\theta_0 \in H$, where $\sum_n \epsilon_n = \infty$, $\epsilon_n > 0$ for all n, and $\epsilon_n \to 0$. Then the shifted interpolation process $x_n(t) = \vartheta^\epsilon(t_n + t), t \geq 0$ and $\vartheta^\epsilon(0) = \theta_0$, converge as $n \to \infty$ to the solution of the ODE given by G.

Proof. The arguments for coercivity of G_c are as presented in Example 2.12 for the special case of the negative gradient. Since G is coercive for the NLP, all KKT points of the NLP are asymptotically stable points of G and, by construction, of G_c as well. Furthermore, G_c is locally Lipschitz continuous as G is, and bounded by construction. Hence, G_c is coercive for the NLP. The proof then follows from Theorem 2.6 for decreasing gain size and Theorem 2.6 for fixed gain size. □

Projection is an alternative to clipping for adjusting a given NLP by constraining the NLP to some large convex and compact set such that the essence of the problem remains untouched while establishing convergence of the ODE. This *truncation principle* is used frequently in the literature for simplifying convergence analysis and for extending the applicability of the theory. In particular, one uses the argument that, when applying projection on, say, a large enough hypercube, the trajectory of the truncated algorithm is identical to the original one and thus the nominal trajectory inherits the properties of boundedness of the projected one. Next to this analytical application, the truncation principle can be also used for modifying the actual algorithm for reasons of numerical stability. In this case the resulting trajectory is different from that of the nominal trajectory. The precise statement is as follows (the extension to the biased case is left to the reader).

Theorem 2.10 (Truncation Principle). *Given is a NLP $\min_{\theta \in \Theta} J(\theta)$ with feasible region Θ and $G : E \mapsto E$, $E \subseteq \mathbb{R}^d$, that is coercive for this NLP on some set H. Choose $\Theta_0 \subseteq E$ convex and compact such that all KKT points of the NLP are inner points of Θ_0. Then all KKT points of the NLP are asymptotically stable points of $(\Pi_{\Theta_0} G)(\cdot)$ with initial values in $H \cap \Theta_0$ as well, and the additional KKT points stemming from the conditions imposed by Θ_0 are neither globally nor asymptotically stable points of $(\Pi_{\Theta_0} G)(\cdot)$.*

- *Consider the recursion*

$$\theta_{n+1}^\epsilon = \Pi_{\Theta_0}\left(\theta_n^\epsilon + \epsilon G(\theta_n^\epsilon)\right), \quad \theta_0^\epsilon \in H \cap \Theta_0,$$

with constant stepsize $\epsilon > 0$ and let $x_\epsilon(t) = \vartheta^\epsilon(t), 0 \le t$, denote the interpolation process of $\{\theta_n^\epsilon\}$, with $x_\epsilon(0) = \theta_0^\epsilon = \theta_0 \in H \cap \Theta_0$. Let $\{x(t) : t \ge 0\}$ denote the limit of some convergent ϵ-subsequence of $\{x_\epsilon(t) : t \ge 0\}$. Then $x(t)$ is the unique solution of the corresponding projected ODE with projected vector field $(\Pi_{\Theta_0} G)(\cdot)$. Moreover, the convergence of $x_\epsilon(t)$ happens uniformly over finite time-intervals, i.e., for any $0 \le a, b < \infty$ it holds that $\sup_{a \le t \le b} |x_\epsilon(t) - x(t)| \to 0$ as ϵ tends to zero.

- *Alternatively, consider the recursion*

$$\theta_{n+1} = \Pi_{\Theta_0}\left(\theta_n + \epsilon_n G(\theta_n)\right), \quad \theta_0 \in H \cap \Theta_0,$$

where $\sum_n \epsilon_n = \infty$, $\epsilon_n > 0$ for all n, and $\epsilon_n \to 0$. Then the shifted interpolation process $x_n(t) = \vartheta^\epsilon(t_n + t), t \ge 0$ and $\vartheta^\epsilon(0) = \theta_0 \in H \cap \Theta_0$, converge to the solution of the corresponding projected ODE with projected vector field $(\Pi_{\Theta_0} G)(\cdot)$ in (2.26).

Proof. Convergence of the interpolation process to the projected vector field follows from Theorem 2.7. Following the arguments put forward in Remark 2.4, we may conclude that the ODE driven by the projected vector field $(\Pi_\Theta G)(\cdot)$ has continuous and bounded trajectories. As the trajectories of G go in direction of the KKT points of the NLP, G necessarily has a drift toward the inside of Θ_0. This shows that the potential equilibrium points on the boundary of Θ_0 are not asymptotically stable (and thus also not globally stable). The convergence result is an immediate consequence of Theorem 2.7. □

Remark 2.9. Consider the setting in Theorem 2.10. Fix the initial value of the algorithm to θ_0, and choose a bounded set H that contains all stationary points of G. Provided that G is coercive, there is a hypercube $\tilde{H}$, with $H \subset \tilde{H} \subset \mathbb{R}^d$, such that the trajectory of the ODE driven by G started at θ_0 stays inside $\tilde{H}$. Choosing now $\tilde{H} = \Theta_0$ in Theorem 2.10, it follows that the trajectory of the ODE driven by G and by $(\Pi_{\Theta_0} G)(\cdot)$ coincide provided both are initially started in θ_0. The trajectory of the projected ODE is thus equivalent to that of the unprojected one. Therefore the limit ODE is independent of the truncation of the algorithm and we can argue that we work with original algorithm and thereby avoiding projection at all.

2.7 SPECIFIC ALGORITHMS FOR CONSTRAINED OPTIMIZATION

Theorem 2.7 provides projection based results for either constant or decreasing stepsize for constrained optimization. In the following we discuss algorithms that avoid using numerically slow projections on the admissible set given by the constraints, namely, the Arrow-Hurwicz algorithm and the penalty method.

In their original work, Arrow and Hurwicz [8, 9] used an ODE approach to establish the optimality of (1.50). They posed the problem in terms of a competition game, and they added a penalty to their objective function to ensure convexity. The following theorem assumes strict convexity, so no penalty is required. Its proof is stated within the framework of this chapter. Exercise 2.7 focuses on a simple one-dimensional case that can help to illustrate the main ideas of this proof.

Theorem 2.11. *Consider a strictly convex NLP, and assume that $\nabla J(\theta), \nabla g(\theta)$ are continuous, and let $x_0 \in \mathbb{R}^d, y_0 > 0 (\in \mathbb{R}^p), \eta_0 \in \mathbb{R}^q$ be an initial point of the Arrow-Hurwicz algorithm (1.50) such that all inactive constraints at this point are the same as those for the optimal point θ^*. Then the limit ODE is*

$$\begin{aligned}
\frac{dx(t)}{dt} &= -\nabla_\theta \mathcal{L}(x(t), y(t), z(t)) \\
\frac{dy(t)}{dt} &= g(x(t)) \mathbf{1}_{\{(x(t), y(t)) \in \Lambda\}} \\
\frac{dz(t)}{dt} &= h(x(t)),
\end{aligned} \qquad (2.30)$$

where

$$\Lambda^c \stackrel{def}{=} \{(\theta, \lambda) : \lambda \leq 0, g(\theta) < 0\}.$$

In addition, the stable points of (2.30) are saddle points of the Lagrangian.

Proof. To apply Theorem 2.3, we study a simplification of the vector field associated with (2.30). We observe first that inactive constraints do not affect local behavior (their multipliers are zeroes) so they can be ignored from the analysis in a neighborhood of any interior point (θ, λ, η). Call $\tilde{g}(\theta)$ the corresponding vector with only those components j for which $\lambda_j > 0$. Once the total dimension has thus been reduced, the corresponding matrix ∇G is given by

$$\mathbb{A} = \begin{pmatrix} -H(\theta, \lambda, \eta) & -B(\theta)^\top \\ B(\theta) & 0 \end{pmatrix},$$

where $H(\theta, \lambda, \eta) = \nabla_\theta^2 \mathcal{L}(\theta, \lambda, \eta) \in \mathbb{R}^{d \times d}$, and $B^\top(\theta) = (\nabla \tilde{g}(\theta), \nabla h(\theta)) \in \mathbb{R}^{d \times (p+q)}$ is the gradient of the active constraints at θ. For a strictly convex non-linear problem, $H(\theta, \lambda, \eta)$

THE ITERATIVE METHOD

is positive definite, and if $\nabla B(\theta)$ is full rank (i.e., the constraint qualifications hold at θ), then the matrix $\mathbb{A}$ is a Hurwitz matrix, driving the ODE toward its asymptotically stable point, as we now show.

Let v be a eigenvector of unit norm of $\mathbb{A}$ with eigenvalue ρ. Then

$$\mathbb{A}v = \begin{pmatrix} -H(\theta, \lambda, \eta) & -B(\theta)^\top \\ B(\theta) & 0 \end{pmatrix} \begin{pmatrix} v_1 \\ v_2 \end{pmatrix} = \begin{pmatrix} -H(\theta, \lambda, \eta)v_1 - B(\theta)^\top v_2 \\ B(\theta)v_1 \end{pmatrix} = \rho \begin{pmatrix} v_1 \\ v_2 \end{pmatrix},$$

where v_1 is a d-dimensional vector and v_2 is a $(p+q)$-dimensional vector. The first claim is that v_1 cannot be the null vector. Proceeding by contradiction, if $v_1 = 0$ then the above equation would imply that $B(\theta)^\top v_2 = 0$, which has the unique solution $v_2 = 0$ because $B(\theta)$ is full rank. By assumption v is a unit norm eigenvalue and thus it cannot be the zero vector.

We use now some results from matrix algebra: (a) for a symmetric matrix the eigenvalues are real; (b) a real matrix has eigenvalues that are either real-valued, or appear in pairs of complex conjugates; and (c) $\mathbb{A} + \mathbb{A}^*$ is symmetric, where $\mathbb{A}^*$ is the conjugate transpose of $\mathbb{A}$.

The identity $\mathbb{A}v = \rho v$ implies that $v^\top \mathbb{A}^* = \rho^* v^\top$, where ρ^* is the complex conjugate of ρ. Then the real part of ρ is

$$\Re(\rho) = v^\top (\mathbb{A} + \mathbb{A}^*)v = \frac{1}{2}(v_1^\top, v_2^\top) \begin{pmatrix} -H(\theta, \lambda, \eta) & 0 \\ 0 & 0 \end{pmatrix} \begin{pmatrix} v_1 \\ v_2 \end{pmatrix} = -v_1^\top H(\theta, \lambda, \eta) v_1$$

with $H = -[\nabla^2 J(\theta) + \lambda \nabla^2 \tilde{g}(\theta) + \eta \nabla^2 h(\theta)]$ being a positive definite matrix. Because $v_1 \neq 0$ then the real part of all eigenvalues is strictly negative, which shows that ∇G satisfies the Hurwitz condition. The claim now follows from Theorem 2.3. □

Example 2.13. This example illustrates the use of Theorem 2.11 and the interpretation of the limit ODE for a constrained optimization problem using the Arrow-Hurwicz algorithm. Consider

$$\min_{\theta \in \mathbb{R}} J(\theta) \stackrel{\text{def}}{=} \frac{1}{2}\theta^2$$
$$\text{s.t.} \quad c - \theta \leq 0,$$

for $c > 0$, and suppose that we use the Arrow-Hurwicz algorithm (1.50) with constant stepsize:

$$\theta_{n+1} = \theta_n - \epsilon (\nabla_\theta \mathcal{L}(\theta_n, \lambda_n))^\top$$
$$\lambda_{n+1} = \max(0, \lambda_n + \epsilon g(\theta_n)), \quad (2.31)$$

where the Lagrangian and constraint is given by

$$\mathcal{L}(\theta, \lambda) = \frac{1}{2}\theta^2 + \lambda(c - \theta); \quad g(\theta) = c - \theta.$$

The vector field is

$$G(\theta, \lambda) = \begin{pmatrix} -(\nabla_\theta \mathcal{L}(\theta, \lambda))^\top \\ g(\theta) \end{pmatrix} = \begin{pmatrix} -\theta + \lambda \\ c - \theta \end{pmatrix} \quad (2.32)$$

and by simple inspection, the solution to the constrained problem must satisfy $\theta^* = c$ and the corresponding Lagrange multiplier can be evaluated as the solution to the KKT point $-\theta^* + \lambda^* = 0$ so that $\lambda^* = c$.

In this case, the Hessian matrix related to the limit ODE in the region $\lambda > 0$ is

$$\mathbb{A} = \nabla_{(\theta,\lambda)} G(\theta, \lambda) = \begin{pmatrix} -1 & 1 \\ -1 & 0 \end{pmatrix}.$$

The eigenvalues satisfy $(-1 - \rho)(-\rho) + 1 = 0$, with roots $\rho = -\frac{1}{2} \pm i\frac{\sqrt{3}}{2}$, and $G(\theta)$ there for satisfies the Hurwitz condition. By Theorem 2.3, the G is bounded along the trajectories.

In the following we show how a Lyapounov stability argument can be applied for showing that for any finite $x(0)$, the corresponding trajectory $x(t)$ never leaves a compact set. Let $x(t)$ be a trajectory of the ODE corresponding with vector field (2.32), i.e.,

$$\frac{dx_1(t)}{dt} = -x_1(t) + x_2(t) \quad \text{and} \quad \frac{dx_2(t)}{dt} = c - x_1(t),$$

and define the function $V(t) = \frac{1}{2}\left((x_1(t) - c)^2 + (x_2(t) - c)^2\right)$ as the distance from the optimal value. Then

$$\frac{d}{dt} V(t) = (x_1(t) - c)\frac{dx_1}{dt} + (x_2(t) - c)\frac{dx_2}{dt}$$
$$= (x_1(t) - c)(-x_1(t) + x_2(t)) + (x_2(t) - c)(c - x_1(t))$$
$$= -x_1^2(t) + 2cx_1(t) - c^2 = -(x_1(t) - c)^2 \leq 0,$$

that is, starting at a point $x(0)$, the distance to the optimal value is strictly decreasing unless $x = (\theta^*, \lambda^*)$. Following this, for all starting points inside a circle around c the entirety of the trajectory lies within that circle, and therefore $G(x(t))$ is bounded. Hence, we may replace the ODE in Theorem 2.11 by, e.g., its clipped version and thus can guarantee boundedness of the vector field as required in the theorem.

Once the limit ODE has been established, its long-term behavior can be studied around the equilibrium point (c, c), and the corresponding vector field is shown in Figure 2.10.

As shown in the plot, every trajectory that follows this vector field will spiral inward toward the optimal point (c, c). The spiraling effect is due to the complex component of the eigenvalues. It is typical of the Arrow-Hurwicz algorithm and other differential games, where one "player" tries to minimize and the other to maximize the same economic function. Indeed, the major driving component of the algorithms depends on the exponential of the eigenvalues, and for an imaginary number $z = i\alpha$, $e^z = \sin(\alpha) + i\cos(\alpha)$ has oscillatory behavior, as shown in Figure (2.10)(b). ❋❋❋

We now turn to the two-time scale methods for dealing with constraints; see Section 1.3. The main idea is that the penalty parameter changes "very slowly," so it is "seen" by the θ updates as a constant over given time intervals. Then Theorem 1.8 can be used because the algorithm will evolve very close to the exact minimizers $\theta^*(\alpha) = \arg\min J_\alpha(\theta)$.

Theorem 2.12. *Consider the two-time scale penalty algorithm:*

$$\theta_{n+1} = \theta_n - \epsilon \nabla J(\theta_n, \alpha_n) \tag{2.33}$$
$$\alpha_{n+1} = \alpha_n + \delta_\epsilon, \tag{2.34}$$

where $\nabla J(\theta, \alpha) = \nabla (J_\alpha(\theta))^\top$, *for* $\theta \in \mathbb{R}^d$ *and* $\alpha \in \mathbb{R}$, *defined in (1.30), and choose* $\epsilon, \delta_\epsilon \to 0$ *such that*

$$m_\epsilon \stackrel{\text{def}}{=} \frac{\epsilon}{\delta_\epsilon} \to \infty; \quad \frac{\epsilon^2}{\delta_\epsilon} \to 0. \tag{2.35}$$

THE ITERATIVE METHOD

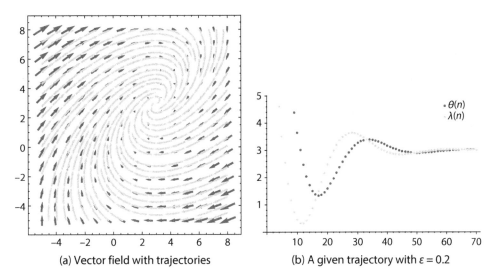

(a) Vector field with trajectories (b) A given trajectory with $\varepsilon = 0.2$

Figure 2.10. (a) The vector field of Example 2.5 with some trajectories streamlined; (b) a typical trajectory shows the oscillating behavior.

Then under the assumptions of Theorem 1.8, $\theta^* = \lim_{t \to \infty} \lim_{\epsilon \to 0} \theta^{\epsilon}_{\lfloor t/\epsilon \rfloor}$ is a KKT point of the constrained problem:

$$\min_{\theta \in \Theta} J(\theta), \qquad \Theta = \{\theta \in \mathbb{R}^d : g(\theta) \leq 0, h(\theta) = 0\}.$$

Proof. We start by studying the limit behavior of the "slower" changing process $\alpha^\epsilon(t) = \lfloor t/\delta_\epsilon \rfloor$. Using learning rate $\delta_\epsilon \to 0$ the drift is simply 1, and the limit ODE $\tilde{\alpha}'(t) = 1$ has solution $\tilde{\alpha}(t) = \tilde{\alpha}(0) + t$. For simplicity we will take $\tilde{\alpha}(0) = 0$.

Let $u \in \mathbb{R}^+$ be any fixed value, and let $n_\epsilon(u) = \lfloor u/\delta_\epsilon \rfloor$, so that $\alpha^\epsilon(u) = \alpha_{n_\epsilon} \to u$. Now consider the dynamics of the "inexact" optimization on θ:

$$\theta_{n_\epsilon + m} - \theta_{n_\epsilon} = -\epsilon \sum_{k=n_\epsilon+1}^{n_\epsilon+m} \nabla J(\theta_k, \alpha_k).$$

The idea is now to use the fact that α_n "updates much slower than" θ_n to replace its value by a constant one. This we do as follows. First, because $J_\alpha(\cdot)$ is linear in α it satisfies the Taylor expansion $\nabla J(\theta_k, \alpha_k) = \nabla J(\theta_k, \alpha_{n_\epsilon}) + L\delta_\epsilon(k - n_\epsilon)$ with bounded Lipschitz constant L. Now evaluate the difference above using m_ϵ iterations of the algorithm to get

$$\theta_{n_\epsilon + m_\epsilon} - \theta_{n_\epsilon} = -\epsilon \sum_{k=n_\epsilon+1}^{n_\epsilon+m} \left(\nabla J(\theta_k, \alpha_{n_\epsilon}) + \rho(\epsilon)\right),$$

where $\rho(\epsilon) = O(m_\epsilon^2 \delta_\epsilon) = O(\epsilon^2/\delta_\epsilon) \to 0$. By construction, the number of terms in the interval $m_\epsilon \to \infty$, and we have

$$\theta_{k+1} \approx \theta_k - \epsilon \nabla J_u(\theta_k)^\top; \quad k = n_\epsilon, \ldots, n_\epsilon + m_\epsilon,$$

so that, as $\epsilon \to 0$, $\theta_{n_\epsilon + m_\epsilon} \to \theta^*(\alpha(u))$, where $\theta^*(\alpha) = \arg\min J_\alpha(\theta)$. Use now Theorem 1.8 to reach the desired conclusion.

Notice that the condition on the stepsizes is satisfied, for example, if $\delta_\epsilon = \epsilon^c, c > 2$.

2.8 PRACTICAL CONSIDERATIONS

Well-Posed Problems

In practice, it is essential to model the cost function to be minimized and the constraints in terms of a well-posed problem. This is not always straightforward. However, a careful pre-analysis of the optimization problem is important to understand what, if anything, can go wrong for the numerical algorithms. We have already introduced the "vanishing gradient" problem as an example, where numerical instabilities may occur, or the "vanishing update." Inspection of the problem and initial numerical results may provide insight into the methodology. It is necessary to "go back to the modeling table" many times when dealing with real-life problems until a satisfactory model is defined.

Coercive Fields for the Problem at Hand

Once the problem has been defined, you must build a suitable target vector field G. Focus on coerciveness of the field for the problem at hand. It is also a modeling issue of how to build this function. Go back to Chapter 1 and familiarize yourself with the constrained optimization methods discussed there. In our examples, we have presented elementary problems, and we have illustrated vector fields that are either functions of the gradients or suitable functions for root finding. In most interesting cases, you will have to use more complex models and incorporate perhaps a combination of constrained methods for efficient algorithms.

Experimentation

Get to know your problem. Write simple code that will help you get acquainted with the behavior of the algorithms. If the dimension is large, you may wish to start with lower-dimensional instances to explore how the algorithms behave.

The limit results presented in this chapter cannot be straightforwardly translated into termination conditions for algorithms. Indeed, the limits as defined in Theorem 2.8 are somewhat elusive as they are unachievable by computational algorithms. Of course, a first hands-on conclusion can be drawn from the theory developed so far: (i) the decreasing ϵ algorithm will produce a valid approximation for a true solution of the original optimization problem if n taken very large, and (ii) the for fixed ϵ algorithm one should run the algorithm for some "small" ϵ up to some large iteration n and take the resulting θ-value as approximation of the solution original optimization problem. Assessment of the quality of the "solution" has then to be done on an ad hoc basis. In Chapter 6 we investigate the convergence of the algorithms and show how a sound output analysis can be performed. However, in this book, we have not provided explicit stopping criteria. This is an issue that depends much on the situation that you will be presented with, so getting to explore the algorithms is an important step together with the theoretical understanding of the methodology.

Fulfilling the Key ODE Conditions on Coervivity

Provided that G is locally Lipschitz continuous and the solution to the initial value problem does not blow up in finite time, condition (C2) holds; see Theorem 2.2.

Condition (C1) in Theorem 2.5 offers two approaches for establishing equicontinuity in the extended sense. As in applications G is typically neither (global) Lipschitz continuous nor bounded, both approaches require adjustments to the original G, where the new vector field $\tilde{G}$ is such that (i) the equilibrium points of G and $\tilde{G}$ are the same, and (ii) the behavior of G and $\tilde{G}$ are similar in the vicinity of the equilibrium.

A first approach is to replace G by a clipped vector field G_c, see Theorem 2.9. Alternatively, one can replace G by a surrogate vector field, see Exercise 2.12. Then, if G is locally Lipschitz (though possibly unbounded), the surrogate G, as defined in Exercise 2.12, becomes globally Lipschitz and the trajectories do not blow up in finite time, which implies that conditions (C1) and (C2) in Theorem 2.5 hold.

An obvious question that arises here is whether these two set of conditions are actually different, i.e., does boundedness of a locally Lipschtiz G not already imply that G is Lipschitz? The answer is no, and the following example illustrates this. Let $G(\theta) = \cos(\theta^2)$. Then $G(\theta)$ is bounded and locally Lipschitz but not globally Lipschitz. To see the latter, note that G is differentiable with unbounded derivative which implies that G cannot be globally Lipschitz as a differentiable function is globally Lipschitz if and only if its derivative is globally bounded.

The second approach is to replace G by a projected version. When projecting onto a compact set, the projected version of any continuous G is bounded. Moreover, the trajectories are bounded and cannot blow up by construction, and the projection of a locally Lipschtiz mapping stays locally Lipschitz on the set of projection, see Theorem 2.7.

We conclude with noting that in machine learning projection and clipping are applied to deal with the problem of the so-called *exploding gradient*, which describes the phenomenon that the norm of the gradient varies significantly in θ throughout the search space. Specifically, the objective function for highly non-linear deep neural networks or for recurrent neural networks often contains cliff-like shaped regions, which give rise to very high derivatives in some areas of $\mathbb{R}^d$. When θ_n gets close to such a cliff region, a plain descent update can send the trajectory very far, possibly losing most of the optimization work that had been achieved up to θ_n.

Bounded Trajectories

Generally speaking, boundedness of the trajectories of an ODE is implied by a center of mass condition; see (2.16). Alternatively, boundedness of trajectories can be enforced by using the projected ODE on some convex and compact set. In an optimization setting, one may use a Lyapunov function argument for establishing boundedness of the trajectories of the target ODE; see Example 2.8.

2.9 EXERCISES

Exercise 2.1. Let $J(\theta) = 2\theta^2 + \theta$. Consider the ODE

$$\frac{d}{dt}x(t) = -J'(x(t)), \quad x(0) \in \mathbb{R},$$

on $\mathbb{R}$. Show that for any $x(0)$ it holds that

$$\sup_t |x(t)| < \infty.$$

Exercise 2.2. Let $f_n(x) = \sin(x) + \frac{x}{n}$, and $f(x) = \sin(x)$. Show that:

(a) for each compact set B and $x \in B$, $\lim_{n \to \infty} f_n(x) = f(x)$, but
(b) $f_n \not\to f$ in the sup-norm.

Exercise 2.3. Use boundedness of G to show that as $\epsilon \to 0$, $\{x_\epsilon(\cdot)\}$ in the proof of Theorem 2.5 is equicontinuous in the extended sense.

Exercise 2.4. Show Theorem 2.6, assuming that G is bounded. Use the following steps. First consider the recursion (2.19) (bias term zero):

$$\theta_{n+1} = \theta_n + \epsilon_n G(\theta_n), \quad \theta \in \mathbb{R}^d.$$

Let $\vartheta^\epsilon(\cdot)$ denote the interpolation process in Definition 2.9. Assume that $\sum_n \epsilon_n = \infty$, $\epsilon_n > 0$ for all n, $\epsilon_n \to 0$ and let $x_n(t) = \vartheta^\epsilon(t_n + t), t \geq 0$, where $t_n = \sum_{k=1}^n \epsilon_k$ as before.

(a) Write the telescopic sum for $x_n(t+s) - x_n(t)$, and express this through an integral approximation.
(b) Show that $\{x_n\}$ is equicontinuous in the extended sense.
(c) Use the Theorem of Ascoli-Arzelà to prove that, when $n \to \infty$, x_n converges (in the sup-norm) to the solution of the ODE:

$$\frac{dx(t)}{dt} = G(x(t)). \tag{2.36}$$

(d) Argue like in the proof of Theorem 2.5 and extend your result to the biased version

$$\theta_{n+1} = \theta_n + \epsilon_n (G(\theta_n) + \beta_n(\theta_n)), \quad \theta \in \mathbb{R}^d,$$

where you assume that $\sum_n \epsilon_n \|\beta_n(\theta)\| < \infty$.

Exercise 2.5. Consider the case $G(\theta) = -\nabla J(\theta)$ in (2.19). Discuss the differences in assumptions and conclusions between Theorem 1.3 and the result in Exercise 2.4.

Exercise 2.6. Consider again the problem of Example 1.1 of the surfer at the beach who wishes to rescue a drowning victim. We wish to minimize $J(\theta)$, but the stationary points are only given in the implicit equation (1.7). Consider the gradient search method:

$$\theta_{n+1} = \theta_n - \epsilon J'(\theta_n).$$

(a) Show that as $\epsilon \to 0$ the interpolation processes converge to the ODE:

$$\frac{dx(t)}{dt} = \frac{\sin(\alpha_2(x(t)))}{v_2} - \frac{\sin(\alpha_1(x(t)))}{v_1}.$$

(b) Show that θ^* is stable and argue that a solution $x(t)$ of the above ODE must then satisfy $\lim_{t \to \infty} x(t) = \theta^*$.
(c) Program the procedure and plot the results, using $a = 2, b = 5, d = 10, v_1 = 3, v_2 = 1$, and $\epsilon = 0.05$. Hint: the derivative can be re written as

$$J'(\theta) = \frac{1}{v_1} \frac{\theta}{\sqrt{\theta^2 + a^2}} - \frac{1}{v_2} \frac{d - \theta}{\sqrt{(d-\theta)^2 + b^2}}.$$

Exercise 2.7. Consider the one-dimensional constrained optimization problem where $d = p = 1$ and $q = 0$ (no equality constraints). For the Arrow-Hurwicz algorithm (2.30) evaluate the eigenvalues of $\mathbb{A}$ and prove that if $J(\cdot)$ is strictly convex and $g(\cdot)$ is convex, then $\mathbb{A}$ is Hurwitz, provided that $g'(\theta^*) \neq 0$. This condition follows if we require that the constraint qualifications hold for this problem.

THE ITERATIVE METHOD

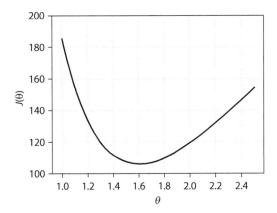

Figure 2.11. The graph of J.

Exercise 2.8. Prove Theorem 2.11, as follows. Consider the strictly convex NLP:

$$\min_{\theta \in \Theta} J(\theta),$$
$$\Theta = \{\theta \in \mathbb{R}^d : g(\theta) \leq 0, h(\theta) = 0\},$$

and assume that $\nabla J(\theta), \nabla g(\theta)$ are continuous and bounded on bounded intervals (e.g., with uniformly bounded Lipschitz constant), and let $x_0 \in \mathbb{R}^d$, $y_0 > 0 (\in \mathbb{R}^p)$, $\eta_0 \in \mathbb{R}^q$ be an initial point of the Arrow-Hurwicz algorithm. Then this algorithm has a local vector field described by the ODE (2.30), whose stable points are saddle points of the Lagrangian.

Exercise 2.9. Let $J(\theta)$ depicted in Figure 2.11 be a smoth function, and consider the optimzation problem

$$\min_{\theta > 0} J(\theta). \tag{2.37}$$

(a) Argue that the problem is well-posed.
(b) For $G = -J'$, argue that G is coercive for (2.37) on $[1.0, 2.4]$.
(c) Use the figure of J to argue that $\theta = 1.8$ is not a stable point of the ODE:

$$\frac{d}{dt} x(t) = G(x(t)).$$

Exercise 2.10. Let $L(\theta)$ be the temperature in the lecture room and denote by θ how far to the right the heat regulator is turned. Assume that $L(\theta)$ is a smooth function. Suppose you want to reach temperature α.

Consider the target vector field

$$G(\theta) = \alpha - L(\theta), \quad \theta \in \mathbb{R}. \tag{2.38}$$

Provide the necessary assumptions on $L(\theta)$ so that

(a) The following problem is well-posed:

$$\min_{\theta} J(\theta) \stackrel{\text{def}}{=} \frac{1}{2}(\alpha - L(\theta))^2. \tag{2.39}$$

(b) The vector field $G(\theta)$ in (2.38) is coercive for (2.39) on $\mathbb{R}$.

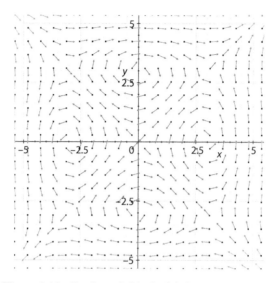

Figure 2.12. Gradient field of $J(\theta)$ for Exercise 2.11.

(c) The algorithm
$$\theta_{n+1} = \theta_n + \frac{1}{n+1}(\alpha - L(\theta_n))$$
converges as $n \to \infty$ to the solution θ^* of (2.39). Mention explicitly the theorems that you use and the reason why the corresponding assumptions are satisfied.

Exercise 2.11. The gradient field of a function $J(\theta)$ is shown in Figure 2.12.

(a) Consider the ODE driven by the negative of the gradient, that is:
$$\frac{d}{dt}x(t) = -\nabla J(x(t))$$
and discuss the nature of the point $(0,0)$ (stable, asymptotically stable, globally stable, or unstable).

(b) Judging from the figure, is this problem well-posed and is the vector-field coercive for the unconstrained optimization problem $\min_\theta J(\theta)$ on $[-5, 5]^2$?

Exercise 2.12. A generic way of obtaining a bounded vector field that is Lipschitz continuous on $\mathbb{R}^d$ is by means of the *surrogate ODE to G*. More specifically, for $G : \mathbb{R}^d \mapsto \mathbb{R}^d$, we define the surrogate vector field by

$$G_r(\theta) = \begin{cases} G(\theta) & \text{if } \theta \in B_r, \\ G_r(r\theta/||\theta||) & \text{if } \theta \notin B_r, \end{cases}$$

for $r > 0$, where $||\theta||$ denotes the Euclidean distance, and B_r the ball of radius r with center at the origin. In a similar way one can define a clipped ODE using the box $[-M, M]^d$.

For G continuous on B_r, show that

(a) $G_r(\theta)$ is bounded on $\mathbb{R}^d$.
(b) If $G(\theta)\theta < 0$ for all θ on the surface of B_r, i.e., for θ such that $\theta/||\theta|| = r$, then G_r has bounded trajectories.
(c) G_r is globally Lipschitz continuous on $\mathbb{R}^d$.

Exercise 2.13. Consider the ODE

$$\frac{d}{dt}x(t) = G(x(t)), \; t \geq 0,$$

for given vector field $G : \mathbb{R}^d \mapsto \mathbb{R}^d$. Show that version of the ODE restricted to the surface of B_r, $r > 0$, is given by

$$\frac{d}{dt}x(t) = G(x(t)) \left(\frac{1}{r} G(x(t)) - \frac{1}{r^3} x(t) \left(x^\top(t) G(x(t)) \right) \right).$$

Note that the surface of B_r fails to be convex as subset of $\mathbb{R}^d$, and the ODE therefore differs from the one in Example 2.7.

Chapter Three

Stochastic Approximation: An Introduction

This chapter contains an overview of stochastic approximation. It is intended to present the ideas via examples and practical situations. The algorithms are discussed without the supporting mathematical theorems, which are deferred to Chapter 4 and Chapter 5. For reference, a "taxonomy" of stochastic approximation is discussed explaining the two main models that will be presented in detail in the following two chapters.

3.1 MOTIVATION

The previous chapters dealt with the analysis of algorithms of the form

$$\theta_{n+1} = \theta_n + \epsilon_n G(\theta_n)$$

designed to "find the zeroes" of a (deterministic) function G, under appropriate assumptions. However, it is often the case that the function $G(\theta)$ is not known with precision, and must be *estimated* using streaming data samples, or generated via *computer simulation*. This is the situation when optimizing real systems in telecommunications, transportation, and robotics; or finding optimal inventory and maintenance policies; or finding economic equilibria, and a number of other important applications. In these cases, the value of G at θ_n is approximated by an observable Y_n, where each n, Y_n should be built as an estimator of the (unknown) function $G(\theta_n)$, in a manner which we will make precise later on. This leads to the following definition.

Definition 3.1. A recursion of the form $\theta_{n+1} = \theta_n + \epsilon_n Y_n$, where $\{Y_n\}$ is a stochastic process, is called a *stochastic approximation* (SA).

Before presenting the theory of stochastic approximations in detail in the following chapters, this chapter provides an overview of stochastic optimization. The leading questions of the theory of stochastic approximation are:

- Under which conditions do stochastic approximation algorithms converge as $n \to \infty$ for the decreasing stepsize algorithm, or (using the approach of Chapter 2) as $\epsilon \to 0$ for the constant stepsize algorithm?
- How can the limits be characterized?
- In which sense does convergence occur, and at what rate, when the sequence $\{Y_n\}$ is a random process, possibly interdependent and dependent on the sequence of values $\theta_m, m \leq n$?

The following example illustrates how stochastic approximation algorithms behave using an academic example. Besides visualizing the result of the algorithms in the presence of noise, we also introduce some important terminology.

STOCHASTIC APPROXIMATION: AN INTRODUCTION

Example 3.1. Suppose that we wish to solve the problem treated in Example 2.5, namely $\min_{\theta \in \mathbb{R}^2} J(\theta)$, where $J(\theta) = 2\theta_1^2 + \theta_2^2$, for $\theta = (\theta_1, \theta_2)^\top$. To solve this problem we would like to follow the iterative algorithm $\theta_{n+1} = \theta_n - \epsilon \nabla J(\theta_n)^\top$, with $\theta_n \in \mathbb{R}^2$, which has a limit behavior as the ODE

$$\frac{dx(t)}{dt} = -4x(t)$$

$$\frac{dy(t)}{dt} = -2y(t).$$

However, suppose that we do not know J or ∇J in closed form. Rather, what we can observe are measurements of the gradient with additive noise. As a simple example, suppose that for each value of $\theta = (\theta_1, \theta_2)^\top \in \mathbb{R}^2$ we can somehow measure the noisy observations of $G(\theta) \stackrel{\text{def}}{=} -\nabla J(\theta)$ given by

$$\hat{G}_1(\theta) \approx -4\theta_1 + \xi_1$$
$$\hat{G}_2(\theta) \approx -2\theta_2 + \xi_2,$$

where ξ_1 and ξ_2 are independent normally distributed, zero-mean random variables (white noise model). Let

$$\phi(\xi, \theta) = (-4\theta_1 + \xi_1, -2\theta_2 + \xi_2)^\top,$$

so that the *observed feedback*, given θ, is the random vector $\phi(\xi, \theta)$. Because the noise is independent of θ and it has a zero mean, it holds that $G(\theta) = \mathbb{E}[\phi(\xi, \theta)]$.

It is assumed that given a value θ (the "input" of the algorithm), $\xi(\theta)$ is available (either observed, or feedback obtained via simulation). When running the SA algorithm, the feedback at the n-th iteration can now be formulated as

$$Y_n := \phi(\xi_n, \theta_n) = \phi(-4\theta_{n,1} + \xi_{n,1}, -2\theta_{n,2} + \xi_{n,2})^\top, \qquad (3.1)$$

where $(\xi_{n,1}, \xi_{n,2}), n \geq 0$ is an iid sequence of independent normally distributed, zero-mean random variables.

To visualize the situation, we have made simulations of the stochastic approximation

$$\theta_{n+1} = \theta_n + \epsilon Y_n,$$

where $\epsilon = 0.01$ and $\xi_{n,i} \sim \mathcal{N}(0, 15)$, independent of the value of θ_n and of previous values of the observations. While this model is known by the simulator, it is assumed that the "controller" who builds the estimates does not know it, and only has access to the observations, or samples. The sequence $\{\xi_n\}$ is called an *underlying process*. There are two "reasonable" approaches to follow:

- We may replace the desired quantity $\nabla J(\theta_n)$ at iteration n by its estimated value using only one observation per update, that is, we use feedback (3.1).
- Because we would like to follow the ODE, we may attempt at sampling "a lot" of observations at each value of θ_n so that the variance of the estimator of $\nabla J(\theta_n)$ is "small." More precisely, we let $\hat{\xi}_n = (\hat{\xi}_{n,1}, \ldots, \hat{\xi}_{n,2K})$, for $K \geq 1$, be a vector of $2K$ iid zero mean normal random variables, and take for the feedback

$$Y_n = \frac{1}{K} \sum_{k=1}^{K} \phi((\hat{\xi}_{n,k}, \hat{\xi}_{n,k+K}), \theta_n))^\top = \left(-4\theta_{n,1} + \frac{1}{K} \sum_{k=1}^{K} \hat{\xi}_{n,k}, -2\theta_{n,2} + \frac{1}{K} \sum_{k=K+1}^{2K} \hat{\xi}_{n,k} \right)^\top,$$

$$(3.2)$$

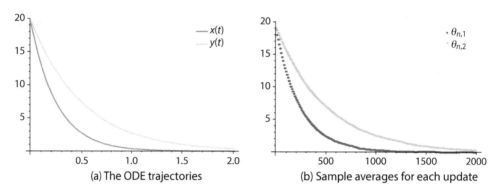

Figure 3.1. Approximations with noise.

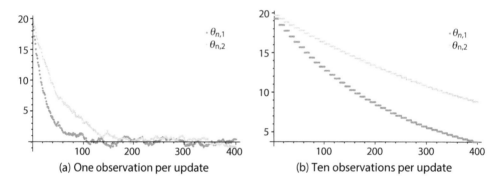

Figure 3.2. Approximations using one or several samples per update.

so that the feedback uses batches of size K for each update. It is worth noting that due to the iid nature of the noise we can interpret Y_n as either obtained from K consecutive observations of the feedback ϕ in (3.1) with fixed θ_n (this is typical in data streaming) or by sampling K realizations of ξ_n at θ_n in parallel (this can be done in a computer simulation).

Figure 3.1(a) shows the limit ODE as we have seen in Chapter 2. The second approach is illustrated in Figure 3.1(b), where each update is built using sample averages with $K = 10$ observations of the noise and $\epsilon = 0.01$. Figure 3.2(a) illustrates a typical trajectory when using only one observation per update; see (3.1). It looks worse as an approximation to the ODE than when several consecutive observations are used per update; see (3.2). However, the sample average approach via batching requires slowing down the algorithm considerably, which affects mainly the first iterations of the algorithm, when it stays at "wrong" values of the estimates for longer than necessary. To compare, Figure 3.2(b) shows the same trajectory as in Figure 3.1(b) that averages ten consecutive values, but only for the same amount of iterations of the algorithm as in Figure 3.2(a). Note that convergence occurs faster in Figure 3.2(a) than in Figure 3.1(b).

This example illustrates an important observation: the time scale. Iterations of the algorithm alone do not provide fair information about the computing effort. Instead, we use time scales that represent computer effort, which is why in Figure 3.2(a) one time step is required per update in (3.1), while in Figure 3.2(b) $K = 10$ time steps are required per update in (3.2).

3.2 ROOT FINDING, STATISTICAL FITTING, AND TARGET TRACKING

Tracking a target's position is an important problem with applications in robotics, defense, estimation, etc. Suppose that we wish to estimate the location $\theta^* \in \mathbb{R}^3$ of a target. Denoting the distance from the target by

$$J(\theta) = \frac{1}{2}(\theta - \theta^*)^\top (\theta - \theta^*),$$

the problem can be stated as the unconstrained optimization problem

$$\min_\theta J(\theta).$$

The gradient-based algorithm in Definition 3.1 would require updating the current estimate via the recursion $\theta_{n+1} = \theta_n - \epsilon_n \nabla J(\theta_n)^\top = \theta_n - \epsilon_n(\theta_n - \theta^*)$, but we don't know θ^*.

Suppose that we can obtain noisy measurements (through radar, satellite, statistics) of θ^*. The n-th observation is denoted ξ_n. In the simplest setting, we have $\mathbb{E}[\xi_n] = \theta^*$, so that the observations are unbiased. Using these estimates, one obtains

$$\theta_{n+1} = \theta_n - \epsilon_n(\theta_n - \xi_n), \tag{3.3}$$

where $\mathbb{E}[\theta_n - \xi_n] = \nabla J(\theta_n)$. The above equation describes the familiar model called regression to the mean. Notice that this particular model $G(\theta) = -(\theta - \theta^*)$ has a unique zero at the point θ^*, as desired.

Solving $\min_\theta J(\theta)$ requires finding the stationary points of the gradient, and is reformulated as rootfinding problem $\mathbb{E}[\theta_n - \theta^*] = 0$. Root-finding problems are also called "target tracking" problems in the control literature, where the idea is that an algorithm has to track a (sometimes moving) target. Here, the random input X is transformed using a known parametrized transformation $v(\theta, X)$ (called the "model" for statistical fitting). The problem is to find the value θ^* such that $L(\theta^*) = \mathbb{E}[v(\theta^*, x)] = \alpha$, where α is a known target. This is also known as an "inverse" problem; see the following example for a motivation. We remark that these problems are examples of *unsupervised learning*.

Example 3.2. A thermostat is to be set up to control a heating device in order to reach a temperature of α degrees in the room. The amount of power θ used by the heating element yields a room temperature $L(\theta)$ in the long run, so we seek θ^* such that $L(\theta^*) = \alpha$, but the function $L(\cdot)$ is unknown (it depends on the room characteristics, the environment, time of day, etc.).

The first approach to this problem would be to "train" the thermostat. In a controlled environment, the device would be set to work under various settings for an *input vector* consisting of the room size, humidity, and some outdoor conditions. A series of statistical tests would estimate the inverse problem, that is, the values of θ^* for each value of α within a reasonable range, as a function of the input vector. This approach would then require the user to choose values for the input vector as well as for the desired room temperature, and the device would simply read the corresponding entry from a preset table. While this static estimation may be accurate for some problems, it is often the case that many important variables that affect the function $L(\theta)$ may not be easy to measure by the regular household user. He or she may prefer a more "intelligent" device that requires the choice of the target room temperature exclusively. Because the thermostat is to be designed to work on different environments and different rooms, a "learning algorithm" can be programmed to adjust the power with minimal input information from the user. One such scheme is precisely a gradient-based optimization algorithm, as we will describe in the following.

Realistically, we must assume that the measurements of the thermometer are imprecise for the actual room temperature $L(\theta)$, so that at value θ, the n-th reading is $\xi_n(\theta) = L(\theta) + w_n$, where the sequence $\{w_n, n \in \mathbb{N}\}$ models the "noise." A common assumption (although not realistic in many cases) is that they are zero mean independent random variables. The consecutive observations $\{\xi_n(\theta_n)\}$ constitute the underlying process for the stochastic approximation model. ✱✱✱

A general, multivariate formulation of the problem in Example 3.2 as an optimization problem is as follows. Let

$$J(\theta) = \frac{1}{2}(L(\theta) - \alpha)^\top (L(\theta) - \alpha), \tag{3.4}$$

for $L: \mathbb{R}^d \mapsto \mathbb{R}^k$ and $\alpha \in \mathbb{R}^k$. The problem is to find

$$\min_\theta J(\theta),$$

with k targets for the tracking problem, and it is assumed that there exists $\theta^* \in \mathbb{R}^d$ such that $L(\theta^*) = \alpha$. The gradient-based algorithm has the deterministic formulation:

$$\theta_{n+1} = \theta_n - \epsilon_n \nabla J(\theta)^\top = \theta_n - \epsilon_n \big((L(\theta_n) - \alpha)^\top \nabla L(\theta_n)\big).$$

In a number of cases, it can be argued that the function $L(\cdot)$ is monotone nondecreasing (or nonincreasing) in each of its components. As we will explain in the following, in this case, a stochastic approximation can be built without any gradient information. In Example 3.2, where $d = k = 1$, more power for the heating element yields a higher room temperature. For cooling elements, more energy is required to decrease the room temperature, so that the function $L(\cdot)$ is monotonic.

We call a mapping $L: \mathbb{R}^d \mapsto \mathbb{R}^d$ *directionally monotone increasing* (resp. decreasing) if, for $1 \le i \le d$, $L(\theta + he_i) \ge L(\theta)$ for any $\theta \in \mathbb{R}^d$ and $h > 0$, with e_i the i-th unit vector. If the inequality can be replaced by a strict inequality, then we call L *strictly directionally increasing* (resp. strictly decreasing). A sufficient condition for L to be strictly directionally increasing is that the Jacobian of L is of the form

$$\nabla L(\theta) = \begin{pmatrix} \frac{\partial}{\partial \theta_1} L_1(\theta) & 0 & \cdots & \cdots \\ 0 & \frac{\partial}{\partial \theta_2} L_2(\theta) & 0 & \cdots \\ & & \ddots & \\ \cdots & & 0 & \frac{\partial}{\partial \theta_d} L_d(\theta) \end{pmatrix}.$$

Denote by $\tilde{G}^\top(\theta) = (L(\theta) - \alpha)^\top M$, where M is the matrix with the partial derivatives above replaced by their sign, and note that M is independent of θ if $L(\cdot)$ is strictly directionally monotone increasing or decreasing. In this case for any $\alpha \in \mathbb{R}^d$ it holds that $\nabla J^\top(\theta) \tilde{G}(\theta) > 0$, which shows that in case of strict directional monotonicity we may replace the actual Jordan matrix by the constant sign matrix in the update and still have a descent direction. This means that the field $\tilde{G}(\theta)$ is coercive for the well-posed optimization problem (3.4). Moreover, the stationary point θ^* is unique and it is also the only stable point of the vector field driving the learning algorithm.

In Example 3.2, it is assumed that *accurate* measurements of the steady temperature $L(\theta)$ are available. Specifically, it is assumed that $\mathbb{E}[\xi(\theta_n)] = L(\theta_n)$. The function $L(\theta)$ here represents the long-term equilibrium temperature reached in the room when the heater has been working at a fixed power setting θ. Suppose now that we use power θ_n for, say, two minutes, during which the thermometer re-adjusts its measurement. After the two minute period, we adjust the power using $\theta_{n+1} = \theta_n - \epsilon_n(\hat{L}(\theta_n) - \alpha)$ and leave the power setting at θ_{n+1} for another two minutes. A problem with the classical (or Robbins-Monro) model (see Theorem 3.1 below) is the assumption that the temperature reading is unbiased for L: is the two-minute period sufficient for the room temperature to reach its equilibrium state? Naturally, this depends on room size, but also on the actual power θ_{n+1}, the initial room temperature ξ_n of the room, and perhaps also on previous values of $\xi_m, m < n$. This measurement problem is an important one: "temperature" is a quantity that cannot be measured as an instantaneous response, because it is an equilibrium quantity. Not only is the temperature of the room not well-defined (until it reaches equilibrium, which may take longer than the updating time), but there may be inaccuracies in the thermometer as well. We will study models with such complexity in Chapters 4 and 5.

Example 3.3 (Linear Regression). Optimization techniques arise naturally in data analysis. Here, a vector of *inputs* ($x_i, i = 1, \ldots, N$) is associated with the corresponding observed *outputs* ($z_i, i = 1, \ldots, N$). The model for the system's response is given by a mapping $h(\theta_1, \theta_2, x)$ and the errors $z_i - h(\theta_1, \theta_2, x_i)$ are assumed to be zero mean, independent random variables. Not knowing the true parameters θ_1^*, θ_2^*, one uses the observations to find the best estimates. In the following, we will discuss linear regression and least-squares approximation in detail.

In a linear regression problem, h is assumed to be given by

$$h(\theta_1, \theta_2, x) = \theta_1 + \theta_2 x, \tag{3.5}$$

and the *linear regression* problem is that of "fitting" the best values of $\theta = (\theta_1, \theta_2)^\top \in \mathbb{R}^2$ for a specific set of observed values. Let $x_1, \ldots, x_N$ be a chosen set of design points, and $z_1, \ldots, z_N$ the corresponding (noisy) observations of the system's response $h(\theta_1, \theta_2, x_i), i = 1, \ldots, N$. Let

$$J(\theta) = \frac{1}{2} \sum_{i=1}^{N} (z_i - \theta_1 - \theta_2 x_i)^2.$$

Then seek $\bar{\theta}$ as the solution to the *deterministic problem*:

$$\min_{\theta \in \mathbb{R}^2} J(\theta).$$

In this problem, the data are seen as fixed (static estimation approach), and the *estimate* $\bar{\theta}$ is the point that minimizes the above expression. It is well known that $\bar{\theta} = (X^\top X)^{-1} X^\top Z$, where X, Z denote the input and output vectors. This solution depends on the design points X and the random vector of output responses Z, and $\bar{\theta}$ is usually not equal to θ^*, although for many models it can be shown that $\mathbb{E}[\bar{\theta}] = \theta^*$. This is commonly referred to as a "training" algorithm (supervised learning). ✸✸✸

Example 3.4 (Learning and Dynamic Model Fitting). Let $Z(X)$ denote the system response to input X, where X is a random variable distributed in the experimentation range: $X \in S$. Like in linear regression, we model the system response by the mapping h given in (3.5). For $\theta = (\theta_1, \theta_2)^\top \in \mathbb{R}^2$, let

$$J(\theta) = \frac{1}{2} \mathbb{E}[(Z(X) - (\theta_1 + \theta_2 X))^2].$$

The problem of *least squares* is to find the parameter values that solve the optimization problem:
$$\min_{\theta \in \mathbb{R}^2} J(\theta).$$

It is left as an exercise (see Exercise 4.4) to verify that for this problem
$$\frac{\partial}{\partial \theta_1} J(\theta) = -\mathbb{E}\left[Z(X) - \theta_1 - \theta_2 X\right], \quad \frac{\partial}{\partial \theta_2} J(\theta) = -\mathbb{E}\left[X Z(X) - \theta_1 X - \theta_2 X^2\right]. \quad (3.6)$$

To minimize this function, Theorem 1.3 suggests use of a gradient procedure of the form
$$\theta_{n+1} = \theta_n - \epsilon_n \nabla_\theta J(\theta_n)^\top = \theta_n + \epsilon_n \mathbb{E}[(Z(X) - \theta_{n,1} - \theta_{n,2} X)(1, X)^\top],$$

but we do not know the (random) function Z. We can use estimates of $Z(x)$ for different (possibly random) values of the input x to construct unbiased estimators of the gradient. This dynamical optimization problem arises naturally when using streaming data for reinforcement learning. Suppose that consecutive design points $\{x_n\}$ are independent and randomly distributed over S. For each x_n we obtain a corresponding random observation $z_n = Z(x_n)$. We identify $\xi_n = (x_n, z_n)$, so that for this example the *underlying process* is just a sequence of iid random pairs that are statistically independent of θ. Call
$$Y_n \stackrel{\text{def}}{=} g(\xi_n, \theta_n) = \begin{pmatrix} z_n - \theta_{n,1} - \theta_{n,2} x_n \\ (z_n - \theta_{n,1} - \theta_{n,2} x_n) x_n \end{pmatrix}.$$

As stated in Exercise 4.4, $\mathbb{E}[Y_n \mid \theta_n] = -\nabla_\theta J(\theta_n)$. The fixed gain size SA algorithm then becomes
$$\theta_{n+1,1} = \theta_{n,1} + \epsilon (z_n - \theta_{n,1} - \theta_{n,2} x_n)$$
$$\theta_{n+1,2} = \theta_{n,2} + \epsilon x_n (z_n - \theta_{n,1} - \theta_{n,2} x_n).$$

This method is often referred to as the *least mean square (LMS)* or *recursive mean square* algorithm. Because the algorithm is designed to improve its estimates as it gathers more and more information, it is often referred to as a *reinforcement learning* algorithm and ϵ is also called the learning rate sequence (a constant stepsize is often used). ✻✻✻

We conclude this section with a historical remark on the important paper [259] wherein Robbins and Monro studied the convergence of this algorithm to θ^* for the one-dimensional case. The following result summarizes their findings.

Theorem 3.1 (Robbins-Monro). *Assume that there is a unique root θ^* to the equation $f(\theta) = \alpha$, and that $f(\cdot)$ is nondecreasing, with $f'(\theta^*) > 0$. Let $\xi_n = \widehat{f}(\theta_n)$ denote the n-th measurement, and assume that $\{\xi_n\}$ are unbiased, that is, $\mathbb{E}[\xi_n \mid \theta_n = \theta] = f(\theta)$, and that $|\xi_n| \leq C < \infty$ is bounded a.s. Let ϵ_n be a sequence satisfying (3.7):*
$$\sum_{n=1}^\infty \epsilon_n = +\infty, \quad \sum_{n=1}^\infty \epsilon_n^2 < \infty, \quad (3.7)$$

with $\epsilon_n > 0$ for all n. Then the stochastic approximation
$$\theta_{n+1} = \theta_n - \epsilon_n (\xi_n - \alpha) \quad (3.8)$$

converges in probability as $n \to \infty$. Specifically, $\mathbb{E}[(\theta_n - \theta^*)^2] \to 0$.

In their paper, Robbins and Monro established more general conditions under which the result holds, including the case where the function $f(\cdot)$ has a jump at θ^*. Let $b_n = \mathbb{E}[(\theta_n - \theta^*)^2]$. The method of proof is based on the expression

$$b_{n+1} = b_n + \epsilon_n^2 \mathbb{E}[f(\theta_n)] - 2\epsilon_n \mathbb{E}[(\theta_n - \theta^*)(f(\theta_n) - \alpha)],$$

which they study using the method of Lemma 1.2 to establish convergence of b_n to zero. For an exhaustive overview of the stochastic root-finding problem we refer to [236].

Remark 3.1. Solving optimization problems leads to finding stationary points and therefore to solving the root-finding problem $\nabla J(\theta) = 0$. Independent of the optimization setting, root-finding problems occur naturally in finding θ such that $L(\theta) = \alpha$ for some response function L, as in Example 3.2. Note that applying the Robbins-Monro root-finding algorithm requires that the dimension of α is not larger than that of θ. Indeed, if $L(\theta)$ and α both are in $\mathbb{R}^k$, and $k < d$, then $\theta \in \mathbb{R}^d$ is not determined through $L(\theta) = \alpha$. To overcome this drawback of direct root finding, the problem is typically cast into a least-squares minimization problem so that solving the identification problem is equivalent to finding the root of the gradient, which then by construction is of appropriate dimension.

3.2.1 Historical Note

B. Delyon [86] mentioned that B. Bru traced back the first known description of stochastic approximation to 1890. The historical anecdote is a result of personal communication of B. Bru with B. Delyon and is based on B. Bru's work on the history of French artillery; see [54]. The problem in question was that of adjusting the angle of a cannon in order to find the optimal angle α for hitting a target at a distance r from the cannon. Finding the correct angle was done in an iterative way by trial and error in the following way. Let θ_n denote the angle of the cannon for firing the n-th shell, resulting in an observed range ξ_n, which yields an error $r - \xi_n$. Experiments showed that $r - \xi_n$ is a random variable, and that an analytical solution for finding the optimal angle was impossible. To overcome this problem, the specialists of the military came up with the algorithm

$$\theta_{n+1} = \theta_n - \frac{\rho}{n}(r - \xi_n),$$

where ρ is some fixed normalization constant. As mentioned by B. Delyon in [86], the important discovery here is the correct choice of the stepsize sequence $1/n$. In the case where exact measurements on $r - \xi_n$ are not available and only the sign is known (either too long or too short), the following algorithm had to be used:

$$\theta_{n+1} = \theta_n - \frac{\rho}{n} \text{sign}(r - \xi_n).$$

Compare with, e.g., Example 3.2.

3.3 A TAXONOMY FOR STOCHASTIC APPROXIMATION

In the presence of noisy measurements or uncertain dynamics, it is important to model the situation of interest in terms of either independent observations of random vectors (e.g., when sensors feed data or when simulations are used to produce synthetic data), or in terms of a stochastic process. Surprisingly, the SA procedure $\theta_{n+1} = \theta_n + \epsilon_n Y_n$ can be shown to approximate the zeroes of the deterministic target vector field $G(\cdot)$ for numerous and very complex models. However, the sense in which the SA procedure "converges" to the limit ODE and the sense in which Y_n "estimates" the vector field G depends on the model of the problem itself. In this section, we first introduce a simple model where consecutive measurements, observations, or simulations of the underlying process only depend on θ_n and not on previous values of the observations. This model will be generalized and studied in detail in Chapter 4. Next, we introduce a more realistic model for many situations where $\{\xi(\theta)\}$ has Markovian dynamics for each value of θ, and the vector field is defined in terms of the long-term behavior, or the stationary measure of the underlying process. The detailed treatment for this model is in Chapter 5.

3.3.1 The General Formulation: The Static Problem

Let $G: \Theta \mapsto \Theta$, with $\Theta \subseteq \mathbb{R}^d$, be a vector field and consider the problem of finding the zeros of G. It is assumed that the function G is not known in closed form, but for every value of the control θ a noisy estimate of the vector $G(\theta)$ can be built (as will be made precise shortly). We will often refer to $G(\theta)$ as the *target vector field*. Given a value of $\theta \in \Theta$, the n-th observation of the *underlying process* $\xi_n(\theta)$ is defined on a probability space $(\Omega, \mathfrak{F}, \mathbb{P}_\theta)$, where $\xi_n(\theta) \in \mathbb{R}^l$, for $l \geq 0$. Often, consecutive values of $\{\xi_n(\theta)\}$ have the same distribution, and the target vector field can be expressed as an expected value of the form

$$G(\theta) = \mathbb{E}[g(\xi(\theta), \theta)].$$

If g is known in explicit form, then g is an obvious choice for the feedback mapping to estimate the target vector field G. As g is typically not available, for estimating $G(\theta)$ we use as the *feedback function* a measurable mapping $\phi(\xi(\theta), \theta))$ as a proxy for $g(\xi(\theta), \theta))$. In formula,

$$G(\theta) = \mathbb{E}[\phi(\xi(\theta), \theta)] + \beta(\theta),$$

where $\beta(\theta)$ is the *bias term*. An SA algorithm has the following steps.

- **Initialization:** Choose initial value θ_0, and use fixed stepsize ($\epsilon = \epsilon_n$) or decreasing stepsize, where we assume that

$$\sum_{n=1}^{\infty} \epsilon_n = +\infty, \quad \sum_{n=1}^{\infty} \epsilon_n^2 < \infty,$$

where $\epsilon_n > 0$ for all n.

- **Running the algorithm:** For $n \geq 0$ repeat the following:
  - **Sampling phase:** For θ_n, sample the underlying randomness $\xi_n(\theta_n)$ according to $\mathbb{P}_{\theta_n}$
  - **θ-update:** Evaluate the feedback and update θ_n:

$$\theta_{n+1} = \theta_n + \epsilon_n \, \phi(\xi_n(\theta_n), \theta_n), \tag{3.9}$$

  where we assume for ease of presentation that $\Theta = \mathbb{R}^d$ so that θ_{n+1} in (3.9) stays feasible, otherwise projection is needed.
  - Set $n : n + 1$ and go back to the sampling phase.

The basic idea of the above algorithm is that under appropriate conditions the limit of $\{\theta_n\}$ obtained via the feedback ϕ in (3.9) traces the behavior of $\theta_{n+1} = \theta_n + \epsilon_n G(\theta_n)$ and thus "finds" the zeros of an appropriate target vector field (refer to Chapter 2 for details).

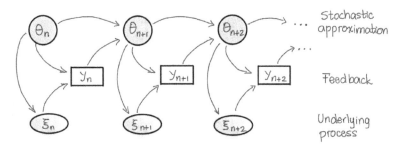

Figure 3.3. Static model.

The information gathered up to the n-th iteration of the SA in (3.9) can be expressed in mathematical language through the filtration built from the σ-fields

$$\mathfrak{F}_n \stackrel{\text{def}}{=} \sigma(\theta_0; \xi_0(\theta_0), \ldots, \xi_n(\theta_n)) \subset \mathfrak{F}, n \geq 0,$$

where we assume that $\mathfrak{F}_0$ contains all $\mathbb{P}$-null sets. Specifically, we assume that all random variables, that is, θ_0 and $\{\xi_n(\theta) : n \geq 0, \theta \in \Theta\}$, are defined on a common probability space $(\Omega, \mathfrak{F}, \mathbb{P})$. Denote the feedback function by

$$Y_n = \phi(\xi_n(\theta_n), \theta_n), \tag{3.10}$$

so that the θ-update step reads $\theta_{n+1} = \theta_n + \epsilon_n Y_n$, where we assume that $\Theta = \mathbb{R}^d$ (otherwise projection is needed). Summarizing our notation we have that

$$\mathbb{E}[Y_n \mid \mathfrak{F}_{n-1}] = \mathbb{E}[\phi(\xi_n(\theta_n), \theta_n) \mid \mathfrak{F}_{n-1}] = G(\theta_n) + \beta_n(\theta_n), \tag{3.11}$$

for $\theta_n \in \Theta$, where we let the bias also depend on the iteration counter n.

It is important to note that given the information up to time $n-1$, i.e., $\mathfrak{F}_{n-1}$, the value of θ_n is known. Moreover, if in the sampling phase the distribution $\mathbb{P}_{\theta_n}$ depends on $\mathfrak{F}_{n-1}$ only through θ_n then this setting is called the *static model*, which is analyzed in Chapter 4. A schematic representation of the process is shown in Figure 3.3.

Batching

As already discussed in Example 3.1, the feedback Y_n may be obtained via batching. We distinguish two kinds of batching, which we will define here properly for ease of reference.

Definition 3.2. We say that Y_n is obtained via *streaming* or *consecutive* uniform batching, if for θ_n we observe a sequence of $K(n)$ consecutive realizations of the underlying process $(\xi_{\tau_n+1}(\theta_n), \ldots, \xi_{\tau_n+K(n)}(\theta_n))$, for $n \geq 0$, with $\tau_n = \sum_{i=0}^{n-1} K(i)$, for $n \geq 1$, and $\tau_0 = 0$, and let Y_n be the sample average over the feedback obtained form reading in the successive observations $\xi_{\tau_n+i}(\theta_n)$, for $1 \leq i \leq K(n)$:

$$Y_n = \frac{1}{K(n)} \sum_{i=1}^{K(n)} \phi(\xi_{\tau_n+i}(\theta_n), \theta_n).$$

We say that Y_n is obtained via consecutive random sized batches when batch n has size $K(n)$, which is a (possibly random) integer.

We say that Y_n is obtained via *independent* or *parallel* batching, if for θ_n we sample $K(n)$ iid replications $\xi_{n,i}(\theta_n)$, where $K(n)$ is a (possibly random) integer, and let Y_n be the sample average over the feedback obtained for the individual observation $\zeta_{n,i}(\theta_n)$, for $1 \leq i \leq K(n)$:

$$Y_n = \frac{1}{K(n)} \sum_{i=1}^{K(n)} \phi(\xi_{n,i}(\theta_n), \theta_n).$$

Streaming batching was discussed in Chapter 1 for two-stage algorithms, where $K(n)$ was supposed to be increasing with n. In practice, random size batches occur naturally when dealing with continuous time process. In those cases, often sensors and local computations are performed every δ units of time, and the number of events observed within each of these time intervals is random and may be dependent on θ_n.

In the following, we discuss how various problems that are typical in applications are cast into our framework. For ease of reading, we present examples with no bias terms.

Example 3.5. Consider again the dynamic model fitting introduced in Example 3.4. The setup satisfies the assumptions of the model with no bias ($\beta_n(\theta) = 0$). Because of the model assumptions, the underlying process consists of the consecutive observations $\xi_n = (x_n, Z(x_n))$ that are actually independent of θ_n. As mentioned in the example, $\mathbb{E}[Y_n \mid \theta_n] = -\nabla J(\theta_n)$ for this model, where ∇J is given in (3.6). A more complex model for statistical learning is given by a neural network for deep learning in a classification problem. The underlying process is modeled here by the consecutive pairs of samples and labels (x_n, z_n), also called the "ground truth." The control parameter θ represents the various parameters of the neural network model (e.g., weights, bias terms, convolution matrices). Proof of unbiasedness of the feedback function requires some more advanced results and will be discussed in Chapter 8 later on. It suffices here to mention that under Lipschitz continuity of the activation functions and from the linearity of matrix multiplication (in θ), it follows that derivative and expectation can be interchanged; thus the *stochastic gradient* (which is the gradient of the samples with respect to θ) can be used as an estimator of the expected gradient of the cost function (commonly a function of the likelihood of the sample, or a direct function of the training error, also called the "loss" function). ✹✹✹

Example 3.6. Consider a service center with one single sever. The server places incoming work orders in a queue and serves the jobs in order of arrival. The server handles N jobs, after which it undergoes maintenance, before starting again for another N jobs. Jobs are completed "on-time" when the average waiting time in queue over N jobs is at most α units of time.

Consecutive service times of each server are denoted by $\{S_n(\theta)\}$ and are independent with a general distribution with $\mathbb{E}[S_n(\theta)] = \theta$. Speeding service times costs money. The problem is to minimize the cost $C(\theta)$, which is a decreasing function of θ, subject to the constraint that $L(\theta) \leq \alpha$, where

$$L(\theta) = \mathbb{E}\left[\frac{1}{N} \sum_{k=1}^{N} W_k(\theta)\right]$$

is the expected average waiting time in queue of the N jobs.

Assume that the iid interarrival times $\{A_n\}$ satisfy $\mathbb{E}[A_1] < \infty$, $\text{Var}(A_1) < \infty$ and that θ is a scale parameter of the service distribution, i.e., $S_n(\theta) = \theta S_n(1)$, with $\{S_n(1)\}$ an iid sequence and assume that $\mathbb{E}[S_1(1)] < \infty$, $\text{Var}(S_1(1)) < \infty$. It is well known that consecutive waiting times of customers constitute a Markov process given by the so-called Lindley recursion. Indeed, let W_n be the waiting time of customer n, then:

$$W_n(\theta) = (W_{n-1}(\theta) + S_{n-1}(\theta) - A_n)_+, \qquad (3.12)$$

where $(x)_+ = \max(0, x)$ is the positive part of the number x, and $W_0(\theta) = S_0(\theta) = 0$, which gives $W_1(\theta) = 0$ as it should because the first customer arriving to an empty system experiences no waiting. We will discuss this example in detail in Part II and will provide a proof of (3.12) in Example 7.16.

Because the cost of service speed is decreasing, and $L(\theta)$ is strictly monotone increasing as a function of θ, it follows from the results in Chapter 1 that the constraint must be active. Thus, the problem can be solved using root finding. As target vector field we take $G(\theta) = -(L(\theta) - \alpha)$. It is easily seen that $G(\theta)$ is coercive for the optimization problem: if $L(\theta) > \alpha$, the vector field points toward smaller values for θ as it should because $L(\theta)$ is monotone increasing in θ, and for $L(\theta) < \alpha$ the vector field points to larger values, again in correspondence with the monotonicity of $L(\theta)$; moreover $G(\theta)$ has unique stable point θ^* satisfying $L(\theta^*) = \alpha$.

For the stochastic approximation we consider the sequence of the first N interarrival and service times as underlying process, that is,

$$\xi_n(\theta) = (A_1, S_0(\theta), A_2, S_1(\theta), \ldots, A_N, S_{N-1}(\theta)),$$

where we let $S_0(\theta) = 0$. There are two possible scenarios: (a) we use simulation to produce copies of the underlying process $\{\xi_n(\theta_n)\}$ given the value of θ_n, or (b) we stream data from the real-time operation of the servers, adjusting their speeds to θ_n for the n-th observation. Either way, we can assume that, given θ_n, the random vector $\xi_n(\theta_n)$ is independent of past values of the underlying process. Using (3.12) to compute the first N waiting times then leads to the feedback mapping

$$\phi(\xi_n(\theta), \theta) = \alpha - \frac{1}{N} \sum_{k=1}^{N} W_k^{(n)}(\theta),$$

where $W_k^{(n)}(\theta)$ is the k-th waiting time in the n-th simulation (or data streaming). It is straightforward to see that $G(\theta) = \mathbb{E}[\phi(\xi_n(\theta), \theta)]$ for all n, so that the bias term vanishes, i.e., $\beta_n(\theta) \equiv 0$, and we may work with the feedback

$$Y_n = \phi(\xi_n(\theta_n), \theta_n) = \alpha - \frac{1}{N} \sum_{k=1}^{N} W_k^{(n)}(\theta_n).$$

For the theoretical development we refer to Chapter 4. ✳✳✳

Example 3.7. This example is related to the field of actuarial science called *risk theory*. It presents an example of a *random horizon* problem. Suppose that you are about to start a company. You have estimated your monthly costs at a random amount $c(\theta)$ that is a decreasing function of the initial endowment θ, with $\theta \geq 0$, that you wish to raise from possible investors. You have determined that contracts will be arriving according to a Poisson process $N(\cdot)$ with rate λ. You will receive a random amount of Z_n dollars per contract, and you assume that $\{Z_n\}$ are iid with a known distribution. The total revenue up to time t is then

$$S(t) = \sum_{k=1}^{N(t)} Z_k - c(\theta)t,$$

and you are interested in estimating the minimal initial endowment θ^* that you would need to operate for T months without bankruptcy. Specifically, let

$$\tau(\theta) = \min(t : S(t) \leq 0),$$

then you seek to estimate θ^* such that $\mathbb{P}(\tau(\theta) \leq T) \leq \alpha$. By assumption, it follows that $\mathbb{P}(\tau(\theta) \leq T)$ is decreasing in θ. By design, this is a root-finding problem.

Denote by $\xi_n = (N(T); Z_1, \ldots, Z_{N(T)})$ be the n-th simulation of the process, so that ξ_n is a process with a random stopping time (not a vector, because the length is random). Using independent simulations of the process, we take as feedback function

$$Y_n = \phi(\xi_n, \theta_n) = \mathbf{1}_{\{\tau(\theta_n) \leq T\}} - \alpha, \quad \theta_n \geq 0.$$

Since $\mathbb{E}[\phi(\xi_n, \theta_n) \mid \theta_n] = \mathbb{P}(\tau(\theta_n) \leq T) - \alpha$ on $\theta_n \in [0, \infty)$, the feedback is unbiased and we have $\beta_n(\theta_n) = 0$ a.s. for all n. For the theoretical development we refer to Chapter 4.

✸✸✸

We conclude this series of examples with a simple setting where we "learn from data" parameters of the objective functions that are not subject to optimization. The example we will present here shows a bias.

Example 3.8. Let X_1 and X_2 be exponentially distributed with mean 1, and let X_3 follow some unknown distribution with finite variance and mean μ for some $\mu > 0$. Consider the target function

$$L(\theta) = \mathbb{E}[\max(\theta X_1 + \mu, X_2)],$$

and find θ^* as the minimizer of

$$\min J(\theta) \stackrel{\text{def}}{=} \frac{1}{2}(L(\theta) - \alpha)^2.$$

Since $L(\theta)$ is strictly monotone increasing with respect to θ, we may choose $G(\theta) = \alpha - L(\theta)$ and solve the root-finding problem

$$L(\theta) - \alpha = 0,$$

with $G(\theta)$ as target vector field, and it follows by standard arguments that G is coercive for the above minimization problem. The samples for X_1 and X_2 are mutually independent and follow the correct distribution, whereas μ is approximated using the sample mean

$$\mu_n = \frac{1}{n} \sum_{i=1}^{n} X_3(i).$$

We take as feedback function

$$Y_n = \phi(\xi_n, \theta_n) = \alpha - \max\left(\theta_n X_1(n) + \frac{1}{n}(X_3(n) + (n-1)\mu_{n-1}), X_2(n)\right),$$

where

$$\xi_n = (X_1(n), X_2(n), X_3(n), \mu_{n-1}).$$

Note that

$$\mu_n = \frac{1}{n} X_3(n) + \frac{n-1}{n} \mu_{n-1} = \frac{1}{n} \sum_{i=1}^{n} X_3(i). \quad (3.13)$$

Inspecting Y_n, we obtain from (3.13) for feasible θ_n, i.e., $\theta_n > 0$,

$$\mathbb{E}[Y_n \mid \mathfrak{F}_{n-1}] = \alpha - \mathbb{E}\left[\max\left(\theta_n X_1(n) + \frac{1}{n}(X_3(n) + (n-1)\mu_{n-1}), X_2(n)\right)\right]$$

$$\neq \alpha - \mathbb{E}[\max(\theta_n X_1(n) + \mu, X_2(n))]$$

$$= G(\theta_n), \quad \theta_n > 0,$$

which shows that $\beta_n(\theta_n) \neq 0$. The bias can be bounded as follows:

$$\beta_n(\theta_n) \leq \left| \mathbb{E}\left[\max\left(\theta_n X_1(n) + \mu, X_2(n)\right)\right] \right.$$
$$\left. - \mathbb{E}\left[\max\left(\theta_n X_1(n) + \frac{1}{n}(X_3(n) + (n-1)\mu_{n-1}), X_2(n)\right)\right]\right|$$
$$\leq \mathbb{E}\left[\left|\max(\theta_n X_1(n) + \mu, X_2(n))\right.\right.$$
$$\left.\left. - \max\left(\theta_n X_1(n) + \frac{1}{n}(X_3(n) + (n-1)\mu_{n-1}), X_2(n)\right)\right|\right]$$
$$\leq \mathbb{E}[|\mu - \mu_n|],$$

and by the Law of Large Numbers we have that $\beta_n(\theta_n)$ tends to 0 as n tends to infinity. Moreover, by the central limit theorem (CLT), $\mu - \mu_n$ is asymptotically normal distributed with zero mean and standard deviation $1/\sqrt{n}$. Then, $|\mu - \mu_n|$ follows a folded normal distribution (for details, see [301]) the mean value of which scales linearly in the standard deviation of the normal variable $\mu - \mu_n$. This shows that $\beta_n(\theta_n) = O(1/\sqrt{n})$. The theory developed in Chapter 4 will provide sufficient conditions for the stochastic approximation to work in presence of (asymptomatically negligible) bias. ✶✶✶

3.3.2 Long-Term Behavior: Markovian Underlying Process

For the static problem discussed in the previous section, the distribution $\mathbb{P}_{\theta_n}$ of the n-th observation $\xi_n(\theta_n)$ depends on $\mathfrak{F}_{n-1}$ only through θ_n. A generalization of this model is to let the distribution of $\xi_n(\theta_n)$ depend on $\xi_{n-1}(\theta_{n-1})$ as well, so that $\xi_n(\theta_n)$ has conditional distribution $\mathbb{P}_{\theta_n}(\cdot|\xi_{n-1}(\theta_{n-1}))$. This extension leads to the Markovian dynamics model.

The *Markovian dynamics model* assumes that for each fixed value of the "control" parameter θ, the underlying process $\{\xi_n(\theta)\}$ is a Markov chain with transition probabilities

$$\mathbb{P}(\xi_{n+1}(\theta) \in A \mid \xi_n(\theta) = x) = P_\theta(x, A),$$

for any Borel set $A \subset \mathbb{R}$. Moreover, the model assumes that each Markov chain P_θ is ergodic with unique stationary measure π_θ. The vector field G is meant to capture the long-term behavior of a given model, which is expressed by assuming that $g(\xi, \theta)$ exists such that

$$G(\theta) = \lim_{N \to \infty} \frac{1}{N} \mathbb{E}\left[\sum_{n=1}^{N} g(\xi_n(\theta), \theta)\right] = \int g(\xi, \theta) \pi_\theta(d\xi). \tag{3.14}$$

In words, $G(\theta)$ is the long-term average of a mapping $g(\cdot, \theta)$ over $\{\xi_n(\theta)\}$. The role of the mapping $g(\cdot, \theta)$ will be illustrated in Example 3.9 below. The precise assumptions on the model are stated in Chapter 5, which also contains the proof of convergence of the stochastic approximation $\theta_{n+1} = \theta_n + \epsilon_n Y_n$ using the feedback $Y_n = \phi(\xi_n(\theta_n), \theta_n)$. The schematic situation of the SA model is shown in Figure 3.4.

Example 3.9. We revisit Example 3.6. It is well known that the queue process is *stable* only when

$$\theta \in \Theta \stackrel{\text{def}}{=} \{\theta \in \mathbb{R}^+ : \theta < \mathbb{E}[A_1]\}, \tag{3.15}$$

and we will also assume that $\text{Var}(S_1(\theta)) < \infty$ for $\theta \in \Theta$. By stable, we mean that the queueing process is ergodic and has a unique stationary distribution. It is well known that for parameters $\theta \in \Theta$ the Markov chain is Harris recurrent and ergodic, with a unique

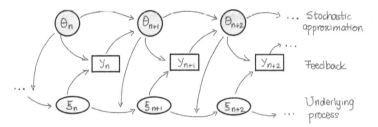

Figure 3.4. Markovian dynamics.

stationary measure. The problem is to adjust the server's speed so that the stationary average wait per client is no higher than a given amount α, while minimizing the cost of operation, which increases as θ decreases. Because it is costly to speed up the service, the solution to the problem is given when the constraint is active. Arguing along the lines put forward in Example 3.6, we see that the minimization problem is well-posed and the vector field

$$G(\theta) = \alpha - \mathbb{E}[W(\theta)], \quad \theta \in \Theta,$$

where $W(\theta)$ denotes the stationary waiting time, is coercive. Apart from the specific case of the $M/M/1$ queue, $W(\theta)$ cannot be computed analytically. However, ergodicity of the waiting time process implies that, w.p.1,

$$\lim_{N \to \infty} \frac{1}{N} \sum_{k=1}^{N} W_k(\theta) = \lim_{N \to \infty} \mathbb{E}\left[\frac{1}{N} \sum_{k=1}^{N} W_k(\theta)\right] = \mathbb{E}[W(\theta)]. \tag{3.16}$$

A straightforward choice for a feedback function would be

$$\phi(\xi(\theta), \theta) = \alpha - \lim_{N \to \infty} \frac{1}{N} \sum_{k=1}^{N} W_k(\theta), \tag{3.17}$$

where $\xi(\theta) = (A_k, S_{k-1}(\theta), k \geq 1)$. From (3.16) is follows that $G(\theta) = \phi(\xi(\theta), \theta)$ w.p.1, and $\beta_n(\theta_n) = 0$, due to unbiasedness. Moreover, due to ergodicity, the problem has become a deterministic problem. The evident catch of this approach is that each update for θ_n requires a time-consuming simulation of the long-run average waiting time. Taking into account that any simulation with finite computation budget will only approximately compute $\mathbb{E}[W(\theta)]$, the averaging inside ϕ will only give $\mathbb{E}[W(\theta)]$ plus some noise, and we are actually back to the setting in Example 3.1.

To overcome the numerical unfeasibility of the naive approach in (3.17) we now present an alternative approach based on the Markovian dynamics model. Use now $\xi_n(\theta)$ as the n-th waiting time to describe the underlying process. It is a Markov process because

$$\xi_n(\theta) = (\xi_{n-1}(\theta) + S_{n-1}(\theta) - A_n)_+,$$

according to Lindley's recursion. For fixed θ, this process is adapted to the filtration with $\mathfrak{F}_n(\theta) = \sigma(A_1, S_0(\theta), A_2, S_1(\theta), \ldots, A_N, S_{N-1}(\theta))$. The feedback mapping in (3.17) is the infinite batch size version of the single-step feedback mapping $\alpha - \xi_n(\theta_n)$. Now suppose that we follow the waiting time sequence and after each update of θ_n, we sample one new observation of the waiting time. Lindley's equation now becomes

$$\xi_n(\theta_n) = (\xi_{n-1}(\theta_{n-1}) + S_{n-1}(\theta_n) - A_n)_+.$$

More formally, the feedback becomes for $\theta_n \in \Theta$

$$Y_n = \alpha - \xi_n(\theta_n) \stackrel{\text{def}}{=} \phi(\xi_n(\theta_n), \theta_n).$$

From the Markov structure, we have

$$\mathbb{E}[Y_n \mid \mathfrak{F}_{n-1}] = \mathbb{E}[\alpha - \xi_{n-1}(\theta_{n-1}) + S_{n-1}(\theta) - A_n)_+ \mid \xi_{n-1}(\theta_{n-1}) = (\cdot)] \neq G(\theta_n),$$

$\theta_n \in \Theta$, where $\xi_{n-1}(\theta_{n-1})$ is the $(n-1)$-th waiting time of θ_{n-1}. While Y_n is biased for the target filed $G(\theta)$ (which refers to the stationary distribution), it does have the important property that Y_n is unbiased for expected deviation of the n-th waiting time given $\xi_{n-1}(\theta_{n-1})$. The bias in Y_n can thus be attributed to the nonstationarity of $\xi_n(\theta_n)$ rather than any issue in the way we compute Y_n. Put another way, the mapping

$$g(x, \theta) = \mathbb{E}[\alpha - x + S_{n-1}(\theta) - A_n)_+ \mid \xi_{n-1}(\theta) = x],$$

satisfies (3.14) on Θ by ergodicity. Moreover, for $\hat{\xi}(\theta)$ a sample of the stationary waiting time at θ, it holds that

$$\mathbb{E}[g(\hat{\xi}(\theta), \theta)] = G(\theta),$$

while

$$\mathbb{E}[\phi(\xi_n(\theta), \theta)] \neq G(\theta),$$

since the expected n-th waiting time for θ is not equal to the expected *stationary* waiting time. The theory provided in Chapter 5 will provide conditions under which this nonstationarity bias does not affect the stochastic approximation. ✻✻✻

3.4 OVERVIEW ON STOCHASTIC APPROXIMATION

Chapters 4 and 5 present the necessary theoretical tools in order to understand the behavior of iterative algorithms in the presence of noise. Emphasis is on understanding under which conditions the noisy iterations "converge" (and in what sense) to the target limit ODE. In particular, Chapter 4 addresses the generic case of feedback mappings for the static problem. In Chapter 5, we discuss the ergodic case as described in the long-term behavior problem formulation. In this case, the noise of observation n is affected by the previous history of the trajectory, and a more complex mathematical model for the underlying processes has to be used. Finally, Chapter 6 discusses the efficiency of the various estimators.

The field of SA is very widespread and distributed over many disciplines from control to statistics and machine learning. It is therefore inevitable that any presentation of SA will in some aspects be incomplete or unbalanced. We refer the interested reader to some recent overviews on SA that complete the results presented here. Accounts on SA with an emphasis on statistical applications can be found in [27, 199]. For SA in the context of root finding we refer to [236]. SA plays also a prominent role in machine learning and overviews can be found in [41, 42, 271]. For SA in simulation based optimization, we refer to [71, 101, 102].

3.5 THE SAMPLE AVERAGE APPROACH

The sample average approach (SAA) also goes in the literature under the name of stochastic counterpart (see [95, 269, 270]) and sample path optimization (see [247, 260]). The idea of the SAA approach is to replace $J(\theta) = \mathbb{E}[L(\theta,\xi)]$, where ξ represents the underlying noisy observation, with its sample average. Note that for SAA the random observation ξ needs to be statistically independent of θ and thus it represents exogenous noise. More specifically, let $L(\theta, \xi_i)$ denote the i-th sample of $L(\theta, \xi)$, and let

$$\hat{J}_n(\theta) = \frac{1}{n} \sum_{i=1}^{n} L(\theta, \xi_i)$$

be the sample average over n iid samples. Using $\hat{J}_n(\theta)$ as a proxy for $J(\theta)$ yields the deterministic problem

$$\min_{\theta} \hat{J}_n(\theta). \tag{3.18}$$

Keeping the sample $\{\xi_i : 1 \le i \le n\}$ fixed, this problem can be solved with deterministic gradient descent, see Chapter 1, using the sample-path dependent derivative

$$\frac{d}{d\theta} \hat{J}_n(\theta) = \frac{1}{n} \sum_{i=1}^{n} \frac{\partial}{\partial \theta} L(\theta, \xi_i).$$

A solution θ_n^* to the above (proxy) problem is then taken as estimator for the solution θ^* of the original problem $\arg_\theta \min J(\theta)$.

Provided that convergence of $\hat{J}_n(\theta)$ toward $J(\theta)$ implies convergence of the minimizers as well, SAA allows the application of standard deterministic optimization algorithms to solve the surrogate problem in (3.18). This precondition for SAA is typically violated if convergence of $\hat{J}_n(\theta)$ toward $J(\theta)$ does not imply convergence of $\nabla \hat{J}_n(\theta)$ toward $\nabla J(\theta)$; see, for example, pp. 152–154 in [272]. For background and details on SAA we refer to [260, 268, 283, 284]. Form a practical point of view, SAA can be only applied if $\hat{J}_n(\theta)$ has an analytically tractable form. Unfortunately, SAA is not suitable when streaming data in real time is considered, because every time a new observation ω_i becomes available the entire procedure has to be restarted. For an SAA approach to stochastic discrete optimization we refer to [185].

Example 3.10. We illustrate SAA in more detail by means of the minimization problem

$$\min_{\theta > 0} \left(\mathbb{E}[X^2(\theta)] + \theta^2 \right),$$

where $X(\theta)$ is exponentially distributed with rate θ. Since $\mathbb{E}[X^2(\theta)] = 2\theta^{-2}$, we have

$$\min_{\theta > 0} \left(\mathbb{E}[X^2(\theta)] + \theta^2 \right) = \min_{\theta > 0} \left(\frac{2}{\theta^2} + \theta^2 \right). \tag{3.19}$$

Stationary points can be found through $-\frac{4}{\theta^3} + 2\theta = 0$, which implies that $\theta^* = \sqrt[4]{2}$. It is straightforward to check that θ^* is indeed the solution of the minimization problem yielding $2\sqrt{2}$ as a minimal value.

We now apply SAA to the pathwise optimization using the sample average. The SAA minimization problem becomes

$$\min_{\theta > 0} \left(\frac{1}{n} \sum_{i=1}^{n} \left(-\frac{1}{\theta}(1 - U_i) \right)^2 + \theta^2 \right) \tag{3.20}$$

and solves as

$$\theta_n^* = \left(\frac{1}{n}\sum_{i=1}^n (-(1-U_i))^2\right)^{\frac{1}{4}}.$$

By the Strong Law of Large Numbers, the sum converges to 2 (the second moment of an exponential random variable with rate 1), and consequently θ_n^* converges to $\sqrt[4]{2}$, which is the correct answer. In realistically sized problems this translation to a deterministic problem offers the great advantage to apply advanced deterministic optimization algorithms for solving (3.20).

It is worth noting that the sample average is essential for SAA. Indeed, solving $\min_{\theta>0}(X^2(\theta)+\theta^2)$ directly using that $X(\theta)=X(1)/\theta$ has solution $\sqrt{X(1)}$. Taking expected values gives

$$\mathbb{E}\left[\sqrt{X(1)}\right] = \frac{1}{2}\sqrt{\pi}.$$

which differs from the actual solution $\sqrt[4]{2}$. The reason for this discrepancy is that the pathwise optimization solves

$$\mathbb{E}\left[\min_{\theta>0}(X^2(\theta)+\theta^2)\right]$$

as opposed to (3.19). ✷✷✷

The SAA approach is very common in statistical learning when using data to fit the parameters of a model. In [309], a model for a bike share station leads a solution that requires a simulation of part of the process, while the rest of the objective function can be found analytically. The simulation was used to evaluate the SAA for a random variable that is independent of the control variable θ, thus enabling a deterministic approach for the optimization. This reference includes an analysis of the error term in the estimated optimal value, proving that under common assumptions, the error is asymptotically normally distributed.

3.6 PRACTICAL CONSIDERATIONS

The type of analysis depends on the chosen model, and often there is flexibility in the interpretation of the problem at hand in terms of a stochastic approximation. Indeed the interrelation between target vector field, feedback function, underlying process, and feedback allows for great flexibility.

In developing stochastic approximation algorithms, we have to distinguish between streaming applications and simulation-based solutions. While simulation-based solutions offer great flexibility in methods and implementation, in streaming applications, the underlying process is restricted to data actually available. This makes the ergodic model very attractive in streaming applications as the feedback only requires one set of new observations for the next update.

Choosing the Gain Sequence

The gain size effects the performance of the algorithm in two ways: the speed of convergence to a local minimizer θ^*, and the statistical properties of the approximation θ_n for θ^* provided by the algorithm. In Chapter 6 we will discuss this in more detail. Here, we will only mention that the optimal (matrix valued) gain size, in the sense that it leads to a maximal speed of convergence of θ_n toward θ^* and a minimal variance of θ_n as an estimator for θ^*, is known to be

$$\epsilon_n = \frac{1}{n+1}(\nabla^2 J(\theta^*))^{-1}, n \geq 0,$$

where θ^* is the true minimizer. For details we refer to Remark 6.1 and [26, 334], or Chapter 8 in [11]. Note that for θ^* being the location of a local minimum of a function in C^2, the Hessian is positive definite, and its inverse exists and is also invertible. Hence, applying the above gain size, we have

$$\epsilon_n \nabla J(\theta_n) = \frac{1}{n+1}(\nabla^2 J(\theta^*))^{-1} \nabla J(\theta_n),$$

which gives a standard decreasing gain size algorithm (with $\epsilon_n = 1/(n+1)$) for descent direction $(\nabla^2 J(\theta^*))^{-1} \nabla J(\theta_n)$; see Lemma 1.1 for details. Generally speaking, the above gain sequence is only of theoretical interest as θ^* is not known. Heuristic approaches that elaborate on the above optimal rate, such as [177, 273, 317], use finite-difference approximations for the Hessian. Specifically, these heuristics periodically estimate $(\nabla^2 J(\theta))^{-1}$ for $\theta = \theta_n$ at selected transition n, and use this estimate as approximation for the optimal rate. Under appropriate smoothness, as θ_n tends to θ^*, the inverse Hessian converges to the optimal rate. For applications of this approach, we refer to [289, 290]. For an instance of an adaptive stepsize algorithm elaborating on the inverse Hessian, we refer to Section 11.1.2 in Part III.

Given the desirability to balance stability of the algorithm in the early iteration (preferring a "small" gain size for small n) with non-negligible steps in the later iteration (preferring a "significant" gain size for large n), a standard for the of the gain size found in the literature is

$$\epsilon_n = \frac{a}{((n+1)A)^\alpha}, \tag{3.21}$$

with $A, a \geq 0$ and $(1/2) < \alpha \leq 1$. The basic idea of this kind of gain sequence is that a larger value of A allows for a larger value of a and thus for small n the term A leads to "small" gains, whereas for n large, the larger value of a avoids the gains and tends to zero too fast. For more details on this, we refer to [233, 275]. Alternative gain sizes, decreasing more slowly than (3.21), are given by

$$\epsilon_n = \frac{\ln(n)}{n}, \ n \geq 2, \quad \epsilon_0 = \epsilon_1 = 1,$$

and

$$\epsilon_n = \frac{1}{n \ln(n)}, \ n \geq 2, \quad \epsilon_0 = \epsilon_1 = 1.$$

A popular adaptive gain sequence rule in case of unbiased gradient updates, called Kesten's rule, only decreases ϵ_n when there is a change in the direction of two consecutive updates, i.e., $(\theta_{n+1} - \theta_n)(\theta_n - \theta_{n-1}) < 0$; see [181]. Extensions of Kesten's rule use linear reinforcement learning techniques to let the stepsize depend on observed consecutive updates; see [277]. These kinds of adaptive algorithms have been extended to biased gradient updates in [51]. An alternative, ad hoc way of defining the gain sequence is to periodically reset the SA after a given number of iterations, where the initial value for the reset SA is the final value achieved by the previous SA; see [275].

Balancing the Outputs

Updates θ_n fluctuate in SA according to their stochastic nature. For n large it is conceivable that θ_n is close to the minimum location θ^* and in effect is hovering around θ^*. This makes it reasonable to use

$$\theta_N^* = \frac{1}{N} \sum_{n=1}^{N} \theta_n$$

as a predictor for θ^*. This averaging approach in SA was introduced for fixed stepsize SA in [249, 274], and is called *Polyak-Rupert averaging* in the literature. Compensating for the fact that the first updates are typically far from the optimum, the "moving window" average

$$\theta_N^* = \frac{1}{m} \sum_{n=N-m+1}^{N} \theta_n,$$

for $N > m$, may be used. For an increasing window size, the asymptotic results are provided in [196, 198]. For details on output analysis of fixed gain size SA, we refer to Section 6.4.

The Polyak-Rupert average can be generalized to

$$\theta_N^* = \frac{\sum_{n=1}^{N} \epsilon_n \theta_n}{\sum_{n=1}^{N} \epsilon_n}.$$

The output is identical to the Polyak-Rupert average for fixed stepsize, i.e., $\epsilon_n = \epsilon$ for all n. For decreasing stepsize the above iterates put less emphasis on the last updates, and the underlying idea is that larger stepsize can be used based on the more "robust" iterates. Hence, the name *robust stochastic approximation* for this method; see [226]. The optimal rate for ϵ_n can be solved for convex $J(\theta)$ and Θ bounded.

For (strongly) convex cost functions $J(\theta)$ the *accelerated* SA methods have been developed; see [111, 112] for details. Accelerated SA uses weighted updates and a combination of the gradient and approximate distance mapping to prevent too large steps in the iterates.

The Role of Projections for Convergence Analysis

One of the main difficulties in analyzing SA is that the sequence θ_n potentially can reach any point in $\mathbb{R}^d$ making convergence analysis challenging. Intuitively, an SA with decreasing stepsize will stay in some bounded region H as θ_n traces some solution point θ^* and will, in the long run, hover around θ^*. Without pre-experiments that explore the global behavior of the objective function, it is impossible to identify a priori a suitable region H. Indeed, in applying a projection on some fixed predefined set, one has to work carefully so as not to exclude the solution θ^*. We refer to Theorem 2.10 and Remark 2.9 for details.

For this reason, variations of SA have been developed that use sequences of sets Θ_m, $m \geq 1$, to project on. Specifically, Θ_m is such that $\Theta_m \subset \Theta_{m+1}$ and Θ_m tends to $\mathbb{R}^d$. Then, the SA is run projected on Θ_1. Once θ_n leaves Θ_1, the projection set is enlarged to Θ_2 and the algorithm is reset to some value inside of Θ_1. One then continues this way until convergence of SA occurs. This way we trace the "correct" projection set alongside running the SA, and convergence analysis can be carried out using this projection set. There are various choices for re-initializing the SA after leaving a Θ_m set, and for choosing the sequence Θ_m. See [74, 75] for early references, and [4, 5, 6] for more recent work. Recently, projections have been proposed for speeding up convergence rate of SA in statistical applications. The basic idea is that, as SA proceeds, more is learned on the location of the solution and informed projections can be constructed so that large variance in the update Y_n (and thus θ_n) can be counter parted by projection. We refer to [285, 286] for details.

An alternative to projections is clipping, and we refer to Theorem 2.10 for details.

A Comment on Global Optimization

Related to the previous discussion. SA generally converges to a local solution. To achieve convergence to the global solution it has been proposed to add additional noise W_n to the standard SA resulting in

$$\theta_{n+1} = \theta_n + \epsilon_n Y_n + \delta_n W_n.$$

Typically, W_n is taken to be an iid sequence of a d-dimensional standard normal distribution. Kushner and Yin [198] propose

$$\epsilon_n = \delta_n = \frac{a}{\log(n+1)}.$$

Many other choices exists, and we refer to the discussion in Section 8.4 of [295] for more details. That adding noise to the gradient has a positive effect on its convergence properties has also been observed in [225] for deep and complex networks, where the updates are taken as

$$\theta_{n+1} = \theta_n + \epsilon_n Y_n + Z_n,$$

with Z_n a vector of independent normal random variables with mean zero and variance $\eta/(1+n)^\gamma$, independent of everything else, for appropriate choices for η and γ. For a pseudo-gradient SA algorithm for global optimization, based on the above approach, with provable convergence properties we refer to [213]. When gradients as well as the Hessian are estimated from simulation, then the stochastic trust-region response surface method in [68, 69] may be used.

3.7 EXERCISES

Exercise 3.1. Let X a real-valued random variable with finite mean value and finite variance. Consider the problem

$$\min_{\theta} \mathbb{E}\left[\frac{1}{2}(\theta - X)^2\right]$$

and let $\{X_n\}$ be a sequence of iid of samples of X.

(a) Show that $Y_n = (\theta_n - X_n)$ is unbiased estimator for

$$\left.\frac{d}{d\theta}\right|_{\theta=\theta_n} \mathbb{E}\left[\frac{1}{2}(\theta - X)^2\right].$$

(b) Let $\epsilon_n = 1/n$ and show that

$$\theta_{n+1} = \theta_n - \frac{1}{n+1}Y_n$$
$$= \frac{1}{n+1}(X_n + X_{n-1} + \cdots + X_0).$$

(c) Write the general model as given in (3.11) for this problem. What is $G(\theta_n)$? What is $\beta_n(\theta_n)$? Discuss convergence of θ_n.

Exercise 3.2. This is a simple version of a *multi-armed bandit problem*. A slot machine has two arms and the winning probability of the two arms, denoted by p_A and p_B, respectively, are not known. It is assumed that the event of winning on each arm is independent of past history and also independent of the other arm's results. Let θ_n denote the belief that we suppose that arm A has higher winning probability. Consequently, we choose arm A at step n with probability θ_n (called "exploitation" in machine learning). The goal is to learn the correct value $\theta^* = 1$ if $p_A > p_B$, and $\theta^* = 0$ otherwise (assuming that $p_A \neq p_B$).

A straightforward learning approach is

$$\theta_{n+1} = \begin{cases} \theta_n & \text{if } n\text{-th game was a loss,} \\ \theta_n + \epsilon_n (1 - \theta_n) & \text{if win on arm A,} \\ \theta_n - \epsilon_n \theta_n & \text{if win on arm B.} \end{cases}$$

(a) Write the above recursion as a stochastic approximation, specifying the feedback Y_n.
(b) Provide the target vector field, assuming no bias term is involved.
(c) Argue that the target vector field is coercive for the learning problem.
(d) *Optional:* Program the procedure to test various schemes for the learning rates $\{\epsilon_n\}$, including the case of constant ϵ. Discuss your results.

Exercise 3.3. Two agents, $k = 1, 2$, play a simultaneous game. At time n each agent chooses an action $a_n^k \in \mathcal{A} = \{a_1, \ldots, a_l\}$, for $k = 1, 2$, and some $l < \infty$. The reward the agents receive for round n depends on the combination of actions the agents played, and it is given by

$$A^k = \begin{pmatrix} r^k(a_1, a_1) & r^k(a_1, a_2) & \ldots & r^k(a_1, a_l) \\ r^k(a_2, a_1) & r^1(a_2, a_2) & \ldots & r^k(a_2, a_l) \\ \vdots & \vdots & & \vdots \\ r^k(a_l, a_1) & r^k(a_l, a_2) & \ldots & r^k(a_l, a_l) \end{pmatrix},$$

for $k = 1, 2$, where $r^k(a_i, a_j)$ is the reward for player k if player 1 plays a_i and player 2 plays a_j. The players have no knowledge on the others motivation of choosing a certain action, and the choice of action of player 1 appears to by random for player 2 and vice versa. However, the players learn their opponents distribution over the actions from played games. This information is publicly available to both players through the sequence of their actions $a_1^k, a_2^k, \ldots$, $k = 1, 2$. From the data, the empirical distribution of player k over the actions set can be computed as follows:

$$p^k(a, n) = \frac{1}{n} \sum_{i=1}^n \mathbf{1}(a_i^k = a), \quad a \in \mathcal{A},$$

and we write $p^k(n)$ for the distribution. Each agent wants so maximize her own reward, so after having played n rounds, agent $k = 1$ computes her mixed strategy p^1 so that

$$p^1(n+1) = g^1(p^2(n)) := \max_p \sum \sum r^1(a_i, a_l) p(a_i) p^2(a_l, n)$$

and agent $k = 2$ computes her mixed strategy p^2 so that

$$p^2(n+1) = g^2(p^1(n)) := \max_p \sum \sum r^2(a_l, a_i) p(a_i) p^1(a_l, n).$$

Let $\theta = (p^1, p^2)$, with p^1, p^2 distributions over the action set, and set $g(\theta) = g(p^1, p^2) = (g^1(p^2), g^2(p^1))$. The game reaches an equilibrium if no player has an incentive to adjust her strategy. In formula, if $g(\theta) = \theta$, such a fixed point is called a *Nash equilibrium*, and a Nash equilibrium is the solution of the root-finding problem

$$g(\theta) - \theta = 0.$$

Note that $g(\theta)$ is by construction a pair of proper distributions. This type of game was introduced by [53] and is called a *fictitious game*. For more details and economic motivation we refer to [105]. Fictitious games attracted much attention in the literature as they offer an explanation of equilibria as a result of learning rather than rationalistic analysis. For mathematical background, we refer to [25] and [176].

(a) Show that the empirical distributions for $k = 1, 2$ follow the recursion

$$p^k(a, n+1) = p^k(a, n) + \frac{1}{n+1}(\mathbf{1}(a^k_{n+1} = a) - p^k(a, n)).$$

(b) Under appropriate conditions, a Nash equilibrium for the game can be found iteratively via

$$\theta_{n+1} = \theta_n + \epsilon_n G(\theta_n),$$

where $G(\theta) = g(\theta) - \theta$. Show that the stochastic version of this algorithm is given by

$$\theta_{n+1} = \theta_n + \epsilon_n Y_n(\xi(n)),$$

where

$$Y_n(\xi(n)) = \Big(\mathbf{1}(a^1_{n+1} = a_1) - p^1(a_1, n), \mathbf{1}(a^1_{n+1} = a_2) - p^1(a_2, n), \ldots,$$
$$\mathbf{1}(a^2_{n+1} = a_I) - p^2(a_I, n)\Big),$$

for $\xi(n) = (a^1_{n+1}, a^2_{n+1})$.

Chapter Four

Stochastic Approximation: The Static Model

The static model for SA was introduced in Chapter 3. It starts with the case where the observations of the processs lead to an unbiased conditional drift for a feedback. This chapter presents rigorous definitions of the model and detailed proofs of convergence of the ensuing SAs. Discussion of the required assumptions are presented via detailed examples.

4.1 MARTINGALE DIFFERENCE NOISE MODEL

The assumptions of the Robbins-Monro procedure are quite restrictive: they require unbiasedness of an additive bounded noise, which excludes many useful models such as Gaussian noise models. To account for more realistic situations this chapter presents the theory of stochastic approximations for the static model in more detail, following the approach of [198]. The method of proof uses properties of a general noise model, called the *martingale difference noise model*. We present the model terminology and notation here.

Let $G(\theta)$ be a vector field defined on $\Theta \subseteq \mathbb{R}^d$, and consider the problem of finding the zeros of G. Furthermore, we assume here that these zeroes are the only asymptotically stable points of the ODE driven by G. It is assumed that the function G is not known in closed form, but for every value of the control θ a noisy estimate of the vector $G(\theta)$ can be built (as will be made precise shortly). We will refer to $G(\theta)$ as the *target vector field*.

For each fixed value of θ, there is a random vector $\xi(\theta)$ whose distribution may depend on θ. An "instantaneous" cost function $g(\xi(\theta), \theta)$ provides the noisy estimate of the target vector field. Specifically, we consider the model where $G(\theta) = \mathbb{E}[g(\xi(\theta), \theta)]$, $\theta \in \Theta$.

The feedback function is defined as $Y_n = \phi(\xi_n(\theta_n), \theta_n)$, where ϕ is a proxy for g (which may not be known explicitly), and $\{\xi_n(\theta_n)\}$ is the corresponding sequence of observations that define the underlying process on a common measurable space $(\Omega, \mathfrak{F})$. In this chapter we study the analysis of the SA algorithm:

$$\theta_{n+1} = \theta_n + \epsilon_n Y_n, \qquad (4.1)$$

where the stepsize sequence is assumed either constant, or that it satisfies

$$\sum_{n=1}^{\infty} \epsilon_n = +\infty, \quad \sum_{n=1}^{\infty} \epsilon_n^2 < \infty, \qquad (4.2)$$

with $\epsilon_n > 0$ for all n, for the basic setting and notation see Section 3.3.1. Refer to Figure 3.3 for a schematic visualization of the model. The information structure of the SA as a stochastic process is defined as the natural filtration of the process, using the increasing sequence of σ-algebras:

$$\mathfrak{F}_n \stackrel{\text{def}}{=} \sigma(\theta_0; \xi_0(\theta_0), \ldots, \xi_n(\theta_n)) \subset \mathfrak{F}, n \geq 0,$$

where we assume that $\mathfrak{F}_0$ contains all $\mathbb{P}$-null sets.

Given the information available up to "time" $n-1$, the expected outcome of Y_n is given by $\mathbb{E}[Y_n \mid \mathfrak{F}_{n-1}]$. Put differently, $\mathbb{E}[Y_n \mid \mathfrak{F}_{n-1}]$ is the trend of Y_n in the next iteration. Specifically for our model, $\mathbb{E}[Y_n \mid \mathfrak{F}_{n-1}] = \mathbb{E}[\phi(\xi_n(\theta_n)) \mid \mathfrak{F}_{n-1}]$ which is not necessarily the same as $G(\theta_n)$, because there may be an estimation bias from the proxy. We denote this bias by

$$\beta_n \stackrel{\text{def}}{=} \mathbb{E}[\phi(\xi_n(\theta_n), \theta_n) \mid \mathfrak{F}_{n-1}] - \mathbb{E}[g(\xi_n(\theta_n), \theta_n) \mid \mathfrak{F}_{n-1}]$$

and we remark that this bias is a random variable, measurable w.r.t $\mathfrak{F}_{n-1}$. The difference between the trend and the realization is denoted by

$$\delta M_n \stackrel{\text{def}}{=} Y_n - \mathbb{E}[Y_n \mid \mathfrak{F}_{n-1}], \tag{4.3}$$

and δM_n thus measures the deviation of the realization Y_n from its (conditional) mean value. Notice that by construction, $\mathbb{E}[\delta M_n] = 0$. Put differently, δM_n is the noise in the measurement of $G(\theta_n) + \beta_n$. In the same way

$$V_n = \mathbb{E}[(\delta M)^2] = \mathbb{E}\left[(Y_n - \mathbb{E}[Y_n \mid \mathfrak{F}_{n-1}])^2\right] \tag{4.4}$$

expresses a measure of the variation of Y_n around its conditional mean given information $\mathfrak{F}_{n-1}$. The cumulative process $\sum_{k=1}^{n} \delta M_k$ is, by construction, a martingale and $\{\delta M_n\}$ is called the *martingale difference process*;[1] see Definition B.34 in Appendix B. With the above notation Y_n can be decomposed into the following three components:

$$Y_n = \underbrace{G(\theta_n)}_{\text{target vector field}} + \underbrace{\delta M_n}_{\text{noise}} + \underbrace{\beta_n}_{\text{error of vector field}}, \quad \text{for } \theta_n \in \Theta. \tag{4.5}$$

Following [198], we call the above model the *martingale difference noise model* for G on Θ and the material presented in this chapter follows the approach in [198]. Note that in (4.5) the noise term δM_n is separated from the target vector field $G(\theta)$, and we express this by saying that the noise in the martingale difference noise model is exogenous. In particular, in the proof of our main theorem we will elaborate on the fact that the limit of Y_n can be studied though the analysis of the limit of $G(\theta_n)$ whereby the noise component can be treated separately. In case $\Theta = \mathbb{R}^d$ feasibilty of θ_n in (4.5) is of no concern and we simply refer to (4.5) as martingale difference noise model for G, or martingale difference noise model for short.

The simplest stochastic setting is that of iid noise. In Example 3.1 the noise was given by exogenous iid random variables. Unfortunately, the iid assumption is often unrealistic as the distribution of the noise depends on θ_n, which in turn depends on the noise in the previous state through (4.1). The martingale difference noise model considers the wider applicable condition that the noise in the next state is unpredictable given the information field $\mathfrak{F}_{n-1}$. This setting is very well suited to SA and occurs naturally in many applications, while the weaker condition of uncorrelated noise will require stronger conditions for enforcing convergence of the algorithm; see Remark 4.1 below.

---

[1] Let $\{\delta M_n\}$ be adapted to filtration $\{\mathfrak{F}_n\}$. Then, $\{\delta M_n\}$ is called a martingale difference process if $\mathbb{E}[\delta M_n \mid \mathfrak{F}_{n-1}] = 0$ a.s.

STOCHASTIC APPROXIMATION: THE STATIC MODEL

4.2 ANALYSIS OF DECREASING STEPSIZE SA

We will now apply the basic ideas of the ODE method established in Chapter 2. Before we can do that, we present a technical result that will be used in the proof of the main theorem below. This result establishes sufficient conditions for the noise in the tail of $\delta M_n + \beta_n$ to tend to zero. We use notation and defintions of the martingale difference noise model.

Proposition 4.1. *Suppose that the feedback sequence $\{Y_n\}$ satisfies the martingale difference noise model in (4.5), so $\mathbb{E}[Y_n \mid \mathfrak{F}_{n-1}] = G(\theta_n) + \beta_n$. Let $\{\epsilon_n\}$, with $\epsilon_n > 0$ for all n, and $\epsilon_n \to 0$, be a deterministic stepsize sequence and assume:*

(a1) $\sum_{i=0}^{\infty} \epsilon_i = \infty$,
(a2) the error terms are asymptotically negligible in the sense that $\sum_{i=0}^{\infty} \epsilon_i \|\beta_i\| < \infty$ w.p.1,
(a3) the variance term (4.4) satisfies: $\sum_{i=0}^{\infty} \epsilon_i^2 V_i < \infty, w.p.1$.

Define the processes

$$Z_m^{(n)} = \sum_{i=n}^{n+m} \epsilon_i \delta M_i, \qquad B_m^{(n)} = \sum_{i=n}^{n+m} \epsilon_i \beta_i.$$

Then,

$$\lim_{n \to \infty} \left(\sup_{m \geq 0} |Z_m^{(n)}(i)| \right) = 0 \quad a.s., \ for \ 1 \leq i \leq d, \tag{4.6}$$

$$\lim_{n \to \infty} \left(\sup_{m \geq 0} |B_m^{(n)}(i)| \right) = 0 \quad a.s., \ for \ 1 \leq i \leq d. \tag{4.7}$$

Proof. Result (4.7) follows directly from assumption (a2). We now show (4.6). The goal is to show that the shifted martingale difference noise converges to 0 with probability 1 as $n \to \infty$. By construction, for each n the process $Z_m^{(n)}$ for $m \geq 0$, is also a martingale. From Proposition B.3 in Appendix B, for a non-negative convex function $q(\cdot)$ a martingale W_m satisfies

$$\mathbb{P}\left(\sup_{n \leq \ell \leq n+m} |W_\ell| \geq \Delta \right) \leq \frac{1}{q(\Delta)} \mathbb{E}[q(W_{n+m})],$$

for $m > 0$. For ease of presentation we assume in the following that $d = 1$. We now apply this property to the martingale $Z_m^{(n)}$ for $q(x) = x^2$ noticing that $\mathbb{E}[Z_n^{(n)}] = 0$, which yields for any $\Delta > 0$ and $\ell \geq 0$

$$\mathbb{P}\left(\sup_{0 \leq \ell \leq m} |Z_\ell^{(n)}| \geq \Delta \right) \leq \frac{1}{\Delta^2} \mathbb{E}\left[\left(\sum_{i=n}^{n+m} \epsilon_i \delta M_i \right)^2 \right]$$

$$= \frac{1}{\Delta^2} \sum_{i=n}^{n+m} \epsilon_i^2 \mathbb{E}[(\delta M_i)^2]$$

$$= \frac{1}{\Delta^2} \sum_{i=n}^{n+m} \epsilon_i^2 V_i \leq \frac{1}{\Delta^2} \sum_{i=n}^{\infty} \epsilon_i^2 V_i =: \frac{K_n}{\Delta^2},$$

where the first equality follows from the martingale property $\mathbb{E}[\delta M_n \delta M_m] = 0$ (see Proposition B.1 in Section B.6) and the second equality follows from (4.4). This implies that for every n,

$$\lim_{m \to \infty} \mathbb{P}\left(\sup_{\ell \leq m} |Z_\ell^{(n)}| \geq \Delta \right) \leq \frac{K_n}{\Delta^2},$$

and we now show that this implies

$$\mathbb{P}\left(\sup_{0\leq m} |Z_m^{(n)}| > \Delta\right) \leq \frac{K_n}{\Delta^2}. \qquad (4.8)$$

Let A_m be the event $A_m = \{\sup_{\ell \leq m} |Z_\ell^{(n)}| \geq \Delta\}$. By construction, $A_m \subset A_{m+1}$. Use now the continuity theorem for increasing events (see Theorem B.1 in Appendix B) to establish that $\mathbb{P}(\lim_{m\to\infty} A_m) = \lim_{m\to\infty} \mathbb{P}(A_m) \leq K_n/\Delta^2$, which establishes (4.8).

Assumption (a3) implies that $\lim_{n\to\infty} K_n = 0$, so that

$$\lim_{n\to\infty} \mathbb{P}\left(\sup_{0\leq m} |Z_m^{(n)}| > \Delta\right) = 0.$$

Call A'_n the event $A'_n = \{\sup_{0\leq m} |Z_m^{(n)}| > \Delta\}$. What we have shown so far is that $\lim_{n\to\infty} \mathbb{P}(A'_n) = 0$. What we wish to show is that $\mathbb{P}(\lim_{n\to\infty} A_n) = 0$. By construction, the sequence $\{A_n\}$ is monotone decreasing. Use again Theorem B.1 and the fact that this holds for any $\Delta > 0$ to establish the desired result:

$$\mathbb{P}\left(\lim_{n\to\infty} \sup_{0\leq m} |Z_m^{(n)}| = 0\right) = 1.$$

□

The following result extends the result of Theorem 2.6 for the stochastic setting given by the martingale difference noise model (the version for the case of Θ convex and bounded is left to the reader).

Theorem 4.1. *Consider the stochastic approximation*

$$\theta_{n+1} = \theta_n + \epsilon_n Y_n, \quad n \geq 0,$$

and suppose that the feedback sequence $\{Y_n\}$ satisfies the martingale difference noise model for vector field G on $\mathbb{R}^d$, so that $\mathbb{E}[Y_n \mid \mathfrak{F}_{n-1}] = G(\theta_n) + \beta_n$. Assume

(a1) $\sum_{i=0}^\infty \epsilon_i = \infty$, with $\epsilon_n > 0$ for all n, and $\epsilon_n \to 0$,
(a2) the error terms are asymptotically negligible, in the sense that $\sum_{i=0}^\infty \epsilon_i \|\beta_i\| < \infty$ w.p.1,
(a3) the variance term (4.4) satisfies: $\sum_{i=0}^\infty \epsilon_i^2 V_i < \infty$, w.p.1.
(a4) G is bounded or Lipschitz continuous, and the trajectories of ODE

$$\frac{dx(t)}{dt} = G(x(t)), x(0) \in \mathbb{R}^d, \qquad (4.9)$$

have a unique limit (as $t \to \infty$) for each initial condition. Let S denote the corresponding set of asymptotically stable points and assume that $S \neq \emptyset$.

Then the shifted interpolation process $x_n(t) = \vartheta^\varepsilon(t_n + t), t \geq 0$ and $\vartheta(0) = \theta_0$, converge as $n \to \infty$ to the solution of the ODE (4.9). This implies that $\theta_n \to S$ a.s. In particular, if θ^* is the only asymptotically stable point of the ODE (4.9), then $\theta_n \to \theta^*$ a.s.

Alternatively, for $\Theta \subset \mathbb{R}^d$ convex and compact, consider the stochastic approximation

$$\theta_{n+1} = \Pi_\Theta(\theta_n + \epsilon_n Y_n), \quad n \geq 0, \text{ and } \theta_0 \in \Theta,$$

and suppose that the feedback sequence $\{Y_n\}$ satisfies the martingale difference noise model for G on Θ, so that $\mathbb{E}[Y_n \mid \mathfrak{F}_{n-1}] = G(\theta_n) + \beta_n$, for $\theta_n \in \Theta$. Assume (a1) to (a3) and *(a4)'* *G is continuous on Θ, and the trajectories of the projected ODE*

$$\frac{dx(t)}{dt} = (\Pi_\Theta G)(x(t)), \quad x(0) \in \Theta, \tag{4.10}$$

have set of asymptotically stable points $S \neq \emptyset$.

Then the shifted interpolation process $x_n(t) = \vartheta^\varepsilon(t_n + t), t \geq 0$ and $\vartheta(0) = \theta_0$, converge as $n \to \infty$ to the solution of the ODE in (4.10). This implies $\theta_n \to S$ a.s. In particular, if θ^* is the only asymptotically stable point of the projected ODE, (4.10) then $\theta_n \to \theta^*$ a.s.

Proof. We prove the first part of the theorem. This proof is done in three parts: (i) telescopic sum and integral representation; (ii) characterizing the limit; and (iii) analyzing the asymptotic behavior.

Telescopic Sum and Integral Representation: We define the shifted interpolated processes $\vartheta^n(t), M^n(t), B^n(t)$:

$$\vartheta^n(t) = \theta_n + \sum_{i=n}^{m(t_n+t)-1} \epsilon_i Y_i,$$

$$M^n(t) = M_n + \sum_{i=n}^{m(t_n+t)-1} \epsilon_i \delta M_i,$$

$$B^n(t) = B_n + \sum_{i=n}^{m(t_n+t)-1} \epsilon_i \beta_i.$$

Following the standard form (4.1) and using (4.5) we have

$$\vartheta^n(t) = \theta_n + \sum_{i=n}^{m(t_n+t)-1} \epsilon_i G(\theta_i) + M^n(t) + B^n(t),$$

where $M^n(t) \to 0, B^n(t) \to 0$ as $n \to \infty$ for all $t > 0$ a.s., which follows directly from Proposition 4.1. Because $\vartheta^n(\cdot)$ is piecewise constant, then the sum of the terms $\epsilon_i G(\theta_i)$ is an integral and

$$\vartheta^n(t) - \vartheta^n(0) = \int_0^t G(\vartheta^n(u)) \, du + M^n(t) + B^n(t) + \rho^n(t),$$

where $\rho^n(t)$ accounts for "end point" errors in the integral approximation and $\rho^n(t) \to 0$ a.s. uniformly in t as $n \to \infty$ (see Exercise 4.6).

Characterizing the Limit ODE: From the a.s. convergence of $M^n(\cdot), B^n(\cdot)$ and $\rho^n(\cdot)$ to zero, there exists a null set N such that for all $\omega \notin N$ these sequences converge to zero as $n \to \infty$, uniformly in t. Pick $\omega \notin N$. For this fixed ω, the rest of the proof is for a deterministic sequence and we can apply Theorem 2.6 to establish that there is a subsequence of functions $\{\vartheta^{n_k}(\cdot, \omega)\}$ with a continuous limit and call $\vartheta(t, \omega)$ the particular limit (we now make it explicit that ω is a fixed trajectory). Then $\vartheta(t, \omega)$ satisfies

$$\vartheta(t, \omega) = \vartheta(0, \omega) + \int_0^t G(\vartheta(u, \omega)) \, du,$$

which means that each convergent subsequence (in n) has a limit that satisfies the ODE (4.16). By Assumption (a4), however, there is a unique solution for each initial condition so all subsequences share the same limit.

Asymptotic Behavior: The final part of the proof uses stability of the limiting ODE to characterize the limit behavior (in t) of the algorithm. Indeed, as $t \to \infty$, for each $\omega \notin N$,

$$\lim_{t \to \infty} \vartheta(t, \omega) = \lim_{t \to \infty} \left(\lim_{n \to \infty} \vartheta^n(t, \omega) \right) = \lim_{n \to \infty} \theta_n(\omega) \in \mathcal{S},$$

which follows because $\mathcal{S}$ is the (invariant) set of stable points of the ODE. Therefore all limits of the algorithm will be stable points. If, in particular, $\theta^* \in \mathcal{S}$ is unique, then $\theta_n \to \theta^*$ a.s. This concludes the proof of the first statement.

For the proof of the second statement, we note that Theorem 2.7 shows that under the conditions put forward in the theorem the interpolation process of the projected SA

$$\theta_{n+1} = \Pi_\Theta(\theta_n + \epsilon_n Y_n)$$

converges to the unique solution of the projected ODE. This shows that the first part of the theorem straightforwardly extends to the projected ODE. $\square$

Remark 4.1. In the case of uncorrelated noise, condition (a3) in Proposition 4.1 and Theorem 4.1 has to be replaced by the stricter condition

$$\sum_{i=0}^{\infty} \epsilon_i^2 V_i (\log i)^2 < \infty.$$

This stems from the fact that uncorrelated noise is a weaker model than the martingale difference noise model. Indeed, denote the pdf of X and Y by f_X and f_Y, respectively, and the joint density by $f_{X,Y}$, then we have three different ways of expressing the relation between two continuous random variables X and Y:

(a) If $f_{X,Y}(x, y) = f_X(x) f_Y(y)$, then X, Y are called *independent*.
(b) X is called *unpredictable by Y* if $\mathbb{E}[X|Y] = \mathbb{E}[X]$ a.s.
(c) X and Y are called uncorrelated (orthogonal) if $\mathbb{E}[XY] = \mathbb{E}[X]\mathbb{E}[Y]$.

It holds that (a) implies (b) and (c), and (b) implies (c). However, (c) does not imply (b), which we illustrate with the following example. Consider the triangle between the points $(-1, 0), (0, 1), (0, 1)$. Let (X, Y) be uniformly distributed on this triangle. Then, by symmetry, we have $\mathbb{E}[X|Y] = 0 = \mathbb{E}[X]$, whereas $\mathbb{E}[Y|X]$ varies in X. But by

$$\mathbb{E}[XY] = \int y\, \mathbb{E}[X|Y = y] f_Y(y)\, dy = 0,$$

so that $\mathbb{E}[XY] = \mathbb{E}[X]\mathbb{E}[Y] = 0$, which implies that $\text{Cov}(X, Y) = 0$. This shows that for achieving the same result in the uncorrelated noise model as in the martingale difference noise model (which resembles (b)), we have to replace the martingale noise inequality in the proofs with an application of the Theorem of Rademacher-Mensov, see [245] and [257], leading to the above stronger condition on the growth of the variance term.

Example 4.1. This example illustrates verification of assumptions (a2) and (a3) in Theorem 4.1, where we use the descent version of the SA algorithm given by $\theta_{n+1} = \theta_n + \epsilon_n Y_n$. In [182], the Robbins-Monro procedure was modified to approximate the solution to

$$\min_{\theta \in \mathbb{R}} J(\theta), \qquad (4.11)$$

when (independent) experiments can be used to estimate the value of the cost function J at any design point θ, but the function may not be differentiable, or an estimate of its gradient may not be available. Because only the cost function can be estimated, and not its gradient, a finite difference approach can be used. Specifically, suppose that $J(\theta) = \mathbb{E}[h(\xi, \theta)]$ for some observable function h, and let $\{\xi_n\}$ be the underlying process of the model. The finite difference leads to the scheme

$$Y_n(\theta_n) = -\frac{h(\xi_n, \theta_n + c_n) - h(\xi'_n, \theta_n - c_n)}{2c_n},$$

where the observations of the "experiments" at $\theta_n + c_n$ and $\theta_n - c_n$ are statistically independent, that is, ξ_n is independent of ξ'_n. By assumption $\mathbb{E}[h(\xi_n, \theta_n) | \mathfrak{F}_{n-1}] = J(\theta_n)$. This procedure is now known as the Kiefer-Wolfowitz procedure, and we sometimes will refer to it as the K-W method for short. Recall from Example 1.6 in Chapter 1 that approximations of gradients via finite differences can lead to the correct limit, provided that the finite difference scheme uses $c_n = n^{-c}$ for a positive constant c.

When $J(\cdot) \in C^3$ admits a Taylor approximation, the order of the bias term β_n of assumption (a2) in Theorem 4.1 can be calculated:

$$\begin{aligned} -\mathbb{E}[Y_n | \mathfrak{F}_{n-1}] &= \frac{J(\theta_n + c_n) - J(\theta_n)}{2c_n} - \frac{J(\theta_n - c_n) - J(\theta_n)}{2c_n} \\ &= \frac{1}{2c_n}\left(J'(\theta_n)c_n + \frac{1}{2}J''(\theta_n)c_n^2 + \frac{1}{6}J'''(x)c_n^3 \right. \\ &\quad \left. - J'(\theta_n)(-c_n) - \frac{1}{2}J''(\theta_n)c_n^2 - \frac{1}{6}J'''(y)(-c_n^3)\right) \\ &= J'(\theta_n) + O(c_n^2), \end{aligned}$$

for x, y appropriately chosen, so that $\beta_n = \mathbb{E}[Y_n | \mathfrak{F}_{n-1}] + J'(\theta_n) = O(c_n^2)$. The term V_n is given by

$$V_n = \frac{v(\theta_n + c_n) + v(\theta_n - c_n)}{4c_n^2},$$

where $v(\theta) = \text{Var}[h(\xi, \theta)]$ is the variance of the observations at parameter value θ. If the variance of the observations is uniformly bounded, then $v(\cdot) \leq V < \infty$ and $V_n = O(c_n^{-2})$. In order for assumptions (a2) and (a3) in Theorem 4.1 to hold, we need to choose the stepsize sequence and the finite difference sequence $\{c_n\}$ simultaneously. First, because of (a2) and (a3), necessarily $c_n \to 0$, which immediately implies that $V_n \to \infty$. For instance, if $\epsilon_n = O(n^{-1})$ and $c_n = n^{-c}$ then $\beta_n = O(n^{-2c})$ and $V_n = O(n^{+2c})$ (the variance blows up). Assumption (a2) follows for any $c > 0$ but for (a3) we need convergence of the series:

$$\sum_n \epsilon_n^2 V_n < \infty \iff \sum_n \frac{1}{n^2} n^{2c} < \infty \implies 2 - 2c > 1,$$

so that $c < 1/2$ for the algorithm to converge almost surely to the correct value. It is worth noting that in case that $h(\xi, \theta)$ is Lipschitz, the variance of the bias can be bounded independently of c_n; see Exercise 4.9. Later in Chapter 6 we will see how to tune the various parameters in order to achieve the best convergence rates.

Recently, a convex combination of a gradient estimator with a two-sided finite difference estimator has been proposed for use in a SA [72]. As detailed in [70], choosing the weight factor in the convex combination achieves updates with low variance, which is favorable for building confidence intervals for the solution of (4.11) as will detail in Chapter 6 later on. ✳✳✳

In the previous example we assumed the variance of the observations to be uniformly bounded, independent of the value of θ_n. In general, the feedback Y_n depends on θ_n and condition (a3) in Theorem 4.1 may be somewhat elusive. In the following examples we illustrate the use of a "perturbation" technique that will yield a method for verifying this assumption for many well-posed problems with coercive fields.

Example 4.2 (Truncation). We revisit Example 3.4 to illustrate how Theorem 4.1 can be applied to show that the procedure converges. First, we must impose some conditions on the problem itself so we have a well-posed optimization problem and a coercive target vector field. Assume that the experimental range $S \subset \mathbb{R}$ is an interval that contains more than one point. Indeed, were S to contain only one point, say x_0, then $\mathbb{E}[Z(x_0)] = y_0$ would be the only corresponding output value, and an infinite number of straight lines could be drawn to pass over this point and the slopes of the lines would not be bounded. At least two different points are required to uniquely define a straight line. Assume also that the observations X have a continuous distribution over S. Finally, assume that consecutive observations of the streaming data $\{\xi_n\} = \{(x_n, z_n)\}$ are iid and have finite variance. This implies that there is a unique point $\theta^* \in \mathbb{R}^2$ with $\|\theta\| < \infty$ that minimizes $J(\theta)$.

The stochastic approximation of Example 3.4 is

$$\theta_{n+1,1} = \theta_{n,1} + \epsilon_n(z_n - \theta_{n,1} - \theta_{n,2} x_n),$$
$$\theta_{n+1,2} = \theta_{n,2} + \epsilon_n x_n(z_n - \theta_{n,1} - \theta_{n,2} x_n),$$

for which $\mathbb{E}[Y_n \mid \mathfrak{F}_{n-1}] = -\nabla J(\theta_n)$. Because we assume that samples are iid, then $\mathrm{Var}(x_n)$ and $\mathrm{Var}(z_n)$ are (finite) constants and V_n is a second-degree polynomial in θ_1, θ_2. Assumptions (a1), (a2), and (a4) of Theorem 4.1 follow if we choose $\epsilon_n = O(n^{-1})$, for example. The elusive Assumption (a3) is much harder to show, because $V_n = O(\|\theta_n\|^2)$ is not necessarily uniformly bounded, unless we can show that θ_n remains in a compact set almost surely, or other similar conditions.

Because the field $-\nabla J$ is coercive on $\mathbb{R}^2$, then for given initial value $x(0) = x_0$, the solution of

$$\frac{dx}{dt} = -\nabla J(x(t)); \qquad (4.12)$$

stays inside a hypercube $H \stackrel{\mathrm{def}}{=} H(x(0)) = [-M, M]^d$; for a direct proof, apply the Lyapunov function argument put forward in Example 2.8. We now apply the truncation principle: let $\tilde{\theta}_n$ be the truncated algorithm that follows the recursion $\tilde{\theta}_{n+1} = \Pi_{[-M,M]^d}(\tilde{\theta}_n + \epsilon_n Y_n)$, so that $\tilde{\theta}_n \in [-M, M]^d$ a.s. Then (a3) is satisfied for the truncated algorithm. Applying the second part of Theorem 4.1 shows that the shifted interpolated processes converge to the solution of the corresponding projected ODE:

$$\frac{dx}{dt} = \Pi_{[-M,M]^d}\left(-\nabla J(x(t))\right); \quad x(0) = x_0.$$

By the observations above, this ODE is equivalent to the unprojected one in (4.12). Therefore, the limit ODE is independent of the truncation of the algorithm and we can make $M \to \infty$ to conclude that the original algorithm also converges to (4.12). ✻✻✻

Example 4.3 (Update Clipping). Alternative to the projection introduced in Example 4.2 we may use clipping; see Example 2.12. For this, let

$$\tilde{\phi}((x_n, z_n), \theta_n) = \begin{pmatrix} z_n - \theta_{n,1} - \theta_{n,2} x_n \\ \theta_{n,2} + \epsilon_n x_n(z_n - \theta_{n,1} - \theta_{n,2} x_n) \end{pmatrix}$$

and, for c sufficiently large, define

$$\phi((x_n, z_n), \theta_n)$$
$$= \tilde{\phi}((x_n, z_n), \theta_n) \left(\mathbf{1}_{\{||\tilde{\phi}((x_n,z_n),\theta_n)||<c\}} + \frac{c}{||\tilde{\phi}((x_n, z_n), \theta_n)||} \mathbf{1}_{\{||\tilde{\phi}((x_n,z_n),\theta_n)||\geq c\}} \right).$$

Then, the stochastic approximation on $\mathbb{R}^d$ is given by

$$\theta_{n+1} = \theta_n + \epsilon_n \phi((x_n, z_n), \theta_n)$$
$$= \theta_n + \epsilon_n \tilde{\phi}((x_n, z_n), \theta_n) \left(\mathbf{1}_{\{||\tilde{\phi}((x_n,z_n),\theta_n)||<c\}} + \frac{c}{||\tilde{\phi}((x_n, z_n), \theta_n)||} \mathbf{1}_{\{||\tilde{\phi}((x_n,z_n),\theta_n)||\geq c\}} \right).$$

By construction, the updates have zero variance outside the norm-ball of radius c. Assumption (a3) thus readily follows as V_n is uniformly bounded. Assumptions (a1), (a2), and (a4) of Theorem 4.1 follow if we choose $\epsilon_n = O(n^{-1})$, for example. Note that while the target vector field $G(\theta) = \mathbb{E}[\phi((x_n, z_n), \theta)]$ is a descent direction, clipping introduces bias, i.e, $G(\theta) \neq -\nabla J(\theta)$. Similar to the truncation argument, one now observes that the trajectories of the ODE using a clipped version of the vector field are the same as the original trajectories, provided that c is large enough. Then the limit ODE is the same as (4.12). Finally, letting $c \to \infty$ one concludes that it is not necessary to clip the SA at all. ✳✳✳

Example 4.4 (Variance Control). Let $\{Y_n(i) : 1 \leq i \leq k\}$ be a collection of iid copies of Y_n conditioned on $\mathfrak{F}_{n-1}$, and denote by

$$\bar{Y}_n^k = \frac{1}{k} \sum_{i=1}^{k} Y_n(i)$$

the sample average of $\bar{Y}_n$. By construction,

$$\mathbb{E}\left[\bar{Y}_n^k \mid \mathfrak{F}_{n-1}\right] = \mathbb{E}\left[Y_n \mid \mathfrak{F}_{n-1}\right]. \tag{4.13}$$

Note that

$$\lim_{k \to \infty} \bar{Y}_n^k = \mathbb{E}[Y_n \mid \mathfrak{F}_{n-1}] = G(\theta_n) \quad a.s.,$$

which shows that taking limit for k toward ∞ the target field is driving the feedback.

Using (4.13), we can relate the variance of $\bar{Y}_n^k$ and Y_n as follows:

$$\frac{1}{k} V_n = \frac{1}{k} \mathbb{E}\left[(Y_n - \mathbb{E}[Y_n \mid \mathfrak{F}_{n-1}])^2\right] = V_n^k = \mathbb{E}\left[(\bar{Y}_n^k - \mathbb{E}[\bar{Y}_n^k \mid \mathfrak{F}_{n-1}])^2\right].$$

Choose now $M \gg 0$, and consider the modified algorithm $\tilde{\theta}_{n+1} = \tilde{\theta}_n + \epsilon_n \tilde{Y}_n$, for $\tilde{\theta}_n \in \mathbb{R}^d$, where

$$\tilde{Y}_n = \begin{cases} Y_n & \text{if } \theta_n \in [-M, M]^d, \\ \frac{1}{k(\theta_n)} \sum_{i=1}^{k(\theta_n)} Y_n(i) & \text{otherwise,} \end{cases}$$

for $k(\theta_n)$ is such that $\text{Var}(\tilde{Y}_n) \leq c$, for some finte constant c. Then, by construction, $\text{Var}(\tilde{Y}_n)$ is uniformly bounded, and $\mathbb{E}[\tilde{Y}_n \mid \mathfrak{F}_{n-1}] = G(\theta_n)$. Hence, (a3) in Theorem 4.1 holds for the modified algorithm. Because the limit is independent of M, we can now argue that M can be made arbitrarily large, yielding the desired result.

Note that $\bar{Y}_n$ is obtained from independent batching, which shows that the variance control scheme cannot be applied to streaming data as this would require consecutive batching. ✳✳✳

Remark 4.2. The truncation argument in Example 4.2, update clipping in Example 4.3, and the variance control scheme in Example 4.4 are theoretical arguments to show convergence of the original algorithm. Notice, however, that in practice we often do not need to code either one in the computer program. Moreover, while truncation and clipping are usually natural in programming (they help to avoid numerical instabilities), the variance control scheme requires sampling outside a large box, which is actually not a good idea from an algorithmic perspective, because it forces the values of θ_n to remain in undesirable areas of potential numerical instabilities for a long time in order to generate or evaluate large samples for the controlled variance.

SAs can be applied to problems that do not necessarily update using the gradient direction. As mentioned in Section 2.5, iterative numerical methods can be used in general to find the zeroes of functions, which can approximate the solution to optimization problems, or can be used to approximate equilibrium solutions, or to approximate fixed points of mappings. These latter problems are important problems in economic theory and mathematical biology, among other fields. The following example illustrates this scenario, and also shows how to verify assumption (a4) for general formulations.

Example 4.5. Consider a continuously differentiable contraction mapping $T: \mathbb{R}^d \to \mathbb{R}^d$ with unique fixed point θ^*. We let $G(\theta) = T(\theta) - \theta$, and finding the fixed point of T then comes down to finding the zero of $G(\theta)$. Suppose that, given a value of θ, a noisy observation or simulation can be performed to produce a random variable $\xi(\theta)$ such that $\mathbb{E}[\xi(\theta)] = T(\theta)$ is Lipschitz continuous, and assume that $\sup_\theta \text{Var}[\xi(\theta)] < \infty$. Consider the SA procedure:

$$\theta_{n+1} = \theta_n + \epsilon_n(\xi(\theta_n) - \theta_n), \qquad (4.14)$$

where $\xi(\theta_n)$ is independent of previous observations $\xi(\theta_0), \ldots, \xi(\theta_{n-1})$. Then this procedure fits the martingale difference noise model, where $Y_n = \xi(\theta_n) - \theta_n$ and $\mathbb{E}[Y_n \mid \mathfrak{F}_{n-1}] = T(\theta_n) - \theta_n = G(\theta_n)$. Assumptions (a1), (a2), and (a3) follow from the usual conditions on ϵ_n and the assumption that the variance of $Z(\theta)$ is uniformly bounded in θ. It only remains to verify condition (a4) of Theorem 4.1. To do this, use Theorem 2.3 to show that the vector field driving the limiting ODE

$$\frac{dx(t)}{dt} = G(x(t) = T(x(t)) - x(t).$$

In the following we show that G is coercive for $\min_\theta J(\theta)$, where we have here the special case that $G = J$. By assumption, there is only one stationary point of this ODE. We now show that it is asymptotically stable, which in turn establishes that $\theta_n \to \theta^*$ a.s.

To show stability, using Lemma 2.1, it suffices to show that $\mathbb{A} = \nabla T(\theta^*)^\top - I$ is a Hurwitz matrix. Call $\mathbb{B} = \nabla T(\theta^*)^\top$. Then every eigenvalue of $\mathbb{A}$ has the form $\rho - 1$, where ρ is an eigenvalue of $\mathbb{B}$. We now show that if ρ is any eigenvalue of $\mathbb{B}$ then its real part (denoted by $\mathfrak{R}(\rho)$) satisfies $\mathfrak{R}(\rho) < 1$, which will complete the proof.

Let ρ be an eigenvalue of $\mathbb{B}$ with corresponding eigenvector v, and use a Taylor approximation of $T(\theta)$ for $\theta = \theta^* + \delta v$, $\delta \approx 0$ to establish that:

$$\frac{T(\theta) - T(\theta^*)}{\|\theta - \theta^*\|} = \frac{\delta \mathbb{B} v}{\delta \|v\|} + O(\delta) = \rho \frac{v}{\|v\|} + O(\delta).$$

Because T is a contraction mapping, then there is a number $0 < \tau < 1$ independent of δ such that
$$\frac{\|T(\theta) - T(\theta^*)\|}{\|\theta - \theta^*\|} < \tau < 1.$$

Choosing a small enough δ, this implies that $|\rho| < 1$, which proves the claim. ✱✱✱

We finalize this section with an alternative to Theorem 4.1 that uses the same ideas as in Example 4.3. It can be very useful when assumption (a3) is not directly provable, but it requires the vector field to be coercive (see Definition 2.13) for a well-posed optimization problem (see Definition 2.12).

Theorem 4.2. *Consider the well-posed NLP $\min_{\theta \in \Theta} J(\theta)$. Suppose that the vector field G is coercive for this NLP on $H \subset \mathbb{R}^d$.*

Consider stochastic approximation algorithm

$$\theta_{n+1} = \theta_n + \epsilon_n Y_n, \quad \theta_0 \in H,$$

where $Y_n = \phi(\xi_n, \theta_n)$ satisfies the martingale difference noise model for G on $\mathbb{R}^d$, so that $\mathbb{E}[Y_n \mid \mathfrak{F}_{n-1}] = G(\theta_n) + \beta_n$. Assume

(a1') $\sum_{i=0}^{\infty} \epsilon_i = \infty$, $\sum_{i=0}^{\infty} \epsilon_i^2 < \infty$, $\epsilon_n > 0$ for all n,
(a2) $\sum_{i=0}^{\infty} \epsilon_i \|\beta_i\| < \infty$ w.p.1,
(a3') For every compact set $M \subseteq \Theta$ there is a finite number V such that for all n $\sup_{\theta \in M} \mathsf{Var}(\phi(\xi_n, \theta)) \leq V$ (uniformly bounded variance on compact sets).

Then $\{\theta_n\}$ converges a.s. and all accumulation points are KKT points of the optimization problem. In particular, if the global optimum θ^* is the only asymptotically stable point of the ODE $dx(t)/dt = G(x(t))$, then $\theta_n \to \theta^*$ a.s.

Alternatively, for Θ convex, compact and $G(\cdot) = (\Pi_\Theta \hat{G})(\cdot)$, for some continuous vector field $\hat{G}$ on Θ, the convergence statement above holds for the projected SA

$$\theta_{n+1} = \Pi_\Theta(\theta_n + \epsilon_n Y_n), \quad \theta_0 \in \Theta \cap H,$$

where $Y_n = \phi(\xi_n, \theta_n)$ satisfies the martingale difference noise model for $\hat{G}$ on Θ (so that $\mathbb{E}[Y_n \mid \mathfrak{F}_{n-1}] = \hat{G}(\theta_n) + \beta_n$, for $\theta_n \in \Theta$), provided that (a1'), (a2), and (a3') hold.

Proof. The first statement of the theorem follows from Theorem 4.1 applied to the clipped version G_c of G, where we make use of Theorem 2.9 for showing that G_c is coercive for the NLP as well. Then, for any given c, continuity of G implies that the set $\Theta^c := \{\theta : \|G(\theta)\| \leq c\}$ is compact and we use Θ^c in (a3'). Because the conditional variance is always bounded by the variance of the random variable, assumption (a3') implies (a3) for the clipped version of the feedback. For c large enough, it holds that the trajectories of the ODEs driven by G and G_c with the same initial value θ_0 coincide, and we may thus run the algorithm without executing the clipping operation. The second statement follows the same way where we make use of Theorem 2.10 applied for $\Theta_0 = \Theta$. □

4.3 ANALYSIS OF CONSTANT STEPSIZE SA

It is common to program the iterative algorithms with constant stepsize ϵ. But in order to analyze the behavior of the ensuing algorithms, one must resort to a different concept of convergence. Recall the framework of Chapter 2, where we studied Euler's method for approximating solutions of ODEs from a functional analysis point of view. The rigorous framework and analysis are presented in Chapter 5, but here we provide a motivation for such methodology.

Notice first that when ϵ is constant and the feedback function $Y_n = \phi(\xi_n, \theta_n)$ is random with non-zero variance, then even if $\theta_n = \theta^*$ is a stable point of the vector field (that is, $\mathbb{E}[\phi(\xi,\theta)] = G(\theta^*) = 0$) it holds with probability one that $\theta_{n+1} \neq \theta^*$ although $\mathbb{E}[\theta_{n+1} \mid \theta_n] = \theta^*$. Indeed, the algorithm will fluctuate, even if started at the optimal value. Of course it is possible to conceive of a problem where the variance terms V_n decrease such that the conditions of Theorem 4.1 are satisfied for constant ϵ. Although this is mathematically possible, this situation almost never occurs in real-life problems and applications of interest.

Instead of analyzing the sequence $\{\theta_n\}$ as $n \to \infty$, the case for constant stepsize looks at the interpolated process (as in Chapter 2) and the "convergence" as $\epsilon \to 0$ is interpreted in terms of consecutive trajectories, or "experiments" that each use a constant ϵ, but where ϵ decreases from one experiment to the next. This sense of convergence (of functions) is akin to the sense in which the value of an integral can be approximated by the areas under the rectangles:

$$\int_0^t G(u) \, du \approx \sum_{n=1}^{\lfloor t/\epsilon \rfloor} \epsilon \, G(n\epsilon),$$

and one shows that the limit as $\epsilon \to 0$ is the true value of the integral. When coding the numerical method, of course one never makes ϵ decrease to zero but chooses ϵ "sufficiently small."

The methodology of Chapter 2 relied on a "compactness argument" for the proof. First, the (deterministic) sequence of interpolated processes was assumed to be equicontinuous in the extended sense. The interpretation of this condition is that it limits the possible growth rate of the sequence, and then the Ascoli-Arzelà Theorem is used to establish that all subsequences have a further convergent subsequence with continuous limits. The reader is encouraged to review Chapter 2 before engaging in the extension of the method to stochastic processes.

In the stochastic setting, Theorem B.13 is essential, and it works as the analogue to the Ascoli-Arzelà Theorem in Chapter 2. It shows that all sequences of stochastic process have convergent subsequences. Instead of equicontinuity, it is required here that the feedback $\{Y_n\}$ be uniformly integrable (see Definition B.30). For stochastic processes, "tightness" is the equivalent of compactness (see Definition B.29), and convergence is established with respect to the weak topology of a space denoted $D^d[0, \infty]$ (see Definition B.31). What this means is that we will work with convergence *in distribution* of the sequence of interpolated processes.

For pedagogical reasons, we state first the result for a model without bias and with bounded vector field. The proof follows the logic of the deterministic method of proof in Theorem 2.5 that presents much of the required ground work, detailing the additional steps required to apply the method to the stochastic case. Introduction of noise creates much complexity in the treatment of the algorithm, and it is somewhat surprising that one can recover the same result as for the deterministic case with relatively general conditions on the stochastic model. The reader is encouraged to revise the proof of Theorem 2.5 in preparation for Theorem 4.3.

In fact, the following theorem is a corollary of Theorem 5.3 in the following chapter because the martingale difference noise model for a static problem is a particular (and simpler) case of the stationary (long-term) problem, so understanding this proof will help the reader to follow the more complex technical proofs in Chapter 5. For a defintion of uniform integrability of a sequence of random variables, we refer to Definition B.30, and we recall the definition of the interpolated process and that of $m(t) = \lfloor t/\epsilon \rfloor$ in Definition 2.9.

STOCHASTIC APPROXIMATION: THE STATIC MODEL

Theorem 4.3. *Let $G: \mathbb{R}^d \mapsto \mathbb{R}^d$ and assume*

$$G(\theta) = \mathbb{E}[g(\xi(\theta), \theta)], \tag{4.15}$$

where $\xi(\theta)$ is a random vector with distribution possibly dependent on θ. For G bounded or Lipschitz continuous, assume that the trajectories of ODE

$$\frac{dx(t)}{dt} = G(x(t)) \tag{4.16}$$

have a unique limit (as $t \to \infty$) for each initial condition. Consider the stochastic approximation algorithm

$$\theta_{n+1}^\epsilon = \theta_n^\epsilon + \epsilon Y_n, \quad \theta_0 \in \mathbb{R}^d, \tag{4.17}$$

with $Y_n = g(\xi(\theta_n^\epsilon), \theta_n^\epsilon)$, where the random vector $\xi(\theta_n^\epsilon)$ has a distribution possibly dependent on θ_n^ϵ but statistically independent of previous values $\{\xi(\theta_k^\epsilon), k < n\}$. Provided that $\{Y_n\}$ is uniformly integrable, the interpolated process $\{\vartheta^\epsilon(t) = \theta_{m(t)}^\epsilon : t \geq 0\}$ converges in distribution, as $\epsilon \to 0$; the limit process is a.s. continuous and satisfies the ODE (4.16).

Under the same statistical assupmtions as above, it holds for Θ convex and compact that the interpolated process $\{\vartheta^\epsilon(t) = \theta_{m(t)}^\epsilon : t \geq 0\}$ of the projected version of (4.17), given by

$$\theta_{n+1}^\epsilon = \Pi_\Theta(\theta_n^\epsilon + \epsilon Y_n), \quad \theta_0 \in \Theta,$$

converges in distribution, as $\epsilon \to 0$; the limit process is a.s. continuous and satisfies the corresponding projected ODE, given by replacing $G(\cdot)$ by $(\Pi_\Theta G)(\cdot)$ in (4.16).

Proof. First we establish the compactness argument for the stochastic case. Using Theorem B.13 it follows that $\{\vartheta^\epsilon(\cdot)\}$ is a tight sequence, so every sequence has a subsequence that converges in distribution, and all limits are Lipschitz continuous processes w.p.1. Thus, from now until the end of the proof, pick a weakly convergent subsequence $\lim_k \epsilon_k = 0$, with limit process ϑ:

$$\vartheta^{\epsilon_k}(\cdot) \stackrel{d}{\Longrightarrow} \vartheta(\cdot) \quad \text{as } k \to \infty.$$

For this proof we will use the same technique as in the proof of Theorem 2.5. From the definition of the interpolated process, we express the telescopic sum:

$$\vartheta^{\epsilon_k}(t+s) - \vartheta^{\epsilon_k}(t) = \sum_{n=m(t)}^{m(t+s)-1} \epsilon_k Y_n = \epsilon_k \sum_{n=m(t)}^{m(t+s)-1} g(\xi_n, \theta_n^{\epsilon_k}).$$

Given ϵ_k, the information structure of the interpolated process is related to the original sigma algebra for the sequence $\theta_n^{\epsilon_k}$, so that $\tilde{\mathfrak{F}}_t^{\epsilon_k} = \sigma(\vartheta^{\epsilon_k}(s); s \leq t) = \mathfrak{F}_{m(t)}^{\epsilon_k} = \sigma(\theta_n^{\epsilon_k}; n \leq m(t))$. From this expression, conditioning on the initial sigma algebra $\mathfrak{F}_{m(t)}^{\epsilon_k}$, it follows (using the *Law of the iterated expectation*; see Theorem B.6) in Appendix B that

$$\mathbb{E}\left[\vartheta^{\epsilon_k}(t+s) - \vartheta^{\epsilon_k}(t) \big| \tilde{\mathfrak{F}}_t^{\epsilon_k}\right] = \epsilon_k \sum_{n=m(t)}^{m(t+s)-1} \mathbb{E}\left[\mathbb{E}[g(\xi_n, \theta_n^{\epsilon_k}) | \mathfrak{F}_{n-1}^{\epsilon_k}] \big| \mathfrak{F}_{m(t)}^{\epsilon_k}\right] \tag{4.18}$$

$$= \mathbb{E}\left[\epsilon_k \sum_{n=m(t)}^{m(t+s)-1} G(\theta_n^{\epsilon_k}) \Big| \tilde{\mathfrak{F}}_t^{\epsilon_k}\right],$$

and the expression inside the conditional expectation corresponds to an integral:

$$\epsilon_k \sum_{n=m(t)}^{m(t+s)-1} G(\theta_n^{\epsilon_k}) = \int_t^{t+s} G(\vartheta^{\epsilon_k}(u))\,du + \rho(\epsilon_k), \qquad (4.19)$$

where the error in the approximation of the integral is of order $O(\epsilon_k)$. We have used here the assumption that given $\theta_n^{\epsilon_k}$ the distribution of ξ_n is independent of the past values, and the model assumption (4.15), which states that $g(\xi(\theta), \theta)$ is an unbiased estimator for $G(\theta)$.

Define now the process of the difference:

$$M^{\epsilon_k}(t) = \vartheta^{\epsilon_k}(t) - \vartheta^{\epsilon_k}(0) - \int_0^t G(\vartheta^{\epsilon_k}(u))\,du.$$

The proof is completed as follows. Using that $\rho(\epsilon_k) \stackrel{d}{\Longrightarrow} 0$, $M^{\epsilon_k}(\cdot)$ converges in distribution to a limit process $M(\cdot)$ with natural filtration $\{\mathfrak{F}_t\}$ that satisfies $\mathbb{E}[M(t+s) - M(t) \mid \mathfrak{F}_t] = 0$ (for a proof use (4.18) and (4.19)). Since the distribution of ξ_n for $n > m(t)$ is independent of the past, we can evoke Theorem B.17 to conclude that that $M(\cdot)$ is a martingale. Because $\vartheta^{\epsilon_k}(\cdot) \stackrel{d}{\Longrightarrow} \vartheta(\cdot)$ as $k \to \infty$ along the chosen subsequence, it follows that

$$M(t) = \vartheta(t) - \vartheta(0) - \int_0^t G(\vartheta(u))\,du,$$

and this is a Lipschitz continuous martingale with $M(0) = 0$, which by Theorem B.16 implies that $M(t) \equiv 0$ w.p.1. The proof that the second part in the above expression is Lipschitz continuous is given in Exercise 4.8.

This establishes that the limit process $\vartheta(\cdot)$ along the chosen subsequence satisfies the ODE $x'(t) = G(x(t))$, but by assumption on the vector field, the solution to this ODE is unique, thus any such weakly convergent subsequences have the same limit, which establishes the result.

For the proof of the second part of the theorem, we note that Theorem B.13 does not apply to the projected version. Arguments for showing weak convergence of the interpolation process in the presence of projections are provided in Section 8.1, Part 1, on page 251 of [198]. □

In parallel with the decreasing stepsize section, we finalize this section with the result including a possibly decreasing bias, when the field is coercive for a well-posed problem, so that the condition that G is bounded is no longer required. In addition, a variance condition is used instead of the uniform integrability condition, which is usually harder to verify directly.

Theorem 4.4. *Consider the well-posed NLP $\min_{\theta \in \Theta} J(\theta)$. Suppose that the vector field G is coercive for this problem on $H \subseteq \mathbb{R}^d$.*

Let $\xi(\theta)$ be a random vector with distribution possibly dependent on θ and $g(\xi(\theta), \theta)$ be a function such that $G(\theta) = \mathbb{E}[g(\xi(\theta), \theta)]$, for $\theta \in \mathbb{R}^d$. Consider the stochastic approximation algorithm:

$$\theta_{n+1}^\epsilon = \theta_n^\epsilon + \epsilon Y_n^\epsilon, \qquad \theta_0^\epsilon \in H,$$

with

$$Y_n = \phi^\epsilon(\xi_n, \theta_n^\epsilon) \stackrel{def}{=} g(\xi_n, \theta_n^\epsilon) + b^\epsilon(\xi_n, \theta_n^\epsilon), \qquad (4.20)$$

where the random vector $\xi_n = \xi(\theta_n^\epsilon)$ has a distribution possibly dependent on θ_n^ϵ but statistically independent of previous values $\{\xi_k, k < n\}$. Consider the following properties:

(i) *The error is decreasing (in ϵ): for every compact set $M \subseteq \Theta$, if $\beta^\epsilon(\theta) \stackrel{def}{=} \mathbb{E}[b^\epsilon(\xi(\theta), \theta)]$ then*

$$\sup_{\theta \in M} \|\beta^\epsilon(\theta)\| = O(\epsilon).$$

(ii) *The variance is locally bounded: for every compact set $M \subset \Theta$ there is a finite number V such that $\sup_{\theta \in M} \text{Var}(\phi^\epsilon(\xi(\theta), \theta)) \leq V$.*

Provided that (i) and (ii) hold, the interpolated processes $\vartheta^\epsilon(t) = \theta_{m(t)}$ converge in distribution, as $\epsilon \to 0$, to a limit process which is continuous w.p.1 and satisfies the ODE:

$$\frac{dx(t)}{dt} = G(x(t)), \tag{4.21}$$

and $\lim_{t \to \infty} \lim_{\epsilon \to 0} \theta^\epsilon_{\lfloor t/\epsilon \rfloor}$ is a KKT point of the optimization problem.

Alternatively, for Θ convex, compact and $G(\cdot) = (\Pi_\Theta \hat{G})(\cdot)$, for some continuous vector field $\hat{G}$, so that $\hat{G}(\theta) = \mathbb{E}[g(\xi(\theta), \theta)]$ holds for $\theta \in \Theta$, and (4.20) holds for $\theta^\epsilon_n \in \Theta$, the statement holds for the projected SA

$$\theta^\epsilon_{n+1} = \Pi_\Theta(\theta^\epsilon_n + \epsilon Y^\epsilon_n), \quad \theta^\epsilon_0 \in \Theta \cap H,$$

provided that (i) and (ii) hold for $M = \Theta$.

Proof. The first statement of the theorem follows from Theorem 4.3 applied to the clipped version G_c of G, where we make use of Theorem 2.9 for showing that G_c is coercive for the NLP as well. Then, for any given c, continuity of G implies that the set $\Theta^c := \{\theta : \|G(\theta)\| \leq c\}$ is compact and we use Θ^c in *(ii)*. For c large enough, it holds that the trajectories of G and G_c with the same initial value θ_0 coincide, and we may thus run the algorithm without executing the clipping operation. The second statement follows the same way where we make use of Theorem 2.10 applied for $\Theta_0 = \Theta$. □

Example 4.6. Consider a continuous probability distribution F with support $\mathbb{R}$. For $\alpha \in (0, 1)$ the α-quantile is defined as

$$q^\alpha = F^{-1}(\alpha) = \inf(u \in \mathbb{R} : F(u) \geq \alpha).$$

Estimating quantiles often requires a sequence of observations $\{\xi_1, \xi_2, \ldots, \xi_m\}$ assumed iid according to F. In particular, the standard estimator considers the whole sample:

$$\hat{q}^\alpha = \min_{u \in \mathbb{R}} \left(\frac{1}{m} \sum_{i=1}^{m} \mathbf{1}_{\{\xi_i \leq u\}} \leq \alpha \right).$$

This expression uses the estimator of the distribution function

$$F(u) = \mathbb{P}(\xi \leq u) = \mathbb{E}[\mathbf{1}_{\{\xi \leq u\}}] \approx \frac{1}{m} \sum_{i=1}^{m} \mathbf{1}_{\{\xi_i \leq u\}}.$$

It can be shown that this expression is equivalent to finding the place that approximates the $100(1-\alpha)\%$ of the samples. Let $(t_1, \ldots, t_m)$ be the order statistics of the observations, then

$$\hat{q}^\alpha = t_{\lceil m\alpha \rceil}.$$

An excellent tutorial on quantile estimation can be found in [90], where the convergence properties of the estimators are developed in detail.

Fix α and suppose now that one wishes to compute the α-quantile not knowing the distribution F, but streaming data $\{\xi_1, \xi_2, \ldots\}$ so that the observations are obtained in real time. Using $\hat{q}^\alpha$ as above has the computational problem of complexity, having to reorder the sample every time a new observation is gathered. Example 4.2 in [295] presents an intriguing alternative. We present the idea within our framework and notation.

The goal is to find θ^* such that
$$F(\theta^*) = \alpha.$$

Because it is a continuous distribution, F is strictly monotonically increasing and there is a unique value θ^* satisfying the equality. This problem can then be cast as a tracking problem. Indeed, the vector field
$$G(\theta) \stackrel{\text{def}}{=} \alpha - F(\theta) = \alpha - \mathbb{E}[\mathbf{1}_{\{\xi \le \theta\}}]$$

is coercive for this problem and seeks to minimize the distance $J(\theta) = (\alpha - F(\theta))^2$. Note that this problem is of the root-finding type as discussed in Section 3.2.

The iid observations $\{\xi_n\}$ form the underlying process and one can build a feedback estimate satisfying the martingale difference noise model as follows. Let
$$Y_n = \alpha - \mathbf{1}_{\{\xi_n \le \theta_n\}}, \qquad \theta_{n+1} = \theta_n + \epsilon_n Y_n.$$

Because the samples are independent, $\mathbb{E}[Y_n \mid \mathfrak{F}_{n-1}] = G(\theta_n)$. By construction, it follows that $|Y_n| \le 2$ a.s., so that we can apply Theorem 4.2 for decreasing stepsizes, using $\sum_n \epsilon_n = +\infty$, $\sum_n \epsilon_n^2 < \infty$. This shows that $\theta_n \to \theta^*$ a.s., thus it is a consistent estimator of the quantile q^α. Alternatively, one can apply Theorem 4.4 for constant ϵ to conclude that as $\epsilon \to 0$ the interpolated process converges to the ODE $dx/dt = G(x(t))$, which by construction has a single global asymptotically stable point at θ^*.

In case F has support $\Theta \ne \mathbb{R}$, such that Θ is closed and convex, the resulting SA becomes $\theta_{n+1} = \Pi_\Theta(\theta_n + \epsilon_n Y_n)$. ✷✷✷

4.4 PRACTICAL CONSIDERATIONS

In this chapter we have studied the algorithm
$$\theta_{n+1} = \theta_n + \epsilon_n Y_n,$$

where we assumed that Y_n satisfies the martingale difference noise model with exogenous noise. The key result was that for the martingale difference noise model, we show for decreasing stepsize that the influence of the noise tends to zero as n tends to infinity, and θ_n approaches the path of a deterministic vector field $G(\theta)$. Recall that the link between Y_n and G is that
$$G(\theta_n) = \mathbb{E}[Y_n | \mathfrak{F}_{n-1}] + \beta_n.$$

This type of algorithm finds the zeros of G, which offers a great flexibility in applications of the algorithm. We summarize in the following typical scenarios for choosing G and Y_n for a range of applications.

- In case of a well-posed unconstrained optimization problem, a natural choice for the vector field is $G(\theta) = -\nabla J(\theta)$, and preferably an unbiased feedback estimator

$Y_n = \phi(\xi(\theta_n), \theta_n)$ such that $\mathbb{E}[\phi(\xi(\theta), \theta)] = -\nabla J(\theta)$. Part II of this book is dedicated to the construction of such gradient estimators.

- In case of a target tracking problem $L(\theta^*) = \alpha$ with strictly monotone $L(\cdot)$ it suffices to take the observed deviation from the target. See Section 3.2 where we use $G(\theta) = (L(\theta) - \alpha)^\top \widehat{M}$ as vector field and apply root-finding techniques for solving $G(\theta) = 0$. In this case any unbiased estimator of $L(\theta)$ is a valuable choice for the feedback estimator Y_n.

- Now consider a constrained optimization problem with soft constraints only. As discussed in Chapter 1, a standard approach is to remodel the optimization problem as an unconstrained one by, for example, including a penalty function. Then the vector field $G(\theta)$ becomes the negative gradient of the penalized cost function, see (1.30), plus the penalty, and Y_n is taken to be an unbiased estimate for $G(\theta)$. It is worth noting that in running the algorithm, two time scales have to be taken into account for updating the penalty function along the run of the algorithm. Alternatively, the constraint can be included in G through the Lagrangian function, for example, by using the Arrow-Hurwitz scheme; see (1.50) and Theorem 2.11.

For the projection method, constraints are included in the vector field and the limit is the projected ODE. When projections can be easily handled (e.g., by truncations) this is straightforward to implement. For complex feasible regions, however, the projection can be implemented as discussed in Chapter 1 via a two-time scale algorithm; see Example 1.11.

4.5 EXERCISES

Exercise 4.1. Use the Robbins-Monro procedure in Theorem 3.1 to adjust the power control variables of Example 3.2 with stepsize condition (3.7) so as to attain the target temperature. Pose the model and specify which assumptions you must make about M and w_n for convergence. Does your result imply that $\theta_n \to \theta^*$ almost surely?

Exercise 4.2. Let $\{X_k\}$ be an iid sequence with mean μ and define

$$S_n = \sum_{k=1}^n (X_k - \mu), n \geq 1,$$

Show that S_n is a martingale with respect to the natural filtration of the process, where $\mathfrak{F}_n = \sigma(X_1, \ldots, X_n)$.

Exercise 4.3. Let $\delta M_i = Y_i - \mathbb{E}[Y_i \mid \mathfrak{F}_{i-1}]$ be defined as in the martingale difference noise model. Show that the process $M_n \stackrel{\text{def}}{=} \sum_{i=0}^n \epsilon_i \delta M_i$ is a martingale process on $(\Omega, \mathbb{P}, \{\mathfrak{F}_n\})$. Show that $\mathbb{E}[\delta M_n \delta M_m] = 0$. (For the basic definition and properties of martingale processes we refer to Section B.6 in Appendix B.)

Exercise 4.4. Refer to the model in Example 3.4. Show that for a random variable X with finite variance,

$$\nabla J(\theta) = \begin{pmatrix} -\mathbb{E}\left[Z(X) - \theta_1 - \theta_2 X\right] \\ -\mathbb{E}\left[X Z(X) - \theta_1 X - \theta_2 X^2\right] \end{pmatrix}. \tag{4.22}$$

For each experimental point x_n we obtain a random observation $\xi_n = Z(x_n)$ with $\mathbb{E}(Z(x_n)) = h(x_n)$. The feedback function is $Y_n = (\xi_n - \theta_n(1) - \theta_n(2)x_n)(1, x_n)^T$. Use (4.22) to show the claim that $\mathbb{E}[Y_n \mid \mathfrak{F}_{n-1}] = -\nabla J(\theta_n)$.

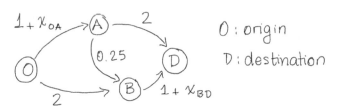

Figure 4.1. Traffic flow equilibrium.

Exercise 4.5. Revisit Exercise 4.3. Explain why the continuity theorem for decreasing sets implies that as $n \to \infty$, $\sup_m |\sum_{k=n}^{m} \delta M_k| \to 0$ with probability 1.

Exercise 4.6. Refer to the proof of Theorem 4.1. Assume that G is bounded and show that $\rho^n(\cdot)$ converges a.s., uniformly in t, and that for each $\omega \notin N$, the sequence $\{\vartheta^n(\cdot)\}$ is equicontinuous.

Exercise 4.7. Consider the following well-known Nash equilibrium problem in transportation. The transportation time along an arc i is denoted by t_i and it depends on the vector of normalized traffic flow x, where $x_{(k,l)}$ is the total traffic on the arc (k,l). In Figure 4.1 there are three possible routes from origin to destination, $\{(OAD), (OABD), (OBD)\}$ and the corresponding travel times per arc are shown. The fraction of traffic on route r is denoted by θ_i, so that $\sum_{i=1}^{3} \theta_i = 1.0$, and $0 \leq \theta_i \leq 1$, for $i = 1, 2, 3$. Therefore the fraction of traffic on the arc (O, A) is given by $x_{(O,A)} = \theta_1 + \theta_2$ and similarly for other arcs. The total time on route i is the sum of the travel times along the arcs that form the route and is denoted by $T_i, i = 1, 2, 3$.

A Nash equilibrium will be an allocation such that if a driver on route i decides to change their path, then they will experience a larger delay than staying at equilibrium. The vector of travel times per route is given by

$$T(\theta) = \begin{pmatrix} 3 + \theta_1 + \theta_2 \\ 2.25 + \theta_1 + 2\theta_2 + \theta_3 \\ 3 + \theta_2 + \theta_3 \end{pmatrix}.$$

(a) Show that there is a unique Nash equilibrium, by showing that there is a unique value θ^* such that $T_i(\theta^*) = $ constant, is independent of i.
(b) In order to attain equilibrium, the flow on path i should aim to "equalize" the delay. Suppose now that we do not know the various constants in the delay function, but can only just estimate the delays by running a simulation, which yields an unbiased estimator $\widetilde{T(\theta)}$.

$$\theta_{n+1,i} = \theta_{n,i} - \epsilon_n \left(\widetilde{T_i(\theta_n)} - \frac{1}{3} \sum_k \widetilde{T_k(\theta_n)} \right), \quad (4.23)$$

Show that the algorithm is mass preserving, that is, $\sum_i \theta_{0,i} = 1$ then $\sum_i \theta_{n,i} = 1$, where you may assume that ϵ_n is sufficiently small so that projection is not required.
(c) Characterize the behavior of the stochastic approximation (4.23) and specify your assumptions. In particular, show that the target vector field is coercive for the equilibrium problem.

By the way, this example is classic to show that the Nash equilibrium does not minimize overall travel time. For the specific model, $\theta^* = (0.25, 0.50, 0.25)^T$, and $T_i(\theta^*) = 3.75$ for all i. However, for $\theta = (0.50, 0.0, 0.50)$, then $T_1(\theta) = T_3(\theta) = 3.5$.

Exercise 4.8. Show that under the conditions put forward in Theorem 4.3 it holds that

$$F(t) = \int_0^t G(\vartheta(u))\, du, \quad t \geq 0,$$

is Lipschitz continuous.

Exercise 4.9. Let $J(\theta) = \mathbb{E}[h(\xi, \theta)]$ for $\theta \in \mathbb{R}$. Show that the bias of the FD estimator in Example 4.1 is independent of c_n provided that h is Lipschitz continuous with modulus L such that $\text{Var}(L(\xi))$ is finite.

Chapter Five

Stochastic Approximation: Markovian Dynamics

The dynamic model for stochastic approximation has been introduced in Chapter 3. The model contemplates the case where the target field is defined as an ergodic, or long-term average of quantities of the observed process, so the model of Chapter 3 is insufficient to describe the situation. In this chapter we present rigorous definitions of the model and detailed proofs of convergence of the ensuing stochastic approximations.

5.1 LONG-TERM STATIONARY DYNAMICS: MARKOVIAN MODEL

The first two chapters dealt with analysis of algorithms of the form

$$\theta_{n+1} = \theta_n + \epsilon_n G(\theta_n)$$

designed to "find the zeroes" of a vector field G, under appropriate assumptions. We will assume here that there is no closed expression for $G(\theta)$, but that

- we can build *estimators* for $G(\theta)$, and that
- the control variable can be changed at will, while the process is operating (or being simulated).

Chapter 4 presented a methodology to study stochastic approximation of the form

$$\theta_{n+1} = \theta_n + \epsilon_n Y_n \tag{5.1}$$

analyzing the behavior and convergence properties. The tools of the analysis relied heavily on the decomposition in (4.5). In this chapter, we provide the basis for convergence analysis of the same algorithm when a decomposition of $G(\cdot)$ like in (4.5) is to no avail. Next, we provide an illustrative example of this phenomenon.

Example 5.1. Suppose that an exercise machine can adjust the resistance $\theta \in \Theta = [0.5, 1]$ while a person is exercising. Were θ to be kept fixed, the heart rate of the person would attain a steady value of $L(\theta)$. The goal is to program an algorithm for the machine's computer such that if the person sets a target heart rate of α, the resistance will adjust to reach the value θ^* such that $L(\theta^*) = \alpha$.

The machine cannot accurately measure the heart rate instantaneously, but instead, the number of heartbeats (or pulses) in a time interval of fixed length may be used to estimate for $L(\theta)$. As an example, suppose that for a given person and environmental conditions, $\xi_n := \xi_n(\theta)$ is the number of heartbeats in the n-th interval of 0.3 sec for the machine operating with resistance θ. In order to work this example, we assume the following oversimplified Markov model for the heart beats:

STOCHASTIC APPROXIMATION: MARKOVIAN DYNAMICS

$$\mathbb{P}(\xi_{n+1} = 1 \mid \xi_n = 0, \theta) = p_{0,1}(\theta),$$

$$\mathbb{P}(\xi_{n+1} = 1 \mid \xi_n = 1, \theta) = p_{1,1}(\theta).$$

For each fixed value of θ, the heart rate is defined as the long-term average number of heart beats per minute, (200 intervals of 0.3 sec each):

$$\lim_{N \to \infty} \frac{200}{N} \sum_{n=1}^{N} \xi_n = \lim_{N \to \infty} \frac{200}{N} \sum_{n=1}^{N} \mathbf{1}_{\{\xi_n = 1\}} = \lim_{N \to \infty} \frac{200}{N} \sum_{n=1}^{N} \mathbb{E}[\mathbf{1}_{\{\xi_n = 1\}} \mid \theta] = 200 \, \pi_\theta(1) \quad \text{a.s.}$$

That is, the heart rate can be conveniently expressed in terms of the stationary distribution $\pi_\theta = (\pi_\theta(0), \pi_\theta(1))^\top$ of the Markov chain, yielding

$$L(\theta) = \pi_\theta(1) \times (60/.3) = 200 \, \pi_\theta(1)$$

as the heart rate (per minute).

Using a root finding formulation we propose the vector field

$$G(\theta) = -(L(\theta) - \alpha) = \alpha - 200 \, \pi_\theta(1).$$

A straightforward choice for Y_n in (5.1) is

$$Y_n = \phi(\xi_n) = -(200 \times \mathbf{1}_{\{\xi_n = 1\}} - \alpha) = -(200 \xi_n - \alpha), \tag{5.2}$$

where we note that the dependence on θ_n is only implicit because $\xi_n = \xi(\theta_n)$. Provided the information up to time $n-1$, the expected outcome of Y_n is

$$\mathbb{E}[Y_n \mid \xi_{n-1}, \theta_n] = -200 \, \mathbb{P}(\xi_n = 1 \mid \xi_{n-1}, \theta_n) + \alpha$$

$$= -200 \, p_{\xi_{n-1}, 1}(\theta_n) + \alpha$$

$$= G(\theta_n) + \beta_n, \tag{5.3}$$

for $\theta_n \in [0.5, 1]$, which implies that

$$\beta_n = 200 \left(\pi_{\theta_n}(1) - p_{\xi_{n-1}, 1}(\theta_n) \right).$$

The exogenous noise approach in Chapter 4 requires that

$$\sum_i \epsilon_i |\beta_i| < \infty,$$

with $\sum_i \epsilon_i = \infty$, which is not satisfied in this example. To see this, notice that $\beta_n/200$ is a binary random variable depending on both θ_n and ξ_n that has the distribution

$$\beta_n/200 = \begin{cases} \pi_{\theta_n}(1) - p_{0,1}(\theta_n) & \text{w.p. } \mathbb{P}(\xi_{n-1} = 0), \\ \pi_{\theta_n}(1) - p_{1,1}(\theta_n) & \text{w.p. } \mathbb{P}(\xi_{n-1} = 1). \end{cases}$$

We observe that the only way that the bias term is null is the case when $p_{0,1}(\theta) = p_{1,1}(\theta) = \pi_\theta(1)$, which corresponds to a "degenerate" Markov chain, since in this case ξ_n is independent of the past. This is the model of independent noise of Chapter 4. Otherwise the values for β_n are non-zero and $|\beta_n| \geq 200 \, b(\theta_n)$, where $b(\theta_n) = \max(|\pi_{\theta_n}(1) - p_{0,1}(\theta_n)|, |\pi_{\theta_n}(1) - p_{1,1}(\theta_n)|) > 0$.

However, this does by no means imply that the algorithm (5.1) fails to converge to the optimal value. In particular, we know by the calculations above that

$$\mathbb{E}_{(\xi_{n-1} \sim \pi_{\theta_n})}\left[\mathbb{E}[Y_n \mid \xi_{n-1}, \theta_n]\right] = G(\theta_n),$$

that is, the expectation w.r.t. the stationary measure π_{θ_n} is unbiased for $\theta_n \in [0.5, 1]$. This is the precise sense in which we say that "Y_n is an estimator of the target vector field" for the (long-term) stationary problem. Indeed, as we will show in this chapter, using a more elaborate method of proof, we can relax the assumption on $\mathbb{E}[Y_n \mid \mathfrak{F}_{n-1}]$ and thereby establish sufficient conditions for convergence. ✷✷✷

We first introduce the model that describes the target vector field $G(\theta)$ for each fixed value of θ, and we invite the reader to recap the generic notation introduced in Section 3.3.

Underlying Process

We begin by assuming that for each fixed value of the control variable $\theta \in \Theta$, there is an *underlying* stochastic process $\{\xi_n(\theta)\}$ on $(\Omega, \{\mathfrak{F}_n(\theta)\}, \mathbb{P}_\theta)$ modeling the randomness inherent in the dynamics of the system of interest, where the subscript θ expresses the θ-dependence of the system. This underlying process $\{\xi_n(\theta)\}$ is assumed to be Markovian, that is

$$P_\theta(\xi_n, \cdot) \stackrel{\text{def}}{=} \mathbb{P}_\theta(\xi_{n+1} \in \cdot \mid \xi_n) = \mathbb{P}_\theta(\xi_{n+1} \in \cdot \mid \mathfrak{F}_n(\theta)),$$

and we assume that there is a unique stationary measure $\pi_\theta(\cdot)$. We will use the notation $\mathbb{P}_\theta, \mathbb{E}_\theta$ to indicate probability and expectations with respect to the fixed-θ process $\{\xi_n(\theta)\}$.

Target Vector Field

We assume that the target vector field satisfies

$$G(\theta) = \lim_{N \to \infty} \mathbb{E}\left[\frac{1}{N} \sum_{n=1}^{N} g(\xi_n(\theta), \theta)\right]$$

$$= \lim_{N \to \infty} \mathbb{E}\left[\frac{1}{N} \sum_{n=1}^{N} \underbrace{\mathbb{E}[g(\xi_n(\theta), \theta) \mid \mathfrak{F}_{n-1}]}_{= \tilde{g}(\xi_{n-1}(\theta), \theta)}\right] = \int g(\xi, \theta) \pi_\theta(d\xi), \quad (5.4)$$

where $g(\xi, \theta)$ is a well-defined function representing the instantaneous feedback information, and $\tilde{g}(\xi, \theta)$ is the one-step ahead expectation of the feedback. As was the case in Chapter 4, we will seek to find the zeroes of $G(\cdot)$ via a stochastic approximation. Before presenting the assumptions on the feedback for (5.1) we come back to Example 5.1 to illustrate the model for the target vector field in that case.

Example 5.2. Going back to the Example 5.1 we can use here $g(\xi, \theta) = -(200\xi - \alpha)$ and calculate

$$\tilde{g}(\xi_{n-1}(\theta), \theta) \stackrel{\text{def}}{=} \mathbb{E}[Y_n \mid \mathfrak{F}_{n-1}(\theta)] = -(200 \, p_{\xi_{n-1}(\theta),1}(\theta) - \alpha),$$

for the process with fixed value of θ, and by definition of the stationary measure,

$$G(\theta) = \int g(x, \theta) \pi_\theta(dx) = -\sum_x \pi_\theta(x)(200 \, p_{x,1}(\theta) - \alpha) = -200 \, \pi_\theta(1) + \alpha = -(L(\theta) - \alpha).$$

(5.5)

Notice that for this type of problem, one may alternatively use estimation "by batches," that is, we may choose

$$Y_n^{(K)} = \frac{1}{K} \sum_{k=nK}^{(n+1)K-1} g(\xi_k(\theta), \theta),$$

so that only any K iterations of the underlying process θ_n is updated; see Example 3.1. As there is no θ-updated during a K-batch, $\mathbb{E}[Y_n^{(K)} \mid \xi_{nK}, \theta] = G(\theta)$ for any value of K provided that ξ_{nK} is stationary for π_θ. Moreover, by ergodicity

$$\lim_{K \to \infty} Y_n^{(K)} = G(\theta_n) \quad \text{a.s.}$$

The batch size K is usually hard to choose in advance, and sometimes we use trial and error experimentation with preliminary calculations in order to establish the better batch size. An implementation of this procedure is included in Exercise 5.3 to illustrate the use of batching. ✼✼✼

The Model for Y_n

Consider now the case where Y_n is constructed as a function of the observed values $(\xi_0, \ldots, \xi_n)$ (the *history* of the process), where we drop the notation of θ for notational convenience. When θ_n varies according to (5.1) we consider the augmented process $\{(\xi_{n-1}, \theta_n)\}$ that satisfies

$$\mathbb{P}(\xi_n \in B \mid \xi_{n-1}, \theta_n) = P_{\theta_n}(\xi_{n-1}, B),$$

for any measurable set B, $\theta_n \in \Theta$, and denote the natural filtration of $\{\xi_n\}$ by $\{\mathfrak{F}_n\}$.

Remark 5.1. In the literature, such a model is a particular case of what is also known as a Markov Decision Process (MDP). In this chapter we do not study the problem in the context of MDPs. Indeed, we seek a time-independent value θ^* being a stable point of $G(\cdot)$, which is different from the typical MDP approach where one allows the control θ_n to be a function of the state ξ_n.

By assumption, the estimator Y_n is measurable with respect to $\mathfrak{F}_n$. Because θ_n is a function of θ_{n-1} and of Y_{n-1} it follows that θ_n is measurable with respect to $\mathfrak{F}_{n-1}$. If known, a natural choice for the feedback is $Y_n = g(\xi_n, \theta_n)$, so that using (5.4), $\mathbb{E}[Y_n \mid \mathfrak{F}_{n-1}] = \tilde{g}(\xi_{n-1}, \theta_n)$. Otherwise, use a proxy $\phi(\xi_n, \theta_n)$ that may lead to a bias term. For our analysis we assume that

$$\mathbb{E}[Y_n \mid \mathfrak{F}_{n-1}] = \tilde{g}(\xi_{n-1}, \theta_n) + \beta_n,$$

where the error terms β_n are well-defined random variables, measurable w.r.t $\mathfrak{F}_{n-1}$.

Example 5.3. Going back to Example 5.2, we can calculate

$$\tilde{g}(\xi_{n-1}, \theta_n) = \mathbb{E}[Y_n \mid \mathfrak{F}_{n-1}] = -(200 \, p_{\xi_{n-1}, 1}(\theta_n) - \alpha),$$

for $n \geq 0$, see (5.5). Comparing with (5.3) we see that replacing $G(\theta_n)$ by $\tilde{g}(\xi_{n-1}, \theta_n)$ leads to zero bias. ✼✼✼

For ease of reference we summarize the above setup in the following definition.

Definition 5.1. The *stationary problem* is determined by a parametrized family of Markov chains $\{\xi_n(\theta)\}$, $\theta \in \Theta \subseteq \mathbb{R}^d$, on some state space Ξ, with a unique stationary measure $\pi_\theta(\cdot)$ for each value of θ. The target vector field satisfies

$$G(\theta) = \int g(\xi,\theta)\pi_\theta(d\xi) = \lim_{N\to\infty} \mathbb{E}\left[\frac{1}{N}\sum_{n=1}^{N} g(\xi_n(\theta),\theta)\right]$$

$$= \lim_{N\to\infty} \mathbb{E}\left[\frac{1}{N}\sum_{n=1}^{N} \underbrace{\mathbb{E}[g(\xi_n(\theta),\theta)\mid \mathfrak{F}_{n-1}]}_{=\tilde{g}(\xi_{n-1}(\theta),\theta)}\right] \tag{5.6}$$

for some *instantaneous feedback* $g(\xi,\theta)$. The function $\tilde{g}(\xi,\theta)$ is its corresponding one-step ahead expectation. The underlying process $\{\xi_n(\theta)\}$ satisfies the Markov property

$$\mathbb{P}(\xi_n(\theta) \in B \mid \xi_{n-1}(\theta) = x) = P_\theta(x, B), \quad \theta \in \Theta,$$

for all measurable Borel sets B and $x \in \Xi$, and it has natural filtration $\{\mathfrak{F}_n\}$. The *feedback*

$$Y_n = \phi(\xi_n(\theta_n), \theta_n) = g(\xi_n(\theta_n), \theta_n) + b(\xi_n(\theta_n), \theta_n),$$

where $b(\xi_n(\theta_n), \theta_n)$ expresses the sample bias, satisfies

$$\mathbb{E}[Y_n \mid \mathfrak{F}_{n-1}] = \tilde{g}(\xi_{n-1}, \theta_n) + \beta_n,$$

for $\theta_n \in \Theta$, where the error terms β_n are well-defined random variables, measurable w.r.t $\mathfrak{F}_{n-1}$. More specifically,

$$\int g(z,\theta)P_\theta(\xi,dz) = \tilde{g}(\xi,\theta) \quad \text{and} \quad \beta_n = \int b(z,\theta_n)P_{\theta_n}(\xi_{n-1}(\theta_{n-1}),dz)$$

for all ξ,θ, and any n.

Refer to Figure 3.4 for a schematic representation of the iterations, which now describe what is known in engineering as a "closed loop control."

The main difference between the static model of Chapter 4 and the dynamic model discussed here lies in the distribution of the underlying process. In Chapter 4, we assume that consecutive samples of the underlying θ-fixed process provide estimates for $G(\theta)$. The stationary problem assumes that $G(\theta)$ satisfies (5.6). Typically, one cannot sample directly from the stationary distribution π_θ, so that the results of Chapter 4 are not applicable for the dynamic model in Definition 5.1. More specifically, the model now has *correlated noise* because $\mathbb{E}[Y_n \mid \mathfrak{F}_{n-1}]$ is no longer just a function of θ_n, as it was the case in the model in Chapter 4. The correlated noise model is also called an *endogenuous noise model* in the literature.

The stochastic approximation still considers building the feedback Y_n using the samples $\phi(\xi_n, \theta_n)$ even when θ_n is changing. It is a very strong result (and perhaps quite surprising for many readers) that this simple update scheme actually provides the correct solution. In what follows we carefully present the proof that indeed, as n increases, the underlying process $\{\xi_n(\theta_n)\}$ does converge (weakly) jointly with the process $\{\theta_n\}$ and the limit corresponds to the stationary process at points θ^* that are stable points of the ODE. As we will see in the proof of Theorem 5.3, the recursion of the stochastic approximation actually provides an estimate of the target vector field $G(\theta)$ by "averaging" consecutive values of the feedback function $\phi(\xi_n(\theta_n), \theta_n)$ (which is the proxy for $g(\xi_n(\theta_n), \theta_n)$).

Algorithm 5.1 shows the main structure of the code in case $\Theta = \mathbb{R}^d$, when batches of K consecutive observations for θ_n are used to build the feedback estimator Y_n. The code

Algorithm 5.1 Stochastic approximation: Markovian dynamics

Define stepsize sequence ϵ_n
Define feedback function / proxy $\phi(\xi, \theta)$
Define transition probability function $\text{Prob}(\theta, \xi) = P_\theta(\xi, \cdot)$ (if simulation)
Initialize $\theta_0, \xi[0, 0]$
for $(n = 0, \ldots N - 1)$ **do** (iterations of the SA algorithm)
    for $(k = 1, \ldots K)$ **do**
        Generate (or Read) sample $\xi[n, k] \sim \text{Prob}(\theta_n, \xi[n, k-1])$ (simulating data—or streaming)
    $Y_n = \text{Mean}(\phi(\xi[n, \cdot])$
    $\theta_{n+1} = \theta_n + \epsilon_n Y_n$
    $\xi[n+1, 0] := \xi[n, K]$

assumes that the compiler has a built-in function $\text{Mean}(v)$ that calculates the sample mean of a vector $v = (v_1, \ldots, v_K)$. The difference between the algorithms for the static model of Chapter 4 and the dynamic model of Chapter 5 is entirely on the line where the generation of the following observation of the underlying process $\xi_k^{(n)}$ is performed. Notice that these algorithms can be implemented for simulation-based optimization, or instead of generating the following observation, one could be streaming data from real sources.

Remark 5.2. Notice that the state of the Markov chain is not reset to the initial value after changing the value of θ_n. This is very important: if we were to use the same state, say $\xi_0^{(n)} \equiv x_0$ for all n then we would be estimating a feedback that corresponds to a *finite* horizon average, rather than the long-term average desired, and clearly

$$\mathbb{E}\left[\frac{1}{K} \sum_{k=1}^{K} \phi(\xi_k(\theta), \theta) \Big| \xi_0(\theta) = x\right] \neq G(\theta),$$

for fixed initial value x.

5.2 ANALYSIS OF THE DECREASING STEPSIZE SA

The following theorem will be stated without proof. The proof is an extension to that of Theorem 4.1, and it uses the methodology of Chapter 6 of [198].

Theorem 5.1. *Consider the setting of the stationary problem forward in Definition 5.1. Assume the following conditions hold:*

(a1) The transition probabilities $P_\theta(\xi, \cdot)$ are continuous in (θ, ξ) on $\mathbb{R}^d \times \Xi$. The stationary measure of the fixed-θ process $\{\xi(\theta)\}$, denoted by π_θ, is unique, and the set $\{\pi_\theta(\cdot), \theta \in M\}$ of stationary measures of the fixed-θ processes is tight for each compact set M, for $M \subset \mathbb{R}^d$.
(a2) The instantaneous feedback $g(\xi, \theta)$ and the one-step ahead expectation of the feedback $\tilde{g}(\xi, \theta)$ are jointly continuous measurable functions (they may depend on n, but we will introduce the simplest model here), such that

$$\mathbb{E}[Y_n \mid \mathfrak{F}_{n-1}] = \tilde{g}(\xi_{n-1}, \theta_n) + \beta_n, \quad \theta_n \in \mathbb{R}^d,$$

and, for every fixed $\theta \in \mathbb{R}^d$,

$$\lim_{N \to \infty} \frac{1}{N} \sum_{n=1}^{N} \mathbb{E}[g(\xi_n(\theta), \theta)] = \lim_{N \to \infty} \frac{1}{N} \sum_{n=1}^{N} \mathbb{E}[\tilde{g}(\xi_{n-1}(\theta), \theta)]$$
$$= \int g(x, \theta) \pi_\theta(dx) = G(\theta).$$

(a3) $V_n = \mathbb{E}[(Y_n - \mathbb{E}(Y_n | \mathfrak{F}_{n-1}))^2]$ satisfies $\sum_{i=0}^{\infty} \epsilon_i^2 V_i < \infty$.
(a4) $\sum_{i=0}^{\infty} \epsilon_i = \infty$, $\epsilon_n > 0$ for all n, $\epsilon_n \to 0$.
(a5) The error terms are asymptotically negligible, that is, $\sum_{i=0}^{\infty} \epsilon_i \|\beta_i\| < \infty$ w.p.1.
(a6) G is bounded or Lipschitz continuous, and the trajectories of the ODE

$$\frac{dx(t)}{dt} = G(x(t)) \tag{5.7}$$

have a unique limit (as $t \to \infty$) for each initial condition. Let $\mathcal{S}$ denote the corresponding set of asymptotically stable points and assume that $\mathcal{S} \neq \emptyset$.

Let

$$\theta_{n+1} = \theta_n + \epsilon_n Y_n, \quad \theta_0 \in \mathbb{R}^d.$$

Then the shifted interpolation process $x_n(t) = \vartheta^\epsilon(t_n + t), t \geq 0$ and $\vartheta(0) = \theta_0$, converge as $n \to \infty$ to the solution of the ODE (5.7). This implies that $\theta_n \to \mathcal{S}$ a.s. In particular, if θ^* is the only asymptotically stable point of the ODE (5.7), then $\theta_n \to \theta^*$ a.s.

Alternatively, for $\Theta \subset \mathbb{R}^d$ convex and compact, assume (a1) to (a5) hold on Θ, and

(a6)' G is continuous on Θ, and the trajectories of the projected ODE

$$\frac{dx(t)}{dt} = (\Pi_\Theta G)(x(t)), \quad x(0) \in \Theta, \tag{5.8}$$

have set of asymptotically stable points $\mathcal{S} \neq \emptyset$.

Let

$$\theta_{n+1} = \Pi_\Theta(\theta_n + \epsilon_n Y_n), \quad \theta_0 \in \Theta.$$

Then the shifted interpolation process $x_n(t) = \vartheta^\epsilon(t_n + t), t \geq 0$ and $\vartheta(0) = \theta_0$, converge as $n \to \infty$ to the solution of the ODE (5.8). This implies that $\theta_n \to \mathcal{S}$ a.s. In particular, if θ^* is the only asymptotically stable point of the projected ODE, (5.8) then $\theta_n \to \theta^*$ a.s.

Example 5.4. Consider again the problem in Example 5.1 of tracking the heart rate while a person is exercising, with parameters given by the following simple model:

$$\mathbb{P}(\xi_{n+1} = 1 | \xi_n = 0, \theta) = \theta,$$
$$\mathbb{P}(\xi_{n+1} = 1 | \xi_n = 1, \theta) = \theta^2.$$

For any $\theta \in [0.5, 1]$, the fixed-θ process is a recurrent Markov chain with state space $\{0, 1\}$. We calculate the stationary measure:

$$\pi_\theta(1) = \pi_\theta(1)\theta^2 + (1 - \pi_\theta(1))\theta \Longrightarrow \pi_\theta(1) = \frac{\theta}{1 + \theta(1 - \theta)}.$$

Thus the function $L(\theta) = 200 \pi_\theta(1)$ is increasing in θ. To see this, it suffices to verify that

$$\frac{d\pi_\theta(1)}{d\theta} = \frac{1+\theta-\theta^2-(\theta-2\theta^2)}{(1+\theta-\theta^2)^2} = \frac{1+\theta^2}{(1+\theta-\theta^2)^2} > 0.$$

Thus, for any $\alpha \in [80, 200]$ there is a unique value θ^* such that $L(\theta^*) = \alpha$. Of course if we do know the model, then we can solve the problem using deterministic root finding.

In general, if the functions $p_{i,1}(\theta)$ are continuously differentiable and increasing in θ, then for any θ such that $0 < p_{i,1}(\theta) < 1$ the ensuing Markov chain will be ergodic, so the invariant measure exists. Assumptions (a1) and (a2) are verified for this problem, as can be seen in Example 5.2. In this case, the measurements Y_n have no bias terms β_n, and they are absolutely bounded by 200, so they have uniformly bounded variance. Any stepsize sequence satisfying the standard conditions will ensure (a3), (a4) and (a5). Because $L(\theta)$ is continuously differentiable and bounded, then (a6) is satisfied. For each $\alpha \in [80, 200]$, there is a unique stable point of (5.5). Therefore our learning algorithm (5.1) will converge to the optimal setting—provided that the person exercises long enough, of course! ✲✲✲

The following lemma shows that Theorem 5.1 extends to streaming/consecutive batching; see Definition 3.2. The proof provides the details of using streaming batching as in Example 5.2.

Lemma 5.1. *Theorem 5.1 can be extended to Y_n^K where Y_n^K is obtained from streaming/consecutive batching of Y_n, that is, if (a2) holds for $K = 1$, then it holds for any $K \geq 1$. Moreover, as K tends to infinity, the variance of Y_n^K tends to zero.*

Proof. By construction

$$\mathbb{E}[Y_n^K | \mathfrak{F}_{n-1}] = \frac{1}{K} \sum_{i=0}^{K-1} \mathbb{E}[\phi(\xi_{nK+i}(\theta_n), \theta_n) | \mathfrak{F}_{n-1}] =: g^K(\xi_{Kn}, \theta_n).$$

Under stationarity of ξ_n under θ, the components of ξ_n^K are distributed according to π_θ, which gives

$$G(\theta) = \int g(x, \theta) \pi_\theta(dx)$$

$$= \frac{1}{K} \sum_{i=0}^{K-1} \int g(x_i, \theta) \pi_\theta(dx_i)$$

$$= \frac{1}{K} \sum_{i=0}^{K-1} \int \left(\int \left(\cdots \int g(x_i, \theta) P_\theta(x_{i-1}, dx_i) \cdots \right) P_\theta(x_0, dx_1) \right) \pi_\theta(dx_0)$$

$$= \int \cdots \int \left(\frac{1}{K} \sum_{i=0}^{K-1} g(x_i, \theta) \right) P_\theta(x_{K-2}, dx_{K-1}) \cdots P_\theta(x_0, dx_1) \pi_\theta(dx_0)$$

$$= \int g^K(x) \pi_\theta(dx).$$

This shows that we replace Y_n in (a2) by Y_n^K without effecting the limiting result. The second part of the lemma follows from ergodicity as Y_n^K tends a.s. toward the constant $G(\theta_n)$. Bias can be ingored due to (a5). □

Condition (a2) in Theorem 5.1 can be achieved by taking streaming batches of appropriate length K; however, this comes at a computational cost. Condition (a3) in Theorem 5.1 may be very difficult to verify for many practical applications. Our next result considers the case where G is coercive for a well-posed problem, in which case it is sufficient to verify bounded variance on compact sets. While weaker, this assumption is usually easier to verify in practice than condition (a3). The theorem is an adaption of Theorem 5.1 to the NLP setting, and the proof is left to the reader as an exercise.

Theorem 5.2. *Conisder the well-posed NLP $\min_{\theta \in \Theta} J(\theta)$. Suppose that the vector field G is coercive for this problem on H. Consider the stationary problem of Definition 5.1.*

Assume the following conditions hold:

(a1) The transition probabilities $P_\theta(\xi, \cdot)$ are continuous in (θ, ξ) on $\mathbb{R}^d \times \Xi$. The stationary measure of the fixed-θ process $\{\xi(\theta)\}$, denoted by π_θ, is unique, and the set $\{\pi_\theta(\cdot), \theta \in M\}$ of stationary measures of the fixed-θ processes is tight for each compact set $M \subset \mathbb{R}^d$.

(a2) The instantaneous feedback $g(\xi, \theta)$ and the one-step ahead expectation of the feedback $\tilde{g}(\xi, \theta)$ are jointly continuous measurable functions (they may depend on n, but we will introduce the simplest model here), such that

$$\mathbb{E}[Y_n \mid \mathfrak{F}_{n-1}] = \tilde{g}(\xi_{n-1}, \theta_n) + \beta_n, \quad \theta_n \in \mathbb{R}^d,$$

and, for every fixed $\theta \in \mathbb{R}^d$,

$$\lim_{N \to \infty} \frac{1}{N} \sum_{n=1}^{N} \mathbb{E}[g(\xi_n(\theta), \theta)] = \lim_{N \to \infty} \frac{1}{N} \sum_{n=1}^{N} \mathbb{E}[\tilde{g}(\xi_{n-1}(\theta), \theta)]$$

$$= \int g(x, \theta) \pi_\theta(dx) = G(\theta).$$

(a3') For every compact set $M \subset \Theta$ there is a finite number V such that for all $n \sup_{\theta \in M} \mathrm{Var}(\phi(\xi_n, \theta)) \leq V$.

(a4') $\sum_{i=0}^{\infty} \epsilon_i = \infty$, $\sum_{i=0}^{\infty} \epsilon_i^2 < \infty$, and $\epsilon_n > 0$ for all n.

(a5) The error terms are asymptotically negligible, that is, $\sum_{i=0}^{\infty} \epsilon_i \|\beta_i\| < \infty$ w.p.1.

Let

$$\theta_{n+1} = \theta_n + \epsilon_n Y_n, \quad \theta_0 \in H,$$

then θ_n converges a.s. to a KKT point of the optimization problem. In particular, if θ^* is the only asymptotically stable point of the ODE given by G, then $\theta_n \to \theta^*$ a.s.

Alternatively, for $\Theta \subset \mathbb{R}^d$ convex, compact and $G(\cdot) = (\Pi_\Theta \hat{G})(\cdot)$, for $\hat{G}$ continuous on Θ, the above convergence statement holds for the projected SA

$$\theta_{n+1} = \Pi_\Theta(\theta_n + \epsilon_n Y_n), \quad \theta_0 \in \Theta \cap H,$$

provided that the above conditions (a1) to (a5) holds on Θ, and $\hat{G}$ in (a2) (and G replaced by $\hat{G}$ in (5.6)).

We conclude this section with an example on optimization that builds on Example 4.6.

Example 5.5. Consider the following stochastic recursive sequence:

$$V_{k+1}(\theta) = U_k (V_k(\theta) + S_k(\theta)), \quad k \geq 0, \tag{5.9}$$

where $\{S_k(\theta)\}$ denotes an iid sequence of exponentially distributed random variables with mean θ and $\{U_k\}$ is an iid sequence of uniformly distributed random variables on $[0, 1]$. The sequence $\{V_k(\theta)\}$ generated by (5.9) generalizes the classical Lindley recursion for waiting times in queueing theory; see Example 3.6. The model has also been used to analyze the window size of the transmission control protcol in computer communication networks; see [15, 234]. Moreover, the sequence $\{V_k(\theta)\}$ can be interpreted as a reflected autoregressive process of order 1, reflected at 0. For $\theta > 0$, $\{V_k(\theta)\}$ is ergodic and the Laplace transform of the stationary distribution can be obtained in explicit form; see [167, 206]. Moreover, the stationary mean of $\{V_k(\theta)\}$ is equal to θ.

In the following we study the (academic) example of adjusting the control parameter θ so that the α-quantile of the stationary $V_k(\theta)$ is equal to some predefined value $\eta > 0$. This is a target tracking example with the added challenge that the target is defined via a stationary characteristic that is not directly observable.

Let q_θ^α be the α-quantile of the stationary distribution of $V_k(\theta)$, which we denote by F_θ, and consider the following quantile fitting problem

$$\min_{\theta \in (0,\infty)} \frac{1}{2}(q_\theta^\alpha - \eta)^2. \tag{5.10}$$

Provided that q_θ^α is stricly monotone as a mapping of θ (for given α), the optimization problem is well-posed. By computation,

$$\frac{d}{d\theta}\frac{1}{2}(q_\theta^\alpha - \eta)^2 = \frac{d}{d\theta}q_\theta^\alpha (q_\theta^\alpha - \eta).$$

Use now the definition of the quantile $F_\theta(q_\theta^\alpha) = \alpha$ and differentiate w.r.t θ to obtain

$$F'_\theta(q_\theta^\alpha) + f_\theta(q_\theta^\alpha)\frac{d}{d\theta}q_\theta^\alpha = 0,$$

which implies that:

$$\frac{d}{d\theta}q_\theta^\alpha = -\frac{F'_\theta(q_\theta^\alpha)}{f_\theta(q_\theta^\alpha)}.$$

Thus, when using the negative gradient as the target vector field for (5.10) it has the expression

$$G^{\text{opt}}(\theta) = \frac{F'_\theta(q_\theta^\alpha)}{f_\theta(q_\theta^\alpha)}(q_\theta^\alpha - \eta).$$

This target field requires evaluating F'_θ and f_θ, which we assume not available to us in a closed form. Only consecutive observations of the underlying process $\{V_n(\theta_n)\}$ are assumed to be available for the updates. Note that $f_\theta(\cdot) > 0$ by definition of a pdf. Moreover, for the exponential random variables $S_k(\theta) \stackrel{d}{=} \theta S_k(1)$, so that under this representation $S_k(\theta)$ is strictly monotone increasing in θ for all k. By induction, it then follows from (5.9) that $V_k(\theta)$ is strictly monotone increasing in θ for all k. This shows that $F_\theta(\cdot)$ is strictly monotone decreasing in θ (higher values for θ lead to higher values for the stationary $V_k(\theta)$, and the mass of $F_\theta(\cdot)$ shifts to the right). Consequently, q_θ^α is striclty monotone and the optimization problem in (5.10) is thus well-posed. Morever, due to the monotonicity of

$F_\theta(\cdot)$ we may replace $G(\theta)$ by the descent direction target field

$$\hat{G}^{\text{opt}}(\theta) = \eta - q_\theta^\alpha,$$

which is also coercive on $(0, \infty)$ because it is a descent direction and has the same zeroes as G^{opt}.

In solving the minimization problem in (5.10), we will combine the recursive learning scheme for q_θ^α put forward in Example 4.6 with the SA updates for solving (5.10). For this we introduce a two-dimensional parameter (θ_n, Q_n), where Q_n denotes the current proxy of the quantile obtained via the scheme in Example 4.6 and θ_n denotes the current approximation of the optimal control parameter. From Example 4.6 it follows that the target vector field for recursive quantile estimation can be written as the mapping

$$G^q(\theta, Q) = \alpha - F_\theta(Q),$$

where θ is a fixed distributional parameter.

We will now introduce a scheme that *simultaneously updates both parameters*. For this, we choose as underlying process $\xi_n(\theta) = V_n(\theta)$. In combining the two update schemes we take Q_n to be the proxy for $q_{\theta_n}^\alpha$. This leads to the following two-dimensional SA:

$$\theta_{n+1} = \Pi_{[\delta, 1/\delta]}(\theta_n + \epsilon_n(\eta - Q_n)),$$
$$Q_{n+1} = Q_n + \epsilon_n\left(\alpha - \mathbf{1}_{\{\xi_n(\theta_n) \leq Q_n\}}\right).$$

where, without loss of generality, we use in the first update rule projection on $[\delta, 1/\delta]$, for δ small, to keep θ_n feasible.

The feedback here is

$$Y_n = \left(\eta - Q_n, \alpha - \mathbf{1}_{\{\xi_n(\theta_n) \leq Q_n\}}\right)^\top,$$

We now argue that Theorem 5.2 applies to Y_n. Condition (a1) follows from [206], condition (a3) holds as the variance of the indicator mapping is bounded, condition (a4) holds for the standard choice of gain size, and condition (a5) holds due to the unbiasedness of Y_n on $[\delta, 1/\delta]$. In the following we turn to establishing condition (a2). Consider the following joint update mapping:

$$\tilde{g}(\xi_{n-1}(\theta_{n-1}), (\theta_n, Q_n)) = \mathbb{E}[Y_n | \mathfrak{F}_{n-1}]$$
$$= \left(\Pi_{[\delta, 1/\delta]}(\eta - Q_n), \alpha - \mathbb{E}[\mathbf{1}_{\{\xi_n(\theta_n) \leq Q_n\}} | \xi_{n-1}(\theta_{n-1})]\right)^\top$$

and

$$g(\xi, (\theta, Q)) = \left(\Pi_{[\delta, 1/\delta]}(\eta - Q_n), \alpha - \mathbf{1}_{\{\xi \leq Q\}}\right)^\top.$$

For fixed θ, Q, such that $\eta - Q_n \in [\delta, 1/\delta]$, ergodicity of $\xi_n(\theta)$ implies

$$\lim_{N \to \infty} \frac{1}{N} \sum_{n=1}^N \mathbb{E}[\tilde{g}(\xi_{n-1}(\theta), (\theta, Q))] = \lim_{N \to \infty} \frac{1}{N} \sum_{n=1}^N \mathbb{E}[g(\xi_n(\theta), (\theta, Q))] = G(\theta, Q),$$

with

$$G(\theta, Q) = \left(\hat{G}^{\text{opt}}(\theta, Q), G^q(\theta, Q)\right)^\top. \qquad \text{✵✵✵}$$

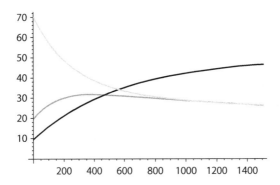

Figure 5.1. Trajectory of the ODE.

5.3 ANALYSIS OF THE CONSTANT STEPSIZE SA

In this section we study the SA with constant stepsize:

$$\theta^\epsilon_{n+1} = \theta^\epsilon_n + \epsilon Y^\epsilon_n, \qquad (5.11)$$

where $\epsilon > 0$ is kept constant for all $n \in \mathbb{N}$, which is the stochastic version of (2.19) for fixed gain size. In light of our previous analysis, it is apparent that method of Chapter 4 is not applicable for the stationary (long-term) problem.

Theorem 2.5 provides sufficient conditions, so that the zeroes of $G(\cdot)$ can be found iteratively by (2.19) with constant stepsize where the interpolation process approaches an ODE whose limits are the desired stationary points. As in Theorem 2.5, the sense in which the algorithm (5.11) converges is in terms of the interpolated processes, as $\epsilon \to 0$ (not as $n \to \infty$). In addition, the stochastic processes converge *in distribution* to the solution of the ODE. In this section, we present an analysis of the constant stepsize algorithm that "averages out" the noise to recover the result of the deterministic case Theorem 2.5.

Example 5.6. To illustrate the concept of convergence, we revisit here the problem of finding an economic equilibrium of Exercise 4.7. In that example, using the data given in the exercise for $N = 100$, the ODE can be approximated by a deterministic recursion $\theta_{n+1} = \theta_n + \epsilon G(\theta_n)$. This example follows the martingale difference noise model, so it is simpler than the stationary problem for the correlated noise model of Definition 5.1. The purpose of this example is to provide a visualization of convergence (in distribution) as $\epsilon \to 0$ of the stochastic approximation to the solution of an ODE. Figure 5.1 demonstrates this result was obtained at $\epsilon = 0.5$.

Suppose that we do not know the exact formulas for the individual travel times given the allocation θ, only that they are affine functions of θ. Estimates can be obtained by simulation, which adds a random non-negative noise with gamma distribution. That is,

$$\widehat{T}_i(\theta) = T_i(\theta) + \gamma_i,$$

where $\gamma_i, i = 1, 2, 3$ are independent with gamma distribution with parameters $(2, 0.5)$, so that $\mathbb{E}[\gamma_i] = 1$. Notice that although this creates biased estimates for T, the noise in the estimation of G has zero mean, using

$$Y^\epsilon_n(i) = -\left(\widehat{T}_i(\theta^\epsilon_n) - \frac{1}{3}\sum_k \widehat{T}_k(\theta^\epsilon_n)\right).$$

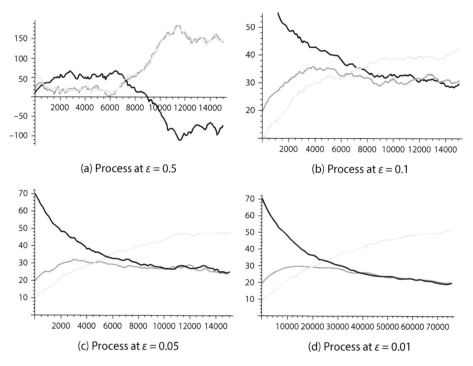

Figure 5.2. Results of constant stepsize processes, as the stepsize is made smaller.

The following graphs were obtained using decreasing $\epsilon = 0.5, 0.1, 0.05, 0.01$. The behavior is very erratic for $\epsilon = 0.5$ compared to the ODE, even for this relatively small error. Because we are not truncating the iterates to ensure that the traffic flow is non-negative, the sequence θ_n^ϵ had negative components in numerous experiments, and when this happens, it does not return to positive values. In contrast, when the deterministic algorithm is started at negative values, the sequence always comes back to the stable point.

This example also illustrates the relationship between the time scale of the limit ODE and the value of ϵ. Because the time is defined in terms of $t = n\epsilon$, the number of iterations needed to attain the same time grows as $1/\epsilon$ (compare the x-axes in Figure 5.2). ✳✳✳

In this section we will follow the common notation adding the superindex ϵ to the various variables to stress the fact that ϵ is the parameter with respect to which we will study the limiting behavior. The process $\{(\xi_n^\epsilon(\theta_n^\epsilon), \theta_n^\epsilon)\}$ defines the structure of the model as a Markov process (and as usual we write ξ_n^ϵ for $\xi_n^\epsilon(\theta_n^\epsilon)$), where $\mathfrak{F}_n^\epsilon = \sigma((\xi_i^\epsilon, \theta_{i+1}^\epsilon); i = 0, \ldots, n)$ is the filtration, the transition probabilities are given by

$$\mathbb{P}(\xi_{n+1}^\epsilon \in B \mid \mathfrak{F}_n^\epsilon) = P_{\theta_{n+1}^\epsilon}(\xi_n^\epsilon, B),$$

for any measurable set $B \subset \Omega$ and θ_{n+1}^ϵ updates according to (5.11), where we assume that Y_n^ϵ is measurable w.r.t. $\mathfrak{F}_n^\epsilon$. We will use the notation

$$\mathbb{E}[Y_n^\epsilon \mid \mathfrak{F}_{n-1}^\epsilon] = \mathbb{E}[\phi(\xi_n^\epsilon, \theta_n) \mid \mathfrak{F}_{n-1}^\epsilon] \approx \tilde{g}(\xi_{n-1}^\epsilon, \theta_n^\epsilon), \qquad (5.12)$$

for well-defined functions $\phi, \tilde{g}$ (it may depend on n, but we will introduce the simplest model here).

As mentioned in Chapter 2, the ODE limit provides more insight into the iterative method than results that characterize only the limit points $\lim_{n\to\infty} \theta_n$. The limit occurs as

$\epsilon \to 0$, which of course is never attained (like $n \to \infty$, which is never attained either), and so we must interpret the limit of θ_n in terms of the algorithms being "close" to a deterministic trajectory when "ϵ is small enough," as illustrated in Example 5.6.

Theorem 5.3 below extends the results of Theorem 4.3 to the stationary model in Definition 5.1. Although not necessary, the reader is encouraged to review that proof in preparation for the following theorem. When the conditions of Theorem 5.3 hold, the stochastic approximation is close to the solution of

$$\frac{dx(t)}{dt} = G(x(t)),$$

where the vector field G in (5.4) is a stationary average mean drift of the algorithm. This is helpful in practice in many ways. First, we may wish to define a target or desired ODE that will have as stable points the optimal values of our problem. Second, we may be able to find formulations such as changes in variables or in a time scale that will make it more convenient for the estimation of G, or that may achieve faster convergence to the stable points. For example, in the root finding version of the problem, instead of using the gradient of the distance, we simply use $L(\theta) - \alpha$, which is easier to estimate and under monotonicity conditions has the same stable points as the gradient. The ODE trajectories not only determine the limit points but provide a global understanding of how the algorithm gets there.

In the following theorem we use the conditions of *uniform integrability* (Definition B.30) and *tightness* (Definition B.29).

Theorem 5.3. *Consider the stationary problem of Definition 5.1, and assume the following conditions hold:*

(a1) The transition probabilities of the fixed-θ processes $P_\theta(\xi, \cdot)$ are weakly continuous (uniformly) w.r.t. (θ, ξ) on $\mathbb{R}^d \times \Xi$. That is, for each bounded and continuous real-valued function f, and any $\rho > 0$, there is a $\delta > 0$ such that

$$\left| \int f(x) P_\theta(\xi, dx) - \int f(x) P_{\theta'}(\xi', dx) \right| \leq \rho$$

whenever $\|(\xi, \theta) - (\xi', \theta')\| < \delta$.

(a2) The stationary measure π_θ is unique, and the set $\{\pi_\theta(\cdot), \theta \in M\}$ of stationary measures of the fixed-θ processes is tight for each compact set M, for $M \subset \mathbb{R}^d$.

(a3) The instantaneous feedback $g(\xi, \theta)$ and the one-step ahead expectation of the feedback $\tilde{g}(\xi, \theta)$ are jointly continuous measurable functions (they may depend on n, but we will introduce the simplest model here), such that for every fixed $\theta \in \mathbb{R}^d$,

$$\lim_{N \to \infty} \frac{1}{N} \sum_{n=1}^{N} \mathbb{E}[g(\xi_n, \theta)] = \lim_{N \to \infty} \frac{1}{N} \sum_{n=1}^{N} \mathbb{E}[\tilde{g}(\xi_{n-1}, \theta)] = \int g(x, \theta) \pi_\theta(dx) = G(\theta).$$

(a4) The feedback sequence $\{Y_n^\epsilon, \epsilon > 0\}$ satisfies $\mathbb{E}[Y_n^\epsilon \mid \mathfrak{F}_{n-1}^\epsilon] = \tilde{g}(\xi_{n-1}^\epsilon, \theta_n^\epsilon) + \beta_n(\xi_{n-1}^\epsilon, \theta_n^\epsilon)$, for $\theta_n^\epsilon \in \mathbb{R}^d$, and it is uniformly integrable, that is,

$$\sup_{n,\epsilon} \lim_{K \to \infty} \mathbb{E}\left[\|Y_n^\epsilon\| \mathbf{1}_{\{\|Y_n^\epsilon\| \geq K\}} \right] = 0.$$

(a5) The set $\{(\xi_n^\epsilon, \theta_n^\epsilon); \epsilon > 0\}$ is tight.

(a6) The bias is asymptotically negligible, that is,

$$\lim_{n,m,\epsilon} \frac{1}{m} \sum_{i=n}^{n+m-1} \mathbb{E}\big[\|\beta_n(\xi_{n-1}^\epsilon, \theta_n^\epsilon)\|\big] = 0,$$

where the limit is taken as $n \to \infty$, $m \to \infty$ and $\epsilon \to 0$ simultaneously in any way.
(a7) G is bounded or Lipschitz continuous, and the trajectories of ODE

$$\frac{dx(t)}{dt} = G(x(t)) \qquad (5.13)$$

have a unique limit (as $t \to \infty$) for each initial condition.

Let

$$\theta_{n+1} = \theta_n + \epsilon_n Y_n,$$

then the interpolated processes $\vartheta^\epsilon(\cdot)$ converge in distribution, as $\epsilon \to 0$, to a limit process which is continuous w.p.1 and satisfies the ODE in (5.13).

Let $\Theta \subset \mathbb{R}^d$ be convex and compact. Provided that the above statistical assumptions hold on Θ, then the interpolated processes $\vartheta^\epsilon(\cdot)$ of the projected stochastic approximation

$$\theta_{n+1} = \Pi_\Theta(\theta_n + \epsilon_n Y_n), \quad \theta_0 \in \theta,$$

converge in distribution, as $\epsilon \to 0$, to a limit process that is continuous w.p.1 and satisfies the projected ODE given by replacing $G(\cdot)$ by $(\Pi_\Theta G)(\cdot)$ in (5.13).

Provided that the above holds, Y_n^ϵ can be replaced by updates obtained via streaming/consecutive batching, that is, Y_n^ϵ can be replaced by $Y_n^{\epsilon,K}$, for any $K \geq 1$, where

$$Y_n^{\epsilon,K} = \frac{1}{K} \sum_{i=0}^{K-1} \phi(\xi_{Kn+i}^\epsilon, \theta_n^\epsilon),$$

refer to (5.12).

Remark 5.3. Assumptions (a1), (a2), and (a3) in Theorem 5.3 refer specifically to the process with fixed θ and are usually easier to verify than assumptions (a4) and (a5). The uniform integrability assumption is often replaced by the weaker condition that $\{Y_n^\epsilon\}$ has a uniformly bounded variance, which implies assumption (a4) and is often easier to verify with most models. For the latter, it is common to use a stochastically dominating process to find uniform bounds. Example 5.7 later on will illustrate this method.

Proof. The method of proof follows three stages and will use the same ideas presented in Chapter 2, as well as adequate treatment of the noise. We first deal with the unbiased case.

Telescopic sum and integral representation: From the SA algorithm,

$$\vartheta^\epsilon(t+s) - \vartheta^\epsilon(t) = \epsilon \sum_{n=m(t)}^{m(t+s)-1} Y_n^\epsilon.$$

From (a4), the processes $\{\vartheta^\epsilon(\cdot), \epsilon > 0\}$ are tight in the space $D^d[0,\infty)$ of cadlag processes under the Skorokhod topology; see Definition B.31. Moreover, using Theorem B.13, any limit process $\vartheta(\cdot)$ as $\epsilon \to 0$ is Lipschitz continuous w.p.1. Theorem B.13 is the stochastic analogue of Ascoli-Arzelà Theorem. We now apply the compactness argument of proof.

STOCHASTIC APPROXIMATION: MARKOVIAN DYNAMICS 133

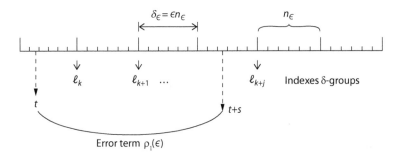

Figure 5.3. Grouping into batches of n_ϵ terms for averages. Batches are indexed by ℓ, $\ell+1, \ldots \ell+j$. Error in the approximation using whole batches is of order $O(\epsilon)$.

From now until further notice, pick a (weakly) convergent subsequence, labeled by $(\epsilon_k, k \to \infty)$, so that
$$\vartheta^{\epsilon_k}(\cdot) \stackrel{d}{\Longrightarrow} \vartheta(\cdot),$$
where $\vartheta(\cdot)$ is Lipschitz continuous w.p.1. Because of our already cumbersome indexing notation, we will omit the subindex k for the chosen ϵ-subsequence.

Rewrite now the telescopic sum regrouping terms into larger subintervals (batches), each containing n_ϵ intervals of size ϵ each (see Figure 5.3), that is,
$$\delta_\epsilon = \epsilon\, n_\epsilon,$$
yielding
$$\vartheta^\epsilon(t+s) - \vartheta^\epsilon(t) = \sum_{l=\lfloor t/\delta_\epsilon \rfloor}^{\lfloor (t+s)/\delta_\epsilon \rfloor} \epsilon \sum_{n=ln_\epsilon}^{(l+1)n_\epsilon - 1} Y_n^\epsilon + \rho_1(\epsilon)$$
$$= \sum_{l=\lfloor t/\delta_\epsilon \rfloor}^{\lfloor (t+s)/\delta_\epsilon \rfloor} \delta_\epsilon \left(\frac{1}{n_\epsilon} \sum_{n=ln_\epsilon}^{(l+1)n_\epsilon - 1} Y_n^\epsilon \right) + \rho_1(\epsilon),$$
where ρ_1 is the end point error from the larger intervals of size δ_ϵ, and it satisfies
$$\|\rho_1(\epsilon)\| \le \|\vartheta^\epsilon(t+s) - \vartheta^\epsilon(\delta_\epsilon \lfloor (t+s)/\delta_\epsilon \rfloor)\| + \|\vartheta^\epsilon(t) - \vartheta^\epsilon(\delta_\epsilon \lfloor t/\delta_\epsilon \rfloor)\|.$$

Choose now the regrouping such that
$$\lim_{\epsilon \to 0} \delta_\epsilon = 0, \quad \lim_{\epsilon \to 0} n_\epsilon = +\infty,$$
as illustrated above, and notice that an obvious choice is $\delta_\epsilon = \sqrt{\epsilon}, n_\epsilon = 1/\sqrt{\epsilon}$.

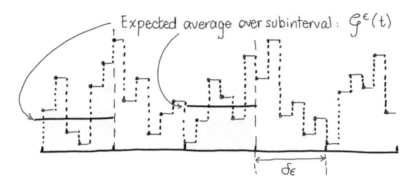

Figure 5.4. Visualization of integral representation.

Under this choice, $\|\rho_1(\epsilon)\| \stackrel{d}{\Longrightarrow} 0$, uniformly in t (see Exercise 5.2). If the updates were iid, we could now use the Law of Large Numbers for the averages of the Y_n^ϵ and (a3) to obtain the desired result as an integral. Because our model is more realistic, it allows for feedback dependencies between the control variables θ_n^ϵ and the updating variables Y_n^ϵ, and we proceed with a more careful analysis. Following the same idea as in the proof of Theorem 4.3, we will now use conditional expectations in order to use the assumption (a4) that relates the feedback Y_n with the function g. For the following expression, use the property known as the *law of iterated expectations*, which is property (b) of Theorem B.6 in Appendix B.

By construction, the process $\{\vartheta^\epsilon(\cdot)\}$ is adapted to the filtration $\{\tilde{\mathfrak{F}}_t^\epsilon\}$, where $\tilde{\mathfrak{F}}_t^\epsilon = \mathfrak{F}_{\lfloor t/\epsilon \rfloor}^\epsilon\}$, and therefore for any indices k, l with $l \geq \lfloor t/\delta_\epsilon \rfloor$ and $k \leq \lfloor t/\epsilon \rfloor$,

$$\mathbb{E}\left[\frac{1}{n_\epsilon}\sum_{n=ln_\epsilon}^{(l+1)n_\epsilon - 1} Y_n^\epsilon \,\Big|\, \mathfrak{F}_k^\epsilon\right] = \mathbb{E}\left[\mathbb{E}\left[\frac{1}{n_\epsilon}\sum_{n=ln_\epsilon}^{(l+1)n_\epsilon - 1} Y_n^\epsilon \,\Big|\, \mathfrak{F}_{ln_\epsilon - 1}^\epsilon\right]\,\Big|\, \mathfrak{F}_k^\epsilon\right]$$

$$= \mathbb{E}\left[\mathbb{E}\left[\frac{1}{n_\epsilon}\sum_{n=ln_\epsilon}^{(l+1)n_\epsilon - 1} \mathbb{E}[Y_n^\epsilon \,|\, \mathfrak{F}_{n-1}^\epsilon]\,\Big|\, \mathfrak{F}_{ln_\epsilon - 1}^\epsilon\right]\,\Big|\, \mathfrak{F}_k^\epsilon\right],$$

$$= \mathbb{E}\left[\mathbb{E}\left[\frac{1}{n_\epsilon}\sum_{n=ln_\epsilon}^{(l+1)n_\epsilon - 1} \tilde{g}(\xi_{n-1}^\epsilon, \theta_n^\epsilon)\,\Big|\, \mathfrak{F}_{ln_\epsilon - 1}^\epsilon\right]\,\Big|\, \mathfrak{F}_k^\epsilon\right], \quad (5.14)$$

where we have used (5.12). To express our telescopic sum as an integral, we now define the piecewise constant function:

$$\mathcal{G}^\epsilon(u) = \frac{1}{n_\epsilon}\sum_{n=ln_\epsilon - 1}^{(l+1)n_\epsilon} \mathbb{E}\left[\tilde{g}(\xi_{n-1}^\epsilon, \theta_n^\epsilon)\,\Big|\, \mathfrak{F}_{ln_\epsilon - 1}^\epsilon\right], \quad \text{for } u \in [l\delta_\epsilon, (l+1)\delta_\epsilon), \quad (5.15)$$

and notice that $\mathcal{G}^\epsilon(l\delta_\epsilon)$ is a random variable depending on $(\theta_{ln_\epsilon}^\epsilon, \xi_{ln_\epsilon}^\epsilon)$. Then we have

$$\mathbb{E}\left[\vartheta^\epsilon(t+s) - \vartheta^\epsilon(t)\,\Big|\, \tilde{\mathfrak{F}}_t^\epsilon\right] = \mathbb{E}\left[\int_t^{t+s} \mathcal{G}^\epsilon(u)\,du \,\Big|\, \tilde{\mathfrak{F}}_t^\epsilon\right] + \rho_1(\epsilon). \quad (5.16)$$

The idea is to characterize the limit function $\vartheta(\cdot)$ along the chosen convergent subsequence.

The Averaging Result: In this stage of the proof, it is shown that assumptions (a1), (a2), (a3), and (a5) imply that for each u,

$$\mathcal{G}^\epsilon(u) \xrightarrow{d} G(\vartheta(u)), \qquad (5.17)$$

as $\epsilon \to 0$ along the chosen weakly convergent subsequence. We give now the idea of the proof. Fix a value $u \in \mathbb{R}^+$. Then for each $\epsilon > 0$ there is a unique index ℓ_ϵ such that $\ell_\epsilon n_\epsilon \leq u < (\ell_\epsilon + 1)n_\epsilon$. Use the Skorohod representation in Theorem B.12 (Appendix B) to define a sequence $\tilde{\vartheta}^\epsilon(\cdot)$ such that it converges to the chosen limit w.p.1. Because the limit is a.s. Lipschitz continuous and $\delta_\epsilon \to 0$, then

$$\sup \left(\|\tilde{\vartheta}^\epsilon(s) - \vartheta(u)\| : s \in [\ell_\epsilon \delta_\epsilon, (\ell_\epsilon + 1)\delta_\epsilon) \right) \to 0.$$

Use Assumption (a1) and continuity of g to replace the process $\{(\xi_{n-1}^\epsilon, \theta_n^\epsilon)\}$ in the definition of $\mathcal{G}^\epsilon$ by the process at fixed value $\vartheta(u)$, that is,

$$\mathcal{G}^\epsilon(u) \approx \frac{1}{n_\epsilon} \sum_{n=\ell_\epsilon n_\epsilon - 1}^{(\ell_\epsilon+1)n_\epsilon} \mathbb{E}\left[\tilde{g}(\xi_{n-1}(\vartheta(u)), \vartheta(u)) \mid \mathfrak{F}_{\ell_\epsilon n_\epsilon - 1}^\epsilon \right]. \qquad (5.18)$$

Assumptions (a2) and (a5) are needed in order to ascertain that the initial value $(\xi_{\ell_\epsilon n_\epsilon - 1}^\epsilon, \theta_{\ell_\epsilon n_\epsilon}^\epsilon)$ has a limit distribution (the mass "does not go away"; see Theorem B.11 in Appendix B) and that this limit distribution is actually the stationary measure $\pi_{\vartheta(u)}$. Finally, use (a3) to establish the claim (5.17).

The Martingale Representation: The proof is now completed by showing that (5.16) and (5.17) imply that the limit of the chosen weakly convergent subsequence satisfies

$$\vartheta(t+s) - \vartheta(t) = \int_t^{t+s} G(\vartheta(u)) \, du, \qquad (5.19)$$

where ϑ is the limit process of the chosen subsequence. The limit is the (unique) solution to the ODE (5.13), and because this limit is the same for any such convergent subsequence, the claim follows.

To show (5.19), define the process:

$$M^\epsilon(t) = \vartheta^\epsilon(t) - \vartheta^\epsilon(0) - \int_0^t \mathcal{G}(\vartheta^\epsilon(u)) \, du.$$

We will now use the martingale characterization Theorem B.17 (Appendix B). Notice that for any $p \in \mathbb{N}$, $s_i \leq t; i \leq p$, and any bounded and continuous function h,

$$\mathbb{E}\left[h(\vartheta^\epsilon(s_i), i \leq p) \left(\vartheta^\epsilon(t+s) - \vartheta^\epsilon(t) - \int_t^{t+s} \mathcal{G}^\epsilon(u) \, du \right) \right] =$$

$$\mathbb{E}\left[h(\vartheta^\epsilon(s_i), i \leq p) \mathbb{E}\left[\vartheta^\epsilon(t+s) - \vartheta^\epsilon(t) - \int_t^{t+s} \mathcal{G}^\epsilon(u) \, du \bigg| \mathfrak{F}_{\lfloor t/\epsilon \rfloor}^\epsilon \right] \right] = O(\mathbb{E}[\rho_1(\epsilon)]),$$

for all ϵ such that $s_p \leq \epsilon \lfloor t/\epsilon \rfloor < t$. Along the chosen subsequence, $\mathbb{E}[\|\rho_1(\epsilon)\|] \to 0$, and $\vartheta^\epsilon \xrightarrow{d} \vartheta$, so that

$$\mathbb{E}\left[h(\vartheta(s_i), i \leq p) \left[\vartheta(t+s) - \vartheta(t) - \int_t^{t+s} G(\vartheta(u)) \, du \right] \right] = 0,$$

thus $M^\epsilon \xrightarrow{d} M$ where

$$M(t) = \vartheta(t+s) - \vartheta(t) - \int_t^{t+s} G(\vartheta(u)) \, du$$

is a martingale. Finally, Lipschitz continuity of the limit process $\vartheta(\cdot)$ implies Lipschitz continuity of $M(\cdot)$. Using now Theorem B.16 (Appendix B), because $M(\cdot)$ is a martingale which is Lipschitz continuous a.s. and which satisfies $M(0) \equiv 0$, then $M(t) \equiv 0$ a.s., which establishes that the limit process $\vartheta(\cdot)$ satisfies the ODE:

$$\frac{dx(t)}{dt} = G(x(t)).$$

The generalization to batches is a direct consequence of Lemma 5.1.

We now consider the case of a biased algorithm. Introducing bias in (5.14) leads to an additional term in (5.15). By virtue of (a6) the error introduced is of the form $\rho_2(\epsilon)$ where $\mathbb{E}[\rho_2(\epsilon)]$ tends to zero as $\epsilon \to 0$. Hence, we may apply the same lien of argument as before while including bias.

For the extension of the proof to the projected case involving the correction force, we refer to Theorem 4.4 in Section 8.4.3 in [198]. □

The statement of assumption (a2) in Theorem 5.3 is much stronger than needed in practice. One evident case is when a projected algorithm is used over a compact set Θ, so that all values of the parameter $\theta_n^\epsilon \in \Theta$ are in a compact set. In that case, it is enough that the invariant measure exists for $\theta \in \Theta$. For other values of $\theta \notin \Theta$ it may be that the invariant measure does not exist. This is the subject of our following examples.

Example 5.7. Consider a $GI/G/1$ queueing system (a setting similar to Example 3.6): inter arrival times are iid $\{A_n\}$, and there is a single server that uses an amount $S_n(\theta)$ of time to serve the n-th client. Clients queue up until they find the server free (first come, first serve discipline). Assume that $\{S_n(\theta)\}$ is an iid sequence with $\theta = \mathbb{E}[S_n(\theta)]$, and also assume that $\text{Var}(S_1(\theta)) < \infty$ for $\theta \geq 0$. Assume that $\mathbb{E}[A_1] < \infty$, $\text{Var}(A_1) < \infty$, and that the service time distribution has a Lebesgue density $f_\theta(s)$, which is continuously differentiable in θ for all s. It is well known that the queue process is *stable* only when

$$\theta \in \Theta = \{\theta \in \mathbb{R}^+ : \theta < \mathbb{E}[A_1]\}. \tag{5.20}$$

By stable, we mean that the queue process is ergodic and has a unique stationary distribution. For this type of problem, the stationary problem is well-posed only for $\theta \in \Theta$.

The consecutive waiting times of customers constitute a Markov process given by the so-called Lindley recursion. Let ξ_n be the waiting time of customer n, then

$$\xi_n^\epsilon = (\xi_{n-1}^\epsilon + S_{n-1}(\theta_{n-1}^\epsilon) - A_n)_+, \tag{5.21}$$

where $(x)_+ = \max(0, x)$ is the positive part of the number x. It is well known that for stable parameters $\theta \in \Theta$ the Markov chain is ergodic, with a unique stationary measure. We will discuss this example in detail in Part II and will provide a proof of (5.21) in Example 7.16.

The problem is to adjust the server's speed so that the stationary average wait per client is no higher than a given amount α, while minimizing the cost of operation, which increases as θ decreases. Because it is costly to speed up the service, the solution to the problem is given when the constraint is active. For $\theta \in \Theta$, let

$$L(\theta) = \lim_{n \to \infty} \mathbb{E}_\theta[\xi_n(\theta)] = \lim_{N \to \infty} \frac{1}{N} \sum_{k=1}^N \mathbb{E}[\xi_n(\theta)]$$

be the stationary waiting time, which is strictly increasing in θ. Then we seek θ^* such that $L(\theta^*) = \alpha$. For this model, there is a unique value $\theta^* \in \Theta$ that solves the problem.

Therefore we use a root finding scheme to optimize the server. The target vector field is $G(\theta) = \alpha - L(\theta)$.

Thus the server must adjust θ to solve the inverse problem $L(\theta^*) = \alpha$. Following our root finding examples put forward in Section 3.2, we propose to use

$$\theta^\epsilon_{n+1} = \theta^\epsilon_n - \epsilon(\xi^\epsilon_n - \alpha),$$

so that $Y_n = -(\xi^\epsilon_n - \alpha)$, for $\theta_n \in \Theta$ and we use projection otherwise, as we will explain below. Call $\mathfrak{F}^\epsilon_n$ the σ-algebra generated by $(\theta^\epsilon_0; \xi^\epsilon_1, \ldots, \xi^\epsilon_n)$. There are several important differences between this problem and the static problem discussed Chapter 4:

- The feedback has a bias that depends on the distribution of ξ^ϵ_{n-1}, that is,

$$\mathbb{E}[Y_n \mid \mathfrak{F}^\epsilon_{n-1}] = -\mathbb{E}[(\xi^\epsilon_{n-1} + S_n(\theta^\epsilon_n) - A_n)_+ \mid \xi^\epsilon_{n-1}, \theta^\epsilon_n] + \alpha \neq \alpha - L(\theta^\epsilon_n) = G(\theta^\epsilon_n),$$

where the expectation is unbiased when ξ_{n-1} has the stationary distribution.
- Consecutive observations are correlated: the process ξ_n and the control values θ_n are coupled via a Markov process.

We therefore use the stationary problem formulation to assess the convergence of the algorithm. In order to apply Theorem 5.3, we need to verify the assumptions.

For this example, we will assume that the optimal value θ^* can be argued to be in the interval $\bar{\Theta} = [0, \mathbb{E}[A_1] - \delta] \subset S$, for a known small quantity δ, so that we can use the projected version of the algorithm to ensure that $\theta^\epsilon_n \in \bar{\Theta}$ at all times. We now verify assumption (a1) in Theorem 5.3. Let $\theta \in \bar{\Theta}$ and, for ease of notation, remove the dependency on θ of the Markov chain $\{\xi_n\}$. Given $\xi_n = \xi$, the variable ξ_{n+1} has a mixed distribution with a mass at zero. Call $p_\theta(\xi, \xi')$ the density for $\xi' > 0$. Then we can calculate

$$p_\theta(\xi, 0) = \mathbb{E}\left[\mathbb{P}(A_{n+1} > \xi + x \mid S_n(\theta) = x)\right] = 1 - \int_0^\infty F_a(\xi + x) f_\theta(x)\, dx$$

$$p_\theta(\xi, \xi') = \int_0^\infty f_a(x + \xi - \xi')\, f_\theta(x) dx, \qquad \text{for } \xi' > 0,$$

where f_a and F_a are the density and distribution of the inter-arrival times, respectively. The above transition probability is weakly continuous in (ξ, θ).

We need to verify that (a2) holds on $\bar{\Theta}$. We illustrate how a *stochastic domination* argument can be used to this end. It will show that for the stationary measure at any $\theta \in \bar{\Theta}$, the system cannot be "worse" than the system with $\underline{\theta} = \mathbb{E}[A_1] - \delta$. In what follows, we show that $\mathrm{Var}(W(\theta)) \leq \mathrm{Var}(W(\underline{\theta})) \stackrel{\text{def}}{=} V < \infty$, yielding (a2).

Stochastic Domination Argument: Consider the representation of *common random numbers* (CRN), where $S_n(\theta) = F^{-1}_\theta(U_n)$ and $\{U_n; n \in \mathbb{N}\}$ are iid uniform in $(0, 1)$. It follows that $S_n(\cdot)$ is increasing a.s. in θ. Consider any value $\theta \in \Theta$, and let $\tau_1(\underline{\theta}) = \min(n > 1 : \xi_{n+1}(\underline{\theta}) = 0)$. Using CRN for the service times, and (5.21), $\xi_n(\theta) \leq \xi_n(\underline{\theta})$ a.s. Because all customers have smaller service times in the θ-system, it follows that the end of the first cycle for the $\underline{\theta}$-system must also be the end of a cycle for the θ-system (no wait in the slower server system implies no wait for the faster one). That is, there is an integer $m \leq \tau_1(\underline{\theta}) < \infty$ a.s. such that $\tau_m(\theta) = \tau_1(\underline{\theta})$. In words, this means that there are m θ-cycles of the system with faster service times θ within the first cycle of the "worst" system, at $\underline{\theta}$. Use now the regenerative formulas that imply, for any non decreasing, non-negative function f, that

$$\mathbb{E}[f(W(\theta))] = \frac{\mathbb{E}[\sum_{n=1}^{\tau_m(\theta)} f(\xi_n(\theta))]}{\mathbb{E}[\tau_m(\theta)]} \leq \frac{\mathbb{E}[\sum_{n=1}^{\tau_1(\underline{\theta})} f(\xi_n(\underline{\theta}))]}{\mathbb{E}[\tau_1(\underline{\theta})]} = \mathbb{E}[f(W(\underline{\theta}))],$$

where the equality on the left follows from applying the renewal property to $\tau_m(\theta)$ rather than $\tau_1(\theta)$, and the inequality follows from $\mathbb{E}[\tau_m(\theta)] = \mathbb{E}[\tau_1(\underline{\theta})]$ and $f(\xi_n(\theta)) \leq f(\xi_n(\underline{\theta}))$, which is due f being non decreasing and $\xi_n(\theta) \leq \xi_n(\underline{\theta})$ a.s. This yields as a particular case that the stationary variance is bounded by that at $\underline{\theta}$ for all $\theta \in \bar{\Theta}$, which in turn implies tightness of $\{\pi_\theta, \theta \in \bar{\Theta}\}$.

Assumption (a3) follows from the ergodicity of $\{\xi(\theta)\}$ and the definition of $G(\theta)$, so we need now verification of (a4) and (a5). Because we are truncating, the same stochastic domination argument as above can be used to establish that $\text{Var}(Y_n)$ is uniformly bounded, and this, in turn, also yields tightness of the sequence $\{(\xi_n^\epsilon, \theta_n^\epsilon)\}$. ✲✲✲

Example 5.8. In the previous example, we considered a projected version of the algorithm. A first observation is that the target ODE is coercive for $\theta < 0$ as well. Indeed, in this case the expected service would be negative, which can be realized by letting $S_n(\theta) = -S_n(|\theta|)$, for $\theta < 0$, and so that for $\theta < 0$ the traget ODE moves to the right as it should be.

We now turn to upper bound of the projection interval. It is sometimes the case that a bound such as $\mathbb{E}[A_1] - \delta$ cannot be found, but it can be argued that the field is coercive and that the target ODE stays within the stability region. This is a more realistic scenario because the arrival distribution may not be known. Thus, the server will be adjusting θ as it operates, using only the consecutive arrival times, which can be observed.

The first assumption is verified as in the previous example, but tightness and uniform integrability are usually the most difficult conditions to verify in practice.

We illustrate with this example a useful technique that parallels Lyapunov stability ideas and uses convergence of super (sub) martingales to ensure recurrence of the process $\{\xi_n^\epsilon\}$ when θ_n follows (5.11). Propose a stochastic Lyapunov function, given $\mathfrak{F}_n^\epsilon$:

$$V(\theta_n) = \frac{1}{2}(\xi_n^\epsilon - \alpha)^2,$$

which given $\mathfrak{F}_n$, is a function of θ_n and of the exogenous random variable A_{n+1}.

Consider the evolution of the process $\{\xi_n^\epsilon\}$ within a single busy cycle. Let $\tau = \min(n > 0: \xi_n^\epsilon = 0)$ (i.e., for all $n < \tau$, $S_n(\theta_n^\epsilon) > A_{n+1} - \xi_n^\epsilon$) and consider the stopped process $V(\theta_{\tau \wedge n})$. Use a Taylor approximation with remainder for $V(\cdot)$ to obtain, for $n \leq \tau$,

$$\mathbb{E}[V(\theta_{n+1}) - V(\theta_n) \mid \mathfrak{F}_n] = \mathbb{E}\left[(\xi_n^\epsilon - \alpha)(\theta_{n+1}^\epsilon - \theta_n^\epsilon)S'(\theta_n^\epsilon) \mid \mathfrak{F}_n^\epsilon\right] + O(\epsilon^2)$$
$$= -\epsilon \mathbb{E}\left[(\xi_n^\epsilon - \alpha)^2 S'(\theta_n^\epsilon) \mid \mathfrak{F}_{n-1}\right] + O(\epsilon^2),$$

which for small enough ϵ is non positive, proving that $V(\cdot)$ is a non-negative local super martingale within cycles. On the other hand, $V(\theta_\tau) = 0.5\alpha^2$, a constant. Thus, the state $\{\xi_n^\epsilon = 0\}$ must be a positive recurrent state. This shows that $\{\xi_n\}$ are tight, thus uniformly integrable, and also that $\theta_n^\epsilon \in S$ infinitely often. This condition implies (a5), and using the assumption of finite variance for stable mean service times, it also implies (a4).

Because the probability $\mathbb{P}(\theta_n^\epsilon \notin \Theta) \to 0$ as $n \to \infty$ for small ϵ, we need only verify (a2) in compact sets inside Θ, which we did already in the previous example. To justify this claim, we point out that assumption (a2) is required only when establishing that assumption (a3) can be applied to the right-hand side of (5.18). Although we assume here that the stochastic approximation is not truncated, so that it is possible that the values of θ_n^ϵ may wander outside the stable region, it follows from the tightness argument above that this event is not persistent. In order to establish that assumption (a2) is required only inside the stability region, proceed by contradiction. If the limit of the chosen subsequence satisfies that $\vartheta(u) \notin S$, then it would mean that for small enough ϵ the process θ_n^ϵ is outside the stability region for $n \in [\ell_\epsilon n_\epsilon, (\ell_\epsilon + 1) n_\epsilon)$, which would imply that ξ_n^ϵ is a not a recurrent state, contradicting the result above.

Using these arguments, we may program algorithms without the need for truncation, because the vector field "points away" from the undesired regions. ✻✻✻

Our final theorem statement is provided without proof. The proof is based on truncation arguments and stability arguments introduced in the examples above, and it is left as exercise for the reader.

Theorem 5.4. *Conisder the well-posed NLP $\min_{\theta \in \Theta} J(\theta)$. Suppose that the vector field G is coercive for this problem on $H \subset \mathbb{R}^d$. Consider the stationary problem of Definition 5.1 for the stochastic approximation, and assume the following conditions hold:*

(a1) The transition probabilities of the fixed-θ processes $P_\theta(\xi, \cdot)$ are weakly continuous (uniformly) w.r.t. (θ, ξ) on $\mathbb{R}^d \times \Xi$. That is, for each bounded and continuous real-valued function f, and any $\rho > 0$, there is a $\delta > 0$ such that

$$\left| \int f(x) P_\theta(\xi, dx) - \int f(x) P_{\theta'}(\xi', dx) \right| \leq \rho$$

whenever $\|(\xi, \theta) - (\xi', \theta')\| < \delta$.

(a2') The stationary measure π_θ is unique, and the set $\{\pi_\theta(\cdot), \theta \in M\}$ of stationary measures of the fixed-θ processes is tight for each compact set $M \subset \mathbb{R}^d$.

(a3) The instantaneous feedback $g(\xi, \theta)$ and the one-step ahead expectation of the feedback $\tilde{g}(\xi, \theta)$ are jointly continuous measurable functions (they may depend on n, but we will introduce the simplest model here), such that for every fixed $\theta \in \mathbb{R}^d$,

$$\lim_{N \to \infty} \frac{1}{N} \sum_{n=1}^{N} \mathbb{E}[g(\xi_n(\theta), \theta)] = \lim_{N \to \infty} \frac{1}{N} \sum_{n=1}^{N} \mathbb{E}[\tilde{g}(\xi_{n-1}(\theta), \theta)]$$

$$= \int g(x, \theta) \pi_\theta(dx) = G(\theta).$$

(a4') The feedback sequence $\{Y_n^\epsilon, \epsilon > 0\}$ satisfies $\mathbb{E}[Y_n^\epsilon \mid \mathfrak{F}_{n-1}^\epsilon] = \tilde{g}(\xi_{n-1}^\epsilon, \theta_n^\epsilon) + \beta_n(\xi_{n-1}^\epsilon, \theta_n^\epsilon)$, for $\theta_n^\epsilon \in \mathbb{R}^d$, and for every compact set $M \subset \mathbb{R}^d$ there is a finite number V such that $\sup_{\theta \in M} \text{Var}(\phi(\xi(\theta), \theta)) \leq V$.

(a5) The set $\{(\xi_n^\epsilon, \theta_n^\epsilon); \epsilon > 0\}$ is tight.

(a6) The bias is asymptotically negligible, that is,

$$\lim_{n,m,\epsilon} \frac{1}{m} \sum_{i=n}^{n+m-1} \mathbb{E}\big[\|\beta_n(\xi_{n-1}^\epsilon, \theta_n^\epsilon)\| \big] = 0,$$

where the limit is taken as $n \to \infty$, $m \to \infty$ and $\epsilon \to 0$ simultaneously in any way.

Use the stochastic approximation

$$\theta_{n+1}^\epsilon = \theta_n^\epsilon + \epsilon Y_n^\epsilon \quad \theta_0^\epsilon \in H.$$

Then the interpolated processes $\vartheta^\epsilon(t) = \theta_{m(t)}$ converge in distribution, as $\epsilon \to 0$, to a limit process that is continuous w.p.1 and satisfies the ODE $dx(t)/dt = G(x(t))$ and

$\lim_{t\to\infty} \lim_{\epsilon \to 0} \theta^\epsilon_{\lfloor t/\epsilon \rfloor}$ *is a KKT point of the optimization problem.*

Alternatively, for $\Theta \subset \mathbb{R}^d$ *convex, compact and* $G(\cdot) = (\Pi_\Theta \hat{G})(\cdot)$, *for some continuous vector field* $\hat{G}$ *on* Θ, *the statement holds for the projected stochastic approximation*

$$\theta_{n+1} = \Pi_\Theta(\theta_n + \epsilon_n Y_n), \quad \theta_0^\epsilon \in \Theta \cap H,$$

provided that the above conditions hold on Θ, *and* $\hat{G}$ *in (a3) (and G replaced by* $\hat{G}$ *in (5.6)).*

Relation with the Static Model

We conclude this chapter with an example illustrating that the Markovian noise model is a conservative extension of the static model in Chapter 4. For this, let θ be a scaling parameter of F_θ (i.e., it holds that $F_\theta(x) = F_1(x/\theta)$, for $\theta \neq 0$) and let $\{\xi_n(\theta)\}$ be an iid sequence distributed according to F_θ. We now interpret $\{\xi_n(\theta)\}$ as Markov process, which gives $P_\theta(x, \cdot) = F_\theta(\cdot)$ and $F_\theta = \pi_\theta$ as the stationary distribution of P_θ. Consider the optimization problem

$$\min_\theta \frac{1}{2}\mathbb{E}[(\xi(\theta) - \alpha)^2], \qquad (5.22)$$

for some α being an inner point of the support of F_θ, and $\xi(\theta)$ a sample of F_θ. It holds that

$$\frac{d}{d\theta}\left(\frac{1}{2}\mathbb{E}[(\xi(\theta) - \alpha)^2]\right) = \mathbb{E}\left[\frac{\xi(\theta)}{\theta}(\xi(\theta) - \alpha)\right],$$

where we use the fact that $d\xi(\theta)/d\theta = \xi(\theta)/\theta$; for details we refer to Chapter 8 in Part II. Hence, we may choose

$$Y_n = \phi(\xi_n(\theta_n), \theta_n) = g(\xi_n(\theta_n), \theta_n) = \frac{\xi_n(\theta)}{\theta_n}(\alpha - \xi_n(\theta_n)),$$

and apply the SA

$$\theta_{n+1} = \theta_n + \epsilon Y_n$$

for solving (5.22), where we assume for simplicity that $\theta \in \mathbb{R}$. The target vector field is the negative derivative of the objective in (5.22), that is,

$$G(\theta) = -\frac{d}{d\theta}\left(\frac{1}{2}\mathbb{E}[(\xi(\theta) - \alpha)^2]\right) = \mathbb{E}\left[\frac{\xi(\theta)}{\theta}(\alpha - \xi(\theta))\right].$$

For this choice for Y_n we have in the Markovian noise framework, as introduced in this chapter, that

$$\mathbb{E}[Y_n | \mathfrak{F}_{n-1}] = \mathbb{E}[Y_n | \xi_{n-1}(\theta_{n-1}), \theta_n] = \tilde{g}(\xi_{n-1}, \theta_n) = \mathbb{E}\left[\frac{\xi_n(\theta)}{\theta_n}(\alpha - \xi_n(\theta_n))\right].$$

Moreover, since $\xi_n(\theta)$ is independent of $\xi_{n-1}(\theta)$, it follows

$$G(\theta) = \int g(\xi, \theta)\pi_\theta(d\xi) = \mathbb{E}_{\pi(\theta)}[g(\xi(\theta), \theta)] = \tilde{g}(\cdot, \theta),$$

for any θ. This shows that if we identify the problem as a stationary problem, the Cesàro averaging in (5.6) is superfluous. Hence, our formalism leads us to realize that the problem

5.4 PRACTICAL CONSIDERATIONS

This chapter's results show that an SA for a long-run cost function over a Markov chain does require per update only one observation of the underlying process. Hence, while consecutive batching may be applied to produce more stable outcomes, such batching is not required by theory. This allows the algorithm to move quickly, as updating the parameter does not require collecting intermediate observations. This shows that our discussion of efficient computational budget assignment put forward in Example 3.1 carries over to the general setting.

5.5 EXERCISES

Exercise 5.1. Consider the following queueing problem. Consecutive inter arrival times are continuous random variables, iid, with finite moments and unit mean. The mean service time is $\theta > 0$. Let $f_\theta(\cdot)$ be the well-defined density of the corresponding service time $S_n(\theta)$ and assume that $\text{Var}(S_n(\theta)) < \infty$ for all finite values of θ. We wish to minimize the cost of operation $C(\theta) = 1/\theta^2$ while satisfying $L(\theta) \stackrel{\text{def}}{=} \mathbb{P}(W(\theta) > w) \leq \alpha$, for a constant w, where $W(\theta)$ is a random variable with the stationary waiting time distribution and $\alpha \in (0, 1)$.

(a) Argue by using the results of Chapter 1, that at the optimal value θ^* the constraint must be active, that is, $L(\theta^*) = \alpha$.
(b) Let $\{\xi_n\}$ be the sequence of consecutive waiting times. Lindley's equation gives the dynamics of the waiting time process:

$$\xi_n = (\xi_{n-1} + S_n(\theta_n) - A_n)_+, \tag{5.23}$$

where A_n is the inter-arrival time between customers n and $n + 1$, and $S_n(\theta)$ is the service time of customer n, given θ_n. Discuss the validity of the stochastic approximation procedure:

$$\theta_{n+1} = \theta_n - \epsilon_n Y_n,$$

for Y_n an estimator of $L(\theta) - \alpha$ obtained observing the process $\{\xi_n\}$. Specify your model (what will you use for Y_n) and verify the assumptions of Theorem 5.1, assuming that $\theta_n < 1$ infinitely often for this procedure.
(c) Instead of using one observation of the process ξ_n to produce Y_n, consider using an estimation interval with K observations. That is, use $\xi_{nK}, \ldots, \xi_{(n+1)K-1}$ to produce the estimate Y_n of the n-th interval. Write a program to simulate the queue under the service time control and experiment with various values of K. Plot the results and discuss them.

Exercise 5.2. Show that $\|\rho_1(\epsilon)\| \stackrel{d}{\Longrightarrow} 0$ in the proof of Theorem 5.3 as follows. For given t, first establish that $0 < t - \delta_\epsilon \lfloor t/\delta_\epsilon \rfloor < \delta_\epsilon$, then argue that the distribution of the random variable $X^\epsilon(t) = \|\vartheta^\epsilon(t) - \vartheta^\epsilon(\delta_\epsilon \lfloor t/\delta_\epsilon \rfloor)\|$ can be made arbitrarily close to that of $X(t) = \|\vartheta(t) - \vartheta(\delta_\epsilon \lfloor t/\delta_\epsilon \rfloor)\|$ and use the fact that the limit function $\vartheta(\cdot)$ is a.s. Lipschitz continuous. Justify why the result holds uniformly in t to finalize the proof.

Exercise 5.3. Consider the model of the "automatic learning" exercise machine that adjusts the resistance to each user so as to enable them to reach their desired target heart rate (refer to Example 5.1 and Example 5.2), using the Markov chain model of Example 5.4 for consecutive heart beats:

$$\mathbb{P}(\xi_{n+1} = 1 \mid \xi_n = 0, \theta) = \theta,$$
$$\mathbb{P}(\xi_{n+1} = 1 \mid \xi_n = 1, \theta) = \theta^2.$$

(a) Suppose that your algorithm considers batches of K intervals of 0.3 sec in order to get a better estimate of the person's heart rate, so that only every K measurements, θ_n changes

$$\theta_{n+1}^\epsilon = \theta_n^\epsilon - \epsilon \left(200 \sum_{k=nK}^{(n+1)K-1} \xi_k - K\alpha. \right),$$

for $\alpha = 120$. Discuss whether projection is needed to keep θ_n^ϵ feasible. Define the interpolation process $\vartheta^\epsilon(\cdot)$ as usual, that is,

$$\vartheta^\epsilon(t) = \theta_{m(t)}, \quad m(t) = \left\lfloor \frac{t}{\epsilon} \right\rfloor.$$

Using Theorem 5.3, find the limiting ODE for this process. What is the dependency of the behavior of the limiting ODE on K?

(b) Program the procedure for $K = 1, 10, 20$ and show the plots with $\epsilon = 0.001$. Discuss your results.

(c) Program the procedure $\theta_{n+1} = \theta_n + \epsilon_n Y_n$ with decreasing stepsizes $\epsilon_n = O(1/n)$, plot and discuss the results. If you were to patent the algorithm for the fitness industry, what scheme would you choose and why?

Exercise 5.4. Show that for the problem in Example 5.7, assumption (a1) in Theorem 5.3 holds for any compact set in the stability region. This is a $GI/G/1$ queueing system: inter arrival times are iid $\{A_n\}$, there is a single server that uses an amount $S_n(\theta)$ of time to serve the n-th client. Clients queue up until they find the server free (first come, first serve discipline), and $\theta = \mathbb{E}[S_n(\theta)]$ is the mean service time. Assume that $\mathbb{E}[A_1] < \infty$, $\text{Var}(A_1) < \infty$, and that the service time distribution F has a Lebesgue density $f_\theta(s)$ which is continuously differentiable in θ. [*Hint:* using the "inverse function method" one can assume that the service times satisfy $S_n(\theta) = F_\theta^{-1}(U_n)$, where $\{U_n\}$ are iid uniformly distributed in the interval $(0, 1)$.]

Chapter Six

Asymptotic Efficiency

In this chapter we present the notion of algorithmic efficiency that contemplates the trade-off between fast execution and final expected error. The concepts are based on economic theory for loss functions. When applied to stochastic approximations, these concepts have a significant impact in terms of the hyper parameters of the algorithms. Various examples and exercises are included to illustrate application of the results.

6.1 MOTIVATION

This chapter is concerned with the concept of *algorithmic efficiency*. To set the stage, think of the problem of estimating an unknown value θ^*. Were we to choose between two different methods, what criterion should we use? We will provide a definite answer to this question, using a mathematical theory of risk and loss functions.

As an introduction to the problem, let us consider the easiest statistical estimation problem: we seek to estimate $\theta^* = \mathbb{E}[X_n]$, where X_n, $n = 1, 2, \ldots$, are iid random variables with finite variance σ^2. The sample mean is then an *unbiased* estimator:

$$\theta_n = \frac{1}{n} \sum_{i=1}^{n} X_i,$$

that by the strong law of large numbers satisfies $\theta_n \to \theta^*$ a.s, and by the CLT $\sqrt{n}(\theta_n - \theta^*) \approx \mathcal{N}(0, \sigma^2)$. Such results, which concern the "limit" *distribution* of the sequence of "centralized" random variables (thus, the name CLT), allow us to calculate how many samples should be used to attain a given precision. For instance, if we tolerate an error of $\rho = 0.01$, then we need approximately a sample size N such that

$$z_{1-\alpha/2}\sqrt{\sigma^2/N} \approx 0.01,$$

for a confidence level $1 - \alpha$. That is, with probability at least $1 - \alpha$, choosing $N \geq \sigma^2(z_{1-\alpha/2}/\rho)^2$ attains (approximately) the required precision ρ. For a summary of statistical output analysis, please see Appendix D.

An estimator will be more "efficient" than another one if it attains better precision with the *same number of samples*. This, of course, is the well-known motivation for variance reduction in statistical estimation.

The goal when building algorithms to find an "optimal" value θ^* will be to obtain the smallest error at small "cost" N. To generalize the results for general estimation problems, and particularly to stochastic approximations, we will first provide the supporting theorems for the *functional central limit theorem* (FCLT).

6.2 FUNCTIONAL CLT

It should be apparent from the convergence theorems in Chapters 4 and 5 that the behavior of the algorithms, seen as a stochastic process, "approximate" a smooth and deterministic behavior in some limiting sense. In particular, for the constant stepsize algorithm, the interpolation process converges in distribution to a deterministic process when the assumptions of Theorems 4.2, 4.3, or 5.3 are satisfied.

In this section we explore how the processes "hover" around the limit process. As a preliminary motivation, recall that the Law of Large Numbers (LLN) for iid zero-mean random variables states that

$$S_N = \frac{1}{N} \sum_{k=1}^{N} \xi_k \xrightarrow{a.s.} 0,$$

and the CLT establishes that, if $\sigma^2 = \text{Var}(\xi_1) < \infty$, then the "standardized" variable $\sqrt{N} S_N / \sigma$ approaches a standard normal random variable, or

$$\frac{1}{\sqrt{N}} \sum_{k=1}^{N} \xi_k \xrightarrow{d} \mathcal{N}(0, \sigma^2)$$

for large N. This result is at the basis of error estimation for the approximations (e.g., confidence intervals). To prepare for the extension of the CLT to the FCLT, consider the interpolation process of this same partial sum, as follows. For each fixed $\epsilon > 0$, let $m(t) = \lfloor t/\epsilon \rfloor$ and suppose that $\{\xi_k, k \in \mathbb{N}\}$ is an infinite sequence of iid zero-mean random variables. Now let's consider the interpolated process with constant stepsize ϵ, namely $\vartheta^\epsilon(t) = S_{m(t)}$. By construction, for any fixed t the sequence of random variables $\vartheta^\epsilon(t) \to 0$ as $\epsilon \to 0$. Call

$$W^\epsilon(t) = \sqrt{\epsilon} \sum_{k=1}^{m(t)} \xi_k = \sqrt{\epsilon\, m(t)} \left(\frac{1}{\sqrt{m(t)}} \sum_{k=1}^{m(t)} \xi_k \right). \tag{6.1}$$

Notice that $t - \epsilon < \epsilon m(t) \le t$, so that as $\epsilon \to 0$ we have $m(t) \to \infty$ and $\epsilon m(t) \to t$. Fix t and look at the processes $W^\epsilon(t)$ evaluated at that time t. By the usual CLT, as ϵ decreases, the random variables converge: $W^\epsilon(t) \xrightarrow{d} \mathcal{N}(0, \sigma^2 t)$. It is left as an exercise to show that $W^\epsilon(\cdot)$ converges in distribution, as a sequence of processes, to the Wiener process or *Brownian motion* $W(\cdot)$ (see Exercise 6.1). We now briefly recall some definitions and results.

Definition 6.1. A process $\{W(t), t \in \mathbb{R}\}$ on $(\Omega, \{\mathfrak{F}_t\}, \mathbb{P})$ is a vector-valued *Wiener process* (or *Brownian motion*) if there is a matrix Σ, called the *covariance matrix*, such that

(a) $W(0) = 0$, $\mathbb{E}[W(t) \mid \mathfrak{F}_t] = 0$, and $W(\cdot)$ has a.s. continuous paths,
(b) $W(\cdot) \in \mathbb{R}^k$ has *independent increments*, that is, for any set of increasing numbers $\{t_i, i \in \mathbb{N}\}$, $\{W(t_{i+1}) - W(t_i)\}$ are independent random variables,
(c) for any $t \in \mathbb{R}$ and any $s > 0$, the distribution of the increment $W(t+s) - W(t)$ is independent of t, and
(d) for any t, $\mathbb{E}[W(t)^\top W(t)] = \Sigma t$.

When $\Sigma = \mathbb{I}$ is the identity matrix the process is referred to as "standard" Wiener process.

An equivalent definition uses the fact that the increments of a Wiener processes have zero-mean normal distribution, see [44], in lieu of condition (c) above. We now state some results that will be useful when dealing with the FCLT for stochastic approximation methods. For details we refer to Theorem 1.2 in Section 10.1 of [198].

Lemma 6.1. *Let $\{Z_n^\epsilon\}$ and $\{U_n^\epsilon\}$ be sequences of $\mathbb{R}^d$-valued random variables on $(\Omega, \mathfrak{F}, \mathbb{P})$ and call $\mathfrak{F}_n^\epsilon$ the minimal σ-algebra generated by $\{(U_i^\epsilon, Z_i^\epsilon); i \leq n\}$. For $t \geq 0$, define*

$$W^\epsilon(t) = \sqrt{\epsilon} \sum_{i=0}^{\lfloor t/\epsilon \rfloor - 1} Z_i^\epsilon, \quad U^\epsilon(t) = U_n^\epsilon; \ t \in [n\epsilon, (n+1)\epsilon),$$

and suppose that $\mathbb{E}[Z_n^\epsilon \mid \mathfrak{F}_{n-1}^\epsilon] = 0$ w.p.1 for all n and ϵ. Finally, suppose that there exist a matrix Σ and an integer $p > 0$ such that

$$\sup_{n,\epsilon} \mathbb{E}[(Z_n^\epsilon)^{2+p}] < \infty, \quad \text{and} \quad \mathbb{E}\left[Z_n^\epsilon (\delta Z_n^\epsilon)^\top \mid \mathfrak{F}_{n-1}^\epsilon\right] \to \Sigma,$$

in probability, as $n \to \infty, \epsilon \to 0$. Then $W^\epsilon(\cdot)$ converges weakly to a Wiener process with covariance matrix Σ. Suppose that $(U^\epsilon(\cdot), W^\epsilon(\cdot))$ converges in distribution, in the space $D^{2d}[0, \infty)$ to a joint limit $(U(\cdot), W(\cdot))$. Then $W(\cdot)$ is a $\mathfrak{F}_t$-Wiener process, where $\mathfrak{F}_t = \sigma(U(s), W(s); s \leq t)$.

What the above lemma establishes is that the FCLT still holds when the "error terms" are no longer assumed iid, as long as they are zero-mean noise terms, and "well-behaved."

6.2.1 Fixed Stepsize

We present in this section the main result for the fixed stepsize stochastic approximation in the general setting of the Markovian dynamics model. Note that the analysis put forward in this section requires that there is no bias.

Theorem 6.1. *Consider the algorithm*

$$\theta_{n+1}^\epsilon = \theta_n^\epsilon + \epsilon Y_n^\epsilon \tag{6.2}$$

and assume all the conditions of Theorem 5.3 hold for $\Theta = \mathbb{R}^d$. Call $\vartheta(\cdot)$ the limit of the interpolation process $\vartheta^\epsilon(\cdot)$ of the recursion in (6.2), and assume that θ^* is the only stable point of the corresponding limit ODE. As usual, let $\delta M_n^\epsilon = Y_n^\epsilon - \tilde{g}(\xi_{n-1}^\epsilon, \theta_n^\epsilon,)$ be the martingale difference noise (refer to (4.3)) and assume

(a6) There is a neighborhood $B_\rho(\theta^*)$ of θ^* and a symmetric matrix Σ such that

$$\mathbb{E}\left[\delta M_n^\epsilon (\delta M_n^\epsilon)^\top \mathbf{1}_{\{\|\theta_n^\epsilon - \theta^*\| \leq \rho\}} \mid \mathfrak{F}_{n-1}^\epsilon\right] \to \Sigma$$

  in probability, as $n \to \infty, \epsilon \to 0$.
(a7) The function $\tilde{g}(\cdot, \xi)$ admits a Taylor expansion in θ:

$$\tilde{g}(\theta, \xi) = \tilde{g}(\theta^*, \xi) + \nabla_\theta \tilde{g}(\theta^*, \xi)(\theta - \theta^*) + \rho_1(\theta, \xi),$$

  where the error term satisfies $\mathbb{E}[\rho_1(\theta, \xi_n^\epsilon)] = O(\|\theta - \theta^*\|^2)$ as $n \to \infty, \epsilon \to 0$.
(a8) There is a Hurwitz matrix A (i.e., a matrix where all the eigenvalues have a negative real part) such that

$$\lim_{m \to \infty} \frac{1}{m} \sum_{i=n}^{n+m-1} \mathbb{E}\left[\nabla_\theta \tilde{g}(\xi_{n-1}^\epsilon, \theta^*) - A\right] = 0.$$

Define the continuous interpolation cadlag[1] processes:

$$U_n^\epsilon = \frac{\theta_n^\epsilon - \theta^*}{\sqrt{\epsilon}}, \quad U^\epsilon(t) = U_n^\epsilon; \ t \in [n\epsilon, (n+1)\epsilon),$$

$$W^\epsilon(t) = \sqrt{\epsilon} \sum_{i=0}^{\lfloor t/\epsilon \rfloor - 1} \delta M_i^\epsilon.$$

Then the sequence $\{(U^\epsilon(\cdot), W^\epsilon(\cdot))\}$ converges weakly in $D^{2d}[0, \infty)$ to a limit $(U(\cdot), W(\cdot))$ satisfying

$$dU(t) = AU(t)\,dt + dW(t), \tag{6.3}$$

and $W(t)$ is a Wiener process with covariance matrix Σ.

Proof. We will only sketch the steps of the proof, for details the reader is referred to Chapter 10 of [198]. Rewrite the recursion as

$$\theta_{n+1}^\epsilon - \theta_n^\epsilon = \epsilon \tilde{g}(\xi_{n-1}^\epsilon, \theta_n^\epsilon) + \epsilon \delta M_n^\epsilon.$$

The proof now proceeds in three steps: first, we "replace" the random argument θ_n^ϵ by θ^* in the function g. Second, we use the expressions to write the recursion for the sequence U_n^ϵ and express an integral form for it. Third, we characterize the limits using Lemma 6.1.

Expansion around θ^*: Using (a7), we express

$$\theta_{n+1}^\epsilon - \theta_n^\epsilon = \epsilon \big(\tilde{g}(\xi_{n-1}^\epsilon, \theta^*) + \nabla_\theta \tilde{g}(\xi_{n-1}^\epsilon, \theta^*)(\theta_n^\epsilon - \theta^*)\big) + \epsilon \delta M_n^\epsilon + \epsilon \rho_1(\xi_{n-1}^\epsilon, \theta_n^\epsilon),$$

and notice that the error term satisfies $\mathbb{E}[\rho_1(\xi_{n-1}^\epsilon, \theta)] = O(\mathbb{E}[\|\theta_n^\epsilon - \theta^*\|^2])$.

Integral representation: From the definition of U_n^ϵ, we have

$$U_{n+1}^\epsilon - U_n^\epsilon = \sqrt{\epsilon} \tilde{g}(\xi_{n-1}^\epsilon, \theta^*) + \sqrt{\epsilon}\delta M_n^\epsilon + \epsilon A U_n^\epsilon + \epsilon \big(\nabla_\theta \tilde{g}(\xi_{n-1}^\epsilon, \theta^*) - A\big) U_n^\epsilon + \sqrt{\epsilon} \rho_1(\theta_n^\epsilon, \xi_n^\epsilon).$$

Adding up the terms in left and right-hand sides,

$$U^\epsilon(t+s) - U^\epsilon(t) = \int_t^{t+s} A U^\epsilon(u)\,du + W^\epsilon(t) + \sqrt{\epsilon} \sum_{i=m(t)+1}^{m(t+s)} \tilde{g}(\xi_i^\epsilon, \theta^*)$$

$$+ \epsilon \sum_{i=m(t)}^{m(t+s)-1} \big(\nabla_\theta \tilde{g}(\theta^*, \xi_n^\epsilon) - A\big) U_n^\epsilon + \rho(\epsilon),$$

where $m(t) = \lfloor t/\epsilon \rfloor$ also depends on ϵ. Note that the number of terms in the sums is $m(t+s) - m(t) \approx s/\epsilon$. The cumulative error term $\rho(\epsilon)$ converges to zero in mean square error, as $\epsilon \to 0$.

Averaging: The last (and more technical) step carries out the "long term averaging" effects. Indeed, for fixed θ, from the Markovian model assumptions (a1), (a2), and (a3) from Theorem 5.3, the quantities $\tilde{g}(\xi_n, \theta)$ converge geometrically fast to their stationary

---

[1] A continuous time stochastic process $\{\vartheta(t)\}$ on $\mathbb{R}^d$ is called a *cadlag* process if it is right-continuous with left limits at every point t; see Defintion B.31.

expectation, that is, $G(\theta^*) = 0$. This result is used to establish that

$$\sqrt{\epsilon} \sum_{i=m(t)+1}^{m(t+s)} \tilde{g}(\xi_i^\epsilon, \theta^*)$$

converges to zero in (absolute-value) expectation. As well, assumption (a7) will establish that the contribution of the term

$$\epsilon \sum_{i=m(t)+1}^{m(t+s)} \left(\nabla_\theta \tilde{g}(\xi_n^\epsilon, \theta^*) - A\right) U_n^\epsilon$$

also vanishes in the limit. Applying now Lemma 6.1, the processes $(U^\epsilon(\cdot), W^\epsilon(\cdot))$ have a limit $(U(\cdot), W(\cdot))$ where the limit $W(\cdot)$ is a Wiener process with covariance Σ. Under the tightness conditions (a4) and (a5) put forward in Theorem 5.3, averaging of the error terms, the limit must then satisfy

$$\lim_{\epsilon \downarrow 0}(U^\epsilon(t+s) - U^\epsilon(t)) = U(t+s) - U(t) = \int_t^{t+s} AU(u)\, du + W(t),$$

which is the same as (6.3). □

The process $U(\cdot)$ that satisfies (6.3) is called an *Ornstein-Uhlenbeck* process. Theorem 6.1 is important to obtain limiting estimates for confidence intervals. The stationary Ornstein-Uhlenbeck process $U(t)$ defined by $dU(t) = AU(t)dt + dW(t)$ has a normal distribution with variance V satisfying the implicit equation (see [26, 82])

$$AV + VA^\top = -\Sigma. \tag{6.4}$$

This is sometimes called the continuous Lyapunov equation and common high level languages such as Mathematica and Matlab have numerical solvers for it, but we need to know A and Σ.

Let us now discuss how we can use this result for the case of a one-dimensional problem. For $A < 0$ the one-dimensional Ornstein-Uhlenbeck process is of the general form

$$dU(t) = A(U(t) - \mu)\, dt + B dW(t),$$

with W a standard one-dimensional Wiener process, where[2] $B = \sqrt{\Sigma}$, for $\Sigma > 0$, and it satisfies

$$\mathbb{E}[U(t)] = U(0)e^{At} + \mu(1 - e^{At}), \quad \text{which yields } \lim_{t \to \infty} \mathbb{E}[U(t)] = \mu,$$

and by (6.4) it follows that

$$\lim_{t \to \infty} \text{Var}(U(t)) = -\frac{\Sigma}{2A} = V. \tag{6.5}$$

In words, V is the asymptotic covariance of the outcome θ_n^ϵ of the SA for ϵ small and n large.

Recall that we denote the target vector field of the martingale noise difference model by $G(\theta)$. In our case the limit process is centered at $\mu = 0$ and the drift $A = G'(\theta^*) < 0$ (under stability condition (a8)), therefore a given precision can be specified for the limit ODE, by choosing t such that e^{At} is small enough. Accordingly, for sufficiently small ϵ, after $n = \lfloor t/\epsilon \rfloor$

---

[2]In the higher dimensional case, B is a matrix that solves $BB^\top = \Sigma$.

iterations the process will have near stationary values with stationary variance $V = -\Sigma/(2A)$, where Σ is the variance of the noise at the solution θ^*, see (a6). Thus Theorem 6.1 can be used to approximate the distribution of the error:

$$\frac{(\theta_n^\epsilon - \theta^*)}{\sqrt{\epsilon}} \approx \mathcal{N}(0, V) \implies \theta_n^\epsilon \approx \mathcal{N}(\theta^*, -\epsilon\Sigma/(2A)),$$

and this can be used to construct the confidence interval $\theta_n^\epsilon \pm z_{\alpha/2}\sqrt{\epsilon V}$ at significance level α for θ^*, with z_a the a-quantile of the standard normal distribution and V as in (6.5). See Appendix D for details.

Example 6.1. Suppose that we use stochastic approximation for a robot that is supposed to track the location of a target (e.g., in a rescue mission, or deep under the ocean). The target's location is estimated by the robot's sensors via noisy thermal, sonar or optical signals $\{\xi_n\}$, assumed iid, with mean α. The simple root-finding equation

$$\theta_{n+1}^\epsilon = \theta_n^\epsilon - \epsilon(\theta_n^\epsilon - \xi_n)$$

will satisfy the conditions of Theorem 5.3 under regularity assumptions on the noise, for example, under a simple model where each component of the location estimate has constant variance: $\text{Var}(\xi_{n,i}) = \sigma^2 < \infty$. In this case, the model is of exogenous noise, so the values of θ_n do not affect the transition probabilities of ξ_n, which are iid. Here $A = -\mathbb{I}$ is Hurwitz, so that the asymptotic variance is $V = \frac{\sigma^2}{2}\mathbb{I}$. Estimating $\Sigma = \sigma^2\mathbb{I}$ concurrently with the stochastic approximation is straightforward using sample variance of the observations, so that at the end of a large number N of iterates one can provide a confidence interval of the form $\widehat{\alpha_i} = \theta_{N,i} \pm 1.96\sqrt{\frac{\epsilon\widehat{\sigma^2}}{2}}$, for each component α_i of the target vector location. In this case independence of the noisy observations makes it simpler to estimate V using an estimate for Σ rather than estimating V directly with consecutive values of θ_n. Because each component of U_n follows a decrease rate proportional to e^{-t} then given ϵ one can stop at N such that $e^{-N\epsilon} \leq \rho_1$, for a given precision ρ_1. Alternatively, pilot observations $\{\xi_n\}$ can be used to get an estimate $\widehat{\sigma^2}$ and then choose N, ϵ such that $e^{-N\epsilon} \leq \rho_1$ and $\epsilon\widehat{\sigma^2} \leq \rho_2$ for desired precision values ρ_1, ρ_2. ✱✱✱

Except for simple cases, where A is independent of θ and Σ known, V cannot be obtained in a closed form. In these cases one can run many replications of the algorithm in parallel to obtain an iid sample of the end point θ_n^ϵ, and then use the sample variance to build the confidence interval.

6.2.2 Decreasing Gain Size

To finalize the study of the convergence behavior of stochastic approximation algorithms, we now provide the most general results available to date for decreasing stepsize. The following result is in [201], and we will present it without proof. Before stating the theorem, we recall our standard definitions:

$$\beta_n = \mathbb{E}[Y_n \mid \mathfrak{F}_{n-1}] - G(\theta_n) \qquad (6.6)$$

$$V_n = \mathbb{E}[(Y_n - \mathbb{E}[Y_n \mid \mathfrak{F}_{n-1}])^2], \qquad (6.7)$$

for a given target vector field G and SA $\theta_{n+1} = \theta_n + \epsilon_n Y_n$. The following theorem applies to the static as well as the Markovian dynamics model.

Theorem 6.2. *Consider the decreasing stepsize algorithm*

$$\theta_{n+1} = \theta_n + \epsilon_n Y_n$$

for the static problem. Let $G: \mathbb{R}^d \to \mathbb{R}^d$, with $G \in C^1$, be a coercive vector field with a unique point θ^* such that $G(\theta^*) = 0$ and suppose that $\mathbb{A} = \nabla G(\theta^*)$ is a Hurwitz matrix with eigenvalues $\lambda_1, \ldots, \lambda_d$. Call $\lambda_{min} = \min(-\Re(\lambda_i))$. Moreover, assume that a constant K exists such that

$$\|G(\theta)\| \leq K(1 + \|\theta\|).$$

Assume that there are constants γ, β, δ such that

- $\epsilon_n = n^{-\gamma}$, $\|\beta_n\| = O(n^{-\beta})$, $\mathbb{E}[\|\beta_n\|^2] = O(n^{-2\beta})$, $V_n = O(n^{-\delta})$.
- $\gamma + \beta > 1$, $2\gamma + \delta > 1$.

Then $\theta_n \to \theta^*$ w.p.1, and if $\gamma < 1$

$$\mathbb{E}[\|\theta_n - \theta^*\|^2] = O(n^{-\kappa}), \quad \kappa \stackrel{def}{=} \min(2\beta, \gamma + \delta). \tag{6.8}$$

If $\gamma = 1$, then (6.8) is satisfied provided that $\lambda_{min} > \max(2\beta, 1 + \delta, 1)$.

The above result gives the rate of convergence κ of the expected mean square error (MSE) and can be used to design the parameters of the algorithm. The following example illustrates how this result can also be used to assess the benefits of variance reduction through increasing sample sizes.

Example 6.2. In economic theory, the supply of products, denoted by $S(\theta)$, is assumed to be an increasing function of the price θ, while demand, denoted by $D(\theta)$, is a decreasing function of θ. The price dictated by the market is the value θ^* such that $S(\theta^*) = D(\theta^*)$. Many models exist to describe price dynamics (Cournot models and more general models assuming delays and lags in production like in [91]). We propose the following model for finding the price θ^* when the demand function $D(\theta)$ is known but the supply function is unknown. This scenario is in accordance to several views that (particularly for commodities that are very expensive to produce) supplies suffer from unpredictable "shocks," many times due to political and social environments in countries that produce raw materials and/or manage important assembly lines. Suppose that consecutive instances for the supply can be generated through complex simulations and historical data, so that given a price θ, $\{\xi_k(\theta)\}$ are consecutive observations (possibly correlated) of the supply, satisfying that

$$\mathbb{E}[\xi_n(\theta) \mid \xi_{n-1}(\theta_n)] = S(\theta_n).$$

Assume that $\text{Var}[\xi_n(\theta)] = \sigma^2(\theta)$ is a bounded function of θ. Because we assume strict monotonicity of the functions, it follows that $D'(\theta) - S'(\theta) < 0$, where as assume $S(\theta), D(\theta) \in C^1$. Thus, the problem

$$\min_\theta \frac{1}{2}(S(\theta) - D(\theta))^2$$

is well-posed and has a unique solution at θ^* (solving the equation $S(\theta^*) = D(\theta^*)$). Taking as target vector field $G(\theta) = -(S(\theta) - D(\theta))$, strict monotonicity of $S(\theta)$ and $D(\theta)$ implies that $G' < 0$ for all θ, and G is thus Hurwitz. Therefore, the solutions of the ODE

$$\frac{dx(t)}{dt} = G(x(t))$$

do not blow up in finite time; see Theorem 2.3). Provided that $S(\theta)$ and $D(\theta)$ are locally Lipschitz and bounded, G is coercive for the above root-finding problem and has a unique limit point (globally asymptotically stable) $\lim_{t\to\infty} x(t) = \theta^*$. Consider now using the recursion

$$\theta_{n+1} = \theta_n + \epsilon_n(D(\theta_n) - \xi_n(\theta_n)).$$

Under this scheme we have $\beta_n = 0$ (no bias, so $\beta = +\infty$), and $V_n = \text{Var}(\xi_n(\theta_n)) = \sigma^2(\theta_n) = O(1)$. We may therefore apply Theorem 4.1 to establish that the sequence $\theta_n \to \theta^*$ w.p.1., provided that ϵ_n satisfies the usual conditions.

Now suppose that we wish to use Theorem 6.2 to improve the rate of convergence by getting more accurate estimates of the supply. Specifically, assume that for each iteration, we use a number $T_n = n^\delta$ of observations of $\xi_n(\theta_n)$ to build the feedback estimator Y_n. Call $\nu(n) = \sum_{k \le n} T(k)$. Then

$$V_n = \mathbb{E}[(Y_n - G(\theta_n))^2] = \mathbb{E}\left[\left(\left(D(\theta_n) - \frac{1}{T_n}\sum_{k=\nu(n-1)+1}^{\nu(n)} \xi_k(\theta_n)\right) - (D(\theta_n) - S(\theta_n))\right)^2\right]$$

$$= \text{Var}\left[\frac{1}{T_n}\sum_{k=\nu(n-1)+1}^{\nu(n)} \xi_k(\theta_n)\right] = O(T_n^{-1}) = O(n^{-\delta}).$$

According to the result in Theorem 6.2, $\kappa = \gamma + \delta$ so in principle we can make δ as large as we wish to have "infinite" convergence rate. What's wrong with this analysis? Well, it's missing the fact that we are not really using $\|\theta_n - \theta^*\|$ for the analysis, but each iteration requires $\nu(n)$ steps, therefore what Theorem 6.2 ascertains is that

$$\mathbb{E}\|\theta_{\nu(n)} - \theta^*\| = O(n^\kappa).$$

In this example, $\nu(n) = \sum_{k \le n} k^\delta$ is the generalized Harmonic number and it satisfies $n^\delta \le \nu(n) \le n^{\delta+1}$, so $\nu(n) = O(n^{\delta+1})$. Thus

$$\mathbb{E}\|\theta_{\nu(n)} - \theta^*\| = O(n^\kappa) = O\left(\left(\nu(n)^{\frac{1}{\delta+1}}\right)^\kappa\right) = O\left(\nu(n)^{\frac{\kappa}{\delta+1}}\right),$$

where $\kappa = \gamma + \delta$ is maximized at $\gamma = 1$. Therefore the actual convergence rate $\kappa' = \kappa/(\delta+1)$ will be maximized at $\kappa' = 1$ regardless of the value chosen for δ. Recall that $\delta = 0$ is the case when we use only one observation for the feedback Y_n. So we conclude that, from the point of view of asymptotic convergence rate, there is no gain in averaging the partial observations. We will come back to the analysis of computational effort and convergence rate in the next section. ✲✲✲

To motivate the final theorem in this section we first review the usual CLT in the simplest statistical estimation context. Suppose that $\{X_n\}$ is a sequence such that $X_n \to 0$, and that $\mathbb{E}[\|X_n\|^2] = O(n^{-\kappa})$. When using sample averages $X_n = S_n = n^{-1}\sum_{i=1}^n \xi_i$, and $\{\xi_n\}$ are zero-mean iid variables, $V_n = \text{Var}[X_n] = n^{-1}\sigma^2$, so with $\delta = 1$ we have $V_n = O(n^{-\delta})$. Then the standardized sequence $n^{\delta/2} X_n$ will have an approximate normal distribution with variance σ^2. These limiting results are of major importance when estimating the errors in our approximations, confidence intervals and predictions.

Example 6.3. Suppose that a sequence of independent estimators $X_1, X_2, \ldots$ with $X_n \to 0$ w.p.1 has decreasing bias, $\mathbb{E}[X_n] = b_n \to 0$, then the limit distribution depends on how fast

b_n itself decreases. Let $b_n = O(n^{-\beta})$, and $\text{Var}(X_n) = O(n^{-\delta})$ (in particular, if $\text{Var}(X_n) = \sigma^2/n$ then $\delta = 1$).

Case 1: If $\beta > \delta/2$ then the bias terms decrease *very fast* (no bias corresponds to $\beta = +\infty$) because $n^{\delta/2} b_n = O(n^{\delta/2 - \beta})$ and we have

$$\mathbb{E}[n^{\delta/2} X_n] \to 0 \quad \text{and} \quad \text{Var}(n^{\delta/2} X_n) \to \sigma^2,$$

so here $\kappa = \delta$ gives the convergence rate for the MSE. Therefore, in this case the variance dominates the limit behavior: the scaled sequence $n^\delta X_n$ has fluctuations around zero with limiting variance σ^2. The most common situation is when X_n has an approximate normal distribution $\mathcal{N}(0, \sigma^2)$.

Case 2: If $\beta < \delta/2$ then $n^\beta X_n$ has a limiting bias, that is,

$$\bar{B} \overset{\text{def}}{=} \lim_{n \to \infty} n^\beta X_n$$

exists, and of course, here $\mathbb{E}[n^{\delta/2} X_n] \to \infty$. However, scaling with n^β we see that the limiting behavior is controlled now by the bias, because $\text{Var}(n^\beta X_n) \approx n^{-(\delta - 2\beta)} \sigma^2 \to 0$. This is the case where the limiting distribution of the scaled sequence is degenerate, concentrated on $\bar{B}$ and $\kappa = 2\beta$.

Case 3: The case $\beta = \delta/2$ is a mixture of the previous cases, where the scaled sequence has both a limiting bias and a limiting variance. Under normality assumptions, we have $n^{\delta/2} X_n \overset{d}{\Longrightarrow} \mathcal{N}(\bar{B}, \sigma^2)$. ✼✼✼

Theorem 6.3. *Consider the decreasing stepsize SA algorithm*

$$\theta_{n+1} = \theta_n + \epsilon_n Y_n$$

on $\mathbb{R}^d$. Assume that

(a1) $G \in C^1$ has a unique stationary point θ^* such that $G(\theta^*) = 0$ and suppose that $\mathbb{A} = \nabla G(\theta^*)$ is a Hurwitz matrix with eigenvalues $\lambda_1, \ldots, \lambda_d$. Call $\lambda_{min} = \min(-\Re(\lambda_i))$. Moreover, assume that a constant K exists such that

$$\|G(\theta)\| \leq K(1 + \|\theta\|).$$

(a2) There are constants γ, β, δ such that $\gamma + \beta > 1$, $2\gamma + \delta > 1$, and

$$\epsilon_n = n^{-\gamma}, \quad \|\beta_n\| = O(n^{-\beta}), \; \mathbb{E}[\|\beta_n\|^2] = O(n^{-2\beta}), \quad V_n = O(n^{-\delta}).$$

(a3) If $\gamma = 1$, then suppose that $\lambda_{min} > \max(2\beta, 1 + \delta, 1)$.

(a4) There is a symmetric positive definite matrix Σ such that

$$n^\delta \mathbb{E}[(\delta M_n)(\delta M_n)^T] \to \Sigma.$$

Then calling

$$\kappa = \min(2\beta, \gamma + \delta),$$

the scaled sequence $n^{\kappa/2}(\theta_n - \theta^*)$ converges in distribution as follows:

$$n^{\kappa/2}(\theta_n - \theta^*) \overset{d}{\Longrightarrow} \begin{cases} \mathcal{N}(0, V) & \text{if } \gamma + \delta < 2\beta, \\ H^{-1} \bar{B} & \text{if } \gamma + \delta > 2\beta, \\ \mathcal{N}(H^{-1} \bar{B}, V) & \text{if } \gamma + \delta = 2\beta, \end{cases}$$

where $\bar{B} \stackrel{def}{=} \lim_{n\to\infty} -n^\beta b_n$, $H = \mathbb{A} + \beta \mathbb{I}$ and V satisfies

$$V = \int_0^\infty e^{\tilde{H}t} \Sigma e^{\tilde{H}^T t} dt,$$

for $\tilde{H} = \mathbb{A} + \frac{(1+\delta)}{2} \mathbb{I}$, provided that $\tilde{H}$ is negative definite. An equivalent expression for V is given by the implicit equation $\tilde{H}V + V\tilde{H}^T = \Sigma$; see (6.4).

For the proof, we refer to [194, 201].

Remark 6.1. Theorem 6.3 allows for computing the optimal sequence $\{\epsilon_n\}$. For this, let K be a positive definite matrix, such that $K\mathbb{A} + \mathbb{I}/2$ is negative definite, and consider the sequence $\epsilon_n = K/n$. The matrix K can be integrated into g and Σ in the setting of Theorem 6.3, which comes down to replacing $\mathbb{A}$ by $K\mathbb{A}$ and Σ by $K\Sigma K^\top$, respectively. The limiting variance then becomes

$$V = \int_0^\infty e^{(K\mathbb{A} + \mathbb{I}/2)t} K\Sigma K^\top e^{(\mathbb{A}^\top K^\top + \mathbb{I}/2)t} dt.$$

Minimizing V with respect to K gives, $K = -\mathbb{A}^{-1}$, with asympotically optimal variance $V = \mathbb{A}^{-1}\Sigma(\mathbb{A}^{-1})^\top$, see [26, 334] or Chapter 8 in [11].

The following examples illustrate the application of Theorem 6.3.

Example 6.4. Suppose that we use stochastic approximation for a robot that is supposed to track the location of a target (e.g., in a rescue mission, or deep under the ocean). The target's location is estimated by the robot's sensors via noisy thermal, sonar or optical signals $\{\xi_n\}$, assumed iid, with mean α and uncorrelated entries for each ξ_n. The simple root-finding equation

$$\theta_{n+1} = \theta_n - \epsilon_n(\theta_n - \xi_n)$$

will satisfy the conditions of Theorem 5.3 under regularity assumptions on the noise, for example, under a simple model where each component of the location estimate has constant covariance $\text{Cov}(\xi_{n,i}, \xi_{n,j}) = \sigma^2 < \infty$, for $i = j$ and zero otherwise. In this case we have

$$\Sigma = \sigma^2 \mathbb{I},$$

and the model is of exogenous noise, so the values of θ_n do not affect the transition probabilities of ξ_n, which are iid. The target vector field is given by $G(\theta) = -(\theta - \alpha)$, and thus $\mathbb{A} = -\mathbb{I}$, which is Hurwitz. We have

$$\tilde{H} = \mathbb{A} + \frac{1}{2}\mathbb{I} = -\frac{1}{2}\mathbb{I},$$

where we use that $\delta = 0$ since the variance is constant. By Theorem 6.3, the asymptotic variance of the θ_n^ϵ process is given by

$$V = \sigma^2 \int_0^\infty e^{-\mathbb{I}t} dt = \sigma^2 \int_0^\infty e^{-t} \mathbb{I} dt = \sigma^2 \mathbb{I}.$$

Even though we can solve V explicitly for this model, we want to mention that estimating σ^2 (for obtaining V) concurrently with the stochastic approximation is straightforward using sample variance of the observations, so that at the end of a large number N of iterates one can provide a confidence interval for each component α_i of the target vector location. ✻✻✻

Example 6.5. Consider again the case of unconstrained optimization where finite differences are used, and refer to Example 4.1, where we established that $c \in (0, 1/2)$ was necessary for convergence to the optimal value when the noise from different observations are independent. We will now apply Theorem 6.3 to this scheme in order to determine the "best" parameters, that is, those that provide faster convergence rate to the optimal point.

Suppose that $\xi_n(\theta_n) =: \xi_n \stackrel{d}{=} \xi'_n := \xi'_n(\theta_n)$ have the same distribution, and that

$$Y_n = \frac{\phi(\xi_n, \theta_n + c_n) - \phi(\xi'_n, \theta_n - c_n)}{2c_n},$$

where $\mathbb{E}[\phi(\xi_n, \theta_n) \mid \mathfrak{F}_{n-1}] = J(\theta_n)$, and let $c_n = n^{-c}$ for some algorithm parameter c to be determined. As in Example 4.1,

$$\mathbb{E}[Y_n \mid \mathfrak{F}_{n-1}] = J'(\theta_n) + O(c_n^2),$$

so that $\beta = 2c$ and we assume bounded variance: $\mathrm{Var}(\phi(\xi, \theta)) \leq \bar{V}$. We consider the gradient descent SA algorithm

$$\theta_{n+1} = \theta_n - \epsilon_n Y_n = \theta_n + \epsilon_n(-Y_n),$$

which allows us to identify $G(\theta) = -J'(\theta)$ and $\mathbb{A} = -J''(\theta^*)$ in Theorem 6.3, where θ^* is the unique solution of the considered minimization problem.

For the calculation of the conditional variance V_n we will now consider two cases.

Case 1: The noise processes are independent, that is ξ_n is independent of ξ'_n. In this case, as was calculated in Example 4.1, $V_n = O(c_n^{-2}) = O(n^{+2c})$. Thus in this case $\delta = -2c$.

Theorem 6.3 establishes that the convergence rate of the algorithm is $\kappa = \min(2\beta, \gamma + \delta)$, where $\epsilon_n = n^{-\gamma}$ for $\gamma \in (0, 1]$. The design parameter here is c (our variable), and we have determined that $\beta(c) = 2c, \delta(c) = -2c$, which leads to $\kappa(\gamma) = \min(4c, \gamma - 2c)$, and maximizing $\kappa(\gamma)$ is attained at the equality

$$4c = \gamma - 2c, \implies c = \gamma/6,$$

so the maximum rate is achieved when $\gamma = 1$, and this implies that

$$\kappa = \frac{2}{3}, \beta = \frac{1}{3}, c = \frac{1}{6}, \text{ and } \delta = -\frac{1}{3}.$$

Using Taylor expansion as in Example 4.1 we calculate the limit bias:

$$\lim_{n \to \infty} n^\beta b_n = \frac{J'''(\theta^*)}{6},$$

which gives $\bar{B} = -\frac{J'''(\theta^*)}{6}$, and the limiting variance is calculated calling $\mathrm{Var}(\phi(\xi_n, \theta)) = \sigma^2(\theta)$, as follows (using that $c = 1/6$):

$$V_n = \frac{\sigma^2(\theta_n + c_n) + \sigma^2(\theta_n - c_n)}{4c_n^2} \approx \frac{\sigma^2(\theta)}{2c_n^2} = n^{\frac{1}{3}} \frac{\sigma^2(\theta)}{2},$$

so that using the fact that as $\lim_{n \to \infty} \theta_n = \theta^*$ a.s.,

$$\Sigma = \lim_{n \to \infty} n^{-\frac{1}{3}} V_n = \frac{\sigma^2(\theta^*)}{2}.$$

Using the explicit solutions, we find $H = -J''(\theta^*) + 1/3$ and $\tilde{H} = -J''(\theta^*) + 1/3$. Thus,

$$H^{-1}\bar{B} = \frac{J'''(\theta^*)}{6(J''(\theta^*) - 1/3)},$$

and, provided that $J''(\theta^*) > 1/3$,

$$V = \frac{\sigma^2(\theta^*)}{2} \int_0^\infty e^{-2(J''(\theta^*) - 1/3)t} dt = \frac{\sigma^2(\theta^*)}{4(J''(\theta^*) - 1/3)}.$$

To summarize,

$$n^{1/3}(\theta_n - \theta^*) \stackrel{d}{=} \mathcal{N}\left(\frac{J'''(\theta^*)}{6(J''(\theta^*) - 1/3)}, \frac{\sigma^2(\theta^*)}{4(J''(\theta^*) - 1/3)}\right).$$

It is worth emphasizing that there is an asymptotic bias for the scaled error.

Case 2: Consider now the particular case where $\phi(\xi(\omega), \theta)$ is continuous and differentiable in θ for every given ω. Assume that $\sup_\theta \mathbb{E}[\phi'(\xi, \theta)^2] < \infty$. Using *common random numbers* we assume here that the two observations are correlated, that is, we use the same "noise," yielding

$$\phi(\xi_n(\omega), \theta_n + c_n) - \phi(\xi_n(\omega), \theta_n - c_n) = \phi'(\xi_n(\omega), \theta_n) + \kappa_n(\omega),$$

where κ_n is a bounded random variable, and $\mathbb{E}[\kappa_n] = O(c_n^2)$, which now gives $V_n = O(1)$ and $\delta = 0$. This yields (equating $2\beta = \gamma + \delta$)

$$4c = \gamma \implies c = \gamma/4,$$

so the maximum rate is now at $c = 1/4$ and is achieved for $\gamma = 1$, which yields $\kappa = 1$. This is the same convergence rate as if we had an unbiased estimator for the derivative. The calculation of the asymptotic normal distribution is left to the reader. Comparing this convergence rate with the rate $2/3$ for the biased case in Case 1, illustrates the advantage of using an unbiased estimator or a based CRN derivative proxy. ✻✻✻

6.3 ASYMPTOTIC EFFICIENCY

This section presents the main theory introduced in [128], and we refer to that reference for all proofs. Let $\{\theta_n\}$ be a stochastic process on the probability space $(\Omega, \{\mathfrak{F}_n\}, \mathbb{P})$, and suppose that the sequence $\{\theta_n\}$ is used to estimate an unknown parameter θ^*. Call $C(n)$ the computational time required to calculate $\theta_1, \ldots, \theta_n$, which is a $\mathfrak{F}_n$ measurable random variable. We will be mostly concerned with the *loss function*

$$L(\theta) = \|\theta - \theta^*\|^2,$$

representing the square error at value θ. However, other real valued convex functions $L(\cdot)$ can be used as well. Define

$$N(c) = \sup(n \in \mathbb{N} : C(n) \leq c) \tag{6.9}$$

as the (random) number of estimates that can be calculated given a *computational budget* c. The corresponding final estimator is $\theta_{N(c)}$.

Definition 6.2. Given a computational budget c, the *risk function* is

$$R(c) = \mathbb{E}[L(\theta_{N(c)})]. \tag{6.10}$$

The risk function measures the expected mean square error of the final estimator as a function of the computational budget and it is often impossible to evaluate analytically.

Lemma 6.2. *Suppose that $\theta_n \to \theta^*$ wp1 and assume that for all n, $\theta_n \in \Theta$ where Θ is a compact set. Suppose that $\exists \tau > 0$ such that $\lim_{n \to \infty} n^{-\tau} C(n) = K$, where $0 < K < \infty$. Then if $L(x) = (x - \theta^*)^2$ it follows that*

$$\lim_{c \to \infty} R(c) = 0.$$

The proof follows directly from the convergence assumptions, using the dominated convergence theorem.

Definition 6.3. *Suppose that there exists real numbers $0 < \nu < \infty$ and $0 < \mathcal{E} < \infty$ such that*

$$\lim_{c \to \infty} c^\nu R(c) = \frac{1}{\mathcal{E}}. \tag{6.11}$$

Then we call $\mathcal{E}$ the *asymptotic efficiency* of the estimation, and the number ν its corresponding *asymptotic convergence rate*.

Example 6.6. For the simple estimation problem of a sample average of one-dimensional iid observations with variance σ^2, $\theta_n \approx \mathcal{N}(\theta^*, \sigma^2/n)$ for large n and

$$R(c) = \mathbb{E}[(\theta_{N(c)} - \theta^*)^2] \approx \frac{\sigma^2}{N(c)}.$$

If each sample has constant cost λ units (of CPU time, or sampling efforts, or money) then $C(n) = \lambda n$, and $N(c) = \lfloor c/\lambda \rfloor$, which implies

$$\lim_{c \to \infty} \frac{N(c)}{c} = \frac{1}{\lambda},$$

therefore, in this case, called the *canonical* estimation case, $\nu = 1$, because $c^{-1} R(c) \to \sigma^2 \lambda$, yielding

$$\mathcal{E} = \frac{1}{\sigma^2 \lambda}.$$

The basic principle is that the (asymptotic) efficiency of the canonical estimator is inversely proportional to the product of variance and computational effort per sample. This measure of efficiency agrees with the notion of "loss" incurred in estimation, and allows a quantitative comparison between possible estimation or approximation methods. ✽✽✽

Theorem 6.4 (Theorem 1 from [128]). *Assume the following conditions:*

(a1) There is an integer $\kappa \in \mathbb{N}$ such that for each $\epsilon > 0$,

$$U^\epsilon(t) \stackrel{def}{=} \epsilon^{-\kappa/2} (\theta_{\lfloor t/\epsilon \rfloor} - \theta^*)$$

converges in distribution as $\epsilon \to 0$ to a random element $U \in D^d[0, \infty)$, the set of real-valued cadlag processes with the Skorokhod topology,[3] and assume that $U(\cdot)$ is a.s. continuous.

---

[3] For most of the cases that we are interested in, the limit process will be a Wiener or an Orstein-Ulenbeck process.

(a2) There exist $\tau > 0$, CPU > 0 such that w.p.1

$$\lim_{n \to \infty} n^{-\tau} C(n) = \text{CPU}^{\tau}.$$

Then

(i) *The following result (FCLT) holds* $c^{\kappa/2\tau}(\theta_{N(ct)} - \theta^*) \stackrel{d}{\Longrightarrow} U(t^{1/\tau}/\text{CPU})$ *as* $c \to \infty$.
(ii) *The associated CLT result holds, namely,* $c^{\kappa/2\tau}(\theta_{N(c)} - \theta^*) \stackrel{d}{\Longrightarrow} U(1/\text{CPU}) = \text{CPU}^{\kappa/2} U(1)$,
(iii) *The corresponding Weak LLN is satisfied:* $\theta_{N(c)} \to \theta^*$ *in probability, as* $c \to \infty$.

The following lemma provides an explicit characterization of the asymptotic convergence rate and efficiency in the one dimensional case, i.e., $d = 1$.

Lemma 6.3 (Corollary 1 from [128]). *Consider a general loss function* $L(\cdot) \in C^2$, *convex, non-negative and satisfying* $L(\theta^*) = L'(\theta^*) = 0$; $L''(\theta^*) > 0$. *Under the conditions of Theorem 6.4, suppose that* $\{c^{\kappa/\tau} L(\theta_{N(c)}); c > 1\}$ *is uniformly integrable. Then the asymptotic convergence rate and efficiency are given by*

$$\nu = \frac{\kappa}{\tau}, \qquad \mathcal{E} = \frac{2}{L''(\theta^*)} \left(\frac{1}{\text{CPU}^{\kappa} \mathbb{E}[(U(1))^2]} \right).$$

In particular, when $L(\theta) = (\theta - \theta^*)^2$, $L''(\theta^*) = 2$, and the efficiency is again the inverse of a asymptotic variance times the asymptotic cost per sample. This criterion of efficiency provides a way to trade off precision and speed of numerical solutions.

Example 6.7. Let $J : \mathbb{R} \to \mathbb{R}, J \in C^2$ be a convex cost function that we wish to minimize, call θ^* the unique minimum and suppose that $J''(\theta^*) > 1$. Assume that there is a random variable $Z(\theta)$ such that $\mathbb{E}[Z(\theta)] = -J'(\theta)$, with $\sup_\theta \text{Var}(Z(\theta)) < \infty$, and let $Y_n = Z(\theta_n)$. Use a decreasing stepsize SA

$$\theta_{n+1} = \theta_n + \epsilon_n Y_n,$$

where $\epsilon_n = n^{-\gamma}$. Note that $\mathbb{E}[Y_n | \mathfrak{F}_{n-1}] = -J'(\theta_n)$, so that we have no bias. Using Theorem 6.3, here $\delta = 0$, and $\beta = +\infty$ so that

$$n^{\gamma/2}(\theta_n - \theta^*) \stackrel{d}{\Longrightarrow} \mathcal{N}(0, V),$$

where

$$V = \frac{\text{Var}(Z(\theta^*))}{2J''(\theta^*) - 1}.$$

For a proof, note that $\tilde{H} = J''(\theta^*) - 1/2$, $\Sigma = \text{Var}(Z(\theta^*))$, and used the expression of V put forward in Theorem 6.3. Assume that the computational cost for the n-th sample is a $\mathfrak{F}_{n-1}$-measurable random variable t_n with conditional distribution

$$\mathbb{P}(t_n \leq t | \mathfrak{F}_{n-1}) = F_{\theta_n}(t).$$

Let $t(\theta) \sim F_\theta(\cdot)$ and assume that $\sup_{\theta \in \mathbb{R}} \text{Var}(t(\theta)) < \infty$. Assume that $\mathbb{E}[t(\theta)]$ is continuous in the neighborhood of θ^*. By definition, the computational cost up to iteration n is

$$C(n) = \sum_{i=1}^{n} t_i.$$

Using a.s convergence of $\theta_n \to \theta^*$ (as we have a decreasing stepsize SA), continuity of $\mathbb{E}[t(\theta)]$ together with boundedness of the variance, we obtain

$$n^{-1}C(n) = \frac{1}{n}\sum_{i=1}^{n} t_i \to \mathbb{E}[t(\theta^*)] \stackrel{\text{def}}{=} \text{CPU} \quad \text{a.s.,}$$

so that in this example, $\tau = 1$. Using Lemma 6.3 with loss function $L(\theta) = (\theta - \theta^*)^2$, the asymptotic rate and efficiency is given by

$$v = \frac{\gamma}{2}, \quad \mathcal{E} = \frac{1}{\text{CPU}^{2\gamma} V},$$

which supports the known result that the "best" decreasing stepsize rate is obtained setting $\gamma = 1$. ✳✳✳

6.4 PRACTICAL CONSIDERATIONS

The results presented in this chapter may serve as a set of guidelines for designing an SA experiment and in its output analysis. First, unbiased feedbacks are preferred to biased ones. Second, the choice $\epsilon_n = A/(n+1)$ for some finite constant A is analytically justifiable although other choices exist (see Chapter 3). Finally, based the convergence to a multi dimensional Ornstein-Uhlenbeck process, it is reasonable to assume that θ_n is for fixed ϵ and n sufficiently large, approximately (multi)normal, and same convergence to a approximately (multi)normal is to be expected for a decreasing ϵ_n.

For fixed ϵ, convergence of the interpolation to a multi dimensional Ornstein-Uhlenbeck process allows for the following "single-run" output analysis. First, one identifies, for ϵ small, a time index m, from which onward θ_n^ϵ behaves like a stationary sequence with no drift. This can, for example, be assessed by means of the augmented Dickey-Fuller (ADF) test; see [87]. More specifically, it is worth noting that $\{\theta_n^\epsilon\}$ is a Markov chain by construction. This allows to apply methods for the statistical output analysis for stationary Markov chains to θ_n^ϵ, and we refer to Section D.3 in Appendix D, where we discuss methods for identifying the length of the transient period m and explain how the batch means method can be used for providing an statistical output analysis for $\theta^* \approx \mathbb{E}[\lim_n \theta_n^\epsilon]$ (see Theorem D.3), where θ^* is a KKT point of an optimization problem, see Chapter 4 and Chapter 5 for relevant theory.

Alternatively, a statistical output analysis can be build an independent replications. More specifically, set up an SA experiment, where R independent runs of the SA are executed, each update is based on K iid calls to the feedback function, and the SA is evaluated over N iterations, i.e, we obtain R iid copies of θ_N. An algorithm uses *mini-batching* or *batching* in case $K > 1$. Then the total simulation budget C, expressed in terms of the total amount of samples drawn from the underlying is $C = RKN$, and we call (R, K, N) the *budget allocation*. For example, a single run SA with decreasing ϵ_n, would allocate the overall budget as $(1, 1, C)$. We argued extensively that batching the feedback is typically less efficient than using single feedback, i.e., allocate according to $(\hat{R}, 1, \hat{N})$, where $\hat{R}\hat{N} = C$ and the batching comes at the price of reducing N to $\tilde{N}$. Using statistical test for assessing approximate normal distribution of θ_N based on the R samples, like the Jarque-Bera test, see [175], a standard confidence interval for $\mathbb{E}[\theta_N]$ can be obtained. In applications, however, (and we will provide instances in Part II), we may encounter an optimization problem where $K > 1$ is need to compensate for the very large variance of the feedback. To summarize, we advocate to not follow the common practice of using batches for the feedback, but rather invest the budget in producing several SA runs, in formula, $(1, K, N)$, $K > 1$, should rather be $(R, 1, N)$. The former producing smoother pictures, while the latter allows for a proper (!) statistical output analysis. Running some pre-experiments our goal is to find the smallest N for which θ_N is approximately normal, and invest the budget in performing the maximal number of parallel runs R for a powerful statistical output analysis.

Is there an overall recommendation? Obviously there is no one-size-fits all, but as a general guideline we recommend to start with $R \geq 100$, $K = 1$, and $N = \lceil C/R \rceil$. This setting produces R samples $\theta_N(\omega_i)$, $1 \leq i \leq N$. Plotting a histogram of the $\theta_N(\omega_i)$, we can check whether approximate normality is achieved, and if so, an confidence interval for θ_N can be constructed in the standard way for given confidence level. If such an analysis is also carried for, say, $N - \lceil 0.1N \rceil$, and the outcomes are approximately equal, then we have an indication that N is large enough and that the true minimize lies within our confidence interval.

We want to emphasize that the target field $G(\theta)$ is required to be differentiable at the limiting point θ^*. This setting thus rules out the case that G is, for example, the negative projected gradient field and θ^* a point on the boundary of the admissible Θ, i.e., the limiting ODE is $\Pi_\Theta G$. An extension of the weak convergence results presented in this chapter to this case can be found in [231].

As a concluding remark, we note that for five times differentiable, strongly convex cost functions f, and fixed gain size SAs, explicit expressions for $\mathbb{E}[\bar{\theta}_n^\epsilon - \theta^*]$ are obtained in [89], where $\bar{\theta}_n^\epsilon = (1/(n+1)) \sum_{i=0}^n \theta_n^\epsilon$. The results therein show that in this particular case a Richardson extrapolation (see [48]) given by $2\bar{\theta}_n^\epsilon - \theta_n^{2\epsilon}$, will produce better estimates for θ^* but at the price of higher computational cost.

6.5 EXERCISES

Exercise 6.1. Let

$$q(t) = \max\left(i : \frac{1}{i} \leq t, i \in \{1, 2, \ldots\}\right), \quad t \in (0, \infty),$$

and suppose that $\{\xi_n, n \in \mathbb{N}\}$ is an infinite sequence of iid random variables with uniformy bounded variance.

(a) Show that the real-valued processes $\{W_n(\cdot); n \in \mathbb{N}\}$ defined by

$$W_n(t) = \frac{1}{\sqrt{n}} \sum_{n=1}^{q(t)-1} \xi_n$$

are tight in $D^d[0, \infty)$, by showing that for any bounded interval $[t, t+s]$, the set $\{W_n(t+s) - W_n(t)\}$ is tight. (*Hint:* use the fact that $W_n(t+s) - W_n(t) = (1/\sqrt{n}) \sum_{q(t)}^{q(t+s)-1} \xi_n$ and $\text{Var}(\xi_n) = \sigma^2 < \infty$.)

(b) Use the compactness argument to show that the limit process is a Wiener process. What is Σ here?

Exercise 6.2. Usually, weak convergence does not imply other types of convergence. This exercise deals with a particular case where the limit of a random sequence is deterministic. Let $\{X_n\}$ be a sequence of random numbers that converges in distribution to a fixed number

$a \in \mathbb{R}$. Show that $X_n \to a$ in mean square error and in probability. Is it also true that $X_n \to a$ a.s.?

Exercise 6.3. Consider the supply/demand problem of Example 6.2. The demand function $D(\theta) = \theta^{-\eta}, \eta > 0$ is known. However, the supply function $S(\theta)$ is only known to be an increasing function of θ that is analytic (infinitely continuously differentiable). Instead, a complex simulation model is used to produce statistically independent unbiased estimates ξ_n such that

$$\mathbb{E}[\xi_n \mid \theta_n] = S(\theta_n); \quad \text{Var}(\xi_n) = 1. \tag{6.12}$$

(a) Show that
$$\theta_{n+1} = \theta_n + \epsilon Y_n, \quad Y_n = D(\theta_n) - \xi_n$$
satisfies the assumptions of Theorem 6.1.

(b) Use $\eta = 5$ for the demand function. Your economics guru has estimated that $\theta^* \approx 1$ and $S'(\theta^*) \approx 4.5$. With this information, apply Theorem 6.1 to identify the values of a, σ^2 for the (approximate) limit Orstein-Uhlenbeck process $U(t)$, and find T such that $e^{-aT} \approx 0.0001$.

(c) Show that $\epsilon \approx 0.0005$ yields a precision of 0.01 (half the width of the approximate confidence interval after T/ϵ iterations, with confidence level $\alpha = 0.05$).

(d) In this part of the problem you will generate the random observations $\{\xi_n\}$ and run the stochastic approximation. Conditional on θ_n, let $\xi_n \sim \text{LN}(m, v^2)$ have a lognormal distribution. First find the parameters for the m and v such that (6.12) holds, with $S(\theta) = \theta^s, s = 4.3$. Next, run the algorithm and discuss your results.

Exercise 6.4. Use Theorem 4.1 to prove the statement in Theorem 6.2 that θ_n converges almost surely to θ^*.

Exercise 6.5. Show Lemma 6.3. Use a Taylor expansion of $L(\cdot)$ around θ^* to show first that

$$\lim_{c \to \infty} c^{\kappa/\beta} R(c) = (1/2) L''(\theta^*) \lambda^\kappa \mathbb{E}[U(1)^2].$$

Exercise 6.6. An investor wishes to divide her capital in two assets $\{S_i(t), i = 1, 2\}$, which she will access at maturity time T. Assume that her total capital is 1 (by changing monetary units if necessary). Although not known precisely, the corresponding means (μ_1, μ_2) and variances $(\sigma_1^2$ and $\sigma_2^2)$ satisfy $\mu_1 > \mu_2$ and $\sigma_1^2 \gg \sigma_2^2$. In order to balance her profit and risk, the investor wishes to find the proportion θ that solves the problem:

$$\max_{\theta \in [0,1]} \mathbb{E}[X(\theta)]$$

$$\text{s.t. } \mathbb{E}[X(\theta)^2] \leq B,$$

where $\mathbb{E}[S_1(T)^2] > B > \mathbb{E}[S_2(T)^2]$ and $X(\theta) = \theta S_1(T) + (1-\theta) S_2(T)$. Although the exact distributions of the two assets are not known, it is possible to use historical observations or simulations to produce consecutive samples $\xi_n = (\xi_{n,1}, \xi_{n,2}) \stackrel{d}{=} (S_1(T), S_2(T))$.

(a) Argue that the optimal value θ^* of the above problem must satisfy $0 < \theta^* < 1$, and that the constraint will be active at the optimum.

(b) For a particular value $x = (x_1, x_2)$, let $\phi(x, \theta) = -\theta x_1 - (1-\theta) x_2$ so that $J(\theta) = \mathbb{E}[\phi(S_1(T), S_2(T); \theta)]$ is the function that we wish to minimize. Write the Langrangian of the problem and show that it is a convex NLP, so that the Arrow-Hurwicz algorithm for the deterministic problem converges to the optimal solution. Specify the vector field $G(\theta, \lambda)$ ($\lambda > 0$) for the corresponding limit ODE.

(c) Consider the stochastic version of the Arrow-Hurwicz algorithm using one-sided finite differences: for each integer n let $\xi_n = (\xi_{n,1}, \xi_{n,2})$ and $\xi'_n = (\xi'_{n,1}, \xi'_{n,2})$ have the same distribution as $(S_1(T), S_2(T))$ and suppose that these samples are statistically independent, that is, $\xi_n \perp \xi'_n$ for all n. The algorithm is

$$\theta_{n+1} = \theta_n - \frac{\epsilon_n}{c_n}\left(\phi(\xi'_n, \theta_n + c_n) - \phi(\xi_n, \theta_n) + \lambda_n\left(\phi^2(\xi'_n, \theta_n + c_n) - \phi^2(\xi_n, \theta_n)\right)\right)$$

$$\lambda_{n+1} = \left(\lambda_n + \epsilon_n(\phi^2(\xi_n, \theta_n) - B)\right)_+ \quad (6.13)$$

with $\epsilon_n = O(n^{-\gamma})$, $c_n = O(n^{-c})$. Use Theorem 6.2 to establish that the fastest convergence is achieved at $c = 1/4$ and gives $\kappa = 1/2 < 1$.

(d) Let $\Delta_n \stackrel{\text{def}}{=} \xi_{n,1} - \xi_{n,2}$. Show that $J'(\theta) = -\mathbb{E}[\Delta_n]$ and that $g'(\theta) = 2\mathbb{E}[\theta\Delta_n^2 + \xi_{n,2}\Delta_n]$. Use Theorem 4.1 to show that the stochastic Arrow-Hurwitz algorithm

$$\theta_{n+1} = \theta_n - \epsilon_n\left(-\Delta_n + 2\lambda_n(\theta_n\Delta_n^2 + \xi_{n,2}\Delta_n)\right)$$

$$\lambda_{n+1} = \left(\lambda_n + \epsilon_n(\theta_n\Delta_n + \xi_{n,2})^2 - B\right)_+$$

converges to the solution of the optimization problem θ^*. What do you need to assume on the stepsize sequence $\{\epsilon_n\}$? Use Theorem 6.2 to find the convergence rate κ assuming that $\epsilon = n^{-\gamma}$. Specify the values of β and δ. Next, find the asymptotic efficiency and the corresponding rate, using Theorem 6.4.

(e) Assume now that the same observation is used at iteration n in order to calculate $\phi(\xi_n, \theta_n \pm c_n)$. (Notice that in this model the noise is exogenous (the value of the two assets is independent of the proportion that the investor decides to allocate), so that it is straightforward to implement CRNs by using the same random variable $\xi_n = \xi'_n$ a.s.)

$$\theta_{n+1} = \theta_n - \frac{\epsilon_n}{2c_n}\Big(\phi(\xi_n, \theta_n + c_n) - \phi(\xi_n, \theta_n - c_n) + \lambda_n(\phi^2(\xi_n, \theta_n + c_n) \quad (6.14)$$
$$- \phi^2(\xi_n, \theta_n - c_n))\Big)$$

$$\lambda_{n+1} = \left(\lambda_n + \epsilon_n(\phi^2(\xi_n, \theta_n) - B)\right)_+. \quad (6.15)$$

Show that this implementation can achieve $\kappa = 1$ and find the value of c that achieves this maximal rate.[Hint: Use Taylor series for $\phi(\xi, \theta)$ for a fixed value of ξ to express the vector field in terms of the stochastic derivative $\phi'(\xi, \theta)$, and establish that the variance term is $V_n = O(1)$.)

Exercise 6.7. Consider the algorithm

$$\theta_{n+1} = \theta_n + \epsilon_n Y_n,$$

for finding some optimal solution θ^*, for either $\epsilon_n \downarrow 0$ or $\epsilon_n = \epsilon$, for $n \in \mathbb{N}$. Suppose that evaluating Y_n requires one sample from an underlying process, and let your computational budget be sufficient to sample N samples from the underlying process.

(a) Suppose you use your entire simulation budget to simulate one sample of θ_N. What conclusions can be drwan from this in case of decreasing ϵ and in case of fixed ϵ?

(b) Suppose you split your simulation budget to produce k independent runs of the algorithm yielding $\theta_n(\omega_i)$, $1 \leq i \leq k$, for each of the runs, where $kn = N$. What conclusions can be drawn from this data in case of decreasing ϵ and in case of fixed ϵ?

(c) For the given simulation budget N describe the best setup for your optimization algorithm that allows to produce a statistical justifiable assessment on the optimal solution θ^*.

Part II

Gradient Estimation

Chapter Seven

A Primer for Gradient Estimation

In this chapter we introduce the main methods of gradient estimation without proofs. All the supporting theorems and methodologies are found later on in Chapters 8 and 9. To simply the presentation, we consider the one-dimensional case only, that is, we assume $\theta \in \mathbb{R}$ and "derivative estimation" might be a more appropriate term in this chapter. However, as a gradient is easily obtained through its partial derivatives the restriction to the one-dimensional case comes at no loss of generality and we follow the standard literature in referring to the material presented in this part of the monograph as gradient estimation.

7.1 MOTIVATION

Example 7.1. In the mining industry, an investment of θ dollars is used to extract the ore (or "raw material") from the ground, where $\theta \geq 0$. The actual amount of valuable minerals is a random variable $X(\theta)$ that depends on the mining strategy (e.g., excavation, exploration, purification, cut-off grades) as well as on the quality of the soil. For a given geographical area, say an open pit mine for example, $X(\theta)$ is nondecreasing in θ. Following economic theory, the demand curve determines the market price as a function $p(q)$, given that the total quantity supplied is q. The function p is usually modeled as a decreasing convex function. Let S be the aggregate supply from all other competitors.

The net cost of the mining is thus $\theta - X(\theta)\, p(X(\theta) + S)$, which is a random variable (if negative, it means the company has a net profit). Strategic planning for large investments is a decision problem: large investments may yield too much supply with prices lowering below production cost, and investing too little may yield too small a revenue. Suppose that a mining company has a forecast $\hat{S}$ on the quantity supplied by the competitors, then the net cost of mining is given by

$$J(\theta) = \theta - \mathbb{E}[X(\theta)\, p(X(\theta) + \hat{S})].$$

Finding the investment θ that minimizes the net cost leads to

$$\min_{\theta \in [0,\infty)} J(\theta).$$

The stochastic version of the gradient search method of Chapter 1 can be implemented via our results in Part I, using

$$\theta_{n+1} = \Pi_{[0,\infty)}(\theta_n + \epsilon_n Y_n),$$

where Y_n is an estimator of the negative derivative of the objective function, that is, we seek

$$\mathbb{E}[Y_n \mid \mathfrak{F}_{n-1}] = \frac{d}{d\theta}\mathbb{E}[h(X(\theta_n))] - 1, \text{ for } \theta_n \geq 0, \quad \text{where } h(x) \stackrel{\text{def}}{=} x\, p(x + \hat{S}),$$

and we assume that $\hat{S}$ is a random variable independent of x. Notice that the function h is in general a non-linear function of x. Suppose that for a fixed investment amount θ we can simulate the output variables $X(\theta), \hat{S}$. Then the sample average of $h(X(\theta))$ is an unbiased estimator of the profit. However, for the purpose of optimization, we need also to *estimate the gradient* of the profit. ✹✹✹

As we have seen in Part I, many problems in stochastic optimization can be stated as a non linear optimization problem, like the above mining example, where measurements of performance functions, constraints, and their derivatives may be noisy. In particular, we established the conditions under which numerical gradient-based algorithms can be used to approximate the solution to such problems, if we can build appropriate estimators of the gradients.

While finding appropriate gradient estimators can be done in an ad hoc manner, it is preferable to have a calculus of gradient estimation at hand that allows to build gradient estimates for complex problems in a systematic way. Like in calculus, we will first study gradient estimation for simple problems. To transcend these results to problems that are of relevance in practice, we will identify typical stochastic models encountered in applications and then show how the elementary gradient results are related to these problem settings. The analysis of gradient estimators for more elaborate situations will be given in the subsequent chapters.

7.2 ONE-DIMENSIONAL DISTRIBUTIONS

In this section we explain the key issues of gradient estimation by means of simple examples. Let Θ be some non-empty connected subset of $\mathbb{R}$. For $\theta \in \Theta$, let $X(\theta)$ denote some random variable defined on an underlying probability space, and let h be some real-valued measurable mapping such that $\mathbb{E}[h(X(\theta))]$ is well defined for any $\theta \in \Theta$. Estimating the derivative of $\mathbb{E}[h(X(\theta))]$ with respect to θ is of key importance in optimizing $\mathbb{E}[h(X(\theta))]$. In other words, we ask how to estimate

$$\frac{d}{d\theta} \mathbb{E}[h(X(\theta))].$$

Since optimization with unbiased derivative information is preferable to biased estimators, we are seeking for a measurable (possibly random) mapping $\psi(h, X, \theta; \omega)$ such that

$$\frac{d}{d\theta} \mathbb{E}[h(X(\theta))] = \mathbb{E}[\psi(h, X(\theta), \theta)]. \tag{7.1}$$

Suppose now that $X(\theta)$ has density $f_\theta(x)$ and that $X(\theta) = r(\theta, U)$ were $U \in [a, b]$ has density $f(u)$. There are two ways to write the expectation, namely

$$\mathbb{E}[h(X(\theta))] = \int h(x) f_\theta(x)\, dx = \int_a^b h(r(\theta, u))\, f(u)\, du.$$

We will now present three important methods for estimating derivatives that satisfy (7.1). The "pathwise" approach considers the latter expression for the expectation and concentrates on the random variable $h(r(\theta, U))$ in order to estimate the derivative and known as the *infinitesimal perturbation* method. The second approach is to study the first expression, where all dependency on θ is in the distribution. We call this the "distributional" approach and it gives rise to two methods: the *score function* (SF) and the *measure-valued differentiation* (MVD) method. In this chapter we present the basic reasoning for the three methods, postponing all proofs and mathematical formulation to Chapters 8 and 9.

7.2.1 Infinitesimal Perturbation Analysis (Pathwise Gradient)

Representation: A random variable $X(\theta)$ on $(\Omega, \mathfrak{F}, \mathbb{P})$ is, by definition, a function of both ω and θ. It is always possible to *represent* $X(\theta)$ as a function of θ and another random variable U with uniform distribution on $(0,1)$ (see Theorem B.3). In particular, given a distribution function F_θ, the random variable $X(\theta, \omega) = F_\theta^{-1}(U(\omega))$ has distribution F_θ when $U \sim U(0,1)$. Here $F_\theta^{-1}(u) = \inf\{x : F_\theta(x) \geq u\}$ is the generalized inverse of the distribution function (see Exercise 7.1). Other representations are also useful. For example, a normal random variable $X \sim \mathcal{N}(\theta, \sigma^2)$ can be *represented* as a function of the standard normal $Z \sim \mathcal{N}(0,1)$, because $X \stackrel{d}{=} \theta + \sigma Z$. In both examples we can express $X(\theta, \omega)$ explicitly in terms of random variables whose distributions are not dependent on θ.

Given such a representation there is a function r such that the random variable $X(\theta) \stackrel{d}{=} r(\theta, U)$, where U is a random variable whose distribution is *independent* of θ. This is the building block for the analysis, using the *stochastic derivative*, which is the random variable that takes values $\frac{\partial}{\partial \theta} r(\theta, U)$ for each fixed value of U (assuming that it exists).

The derivative estimation problem can be dealt with in a straightforward way provided that the random variable $h(r(\theta, U))$ is differentiable a.s. and interchanging expectation and differentiation is justified (see Chapter 8). Indeed, in this case one obtains

$$\frac{d}{d\theta}\mathbb{E}[h(X(\theta))] = \mathbb{E}\left[\frac{d}{d\theta}h(X(\theta))\right] = \mathbb{E}\left[\left(\frac{d}{d\theta}X(\theta)\right)h'(X(\theta))\right], \qquad (7.2)$$

where $h'(x)$ denotes the derivative of $h(x)$ with respect to x. It is worth noting that stochastic derivatives are measurable, see Exercise 7.2. Alternatively, the random variable $\partial X(\theta) h'(X(\theta))/\partial \theta$ is also called *sample-path derivative*.

Letting

$$\boxed{\psi(h, X(\theta), \theta) = \left(\frac{d}{d\theta}X(\theta)\right) h'(X(\theta))}$$

the expression for the derivative estimator in (7.2) is of the general form (7.1).

For many distributions that are of importance in applications, the sample path (or stochastic) derivative of $X(\theta)$ with respect to θ can be obtained in a simple form, yielding an efficient derivative estimator.

This approach of using stochastic derivatives as estimator for the derivative of an expected value is called *infinitesimal perturbation analysis* (IPA). This terminology is now standard in the field of gradient estimation, and is commonly used in the engineering and control literature.

Example 7.2. Consider an exponential random variable $X(\theta)$ with mean value $\theta > 0$:

$$\mathbb{P}(X(\theta) \leq x) = 1 - e^{-x/\theta},$$

for $x \geq 0$. The exponential distribution is used for modeling life times of items in reliability problems, or service times in production systems. For U uniformly distributed on $(0,1)$ and independent of everything else, let

$$\tilde{X}(\theta) = -\theta \ln(1-U), \qquad (7.3)$$

then $\tilde{X}(\theta)$ is a version of $X(\theta)$, i.e., $\tilde{X}(\theta)$ and $X(\theta)$ are equal in distribution (refer to Exercise 7.1). Without loss of generality we will in the following identify $\tilde{X}(\theta)$ and $X(\theta)$. Taking derivatives yields

$$\frac{d}{d\theta} X(\theta) = -\ln(1-U) = \frac{1}{\theta} X(\theta).$$

Inserting this expression for the derivative into (7.2) yields

$$\frac{d}{d\theta} \mathbb{E}[h(X(\theta))] = \mathbb{E}\left[\frac{1}{\theta} X(\theta) h'(X(\theta))\right]. \qquad (7.4)$$

Observe that above we have paramterized the exponential distribution via the mean θ. While this is a natural parameterization from the point of control theory as θ is a simple scaling of the realizations, in probability theory the exponential distribution is typically parameterized via its rate or inverse scale. More formally, it is common to denote by $\text{Exp}(\lambda)$ the exponential distribution with intensity λ (so that the mean is $1/\lambda$):

$$\mathbb{P}(X \le x) = 1 - e^{-\lambda x},$$

for $x \ge 0$. Let now θ denote the intensity (not the mean) of the exponential random variable $X(\theta) \sim \text{Exp}(\theta)$, then the IPA derivative is given by

$$\frac{d}{d\theta} X(\theta) = \frac{d}{d\theta}\left(-\frac{1}{\theta} \ln(1-U)\right) = -\frac{1}{\theta} X(\theta).$$

Note that the stochastic derivatives of the mean and rate parametrization differ only by sign. ✳✳✳

Remark 7.1. Stochastic derivatives have already been included in Examples 3.4, 4.4, and 6.6, where it was straightforward to show that interchange of derivative and expectation is applicable.

Example 7.3. Refer to Example 7.1. Assume that the price function is of the form $p(x) = x^{-1/d}$. To find the optimal investment level θ, one uses

$$\theta_{n+1} = \theta_n - \epsilon \left(1 - \frac{d}{d\theta} \mathbb{E}\left[X(\theta_n)(X(\theta_n) + \hat{S})^{-1/d}\right]\right).$$

Assume that $X(\theta)$ is exponential with mean 50θ. Then in general, it is necessary to evaluate the expectation numerically, that is, using $d=2$, $\hat{S}=1$ the integral

$$\mathbb{E}\left[\frac{X(\theta)}{\sqrt{X(\theta)+1}}\right] = \frac{1}{50\theta} \int_0^\infty \frac{x e^{-x/(50\theta)}}{\sqrt{x+1}} dx$$

is not in closed form. Figure 7.1 shows the form of the function. We obtained this plot using Mathematica, and it required 190 seconds to execute. Clearly the function is convex and has a unique minimum θ^* (around 10).

A PRIMER FOR GRADIENT ESTIMATION

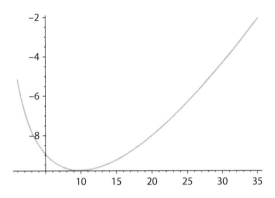

Figure 7.1. Objective function $J(\theta)$.

It is possible to calculate the derivative of this integral expression, but then each step in the recursion will take approximately 200 seconds of execution time, rendering the iterative method very inefficient. We explore now the alternative using the stochastic approximation method and generating consecutive independent samples of $X(\theta_n)$. But, for that we need to use estimation of the derivative of $J(\theta)$. Following Example 7.2,

$$X(\theta) \stackrel{d}{=} -50\,\theta \ln(1-U) \implies \frac{d}{d\theta}X(\theta) = \frac{X(\theta)}{\theta}, \qquad (7.5)$$

and $h(x) = x/\sqrt{1+x}$, so that

$$h'(X(\theta)) = \frac{1}{(1+X(\theta))^{1/2}} - \frac{X(\theta)}{2(1+X(\theta))^{3/2}}. \qquad (7.6)$$

Putting this together, use

$$\theta_{n+1} = \theta_n - \epsilon\left(1 - \frac{X(\theta_n)}{\theta_n} h'(X(\theta_n))\right),$$

where we omit applying a projection to keep θ_n feasible. The corresponding pseudocode is shown in Algorithm 7.1. In the code, the call $\text{Exp}(\lambda)$ returns an exponential random variable with intensity λ. A resulting trajectory of this SA is shown in Figure 7.2.

Algorithm 7.1 Stochastic approximation with IPA: Mining problem

---

H-Prime(x, θ) given in (7.6)
return $1/\sqrt{1+x} - x(1+x)^{-3/2}/2$
Initialize $\theta_0 = 5$, $N = 1000$, $\epsilon = 0.3$
for $(n = 0, \ldots N)$ **do**
    Generate sample: $X = \text{Exp}(1/(50\theta_n))$
    $\theta_{n+1} = \theta_n - \epsilon\left(1 - \frac{X}{\theta_n} \times \text{H-Prime}(X, \theta_n)\right)$

It took Mathematica 0.0238 seconds to run 1,000 iterations of the algorithm using $\epsilon = 0.3$. If more accuracy is desired, one can follow the directives in Chapter 6 to choose appropriate stepsize sequences to achieve better precision. Comparing the execution times it should be apparent that it is beneficial to use derivative estimation rather than numerical

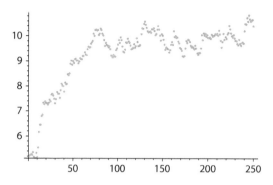

Figure 7.2. A trajectory of the stochastic approximation $\{\theta_n\}$.

integration in cases like this. We deliberately left out projection here, as due to the small variance of the gradient estimator and choice of ϵ, θ_n stays feasible in this experiment and no projection is needed.

Exercise 7.7 asks to verify the conditions under which the stochastic approximation above indeed converges to the optimal value. This requires integrating the relevant results from Part I into this problem. ✳✳✳

Example 7.4. This example provides an overview of IPA derivatives for standard distributions. The derivation of the expressions is left as an exercise to the reader. Let $X(\mu, \sigma)$ have normal distribution with mean μ and standard deviation σ, for $\sigma > 0$. The density is given by

$$\phi_{\mu,\sigma^2}(x) = \frac{1}{\sigma\sqrt{2\pi}} e^{-\frac{1}{2}\frac{(x-\mu)^2}{\sigma^2}}, \quad x \in \mathbb{R}.$$

If Z is a standard normal variable, i.e., $Z \sim \mathcal{N}(0, 1)$, then $X(\mu, \sigma) \stackrel{\text{def}}{=} \mu + \sigma Z$ is a sample of $\mathcal{N}(\mu, \sigma^2)$, which gives the following IPA derivatives

$$\frac{\partial}{\partial \mu} X(\mu, \sigma) = 1,$$

and

$$\frac{\partial}{\partial \sigma} X(\mu, \sigma) = Z = \frac{X(\mu, \sigma) - \mu}{\sigma}.$$

We denote by Weibull(α, λ) the Weibull distribution with shape parameter α and scale parameter λ, and cdf

$$1 - e^{-(\frac{x}{\lambda})^\alpha}, \quad x \geq 0. \tag{7.7}$$

Samples from the Weibull(α, λ) can be obtained using the inverse distribution method:

$$X_{\alpha,\lambda} = \lambda \left(-\ln(1-U)\right)^{\frac{1}{\alpha}},$$

where U is uniformly distributed on $(0, 1)$. The IPA derivative with respect to the scale λ is

$$\frac{\partial}{\partial \lambda} X_{\alpha,\lambda} = (-\ln(1-U))^{\frac{1}{\alpha}} = \frac{1}{\lambda} X_{\alpha,\lambda}.$$

Differentiating with respect to α yields

$$\frac{\partial}{\partial \alpha} X_{\alpha,\lambda} = \frac{\lambda}{\alpha} \left(-\ln(1-U)\right)^{\frac{1-\alpha}{\alpha}}.$$

Note that the above derivative cannot be written in a straightforward way as a transformation of $X_{\alpha,\lambda}$. Later on we will discuss a method for deriving an IPA derivative; see Lemma 8.1.

Let Pareto(α, λ) denote the Pareto type II distribution (also known as Lomax distribution) with scale parameter λ and shape parameter α having cdf

$$1 - \left(1 + \frac{x}{\lambda}\right)^{-\alpha}, \quad x \geq 0.$$

Note that for $\alpha > 1$ the mean value is given by $\lambda/(\alpha - 1)$. Samples of the Pareto(α, λ) distribution can be obtained from

$$X_{\alpha,\lambda} = \lambda((1-U)^{\alpha} - 1).$$

The IPA derivative with respect to the scale parameter is given by

$$\frac{\partial}{\partial \lambda} X_{\alpha,\lambda} = (1-U)^{\alpha} - 1 = \frac{1}{\lambda} X_{\alpha,\lambda},$$

whereas the derivative with respect to α leads to

$$\frac{\partial}{\partial \alpha} X_{\alpha,\lambda} = \lambda(1-U)^{\alpha} \ln(1-U).$$

The above shape derivative cannot be written in a straightforward way as a transformation of $X_{\alpha,\lambda}$. Lemma 8.1 in the following chapter will provide a method for obtaining an IPA derivative. ✻✻✻

Next we present an example of an IPA estimator for the case when θ scales the mean and the standard deviation of a normal random variable.

Example 7.5. Let $Z_{\theta,\theta\sigma}$ follow a normal distribution with mean θ and standard deviation $\theta\sigma$. Then

$$Z_{\theta,\theta\sigma} = \theta\sigma Z_{0,1} + \theta,$$

where $Z_{0,1}$ is a standard normal distribution. We note that θ is a scaling parameter of $Z_{\theta,\theta\sigma}$. The sample path derivative of $Z_{\theta,\theta\sigma}$ exists and equals $\sigma Z_{0,1} + 1$. Hence, the ensuing IPA estimator becomes

$$\frac{d}{d\theta} E\left[h(Z_{\theta,\theta\sigma})\right] = E\left[h'(Z_{\theta,\theta\sigma})(\sigma Z_{0,1} + 1)\right] = E\left[h'(Z_{\theta,\theta\sigma}) \frac{1}{\theta} Z_{\theta,\sigma\theta}\right], \quad (7.8)$$

provided that h is differentiable, and that interchange of differentiation and integration is justified. ✻✻✻

We conclude the discussion of IPA with an example where IPA is applicable only for some of the parameters that characterize the random variable.

Example 7.6. Let $p_i \in (0, 1)$, for $i = 1, \ldots, n$, such that $\sum p_i = 1$, and $\theta_i \in \mathbb{R}$, for $i = 1, \ldots, n$, where $\theta_i \neq \theta_j$ for $i \neq j$. Let X have distribution $\mathbb{P}(X = \theta_i) = p_i$, $1 \leq i \leq n$. We can sample X as follows. Let

$$A_i = \left\{ u : \sum_{k=0}^{i-1} p_k < u \leq \sum_{k=0}^{i} p_k \right\}$$

be a partition of the interval $(0, 1]$, with $p_0 = 0$. Consider now a uniform random variable $U \sim U(0, 1)$. Then the random variable

$$X(\theta) = \theta_i \mathbf{1}_{\{U \in A_i\}} \quad (7.9)$$

has the required distribution, that is, $\mathbb{P}(X(\theta) = \theta_i) = p_i$, with $p_0 = 0$. Using this representation, it is possible to calculate the stochastic derivative (fixing U) as

$$\frac{\partial}{\partial \theta_i} X(\theta) = \mathbf{1}_{\{X(\theta) = \theta_i\}}.$$

Notice that in this case, using the representation (7.9) it follows that X fails to be differentiable with respect to the weights p_i. ✳✳✳

7.2.2 Score Function

It is not always possible to differentiate sample paths or to interchange expectation and differentiation without harming unbiasedness of the estimator. Examples will be provided later on in the text. In the following we discuss alternative methods for obtaining the derivative of $\mathbb{E}[h(X(\theta))]$. To this end, let f_θ denote the pdf of $X(\theta)$. Rewriting the expected value as integral over f_θ, the derivative estimation problem becomes

$$\frac{d}{d\theta} \mathbb{E}[h(X(\theta))] = \frac{d}{d\theta} \int h(x) f_\theta(x)\, dx.$$

Assuming that f_θ is differentiable with respect to θ and that interchanging differentiation and integration is justified (which entails that $h(x)\partial f_\theta(x)/\partial \theta$ is integrable), one obtains

$$\frac{d}{d\theta} \mathbb{E}[h(X(\theta))] = \int h(x) \frac{\partial}{\partial \theta} f_\theta(x)\, dx. \qquad (7.10)$$

The problem with this manipulation is that $\frac{\partial}{\partial \theta} f_\theta(x)$ fails to be a density and we thus cannot sample from it. To see this, note that for a density f_θ it holds that $\int f_\theta(x)dx = 1$ and taking derivatives yields

$$0 = \frac{d}{d\theta} \int f_\theta(x)dx = \int \frac{\partial}{\partial \theta} f_\theta(x)\, dx. \qquad (7.11)$$

Hence, $\partial f_\theta(x)/\partial \theta$ integrates out to zero and is therefore not a density. Often a density can be introduced by means of a simple analytical transformation. This is done as follows:

$$\int h(x) \frac{\partial}{\partial \theta} f_\theta(x)\, dx = \int h(x) \left(\frac{\partial}{\partial \theta} f_\theta(x) \right) \frac{f_\theta(x)}{f_\theta(x)}\, dx$$

$$= \int h(x) \left(\frac{\frac{\partial}{\partial \theta} f_\theta(x)}{f_\theta(x)} \right) f_\theta(x)\, dx. \qquad (7.12)$$

Note that

$$\frac{\partial}{\partial \theta} \log(f_\theta(x)) = \frac{\frac{\partial}{\partial \theta} f_\theta(x)}{f_\theta(x)}$$

and letting

$$\mathsf{SF}(\theta, x) = \frac{\partial}{\partial \theta} \log(f_\theta(x)), \qquad (7.13)$$

the derivative in (7.12) can be written in random-variable language as

$$\int h(x) \frac{\partial}{\partial \theta} f_\theta(x)\, dx = \mathbb{E}[h(X(\theta))\, \mathsf{SF}(\theta, X(\theta))]$$

yielding

$$\frac{d}{d\theta} \mathbb{E}[h(X(\theta))] = \mathbb{E}[h(X(\theta))\, \mathsf{SF}(\theta, X(\theta))],$$

or, in the notation of (7.1)

$$\psi(h, X(\theta), \theta) = h(X(\theta))\, \text{SF}(\theta, X(\theta)).$$

The mapping $\text{SF}(\theta, \cdot)$ is called the *score function* and the estimation approach is called the *score function method*.

Example 7.7. Revisit the exponential distribution with mean $\theta \in (0, \infty) = \Theta$ in Example 7.2. Note that the distribution of $X(\theta)$ has pdf $f_\theta(x) = \exp(-x/\theta)/\theta$, for $x \geq 0$ and $\theta > 0$. The score function can be computed as

$$\text{SF}(\theta, x) = \frac{1}{\theta}\left(\frac{x}{\theta} - 1\right)$$

and the SF estimator becomes

$$\psi(h, X(\theta), \theta) = h(X(\theta))\frac{1}{\theta}\left(\frac{X(\theta)}{\theta} - 1\right). \tag{7.14}$$

Exercise 7.3 calculates the corresponding estimator when θ is the intensity, not the mean, of this distribution. ✳✳✳

Example 7.8. This example provides an overview of the score functions for standard distributions. The derivation of the expressions is left as an exercise to the reader. Let Bernoulli(θ) denote the Bernoulli distribution on $\{0, 1\}$ assigning probability θ to 1 and $1 - \theta$ to 0 with density

$$f_\theta(n) = \theta^n(1-\theta)^{1-n}, \quad n \in \{0, 1\},$$

and note that $X(\theta) = \mathbf{1}_{\{U \leq \theta\}}$ yields a Bernoulli(θ) sample, for U uniform on $[0, 1]$. Then the score function reads

$$\text{SF}(\theta, n) = \frac{n}{\theta} + \frac{n-1}{1-\theta}, \quad n \in \{0, 1\}.$$

For the Poisson(θ) distribution with cdf $\theta^x e^{-\theta}/x!$, for $x \in \mathbb{N}$, the score function can be computed to be

$$\text{SF}(\theta, n) = \frac{x}{\theta} - 1, \quad n \geq 0.$$

For the $\mathcal{N}(\mu, \sigma^2)$ distribution we obtain for the score function with respect to $\theta = \mu$ by

$$\text{SF}_\mu(\theta, x) = \frac{x - \mu}{\sigma^2}, \quad x \in \mathbb{R},$$

and the score function with respect to $\theta = \sigma$ by

$$\text{SF}_\sigma(\theta, x) = -\frac{1}{\sigma} + \frac{1}{\sigma^3}(x - \mu)^2, \quad x \in \mathbb{R}.$$

For the Weibull(α, λ) distribution we obtain as score functions for $\theta = \alpha$

$$\text{SF}_\alpha(\theta, x) = \frac{1}{\alpha} + \left(1 - \left(\frac{x}{\lambda}\right)^\alpha\right)\ln\left(\frac{x}{\lambda}\right), \quad x > 0,$$

and for $\theta = \lambda$

$$\text{SF}_\lambda(\theta, x) = \frac{\alpha}{\lambda}\left(\left(\frac{x}{\lambda}\right)^\alpha - 1\right), \quad x > 0.$$

Note that for $\alpha = 1$ the above recovers the score function of the exponential distribution with mean λ.

Consider the Gamma(α, β) distribution with shape parameter α and scale parameter β. The pdf is given by
$$\frac{\beta^\alpha}{\Gamma(\alpha)} x^{\alpha-1} e^{-\beta x}, \quad x > 0,$$
for $\alpha, \beta > 0$, where $\Gamma(\alpha)$ denotes the gamma function. For $\alpha \in \mathbb{N}$, Gamma(α, β) is the distribution of the sum of α independent exponential variables with mean $1/\beta$. For $\theta = \alpha$ we obtain
$$\mathsf{SF}_\alpha(\theta, x) = \ln(\beta x) - \psi(\alpha), \quad x \in \mathbb{R},$$
where $\psi(\alpha)$ denotes the digamma function, and for $\theta = \beta$ it holds that
$$\mathsf{SF}_\beta(\theta, x) = \frac{\alpha}{\beta} - x, \quad x \in \mathbb{R}.$$

The density of the normal distribution with mean θ and standard deviation σ is given by
$$\phi_{\theta, \sigma^2}(x) = \frac{1}{\sigma\sqrt{2\pi}} e^{-\frac{1}{2}\left(\frac{x-\theta}{\sigma}\right)^2}, \quad x \in \mathbb{R}.$$

For $\theta = \mu$ we obtain
$$\mathsf{SF}_\mu(\theta, x) = \frac{1}{\sigma^2}(x - \theta), \quad x \in \mathbb{R},$$
and for $\theta = \sigma$
$$\mathsf{SF}_\sigma(\theta, x) = \frac{(x - \mu)^2}{\theta^3} - \frac{1}{\theta}, \quad x \in \mathbb{R}. \qquad \text{\textasteriskcentered\textasteriskcentered\textasteriskcentered}$$

Example 7.9. This example is a simplified version of the problem introduced in [155]. The setting is a public train service for local transport such as those found in airports. The time between trains is a random variable $X(\theta)$. Once the train arrives all passengers in the station board the train. In the general setting passengers arrive according to a "bulk" arrival process: a group of P_k passengers arrives at time A_k. We need not make specific assumptions about this arrival process. Call $N(\cdot)$ the counting process for the arrivals, that is, $N(t) = \max(n: A_n \leq t)$. The cumulative waiting time is
$$W(X(\theta)) = \sum_{k=1}^{N(X(\theta))} P_k (X(\theta) - A_k). \qquad (7.15)$$

Assume that $\theta = \mathbb{E}[X(\theta)]$ is a scale parameter of the distribution of $X(\theta)$, where $\theta > 0$. It is in general true that for public transportation systems the cost increases with decreasing θ, so a constraint on the cumulative waiting time is usually imposed. We assume that we wish to minimize the cost subject to $\mathbb{E}[W(\theta)] \leq b$, for given bound b. Following the reasoning of Chapters 1 and 2, it is arguable that the constraint will be active at the optimal value. Thus, we seek θ^* satisfying $\mathbb{E}[W(X(\theta^*))] = b$. If we state the optimization problem as
$$\min_{\theta \in (0,\infty)} J(\theta), \quad \text{where} \quad J(\theta) \stackrel{\text{def}}{=} \frac{1}{2} (\mathbb{E}[W(X(\theta))] - b)^2,$$
then the gradient descent method uses
$$G(\theta) = -(\mathbb{E}[W(X(\theta))] - b) \times \frac{d}{d\theta} \mathbb{E}[W(X(\theta))] \qquad (7.16)$$
for $\theta \in (0, \infty)$. In previous chapters we used a simplified version of the above root-finding problem. Indeed, as θ is a scaling parameter for $X(\theta)$ it follows that $\mathbb{E}[W(\theta)]$ is increasing in

A PRIMER FOR GRADIENT ESTIMATION

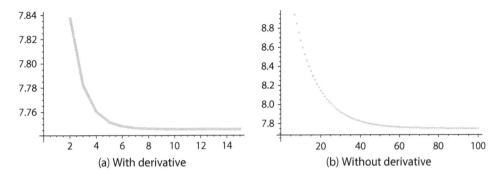

Figure 7.3. Deterministic algorithms (ODE) for root finding with and without the derivative as a factor in the vector field.

θ and $d\mathbb{E}[W(\theta)]/d\theta$ can be replaced by 1, leading to the coercive vector field $-\mathbb{E}[W(X(\theta)) - b]$. But now we can use estimates of gradients, which accelerate the process.

In order to illustrate the increase in efficiency using gradient information for the feedback, consider a very simple model for (7.15) where $P_k \equiv 1$ and $N(\cdot)$ is a homogeneous Poisson process. In that simple case it is possible to calculate $\mathbb{E}[W(X(\theta))]$ explicitly as follows. We first condition on $X(\theta)$. Next we condition on $N(X(\theta))$. It is well known that conditioning on $N(X(\theta)) = n$, the arrival times $\{A_k\}$ have the same distribution as n iid uniform random variables distributed on the interval $[0, X(\theta))$ and ordered. Therefore, $\mathbb{E}[(X(\theta) - A_k) \mid N(X(\theta)), X(\theta)] = X(\theta)/2$ for each $k = 1, \ldots, N(X(\theta))$ and if $N(\cdot)$ is a Poisson process with rate λ then $\mathbb{E}[N(t)] = \lambda t$, which leads to

$$\mathbb{E}[W(X(\theta))] = \mathbb{E}\left[\mathbb{E}\left[\sum_{k=1}^{N(X(\theta))} (X(\theta) - A_k) \,\middle|\, X(\theta), N(X(\theta))\right]\right]$$

$$= \mathbb{E}\left[\mathbb{E}\left[N(X(\theta))\left(\frac{X(\theta)}{2}\right) \,\middle|\, X(\theta)\right]\right] = \lambda \mathbb{E}\left[\frac{X^2(\theta)}{2}\right].$$

In our model, for verification purposes we used $\lambda = 1$ and $X(\theta) \sim \text{Weibull}(2, \theta)$. In this case $\mathbb{E}[X^2(\theta)] = \theta^2$, so we know that for this model

$$\mathbb{E}[W(X(\theta))] = \frac{\theta^2}{2}; \quad \frac{d}{d\theta}\mathbb{E}[W(X(\theta))] = \theta.$$

In particular, this implies that for this specific model, $\theta^* = \sqrt{2b}$, which we will use only as a benchmark to show how the stochastic approximations evolve. Figure 7.3 shows the results of the iterations with and without multiplying $(\mathbb{E}[W(X(\theta))] - b)$ by the derivative of $\mathbb{E}[W(X(\theta))]$. Using derivative information accelerates the process almost ten times.

In real applications, the arrival process of passengers is not a homogeneous Poisson process. The procedures that we develop here can be applied to the general case, only assuming that the distribution of $X(\theta)$ is known. For this problem, we need to estimate the vector field (7.16). Because there is a product of expectations, we will estimate $\mathbb{E}[W(X(\theta))]$ and its derivative independently.

We now present the development of the SF estimator of $d\mathbb{E}[W(X(\theta))]/d\theta$. When $X(\theta) \sim \text{Weibull}(2, \theta)$ the density is

$$f_\theta(x) = \frac{2x}{\theta^2} e^{-(x/\theta)^2}.$$

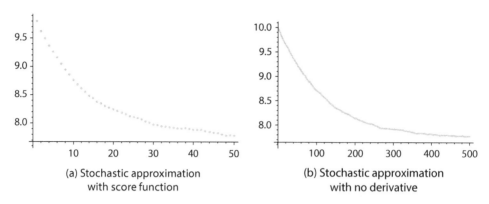

Figure 7.4. Stochastic approximation using $K = 1{,}000$ observations per batch and $\epsilon = 0.001$: (a) with score function; and (b) plain root finding without derivative (notice that this takes 100 times more updates). The bound is set to $b = 30$, and we find $\theta^* = 7.7446$ (true value = 7.7459).

Therefore (see Example 7.8):

$$\text{SF}(\theta, x) = \frac{2}{\theta}\left(\frac{x^2}{\theta^2} - 1\right).$$

In order to implement the stochastic approximation, we use batches of K observations to estimate $\mathbb{E}[W(X(\theta))]$ followed by an independent batch of K observations to estimate its derivative.

Figure 7.4 shows the result of the algorithm using $\epsilon = 0.001$, batch size $K = 1{,}000$ and $n = 50$ updates of θ_n. Here we use the feedback

$$Y_n = -\left(\frac{1}{K}\sum_{k=1}^{K} W(X'_i(\theta_n)) - b\right) \times \left(\frac{1}{K}\sum_{i=1}^{K} \text{SF}(\theta, X_i(\theta_n))W(X_i(\theta_n))\right), \quad (7.17)$$

where $X'_i(\theta)$ is independent of $X_i(\theta)$ and for each generated value of $X_i(\theta)$ or $X'_i(\theta)$ we generate the passenger arrival process $\{(A_k, P_k)\}$. For purposes of verification we used Poisson arrivals with unit rate, so we could use the theoretical values to validate our code. But the simulations can be executed just as easily with more complex data for the passenger arrivals. In the pseudocode for Algorithm 7.2 we show how consecutive batches are used for estimating first the cumulative waiting time and separately the SF derivative. Lines 13 and 22 subtract the very last difference $X - A$, which is negative because of the while condition. This is a common correction factor in simulations with similar random stopping times. We assume that the compiler has predefined functions for generating the random variables and for evaluating the sample mean using the call MEAN(v) for a vector v.

We conclude with a note of caution. The gradient estimation could be simplified by considering the alternative problem $\min_\theta \mathbb{E}[(W(X(\theta)) - b)^2]/2$, with gradient estimator

$$\hat{Y}_n = \frac{1}{2K}\sum_{i=1}^{K}(W(X_i(\theta_n)) - b)^2\,\text{SF}(\theta, X_i(\theta_n)).$$

While a more efficient estimator than Y_n, the SA using $\hat{Y}_n$ will fit the parameter θ using the MSE for $W(\theta) - b$, which is not informative for the original problem of fitting the expected value of $W(\theta)$ to b. Put differently, $\hat{G}(\theta) = -d(\mathbb{E}[(W(X(\theta)) - b)^2]/2)/d\theta$ is coercive for $\min_\theta \mathbb{E}[(W(X(\theta)) - b)^2]/2$, but not for the actual problem $\min_\theta (\mathbb{E}[W(X(\theta))] - b)^2/2$.

※※※

Algorithm 7.2 Stochastic approximation with SF for the train model using (7.17)

1: Define SF function of Weibull$(2, \theta)$ distribution :
2: SCOREFUNCTION(θ, x)
3:   **return** $2/\theta(x^2/\theta^2 - 1)$
4: Initialize: $\theta_0 = 10$, $b = 30$, $\theta_0 = 10$, $\epsilon = 0.001$, $N = 50$, $K = 1000$, $\delta = 10^{-4}$
5: **for** $(n = 0, \ldots, N)$ **do**
6:   **for** $(k = 0, \ldots K)$ **do**
7:     Generate $X \sim$ Weibull$(2, \theta_n)$
8:     Generate $A \sim$ Exp(1)
9:     $W(k) = X - A$
10:     **while** $(A < X)$ **do**
11:       $A \leftarrow A + $ Exp(1)
12:       $W(k) \leftarrow W(k) + (X - A)$
13:     $W(k) \leftarrow W(k) - (X - A)$
14:   MeanWait = MEAN(W)
15:   Initialize: SF $= 0$
16:   **for** $(k = 0, \ldots K)$ **do**
17:     Generate $X \sim$ Weibull$(2, \theta_n)$
18:     Generate $A \sim$ Exp(1)
19:     $W'(k) = X - A$
20:     **while** $(A < X)$ **do**
21:       $A \leftarrow A + $ Exp(1)
22:       $W'(k) \leftarrow W'(k) + (X - A)$
23:     $W'(k) \leftarrow W'(k) - (X - A)$
24:     SF$(k) \leftarrow$ SCOREFUNCTION$(\theta_n, X) \times W'(k)$
25:   MeanDer = MEAN$($ SF $)$
26:   $\theta_{n+1} = \Pi_{[\delta, 1/\delta]}(\theta_n - \epsilon ($MeanWait$- b) \times$ MeanDer $)$

Example 7.10. This example relates the score function to the SAA introduced in Section 3.5 Let $f_\theta(x)$ be the density of $X(\theta)$, and introduce the likelihood ratio

$$\ell(\theta_0, \theta; x) = \frac{f_\theta(x)}{f_{\theta_0}(x)},$$

provided that $f_{\theta_0}(x) \neq 0$ and set it zero otherwise. Then it holds for any integrable cost function h that

$$\mathbb{E}[h(X(\theta))] = \int_\mathbb{R} h(x) f_\theta(x) dx = \int_\mathbb{R} h(x) \ell(\theta_0, \theta; x) f_{\theta_0}(x) dx$$
$$= \mathbb{E}\left[h(X(\theta_0)) \ell(\theta_0, \theta; X(\theta_0))\right],$$

which is a well-known method in statistics known as *change of measure*. Letting $X_i(\theta_0)$, for $1 \leq i \leq N$, be a collection of iid samples of $X(\theta_0)$, we define

$$\hat{J}_n(\theta) = \frac{1}{n} \sum_{i=1}^n h(X_i(\theta_0)) \ell(\theta_0, \theta; X_i(\theta_0))$$

for the sample average over n iid samples. Provided that the likelihood ratio is sufficiently smooth, following the SAA philosophy put forward in Section 3.5, the sample-path function

$\hat{J}_n(\theta)$ can now be used for optimization via deterministic methods, see Chapter 1. Moreover, differentiating $\ell(\theta_0, \theta; X(\theta_0))$ gives

$$\frac{\partial}{\partial \theta}\ell(\theta_0, \theta; x) = \frac{\frac{\partial}{\partial \theta}f_\theta(x)}{f_{\theta_0}(x)},$$

which resembles the score function. Optimization can now be carried out with respect to the likelihood ratio, and for more details we refer to Chapter 14. ✻✻✻

We conclude this section with a discussion of variance reduction techniques for the score function. Since

$$\mathbb{E}[\mathsf{SF}(\theta, X(\theta))] = 0$$

(see (7.11)), we may extend the basic SF estimator by inserting some deterministic, possibly θ-depending, constant β_θ in the following way:

$$\frac{d}{d\theta}\mathbb{E}[h(X(\theta))] = \mathbb{E}\left[\left(h(X(\theta)) - \beta_\theta\right)\mathsf{SF}(\theta, X(\theta))\right].$$

The constant β_θ is an example of a *control variate* used in Monte Carlo simulation for variance reduction; see, for example, [200]. Indeed, finding β_θ^* solving (while keeping θ fixed)

$$\min_{\beta_\theta} \mathbb{E}\left[\left(\left(h(X(\theta)) - \beta_\theta\right)\mathsf{SF}(\theta, X(\theta))\right)^2\right]$$

yields the stabilized SF estimator

$$\left(h(X(\theta)) - \beta_\theta^*\right)\mathsf{SF}(\theta, X(\theta))$$

with a smaller variance. The factor β_θ^* is called *baseline* in the literature (see [326]) and represents the simplest of control variates one can apply to SF. For more details on control variates and their use for variance reduction of the score function, especially in the context of reinforced learning, we refer to [221] and the references therein. As illustrating example consider the SF estimator for the mean of an exponential random variable with mean θ, see Example 7.7. Then,

$$\frac{d}{d\theta}\mathbb{E}[X(\theta)] = \mathbb{E}[\left(X(\theta) - \beta_\theta\right)\mathsf{SF}(\theta, X(\theta))]$$
$$= \frac{1}{\theta}\mathbb{E}\left[\left(X(\theta) - \beta_\theta\right)\left(\frac{X(\theta)}{\theta} - 1\right)\right].$$

For finding β_θ^*, we have to solve

$$\min_{\beta_\theta} \int_0^\infty \left((x - \beta_\theta)\frac{1}{\theta}\left(\frac{x}{\theta} - 1\right)\right)^2 \frac{1}{\theta}\exp(-x/\theta)dx.$$

For example, for $\theta = 1$, the above minimization is solved by $\beta_1^* = 3$, so that

$$(X(1) - 3)\,\mathsf{SF}(1, X(1))$$

is the stabalized SF estimator with minimal variance at $\theta = 1$. More specifically, in this case the variance of the standard SF estimator can be computed to be 13 whereas that of the stabilized estimator is reduced to 4.

7.2.3 Measure-Valued Differentiation

The score function was introduced to deal with the fact that $\frac{\partial}{\partial \theta} f_\theta(x)$ in (7.10) fails to be a density in the expression

$$\frac{d}{d\theta} \mathbb{E}[h(X(\theta))] = \int h(x) \frac{\partial}{\partial \theta} f_\theta(x)\, dx.$$

Specifically, the score function helps to interpret this expression above as an expectation. An alternative way of dealing with this problem stems from measure theory. Under quite general conditions it is possible to write $\frac{\partial}{\partial \theta} f_\theta(x)$ as rescaled difference between two densities. To see this, let

$$c_\theta = \int \max\left(\frac{\partial}{\partial \theta} f_\theta(x), 0\right) dx = \int \max\left(-\frac{\partial}{\partial \theta} f_\theta(x), 0\right) dx, \qquad (7.18)$$

and introduce new densities

$$f_\theta^+(x) = \frac{1}{c_\theta} \max\left(\frac{\partial}{\partial \theta} f_\theta(x), 0\right) \qquad (7.19)$$

and

$$f_\theta^-(x) = \frac{1}{c_\theta} \max\left(-\frac{\partial}{\partial \theta} f_\theta(x), 0\right), \qquad (7.20)$$

where we use (7.11) to show that the normalizing constants for f_θ^+ and f_θ^- are equal. Inserting these densities into the right-hand side of (7.10) yields

$$\frac{d}{d\theta} \int h(x) f_\theta(x)\, dx = c_\theta \left(\int h(x) f_\theta^+(x)\, dx - \int h(x) f_\theta^-(x)\, dx \right), \qquad (7.21)$$

for any h such that interchanging differentiation and integration is justified. The above approach for obtaining the derivative is called *measure-valued differentiation* (MVD), and is realted to weak differentiation. The discussion of details is postponed to Section 8.4 in the next chapter.

Figure 7.5 shows a plot of f_θ and the corresponding $\partial f_\theta / \partial \theta$ for the Weibull$(2, \theta)$ distribution. The constant c_θ corresponds to the area under the positive part of the plot to the right, which equals the area under the negative part, because a density f_θ integrates to 1 for every value of θ. f_θ^+ (f_θ^-) is built as the normalized positive (negative) parts of this derivative and thus they are true densities. Observe that this construction implies that the supports of the two densities are disjoint.

Let $X(\theta) \sim f_\theta$ and $X^\pm(\theta) \sim f_\theta^\pm$ be random variables on the common probability space $(\Omega, \mathfrak{F}, \mathbb{P})$. Then in the notation of (7.1), the MVD estimator is the measurable (random) function

$$\boxed{\psi(h, X(\theta), \theta) = c_\theta (h(X^+(\theta)) - h(X^-(\theta))).}$$

Note that ψ is a random mapping, because we have "enlarged" the information structure when introducing the random variables $X^\pm(\theta)$. Compare with the IPA and SF examples where ψ is a deterministic mapping.

Example 7.11. In the case of the exponential distribution with mean θ (see Examples 7.2 and 7.7) the density is $f_\theta(x) = (1/\theta) e^{-x/\theta}$ and its θ-derivative can be written as follows:

$$\frac{\partial}{\partial \theta} f_\theta(x) = \frac{1}{\theta^3}(x - \theta) e^{-\frac{x}{\theta}} = \frac{1}{\theta}\left(\frac{x}{\theta^2} e^{-\frac{x}{\theta}} - \frac{1}{\theta} e^{-\frac{x}{\theta}} \right). \qquad (7.22)$$

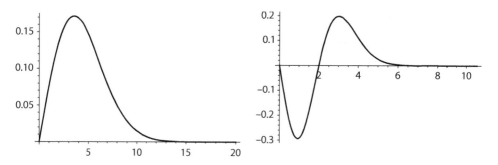

Figure 7.5. *Left*: Weibull$(2, \theta)$ density f_θ. *Right*: $\partial f_\theta / \partial \theta$. Plots show the functions at $\theta = 5$.

Noting that
$$f_\theta^e(x) = \frac{x}{\theta^2} e^{-\frac{x}{\theta}}$$
is the density of the distribution of the sum of two independent exponential random variables with mean θ, known as Gamma$(2, 1/\theta)$ distribution, the derivative expression becomes
$$\frac{d}{d\theta} \int h(x) f_\theta(x) \, dx = \int h(x) \frac{\partial}{\partial \theta} f_\theta(x) \, dx = \frac{1}{\theta} \left(\int h(x) f_\theta^e(x) \, dx - \int h(x) f_\theta(x) \, dx \right). \quad (7.23)$$

Letting $Y(\theta)$ be an exponential random variable with mean θ and independent of $X(\theta)$, the sum $Y(\theta) + X(\theta)$ is Gamma$(2, 1/\theta)$-distributed and the estimator reads in random variable language
$$\frac{d}{d\theta} \mathbb{E}[h(X(\theta))] = \frac{1}{\theta} \mathbb{E}[h(X(\theta) + Y(\theta)) - h(X(\theta))].$$

As an alternative, from (7.22), we obtain the densities corresponding to the positive and negative parts of $\frac{\partial f_\theta(x)}{\partial \theta}$ as
$$f_\theta^+(x) = \frac{(x - \theta) e^{1 - x/\theta}}{\theta^2} \qquad \text{if } x > \theta,$$
$$f_\theta^-(x) = \frac{(\theta - x) e^{1 - x/\theta}}{\theta^2} \qquad \text{if } x \leq \theta,$$

with $c_\theta = e^{-1}/\theta$, which do not belong to an identifiable family of distributions. A decomposition that seperates the positive and negative part is called Hahn-Jordan decomposition. Although the MVD is also unbiased using $X^\pm(\theta) \sim f_\theta^\pm$, in this case we encounter the problem that generating random variables with these distributions is not efficient (inverse function method must be evaluated numerically). This example also shows that in practice, one seeks a representation of the "plus" and "minus" measures that are easy to implement. ✱✱✱

Example 7.12. Consider the normal distribution with mean θ and standard deviation σ, the density of which is given by
$$\phi_{\theta, \sigma^2}(x) = \frac{1}{\sigma \sqrt{2\pi}} e^{-\frac{1}{2}\left(\frac{x - \theta}{\sigma}\right)^2}.$$

A PRIMER FOR GRADIENT ESTIMATION

Taking the derivative of $\phi_{\theta,\sigma^2}(x)$ with respect to θ yields

$$\frac{d}{d\theta}\phi_{\theta,\sigma^2}(x) = \frac{1}{\sqrt{2\pi}}\frac{x-\theta}{\sigma^3}e^{-\frac{1}{2}(\frac{x-\theta}{\theta\sigma})^2} = \frac{1}{\sigma\sqrt{2\pi}}\frac{1}{\sigma}\left(\mathbf{1}_{\{x\geq\theta\}}\frac{x-\theta}{\sigma} - \mathbf{1}_{\{x<\theta\}}\frac{\theta-x}{\sigma}\right)e^{-\frac{1}{2}(\frac{x-\theta}{\sigma})^2}.$$

Assume that interchanging of differentiation and integration is justified for a given function h. Then, substituting $y = (x-\theta)/\sigma$, yields

$$\frac{1}{\sigma}\int_\theta^\infty h(x)\left(\frac{x-\theta}{\sigma}\right)e^{-\frac{1}{2}(\frac{x-\theta}{\sigma})^2}dx = \int_0^\infty h(\sigma y+\theta)\,y\,e^{-\frac{1}{2}y^2}dy$$

(note that $\sigma dy = dx$). As for the second part,

$$\frac{1}{\sigma}\int_{-\infty}^\theta h(x)\left(\frac{\theta-x}{\sigma}\right)e^{-\frac{1}{2}(\frac{x-\theta}{\sigma})^2}dx = \frac{1}{\sigma}\int_{-\infty}^\theta h(x)\left(\frac{\theta-x}{\sigma}\right)e^{-\frac{1}{2}(\frac{-x+\theta}{\sigma})^2}dx$$

$$= \frac{1}{\sigma}\int_{-\theta}^\infty h(-x)\left(\frac{\theta+x}{\sigma}\right)e^{-\frac{1}{2}(\frac{x+\theta}{\sigma})^2}dx.$$

Substituting $y = (x+\theta)/\sigma$ yields

$$\frac{1}{\sigma}\int_{-\theta}^\infty h(-x)\left(\frac{\theta+x}{\sigma}\right)e^{-\frac{1}{2}(\frac{x+\theta}{\sigma})^2}dx = \int_0^\infty h(-\sigma y+\theta)\,y\,e^{-\frac{1}{2}y^2}dy.$$

Hence,

$$\int_{-\infty}^\infty h(x)\frac{d}{d\theta}\phi_{\theta,\sigma^2}(x)dx = \frac{1}{\sqrt{2\pi}}\left(\int_0^\infty h(\sigma y+\theta)\,y\,e^{-\frac{1}{2}y^2}dy - \int_0^\infty h(-\sigma y+\theta)\,y\,e^{-\frac{1}{2}y^2}dy\right)$$

and substituting y by z/σ we arrive at

$$\int_{-\infty}^\infty h(x)\frac{d}{d\theta}\phi_{\theta,\sigma^2}(x)dx$$

$$= \frac{1}{\sigma\sqrt{2\pi}}\left(\int_0^\infty h(z+\theta)\frac{z}{\sigma^2}e^{-\left(\frac{z}{\sqrt{2\sigma^2}}\right)^2}dz - \int_0^\infty h(\theta-z)\frac{z}{\sigma^2}e^{-\left(\frac{z}{\sqrt{2\sigma^2}}\right)^2}dz\right),$$

where we can identify the term

$$\frac{z}{\sigma^2}e^{-\left(\frac{z}{\sqrt{2\sigma^2}}\right)^2}$$

as the density of a Weibull($\alpha = 2, \lambda = \sqrt{2\sigma^2}$) distribution. To summarize, the weak derivative of the normal distribution with respect to the mean is obtained as difference between two shifted Weibull distributions. ✻✻✻

An overview on MVD representations of standard distributions, such as provided in Example 7.4 for IPA and in Example 7.8 for the score functions, is provided in Table 7.1. To simplify the notation, in the table we use the notation f_θ for pdf as well as for probability mass function in the discrete case. The term ds-Maxwell(m,s^2) denotes the double-sided Maxwell distribution with mean μ and shape parameter s having density

$$\frac{1}{s^3\sqrt{2\pi}}(x-\mu)^2 e^{-\frac{1}{2}(\frac{x-\mu}{s})^2}, \quad x \in \mathbb{R}.$$

Table 7.1. Differentiability of common distributions.

f_θ	c_θ	f_θ^+	f_θ^-
Bernoulli(θ) on $\{0, 1\}$	1	Dirac(0)	Dirac(1)
Poisson(θ)	1	Poisson(θ)+1	Poisson(θ)
Normal(θ, σ^2)	$1/\sigma\sqrt{2\pi}$	θ + Weibull(2,$(2\sigma^2)^{1/2}$)	θ - Weibull(2,$(2\sigma^2)^{1/2}$)
Normal(m, θ^2)	$1/\theta$	ds-Maxwell(m,θ^2)	Normal(m,θ^2)
Exp($1/\theta$)	$1/\theta$	Gamma(2, $1/\theta$)	Exp($1/\theta$)
Exp(θ)	$1/\theta$	Exp(θ)	Gamma(2, θ)
Gamma(α, θ)	α/θ	Gamma(α, θ)	Gamma($\alpha + 1, \theta$)
Weibull(α, θ)	α/θ	$[\text{Gamma}(2, \theta^{-\alpha})]^{1/\alpha}$	Weibull(α, θ)

Furthermore, by $[\text{Gamma}(2, \theta)]^{1/\alpha}$ we denote the distribution of the α-root of a Gamma (α, θ) random variable (see Example 7.14 for the full derivation of the last entry in Table 7.1), and "Poisson (θ)+1" denotes the shifted Poisson distribution, i.e., a sample from this distribution is obtained by adding 1 to a sample of the Poisson(θ) distribution. And finally, $\theta \pm$ Weibull(α, λ) denotes the distribution of $\theta \pm W$, where $W \sim$ Weibull(α, λ); see Example 7.12 for a derivation of the derivative representation. Table 7.1 follows Section 4.2.2. of [245] and provides an overview on techniques for jointly sampling the positive and negative part of a weak derivative.

A method for generating double-sided Maxwell distributed variables M_{μ,θ^2} is provided in [156]. It is worth noting that if U is uniformly distributed on $[0, 1]$ and independent of M_{μ,θ^2}, then $U M_{\mu,\theta^2}$ follows a normal distribution with mean μ and standard deviation θ. This leads to the following CRN implementation of the weak derivative of the normal distribution with respect to the standard devation:

$$\frac{d}{d\theta}\mathbb{E}[g(Z_{\mu,\theta^2})] = \frac{1}{\theta}\mathbb{E}\big[g(M_{\mu,\theta^2}) - g(U M_{\mu,\theta^2})\big].$$

As has been observed in [156], the above implementation yields for polynomial g an optimal variance. Here "optimal" means that the corresponding IPA and SF estimators have provably a larger variance.

Remark 7.2. In this monograph we follow the standard form of the Weibull; see, e.g., [258]. An editorial note is in order here. In his monograph [245], Pflug uses for the Weibull distribution the functional form

$$\bar{W}(\alpha, \beta) = 1 - e^{-\beta x^\alpha}, x \geq 0,$$

for $\alpha, \beta > 0$. Comparing with (7.7) it is easily seen that $W(\alpha, \lambda) = \bar{W}(\alpha, \beta(\lambda))$, where

$$\beta(\lambda) = \left(\frac{1}{\lambda}\right)^\alpha.$$

By the chain rule of differentiation, we can transform the weak derivative of normal distribution in Table 7.1 expressed via W into expression in $\bar{W}$. This yields the form proposed by

Pflug, namely,

$$\frac{\partial}{\partial \theta}\mathcal{N}(\theta, \sigma^2) = 1/\sigma\sqrt{(2\pi)}\left(\left[\theta + \bar{W}(2, 1/(2\sigma^2))\right] - \left[\theta - \bar{W}(2, 1/(2\sigma^2))\right]\right).$$

For the Weibull distribution, the transformation is less straightforward due the power $1/\alpha$ in the Gamma distribution. For detailed derivation of the weak derivative of the Weibull, we refer to Example 7.14 below.

We have presented a brief motivation of the main techniques for finding unbiased gradient estimators. We summarize the findings for the exponential distribution in the following example.

Example 7.13. For the case of $X(\theta)$, following an exponential distribution with mean θ we have derived the following IPA, SF, and MVD estimator for $d\mathbb{E}[h(X(\theta))]/d\theta$:

$$\text{IPA} \quad \frac{1}{\theta}\mathbb{E}\left[X(\theta)h'(X(\theta))\right] = \frac{d}{d\theta}\mathbb{E}\left[h(X(\theta))\right]$$

$$= \begin{cases} \frac{1}{\theta}\mathbb{E}\left[h(X(\theta))(X(\theta)/\theta - 1)\theta\right] & \text{SF} \\ \frac{1}{\theta}\mathbb{E}\left[h(X(\theta) + Y(\theta)) - h(X(\theta))\right] & \text{MVD}, \end{cases} \quad (7.24)$$

where $Y(\theta)$ has exponential distribution, independent of $X(\theta)$. Note that the estimator on the left-hand side of (7.24) is based on sample-path analysis whereas the estimators on the right-hand side of (7.24) are based on a distributional analysis. The estimators differ in functional form, complexity and variance. ✳✳✳

Example 7.14. Consider the model in Example 7.9 where passengers arrive at a platform according to a bulk process $N(\cdot)$ with inter arrival times $\{A_k\}$ and corresponding group sizes $\{P_k\}$. Calling $X(\theta)$ the time until the train arrives, the cumulative wait $W(X(\theta))$ is given by (7.15), and $\theta \in (0, \infty)$.

Rather than applying the result from Table 7.1, we fully develop here the MVD estimator for the derivative of $\mathbb{E}[W(X(\theta))]$ to use for the target vector field (7.16), when $X(\theta) \sim$ Weibull$(2, \theta)$. The derivative of the density $f_\theta(x) = (2x/\theta^2)e^{-(x/\theta)^2}$ is given by

$$\frac{\partial}{\partial \theta}f_\theta(x) = \frac{4x^3}{\theta^5}e^{-x^2/\theta^2} - \frac{4x}{\theta^3}e^{-x^2/\theta^2} = \frac{2}{\theta}\left(\frac{2}{\theta^4}x^3 e^{-x^2/\theta^2} - \frac{2x}{\theta^2}e^{-x^2/\theta^2}\right) \quad (7.25)$$

for $\theta \in (0, \infty)$. We recognize the density $f_\theta^-(x) = f_\theta(x)$ as the original Weibull density. To identify the density $f_\theta^+(x)$ we use the fact that if $X \sim F_X$ and $Y = \phi^{-1}(X)$ for some monotone function ϕ, then

$$f_Y(y) = \frac{d}{dy}\mathbb{P}(Y \le y) = \frac{d}{dy}\mathbb{P}(X \le \phi(y)) = \frac{d}{dy}F_X(\phi(y)) = f_X(\phi(y)) \times \phi'(y),$$

where $f_X(\cdot)$ is the density of X. By inspection, we propose $\phi(y) = y^2$ and $X \sim \Gamma(2, \theta^2)$ with density

$$f_X(x) = \frac{x}{\theta^4}e^{-x/\theta^2}$$

and $Y = \sqrt{X}$, which yields

$$f_Y(y) = \frac{y^2}{\theta^4}e^{-y^2/\theta^2}(2y) = \frac{2y^3}{\theta^4}e^{-y^2/\theta^2},$$

which we identify as $f_\theta^+(y)$ as obtained above in (7.25).

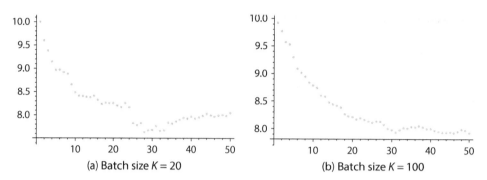

Figure 7.6. Stochastic approximation with MVD using $\epsilon = 0.001$ with batch sizes of $K = 20$ and $K = 100$ observations, respectively.

In summary, $X^+(\theta) \stackrel{d}{=} \sqrt{\text{Gamma}(2, 1/\theta^2)}$ and $X^-(\theta) \stackrel{d}{=} X(\theta)$, for $\theta \in (0, \infty)$. We will now simplify the generation of the "plus" random variable using the inverse function method. First, a Gamma distribution with integer shape parameter k (in our case $k = 2$) is the distribution of the sum of k iid exponential random variables with same scale parameter. Thus

$$X^+(\theta) \stackrel{d}{=} \sqrt{-\theta^2 \ln(1 - U_1) - \theta^2 \ln(1 - U_2)},$$

where U_1, U_2 are independent uniform random variables on $[0, 1]$. On the other hand, $X(\theta) \sim \text{Weibull}(2, \theta)$ so that using again the inverse function method we can express

$$X(\theta) = \theta \sqrt{-\ln(1 - U)}.$$

In order to obtain a positive correlation between $X^+(\theta)$ and $X^-(\theta)$ (which decreases the variance of the MVD difference for a monotone function W) we identify $U = U_1$ and use

$$X^+(\theta) = \sqrt{(X(\theta))^2 + Y(\theta)}; \quad Y(\theta) \sim \text{Exp}(1/\theta^2). \tag{7.26}$$

The corresponding feedback for the stochastic approximation is now

$$Y_n = (W(X'(\theta)) - b) \times \frac{2}{\theta}(W(X^+(\theta)) - W(X(\theta))), \tag{7.27}$$

where $X'(\theta)$ and $X(\theta)$ are iid Weibull$(2, \theta)$ and $X^+(\theta)$ is calculated in (7.26). The feedback can also use estimation by batches, as we did for the SF estimation.

It is interesting to compare Figure 7.4 and Figure 7.6. Although similar, it is noteworthy that MVD only used 20 (left) and 100 (right) observations per batch, while SF used 1,000. In addition, the simulation can be done in an efficient manner as we now explain. While (7.27) seems to require two separate simulations for the derivative, we used the same passengers as follows. First, from (7.26) it follows that $X^+(\theta) > X(\theta)$ w.p.1. Thus we can simplify the difference process:

$$D(\theta) = W(X^+(\theta)) - W(X(\theta)) = \sum_{k=1}^{N(X(\theta))} (X^+(\theta) - X^-(\theta)) + \sum_{k=N(X(\theta))+1}^{N(X^+(\theta))} (X^+(\theta) - A_k)$$

$$= N(X(\theta))(X^+(\theta) - X(\theta)) + \sum_{k=N(X(\theta))+1}^{N(X^+(\theta))} (X^+(\theta) - A_k). \tag{7.28}$$

Algorithm 7.3 shows the corresponding pseudocode, where we used the fact that when the arrival process is stationary, the sum on the right hand side above has the same distribution as the sum from $k = 0$ to $N(X^+(\theta) - X(\theta))$. Line 19 calls for generating $N(X(\theta))$ directly as a Poisson random variable. If the inter arrival times are not exponential, this line of code must be modified accordingly.

Algorithm 7.3 Stochastic approximation with MVD for the train model using (7.28)

1: Initialize: $\theta_0 = 10$, $b = 30$, $\theta_0 = 10$, $\epsilon = 0.001$, $N = 50$, $K = 20$, $\delta = 10^{-4}$
2: **for** $(n = 0, \ldots, N)$ **do**
3:     **for** $(k = 0, \ldots K)$ **do**
4:         Generate $X \sim \text{Weibull}(2, \theta_n)$
5:         Generate $A \sim \text{Exp}(1)$
6:         $W(k) = X - A$
7:         **while** $(A < X)$ **do**
8:             $A \leftarrow A + \text{Exp}(1)$
9:             $W(k) \leftarrow W(k) + (X - A)$
10:        $W(k) \leftarrow W(k) - (X - A)$
11:     MeanWait = $\text{Mean}(W)$
12:     **for** $(k = 0, \ldots K)$ **do**
13:         Generate $X \sim \text{Weibull}(2, \theta_n)$
14:         Generate X^+ using (7.26): $Y \sim \text{Exp}(1/\theta_n^2)$, $X^+ = \sqrt{X^2 + Y}$
15:         $\Delta = X^+ - X$
16:         Generate $N \sim \text{Poisson}(\lambda X)$
17:         Der1 $= N \times \Delta$
18:         Generate $A \sim \text{Exp}(1)$
19:         $W'(k) = \Delta - A$
20:         **while** $(A < \Delta)$ **do**
21:             $A \leftarrow A + \text{Exp}(1)$
22:             $W'(k) \leftarrow W'(k) + (\Delta - A)$
23:         $W'(k) \leftarrow W'(k) - (\Delta - A)$
24:     MeanDer = $\text{Mean}(\text{Der1} + W')$
25:     $\theta_{n+1} = \Pi_{[\delta, 1/\delta]}(\theta_n - \epsilon \, (\text{MeanWait} - b) \times (2\theta_n) \, \text{MeanDer})$

Compare to Algorithm 7.2, it is apparent that the running times are similar, even though the MVD is built from two processes, because we have exploited the "phantom" approach to deal with the difference process directly. The first batch in both algorithms has an expected running time $O(N(X(\theta)))$ for fixed θ, while the second batch in the MVD implementation has expected running time of order $O(N(\Delta(\theta)))$, with $\Delta(\theta) = X^+(\theta) - X(\theta)$. Whether this is smaller or larger than the one for the SF method depends on the various values of θ_n visited by the SA.

Seeking further variance reduction, when $N(\cdot)$ is Poisson with rate λ, it holds that

$$\mathbb{E}[N(X(\theta)) \mid X(\theta)] = \lambda X(\theta),$$

so instead of the above formula one may use the conditional expectation

$$\tilde{D}(\theta) = \lambda X(\theta)(X^+(\theta) - X(\theta)) + \sum_{k=N(X(\theta))+1}^{N(X^+(\theta))} (X^+(\theta) - A_k),$$

that satisfies $\mathbb{E}[\tilde{D}(\theta)] = \mathbb{E}[D(\theta)]$. In Figure 7.6 we used $D(\theta)$. The code using $\tilde{D}(\theta)$ requires even less CPU time because the first passenger arrivals need not be generated at all. ✳✳✳

7.3 A TAXONOMY OF GRADIENT ESTIMATION

In this section we introduce the basic gradient estimation problems. The classification is based on the time horizon of the stochastic experiment, which can be either static (i.e., finite and deterministic) or random (and almost surely finite). Moreover, gradients of long-run characteristics are of interest, and, for Markov processes, gradients of stationary characteristics. In the following we will introduce these estimation problems and we will introduce key examples that will serve as benchmark problems for gradient estimation.

7.3.1 The Finite Horizon Problem

In a *finite horizon* problem the expected value is evaluated over a finite (and nonrandom) collection of random variables. This comprises both the "static" case where the underlying process is a collection of random variables, and the case where the underlying process is a Markov chain and the problem deals with a finite (but not random) horizon cost function. This gives rise to the following definition.

Definition 7.1. For $\theta \in \Theta$, let $X_i(\theta)$, $1 \leq i \leq N$, be a collection of random variables defined on some underlying probability space, such that $X_i(\theta)$ is a measurable mapping on some measurable space $(S, \mathcal{S})$. For a measurable mapping $L : S^N \to \mathbb{R}$, the *finite horizon* gradient estimation problem is to find an unbiased estimator for

$$\frac{d}{d\theta} \mathbb{E}[L_N(X_1(\theta), \ldots, X_N(\theta))],$$

provided the expression exists.

Example 7.15. (Reliability) Consider the reliability network of the figure below. The system has five components whose logical interconnections are depicted in terms of the arcs joining the nodes. Individual components are either working (on) or not (down). For the system to operate, there must be a path between "IN" and "OUT" with components that are all working.

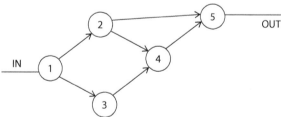

The *lifetime* L of the system is defined as the amount of time that it operates from the moment that all components are new. Each component $i = 1, \ldots, 5$ works for a random time $T_i > 0$ with distribution G_i, after which it breaks down. Suppose that the component lifetimes $\{T_i\}$ are mutually independent with finite expected values. Clearly, $L = T_{i^*}$, where i^* is the index of the last component to brake down that causes the system to break down as well, and it is a random index.

There are three possible paths: $R_1 = \{1, 2, 5\}$, $R_2 = \{1, 2, 4, 5\}$, and $R_3 = \{1, 3, 4, 5\}$. Route i fails at time $L_i = \min(T_j, j \in R_i)$, therefore the lifetime of the system is of the form $L = \max(L_1, L_2, L_3)$. Let I_i be the index of the component that causes the failure of route R_i, that is, $I_i = \arg\min(T_j, j \in R_i)$. Then we can also represent L by the expression $L = \max(T_{I_1}, T_{I_2}, T_{I_3})$, or $L = T_{i^*}$, where $i^* = \arg\max(T_{I_1}, T_{I_2}, T_{I_3})$.

Suppose that $T_3 \sim \text{Exp}(1/\theta)$, so that $\theta = \mathbb{E}[T_3]$ is the parameter of interest, for $\theta \in (0, \infty)$. In order to determine the sensitivity of L with respect to θ, or to find the value of θ that minimizes $\mathbb{E}[L]$ the derivative of $\mathbb{E}[L]$ with respect to θ has to be determined, i.e.,

$$\frac{d}{d\theta}\mathbb{E}\left[\max(T_{I_1}, T_{I_2}, T_{I_3})\right].$$

This is an illustration of a static problem, where there are no dynamics involved and we need only work with a collection of random variables. ✳✳✳

Example 7.16. (**Sojourn Times [Finite Horizon]**) Customers arrive at a service station according to a renewal point process. The interarrival times $\{A_n : n \in \mathbb{N}\}$ are iid, with $\mathbb{E}[A_n] < \infty$ and $\mathbb{P}(A_n = 0) = 0$. Customers are served in order of arrival, and consecutive service times are iid random variables $\{S_n(\theta) : n \in \mathbb{N}\}$. Interarrival times and service times are assumed to be mutually independent. The common distribution of the service times G_θ depends on parameter $\theta = \mathbb{E}[S_n(\theta)]$, where $\theta \in (0, \infty)$. This is what is known as the GI/GI/1 queueing model with the "first come, first served" (FCFS) service discipline. See Example 5.7 for an earlier treatment of this model.

Consider the process of consecutive sojourn (or system) times $\{X_n(\theta)\}$, denoting the total time that the corresponding customer is in the system (from arrival to end of service). The arrival process starts at $T_0 = 0$, and the time of arrival of customer n is $T_n = \sum_{i=1}^n A_i$. Denote the departure time of the n-th customer by $D_n(\theta)$ and let

$$X_n(\theta) = D_n(\theta) - T_n, \quad n \geq 1. \tag{7.29}$$

Then $X_n(\theta)$ denotes the total time that the n-th customer is in the system (from arrival to end of service), also called the *sojourn (or system) time*. As we will explain in the following, the process of consecutive sojourn times $\{X_n(\theta)\}$ forms a Markov chain. To see this, note that if the n-th customer leaves the system prior to the $(n+1)$-th arrival, i.e., $D_n(\theta) - T_{n+1} \leq 0$, then customer $n+1$ has no wait and enters service immediately, in which case the sojourn is equal to the customer's service time:

$$X_{n+1}(\theta) = S_{n+1}(\theta).$$

If, on the other hand, the $(n+1)$-th arrival takes place when the n-th customer is still at the server, i.e., $D_n(\theta) - T_{n+1} > 0$, then customer $n+1$ has to wait until the previous departs, that is, the total wait is

$$D_n(\theta) - T_{n+1} = D_n(\theta) - T_n - A_{n+1} \stackrel{(7.29)}{=} X_n(\theta) - A_{n+1}$$

before service can commence and the sojourn time becomes

$$X_{n+1}(\theta) = S_{n+1}(\theta) + X_n(\theta) - A_{n+1}.$$

To summarize, consecutive sojourn times follow the recursive relation:

$$X_{n+1}(\theta) = \max(0, X_n(\theta) - A_{n+1}) + S_{n+1}(\theta), \quad n \geq 0, \tag{7.30}$$

where we assume that the system starts empty and we formally set $X_0(\theta) = 0$. The above recursive relation is called *Lindley* recursion and shows that $\{X_n(\theta)\}$ is a Markov chain.

Let $W_n(\theta)$ denote the waiting time of the n-th customer (the time the customer spends in the system until start of service), then we obtain along the lines of arguments that led to (7.30) a similar recursion for the waiting times

$$W_{n+1}(\theta) = \max(0, W_n(\theta) + S_n(\theta) - A_{n+1}), \tag{7.31}$$

for $n \geq 0$, and $W_0(\theta) = S_0(\theta) = 0$.

The sojourn time process $\{X_n(\theta)\}$ is adapted to the filtration $\{\mathfrak{F}_n\}$, where $\mathfrak{F}_n$ is the σ-algebra generated by $\{A_1, \ldots, A_n; S_1(\theta), \ldots S_n(\theta)\}$. Suppose that we are interested in evaluating the derivative of the average sojourn times of the first N customers, that is,

$$L_N(X_1(\theta), \ldots, X_N(\theta)) = \frac{1}{N} \sum_{n=1}^{N} X_n(\theta). \tag{7.32}$$

This is a finite horizon gradient estimation problem.

A typical optimization problem involving the finite horizon gradient (8.5) is the following. Let $c(\theta)$ denote the cost/energy needed for operating the server at mean speed θ. Then a stochastic approximation can be applied to minimize

$$J(\theta) = \alpha_1 c(\theta) + \frac{\alpha_2}{N} \sum_{n=1}^{N} \mathbb{E}[X_n(\theta)] = \alpha_1 c(\theta) + \alpha_2 \mathbb{E}\big[L_N(X_1(\theta), \ldots, X_N(\theta))\big],$$

the weighted sum of cost and average sojourn time, for weights $\alpha_1, \alpha_2 > 0$. Note that $J(\theta)$ has no closed-form solution and, with the exception of rather small values of N such as 1, 2, or 3, $J(\theta)$ cannot be solved numerically. Thus, simulation is the only available approach to solving $\min_\theta J(\theta)$. In order to implement stochastic approximation it is required to use an estimator of the derivative of $J(\theta)$. More complex problems involve networks of queues, each with their own service parameter, and in those cases the gradient may have large dimension.

A related problem is that of minimizing, say the cost of operation $c(\theta)$ subject to a constraint in the expected average sojourn time $L_N(X_1(\theta), \ldots, X_N(\theta))$. Or, one may wish to minimize the expected average sojourn time subject to a maximal cost of operation (a budget). The problems are related because the Lagrangian has an expression similar to the function $J(\theta)$ above. If implementing a constrained optimization method, the derivatives of both $c(\theta)$ and $\mathbb{E}[L_N]$ must be estimated. ✳✳✳

7.3.2 The Infinite Horizon Problem

Studying the long-term behavior of a system (commonly referred to as the "infinite horizon" problem in the simulation literature) leads to the following gradient estimation problem.

Definition 7.2. For $\theta \in \Theta$, let $\{X_i(\theta) : i \in \mathbb{N}\}$ be stochastic process defined on some underlying probability space, such that $X_i(\theta)$ is a measurable mapping on some measurable space $(S, \mathcal{S})$. Let $L : S \times \Theta \to \mathbb{R}$ be a measurable mapping. The long-run gradient estimation problem is to find an unbiased estimator for $\nabla J(\theta)$, where

$$J(\theta) := \lim_{N \to \infty} \frac{1}{N} \sum_{n=1}^{N} \mathbb{E}[L(X_n(\theta), \theta)]$$

provided the expression exists.

A PRIMER FOR GRADIENT ESTIMATION

There is a relationship between the finite horizon case and the infinite horizon one. This relationship also links the infinite horizon problem to the methods and algorithms developed in Chapter 5. More specifically, suppose that for any finite N, there is a (possibly random) function $g(X_n(\theta), \theta)$ such that

$$\frac{d}{d\theta}\mathbb{E}\left[\frac{1}{N}\sum_{n=1}^{N}L(X_n(\theta), \theta)\right] = \mathbb{E}\left[\frac{1}{N}\sum_{n=1}^{N}g(X_n(\theta), \theta)\right].$$

Then, Section 9.1 presents conditions under which the finite horizon estimator is a *statistically consistent* estimator for the infinite horizon problem, that is

$$\frac{d}{d\theta}J(\theta) = \lim_{N\to\infty}\frac{d}{d\theta}\mathbb{E}\left[\frac{1}{N}\sum_{n=1}^{N}L(X_n(\theta), \theta)\right] = \lim_{N\to\infty}\frac{1}{N}\sum_{n=1}^{N}\mathbb{E}[g(X_n(\theta), \theta)],$$

i.e., Section 9.1 provides conditions under which the infinite horizon problem can be interpreted as limit of the finite horizon problem. As we will detail in Chapter 9, this allows the application of Theorems 5.2, 5.3, and 5.4 in order to implement stochastic approximation for stochastic gradient search.

Example 7.17. Consider the following inventory problem. A shop stores a quantity of a liquid commodity that is sold by demand (such as petrol, or beer) at a unit price of c dollars. The maximal storage capacity is S, and the commodity is sold during the day as orders are placed. When the daily demand exceeds the stock in inventory, a penalty cost is incurred per unit of unsatisfied demand: we assume that the shop sells all the quantity requested by demands, but it does so by outsourcing the excess amount paying a higher price for it. The penalty cost p is the difference between the amount paid to the external vendor and c. There is also a holding unit cost h for overnight storage. It is well known that for such inventory problems the optimal ordering strategy is of the type called "(s, S)" policy: when the stock on day n falls below a level s, an order is placed to replenish the inventory up to maximal capacity S and this incurs a fixed cost K of delivery plus $k < c$ dollars per unit bought. Suppose that there are no delays in deliveries. We wish to find the optimal value of s, which we now rename θ, so that $\theta \in \Theta = [0, S]$. Let $X_n(\theta)$ be the level of commodity in storage at the start of day n and assume that daily demands are iid $\{\xi_n\}$. Then

$$X_{n+1}(\theta) = \begin{cases} X_n(\theta) - \xi_n & \text{if } \xi_n \leq X_n(\theta) - \theta, \\ S & \text{otherwise.} \end{cases}$$

defines a Markov chain on a continuous state space. The daily cost is defined by

$$L(X_n(\theta), \xi_n) = \begin{cases} h(X_n(\theta) - \xi_n) & \text{if } \xi_n \leq X_n(\theta) - \theta, \\ h(X_n(\theta) - \xi_n) + K + k(S - X_n(\theta) + \xi_n) & \text{if } X_n(\theta) - \theta < \xi_n \leq X_n(\theta), \\ p(\xi_n - X_n(\theta)) + K + kS & \text{if } X_n(\theta) < \xi_n. \end{cases}$$

The profit is simply c dollars per unit sold: $c\,\xi_n$ and it is therefore independent of θ. Under the assumption that the daily demand has a continuous distribution with $\mathbb{P}(\xi_n < S) > 0$ the process is ergodic and has a limit probability. The function $J(\theta)$ to minimize is the long-term daily operating cost at θ:

$$J(\theta) = \lim_{N\to\infty}\frac{1}{N}\sum_{n=1}^{N}L(X_n(\theta), \xi_n).$$

If we wish to use derivative information to find the optimal value θ^* via stochastic approximation then we need to estimate $J'(\theta)$. Chapter 10 treats this type of problem for various different formulations of the inventory optimization. ✳✳✳

7.3.3 The Random Horizon Problem

In the finite horizon setting the performance L is evaluated over a fixed finite number of observations. An extension of this setup is to let L depend on a finite but random number of observations. The precise definition is given in the following.

Definition 7.3. For $\theta \in \Theta$, let $\{X_i(\theta) : i \in \mathbb{N}\}$ be stochastic process defined on some underlying probability space, such that $X_i(\theta)$ is a measurable mapping on some measurable space $(S, \mathcal{S})$. Moreover, let τ_θ be a stopping time adapted to the natural filtration of $\{X_i(\theta) : i \in \mathbb{N}\}$ (see Definition B.9). For $n \in \mathbb{N}$, let $L_n : S^n \to \mathbb{R}$ be a measurable mapping. The random horizon gradient estimation problem is to find an unbiased estimator for

$$\frac{d}{d\theta}\mathbb{E}[L_{\tau_\theta}(X_1(\theta), \ldots, X_{\tau_\theta}(\theta))],$$

provided the expected values exist.

Example 7.18. (Sojourn Times [cycle]) We illustrate the random horizon gradient estimation problem with a classical example from queuing theory, using the sojourn time model in Example 7.16 for a GI/G/1 queue. Since service times and interarrival times are assumed to be mutually independent sequences of independent random variables it holds that whenever the system empties, the sequence of sojourn times starts anew, independent of the past. Suppose that the queue starts initially empty, which implies that the first arriving customer does not have to wait and the first sojourn time equals the service time, i.e., $X_1(\theta) = S_1(\theta)$. Let τ_θ be the first customer after $n = 1$ that finds an empty queue and starts service immediately. That is $\tau_\theta = \min\{n > 1 : X_n(\theta) = S_n(\theta)\}$, or, equivalently,

$$\tau_\theta = \min\{n > 1 : W_n(\theta) = 0\}. \tag{7.33}$$

Define $\tau_\theta = \infty$ if the set on the above right hand side is empty. The collection $(X_1(\theta), \ldots, X_{\tau_\theta - 1}(\theta))$ is called a *cycle* of the sojourn time sequence, and the accumulated sum of sojourn times over a cycle is given by

$$\mathbb{E}\Big[L_{\tau_\theta - 1}(X_1(\theta), \ldots, X_{\tau_\theta - 1}(\theta))\Big],$$

with

$$L_N(X_1(\theta), \ldots, X_N(\theta)) = \sum_{n=1}^{N} X_n(\theta).$$

Note that in the definition of a cycle we only add the first $\tau_\theta - 1$ sojourn times. The reason for this is that the index of the last sojourn before the end of a cycle is not a stopping time, whereas the first sojourn time of a cycle is one. The random horizon problem is that of estimating

$$\frac{d}{d\theta}\mathbb{E}\left[\sum_{n=1}^{\tau_\theta - 1} X_n(\theta)\right] = \frac{d}{d\theta}\mathbb{E}\Big[L_{\tau_\theta - 1}(X_1(\theta), \ldots, X_{\tau_\theta - 1}(\theta))\Big]. \quad\quad\text{✳✳✳}$$

The random horizon problem is related to the finite horizon problem. Suppose that the estimator $\psi(N; X_1(\theta), \ldots, X_N(\theta))$, for $N \geq 1$, solves the finite horizon gradient estimation problem, i.e.,

$$\frac{d}{d\theta}\mathbb{E}[L_N(X_1(\theta),\ldots,X_N(\theta))] = \mathbb{E}[\psi(N;X_1(\theta),\ldots,X_N(\theta))],$$

for all $N < \infty$. Then, as we will show in Section 9.2 under appropriate smoothness conditions and choice of ψ, the following, somewhat surprisingly simple, equation holds

$$\frac{d}{d\theta}\mathbb{E}[L_{\tau_\theta}(X_1(\theta),\ldots,X_{\tau(\theta)}(\theta))] = \mathbb{E}[\psi(\tau(\theta);X_1(\theta),\ldots,X_{\tau(\theta)}(\theta))].$$

Hence, evaluating the finite horizon gradient estimator over the random horizon provides an unbiased gradient estimator.

Relation to the Infinite Horizon Problem

For regenerative processes, infinite horizon problems can be stated using the so-called *regenerative method* for estimation (see Chapter 7 in [263]). To see this, suppose that $\{X_n(\theta) : n \geq 0\}$ is a regenerative process with associated sequence of renewal (resp. regeneration) times $\{\eta_n(\theta)\}$, with $\eta_1 = 0$. From renewal theory it follows that for any mapping $L: S \mapsto \mathbb{R}$, such that the right-hand side of the expression in (7.34) is finite, that

$$L_\infty(\theta) := \lim_{N\to\infty} \frac{1}{N}\sum_{n=1}^N \mathbb{E}[L(X_n(\theta))] = \frac{\mathbb{E}\left[\sum_{n=\eta_k(\theta)}^{\eta_{k+1}(\theta)-1} L(X_n(\theta))\right]}{\mathbb{E}[\eta_{k+1}(\theta) - \eta_k(\theta)]}, \quad (7.34)$$

for any $k \geq 0$, provided that the expected values are finite, see Section B.7 in Appendix B for details. Therefore, for the long-run gradient estimation problem this yields for $k=0$:

$$\frac{d}{d\theta}L_\infty(\theta) = \frac{d}{d\theta}\left(\frac{\mathbb{E}\left[\sum_{n=0}^{\eta_1(\theta)-1} L(X_n(\theta))\right]}{\mathbb{E}[\eta_1(\theta)]}\right)$$

$$= \frac{\frac{d}{d\theta}\mathbb{E}\left[\sum_{n=0}^{\eta_1(\theta)-1} L(X_n(\theta))\right]}{\mathbb{E}[\eta_1(\theta)]} - L_\infty(\theta)\frac{\frac{d}{d\theta}\mathbb{E}[\eta_1(\theta)]}{\mathbb{E}[\eta_1(\theta)]}. \quad (7.35)$$

It is worth noting that the expression on the above right hand side is very challenging for simulation, as it contains fractions of expected values. Indeed, estimators for fractions of this type are studied in the simulation literature and, generally speaking, only asymptotic unbiasedness can be established (as the number of observed cycles tends to infinity). The same holds true for estimating the corresponding confidence intervals. These drawbacks render simulations based on the above right hand side impractical.

A nice trick ([11], Remark 2.3, page 245) that circumvents these drawback (at least to some extent) is the following. Consider

$$J(\theta) = \frac{\mathbb{E}\left[\sum_{k=0}^{\eta_1(\theta)-1} f(X_\theta(k))\right]}{\mathbb{E}[\eta_1(\theta)]}.$$

Let $Z_1(\theta, f)$ be the cycle cost, $Z_2(\theta, f)$ the SF derivative estimator, and let $Z_1(\theta, 1)$ be the cycle length, $Z_2(\theta, 1)$ the corresponding SF derivative estimator. In solving $\min_\theta J(\theta)$, we can trace the stationary points of $J'(\theta)$ by solving the root-finding problem

$$H(\theta) = \mathbb{E}' \left[\sum_{k=0}^{\eta_1(\theta)-1} f(X_\theta(k)) \right] \mathbb{E}[\eta_1(\theta)] - \mathbb{E}\left[\sum_{k=0}^{\eta_1(\theta)-1} f(X_\theta(k)) \right] \mathbb{E}'[\eta_1(\theta)] = 0,$$

see (7.35). Sampling two independent cycles, we apply SA to solving this root-finding problem using as unbiased estimator for H:

$$Y(\theta) = Z_2(\theta, f) Z_1(\theta, 1) - Z_1(\theta, f) Z_2(\theta, 1).$$

7.3.4 Markov Processes: The Stationary Problem

Markov processes are a common tool for modeling and analyzing complex systems. Under appropriate conditions, a Markov process has a unique stationary distribution that is also the unique limiting distribution. Therefore, problem formulation in terms of the cost function $J(\theta)$ in this situation using the stationary expectation is equivalent to the infinite horizon formulation.

Consider a family of general homogeneous Markov process $\mathbf{X}(\theta) = \{X_n(\theta)\}$, where for each $\theta \in \Theta$, $X_n(\theta) \in S$, and the measurable state space $(S, \mathcal{S})$ is a general space (not necessarily discrete or countable). Let $P_{\theta,n}$ denote the Markov kernel of $\mathbf{X}(\theta)$, which is given by

$$P_{\theta,n}(s, A) = \mathbb{P}\big(X_{n+1}(\theta) \in A \mid X_n(\theta) = s\big), \quad n \in \mathbb{N},$$

for $s \in S$ and $A \in \mathcal{S}$. The Markov process and the corresponding Markov kernel are called *homogenous* if $P_{\theta,n}$ is independent of n.

Definition 7.4. Let P_θ be a homogenous Markov kernel and let π_θ be a probability distribution on the state space $(S, \mathcal{S})$ such that

$$\forall A \in \mathcal{S}: \quad \pi_\theta(A) = \int_S P_\theta(s, A) \pi_\theta(s) ds,$$

then π_θ is called a *stationary distribution* for P_θ.

Generally speaking, a Markov process is called *ergodic* if the stationary distribution is the unique limiting distribution. This implies that for an ergodic Markov process the distribution of $X_n(\theta)$ converges independently of the initial distribution to π_θ. Hence, for ergodic Markov processes, the infinite horizon gradient estimation problem can alternatively be formulated in terms of the stationary distribution.

Definition 7.5. For $\theta \in \Theta$, let $\mathbf{X}(\theta) = \{X_i(\theta) : i \in \mathbb{N}\}$ be a homogenous Markov process with unique stationary distribution π_θ. Let L be a measurable, real-valued mapping such that $\int_S |L(s)| \pi_\theta(s) ds$ is finite for $\theta \in \Theta$. The stationary gradient estimation problem is to find an unbiased estimator for

$$\frac{d}{d\theta} \mathbb{E}\left[L(\tilde{X}(\theta)) \right],$$

where $\tilde{X}(\theta)$ is distributed according to π_θ.

Example 7.19. Consider the single server queue put forward in Example 7.16. Provided that service times and interarrival times are iid and mutually independent, it is well known that the *stability condition* $\mathbb{E}[S_1(\theta)] < \mathbb{E}[A_1(\theta)]$ implies that there exists a stationary measure. More specifically, there exists a random variable $W(\theta)$ such that $W(\theta)$ and $\max(0, W(\theta) + S_n(\theta) - A_{n+1})$ are equal in distribution, for $n \geq 1$; see (7.31). This result holds more general for max-plus linear queueing systems, see [14, 145]. In the case of an $M/M/1/\infty$ queue with mean interarrival time $1/\lambda$ and mean service time θ, $W(\theta)$ is known to have the cdf:

$$\mathbb{P}(W(\theta) \leq x) = 1 - \lambda\theta e^{-((1/\theta)-\lambda)x}, \quad x > 0,$$

and $\mathbb{P}(W(\theta) = 0) = 1 - \lambda\theta$ for $\theta \in \Theta = (0, 1/\lambda)$; see [2]. Suppose that operating the server with mean service time θ consumes energy $c(\theta)$. Then solving

$$\min_{\theta} \left(\alpha_1 c(\theta) + \alpha_2 \mathbb{E}[W(\theta)] \right),$$

with $\alpha_i > 0$, $i = 1, 2$, will find the speed of the server that balances the wait of a job in equilibrium and the energy consumption of the server in an optimal way. ✳✳✳

Relation to the Infinite Horizon Problem and the Random Horizon Problem

Under ergodicity of $\{X_n(\theta)\}$ the long-run performance equals the expected stationary performance, which yields a limiting result similar to (7.34):

$$L_\infty(\theta) = \lim_{N \to \infty} \frac{1}{N} \sum_{n=1}^{N} \mathbb{E}[L(X_n(\theta))] = \mathbb{E}\left[L(\tilde{X}(\theta))\right] = \int L(x)\, \pi_\theta(dx), \qquad (7.36)$$

where $\tilde{X}(\theta)$ is distributed according to π_θ. This shows that solving the infinite horizon provides means for the stationary problem.

In an alternative approach to the stationary problem, developed in Chapter 13, we will show how to build a (complex) random mapping $\psi(x)$ so that

$$\frac{d}{d\theta}\mathbb{E}[L(\tilde{X}(\theta))] = \mathbb{E}[\psi(\tilde{X}(\theta))], \qquad (7.37)$$

where $\tilde{X}(\theta)$ is distributed according to π_θ as above. The estimator in (7.37) will thus directly set off in the stationary regime, which is in contrast to the asymptotic approach in (7.36) where one would work under the conditions put forward in Section 9.1 so that

$$\lim_{N \to \infty} \frac{1}{N} \sum_{n=1}^{N} \frac{d}{d\theta}\mathbb{E}[L(X_n(\theta))] = \frac{d}{d\theta}\mathbb{E}\left[L(\tilde{X}(\theta))\right];$$

see Section 7.3.3 for details.

As we will detail later on in Chapter 13, in computer simulation, we can set up an experiment that turns the expression in (7.37) into an unbiased gradient estimator. The basic approach is as follows. We first sample a cycle of $\{X_n(\theta) : n \geq 0\}$ and store the cycle length τ_θ. Next we select σ uniformly out of $\{0, \ldots, \tau_\theta - 1\}$. Then, we set the random number seed back to its initial value (which can be done in computer simulation) and we re-simulate $X_n(\theta)$ up to $X_\sigma(\theta)$. This leads to the following derivative expression:

$$\frac{d}{d\theta}\mathbb{E}[L(\tilde{X}(\theta))] = \mathbb{E}[\psi(\tilde{X}(\theta))] = \frac{1}{\mathbb{E}[\tau_\theta]}\mathbb{E}[\tau_\theta \psi(X_\sigma(\theta))].$$

In words, sampling from the stationary distribution, as required by the gradient estimator in (7.37), can be realized by selecting uniformly a state out of cycle in combination with appropriate rescaling the statistic.

7.4 PRACTICAL CONSIDERATIONS

A Matter of Taste (But Not Always)

Reviewing the examples put forward in this section, it can be seen that in many cases the sensitivity of a model with respect to a parameter θ can be analyzed in various ways: either by establishing a model for which θ acts as a sample-path parameter leading to an IPA estimator or by choosing a model in which θ acts as a distributional parameter leading to either an SF or MVD estimator. To see this, review Example 7.13, and compare Example 7.4, Example 7.8, and Example 7.12. Notably, exceptions are, e.g., the probability weights of a discrete distribution that can only be treated as distributional parameters, and the point masses of a discrete distribution that can only be treated as distributional parameters; see Example 7.6 and Example 7.8 for the special case of the Bernoulli distribution. The choice of the model and ensuing gradient estimator therefore typically depends aspects like ease of implementation, flexibility, and efficiency of the estimator. We will discuss these aspects in detail in the following chapter.

The Ubiquitous Condition $\theta > 0$

Revisit Example 7.1 and Example 7.3. When working on the examples, we have not addressed that θ has to satisfy the hard constraint $\theta > 0$. For ϵ small and the solution being away from the boundary, as indicated by Figure 7.2, it is reasonable to assume that a low variance gradient estimator does produce updates θ_n that do not violate the positivity constraint. Moreover, one could use projection on some interval $[\delta, \infty)$ for δ small (but not too small for reason of numerical stability) to prevent the algorithm from violating the constraint. Alternatively, one can apply the transformation $\theta \to \theta^2$, so that (7.5) becomes

$$X(\theta) = -50\theta^2 \ln(1 - U), \quad \theta \neq 0.$$

The IPA estimator then follows as in Example 7.3, leading to the SA

$$\theta_{n+1} = \theta_n - \epsilon\left(1 - \frac{2X(\theta_n)}{\theta_n^2}h'(X(\theta_n))\right).$$

Since $X(\theta)$ is a continuous random variable, the event $\theta_n = 0$ has probability zero and the above transformed IPA estimator can be applied without projection. As $J(\theta^2)$ tends to zero for $|\theta| \to 0$, we extend $J(\theta)$ to $\theta = 0$ by letting $J(0) = 0$, and obtain $J(\theta^2)$ as differentiable on $\Theta = \mathbb{R}$. This transformation "trick," however, comes at the price of introducing an additional non-informative stationary point. Indeed, comparing $J(\theta)$ with $J(\theta^2)$, it follows that due to the chain rule $J'(\theta^2) = 2\theta J'(\theta)$. Hence, $\theta = 0$ becomes a new stationary point, which is a maximum of $J(\theta^2)$. This is known as the "hairy ball theorem"; see [220].

7.5 EXERCISES

Exercise 7.1. Consider the distribution function $F_\theta(x)$ of some real-valued random variable $X(\theta)$ with parameter θ. The image of the distribution is by definition $[0, 1]$. Suppose that the random variable has a continuous density so that $F_\theta(\cdot)$ is strictly increasing. Let U be a uniform-$[0, 1]$-random variable. Show that $X(\theta, U) = F_\theta^{-1}(U)$ is a random variable with distribution F_θ on $(\Omega, \mathbb{P})$. This is sometimes called the canonical representation and it is a means for generating general distributions from a uniform (pseudo) random number generator.

Exercise 7.2. For Θ a non-empty connected subset of $\mathbb{R}$, let $X(\theta)$, for $\theta \in \Theta$, be a real-valued random variable and let h be a mapping from $\mathbb{R}$ to $\mathbb{R}$ such that $h(X(\theta))$ is measurable for all $\theta \in \Theta$. Show that if h is differentiable and if $X(\theta)$ is differentiable with respect to θ at some point $\theta_0 \in \Theta$, then $dh(X(\theta))/d\theta$ at θ_0 is measurable with respect to $\mathcal{F}$. (Hint: Recall that $(\Omega, \mathcal{F}, \mathbb{P})$ denotes the underlying probability space.)

Exercise 7.3.

(a) Calculate the score function for the exponential distribution $\exp(\theta)$ where θ is the intensity of the distribution.
(b) The pdf of the Pareto II distribution with scale parameter λ and shape parameter α is given by
$$\frac{\alpha}{\lambda}\left(1 + \frac{x}{\lambda}\right)^{-(\alpha+1)}, \quad x \geq 0.$$
Compute the score function for the Pareto II distribution with respect to the scale and shape parameter.

Exercise 7.4. Consider a random variable $X(\theta)$. Assume that $\theta = \mathbb{E}[X(\theta)]$ is a scale parameter of the distribution F_θ; in other words, using the inverse function representation, $X(\theta) = F_\theta^{-1}(U) = \theta F_1^{-1}(U)$. Explain which of the following is a sufficient condition for
$$\mathbb{E}\left[\frac{X(\theta)}{\theta}\right] = \frac{d}{d\theta}\mathbb{E}[X(\theta)].$$

(a) $\mathbb{E}[|X(1)|] < \infty$,
(b) $\mathbb{E}[X(\theta)^2] < \infty$, or
(c) $\mathbb{E}[|\theta X(\theta)|] < \infty$.

Exercise 7.5. Validate the expression for the weak derivative of the normal distribution with respect to the standard deviation as put forward in Table 7.1.

Exercise 7.6. Let θ be a scaling of the mean and the standard deviation of normal distribution, that is, consider $\mathcal{N}(\theta\mu, (\theta\sigma)^2)$. Provide an expression of the weak derivative of $\mathcal{N}(\theta\mu, (\theta\sigma)^2)$ with respect to θ.

Exercise 7.7. Repeat the stochastic approximation method for the mining investment problem of Example 7.3 using IPA, SF, and MVD for derivative estimation. Assume here that the estimators are unbiased.

(a) Using simulation, estimate the corresponding confidence intervals and CPU times for the three derivative estimation methods for $\theta = 5, 8, 10, 12, 15$ and compare.

(b) Apply an appropriate theorem from Part I to establish convergence of the stochastic approximation to the true optimal value θ^*. Specify your choice of stepsize sequence and, for each method, verify the assumptions of the theorems that you use.

(c) Run the stochastic approximations and discuss.

Exercise 7.8. Let
$$X(\theta) = \sum_{i=0}^{N(\theta)} C_i,$$
where $C_0, C_1, \ldots$ are iid, and $N(\theta)$ is a Poisson random variable with mean θ. Define the objective function
$$J(\theta) = \mathbb{E}\bigl[\mathbb{I}\{X(\theta) > x\}\bigr].$$
Derive the SF estimator for $J'(\theta)$.

Chapter Eight

Gradient Estimation, Finite Horizon

In Part I we discussed utilization of finite differences to estimate required gradients for the feedback of SA procedures. It is clear from the results in Chapter 6 that such methods suffer from two sources: they require two evaluations (or simulations) at different values of θ (so they take more CPU time per iteration) and they introduce a bias that slows down the convergence rate. It is for these reasons that *unbiased* gradient estimation can significantly improve efficiency of the SA methods.

Chapter 7 introduced the formulas for the three broad approaches to gradient estimation, and a taxonomy of problems under study. The reader is referred to Example 7.4 and Example 7.8 for a summary of known results that can become handy when solving problems.

In this chapter we present the main technical tools that help establish the conditions under which such gradient estimation methods yield unbiased estimators. We will discuss IPA, SF, and MVD. Moreover, we will present an extension of IPA called smoothed perturbation analysis (SPA).

8.1 PERTURBATION ANALYSIS: IPA AND SPA

The strength of IPA methodology is its ease of use for the static problem. This section presents the mathematical methodology and proofs for unbiasedness of IPA for the finite horizon problem.

8.1.1 Basic Results and Techniques

A Matter of Representation

Consider a family of random variables $\{X(\theta)\}$ parametrized by $\theta \in \Theta \subset \mathbb{R}$ and assume that these random variables are defined on a common probability space $(\Omega, \mathfrak{F}, \mathbb{P})$. A *representation*

$$X(\theta, \omega) \stackrel{d}{=} r(\theta, U(\omega)), \quad \text{with } r: \Theta \times \mathbb{R} \to \mathbb{R},$$

can be used to express $X(\theta)$, where the distribution of U is independent of θ. Note that for fixed ω, $X(\theta, \omega)$ is a mapping from Θ to $\mathbb{R}$, and the derivative of $X(\theta, \omega)$ with respect to θ is well-defined provided that r is differentiable in θ; see Example 7.2. Representations are not unique; Chapter 7 introduced many examples where families of random variables can be represented in this manner, where the common probability space is independent of θ. With this in mind, for any function $h: \mathbb{R} \to \mathbb{R}$ the random variable $h(X(\theta))$ is defined as a composition mapping $h \circ X(\theta)$. In particular, for each fixed ω, $(h \circ X(\theta))(\omega)$ is a function of θ. This representation will be used throughout this section (except for Section 8.1.4). Lipschitz continuity is a key condition for unbiasedness of the IPA estimator.

Definition 8.1. Let $\Theta \subset \mathbb{R}$ be an open connected set, such that for each $\theta \in \Theta$, $X(\theta)$ is a real-valued random variable defined on a common probability space $(\Omega, \mathfrak{F}, \mathbb{P})$. We say that $h(X(\theta))$ is *almost surely Lipschitz continuous* on Θ if a random variable K with $\mathbb{E}[K] < \infty$ exists such that for all $\theta_1, \theta_0 \in \Theta$

$$|h(X(\theta_1; \omega)) - h(X(\theta_0; \omega))| \leq |\theta_1 - \theta_0| K(\omega) \quad \text{a.s.}$$

The r.v. K is called the Lipschitz modulus or just the modulus.

First, we state the key IPA theorem.

Theorem 8.1. *Let $\Theta \subset \mathbb{R}$ be an open connected set, such that $\{X(\theta), \theta \in \Theta\}$ is a family of real-valued random variables on a common underlying probability space $(\Omega, \mathfrak{F}, \mathbb{P})$. Let $\theta_0 \in \Theta$. Assume that*

(i) the sample-path (or stochastic) derivative $dX(\theta)/d\theta$ exists with probability one at θ_0,
(ii) the mapping $h: \mathbb{R} \to \mathbb{R}$ is differentiable at $X(\theta_0)$ with probability one,
(iii) the mapping $h(X(\theta))$ is Lipschitz continuous on Θ with probability one (and has thus by definition an integrable Lipschitz modulus).

Then

$$\left. \frac{d}{d\theta} \mathbb{E}[h(X(\theta))] \right|_{\theta=\theta_0} = \mathbb{E}\left[\left. \left(\left(\frac{d}{d\theta} X(\theta) \right) h'(X(\theta)) \right) \right|_{\theta=\theta_0} \right],$$

where $h'(x)$ denotes the derivative of $h(x)$ with respect to x.

Proof. The proof is an application of the dominated convergence theorem (Theorem B.9 in Appendix B). In terms of random variables, it states that if a sequence of random variables $\{Z_n, n = 1, 2, \ldots\}$ is such that $Z_n \to Z$ a.s. and $|Z_n| < K$ for some random variable K with $\mathbb{E}[K] < \infty$, then $\lim_{n \to \infty} \mathbb{E}[Z_n] = \mathbb{E}[\lim_{n \to \infty} Z_n] = \mathbb{E}[Z]$.

Let $\{\Delta_n\}$ be sequence such that $\Delta_n \to 0$ as $n \to \infty$ and define the sequence of finite differences:

$$Z_n = \frac{h(X(\theta + \Delta_n)) - h(X(\theta))}{\Delta_n}.$$

By conditions (i) and (ii), w.p.1

$$\lim_{n \to \infty} Z_n = h'(X(\theta)) \frac{d}{d\theta} X(\theta)$$

and by condition (iii), using Definition 8.1 the finite differences $\{Z_n(\omega)\}$ are bounded by an integrable random variable K:

$$|Z_n| = \frac{1}{|\Delta_n|} |h(X(\theta + \Delta_n)) - h(X(\theta))| \leq K,$$

with $\mathbb{E}[K] < \infty$. The dominated convergence theorem then implies that

$$\frac{d}{d\theta} \mathbb{E}[h(X(\theta))] = \lim_{n \to \infty} \mathbb{E}[Z_n] = \mathbb{E}[\lim_{n \to \infty} Z_n] = \mathbb{E}\left[h'(X(\theta)) \frac{d}{d\theta} X(\theta) \right],$$

which proves the claim. $\square$

Theorem 8.1 provides the fundamental set of conditions justifying using the stochastic or path-wise derivatives as an unbiased estimator of the derivative of the expectation. Condition (ii) is usually straightforward in applications. We have already used in Chapter 7 the concept of "representation" of random variables as explicit functions both of ω and θ. When these

GRADIENT ESTIMATION, FINITE HORIZON

representations exist condition (i) can be verified. But in some cases we may need alternative ways to establish (i), as well as a convenient expression for the derivative $dX(\theta)/d\theta$ in terms of $X(\theta)$ itself. The existence of the sample path derivative of $X(\theta)$ in condition (i) above can be deduced with the following result in a general way; see [114, 129, 298].

Lemma 8.1. *Let $\Theta \subset \mathbb{R}$ be an open connected set, and let F_θ denote the cumulative distribution function of $X(\theta)$ for $\theta \in \Theta$ having support $S \subset \mathbb{R}$. If $F_\theta(x)$ is continuously differentiable with respect to θ on Θ, and continuously differentiable with respect to x on S, then it holds for any interior point θ of Θ with probability one that*

$$\frac{d}{d\theta}X(\theta) = -\frac{\frac{\partial}{\partial \theta}F_\theta(X(\theta))}{\frac{\partial}{\partial x}F_\theta(X(\theta))}.$$

Proof. Denote the (generalized) inverse of F_θ by F_θ^{-1} and use the canonical representation

$$X(\theta) \stackrel{d}{=} F_\theta^{-1}(U), \qquad U \sim U(0,1).$$

In particular, for Δ such that $\theta + \Delta \in \Theta$

$$F_{\theta+\Delta}(F_{\theta+\Delta}^{-1}(u)) = F_\theta(F_\theta^{-1}(u)), \tag{8.1}$$

for $u \in (0,1)$. Applying Taylor's theorem yields

$$F_{\theta+\Delta}(F_{\theta+\Delta}^{-1}(u)) = F_\theta(F_\theta^{-1}(u)) + \frac{\partial}{\partial \theta}F_\xi(\eta)\Delta + \frac{\partial}{\partial x}F_\xi(\eta)(F_{\theta+\Delta}^{-1}(u) - F_\theta^{-1}(u)),$$

for some point (ξ, η) on the line segment joining $(\theta, F_\theta^{-1}(u))$ and $(\theta+\Delta, F_{\theta+\Delta}^{-1}(u))$. By (8.1), this yields

$$\frac{\partial}{\partial \theta}F_\xi(\eta)\Delta + \frac{\partial}{\partial x}F_\xi(\eta)(F_{\theta+\Delta}^{-1}(u) - F_\theta^{-1}(u)) = 0.$$

Hence,

$$-\frac{\frac{\partial}{\partial \theta}F_\xi(\eta)}{\frac{\partial}{\partial x}F_\xi(\eta)} = \frac{F_{\theta+\Delta}^{-1}(u) - F_\theta^{-1}(u)}{\Delta} = \frac{X(\theta+\Delta, u) - X(\theta, u)}{\Delta}.$$

Letting Δ tend to zero, the claim follows from the continuity of the partial derivatives of F_θ.
□

Example 8.1. For the Weibull and Pareto II distributions sample path differentiation of $X(\theta)$ with respect to the shape parameter of the distribution doesn't yield an expression for the derivative as a function of $X(\theta)$; see In Example 7.4. We now illustrate how to implement Lemma 8.1 in order to find such an expression.

Consider the Weibull(α, λ) distribution

$$F_\theta(x) = 1 - e^{-(x/\lambda)^\alpha}.$$

In Example 7.4 we could not easily identify the IPA derivative as a function of $X(\theta)$. Using Lemma 8.1 the IPA derivative with respect to $\alpha = \theta$ can be obtained by

$$\frac{\partial}{\partial \alpha}X(\theta) = \frac{\left(\frac{X(\theta)}{\lambda}\right)^\alpha \ln\left(\frac{X(\theta)}{\lambda}\right) e^{-\left(\frac{X(\theta)}{\lambda}\right)^\alpha}}{\frac{\alpha}{\lambda}\left(\frac{X(\theta)}{\lambda}\right)^{\alpha-1} e^{-\left(\frac{X(\theta)}{\lambda}\right)^\alpha}} = \frac{X(\theta)}{\theta}\ln\left(\frac{X(\theta)}{\lambda}\right).$$

In the same vein one can find the IPA derivative with respect to the shape parameter of the Pareto II distribution ($\alpha = \theta$) by

$$\frac{\partial}{\partial \alpha} X(\theta) = \frac{\left(1 + \frac{X(\theta)}{\lambda}\right)^{-\alpha} \ln\left(1 + \frac{X(\theta)}{\lambda}\right)}{\frac{\alpha}{\lambda}\left(1 + \frac{X(\theta)}{\lambda}\right)^{-\alpha-1}} = \frac{\lambda}{\theta}\left(1 + \frac{X(\theta)}{\lambda}\right) \ln\left(1 + \frac{X(\theta)}{\lambda}\right).$$

※※※

Example 8.2. We consider here the estimation of the derivative of a quantile. For a family of distributions F_θ that are continuously differentiable with respect to both θ and x, the α-quantile of F_θ is defined by

$$q^\alpha = F_\theta^{-1}(\alpha) = \inf_u \{u \in \mathbb{R} : F_\theta(u) \geq \alpha\},$$

for $\alpha \in (0, 1)$. Provided that the density $f_\theta(x)$ is larger than zero on a neighborhood of q^α, differentiation the implicit equation $\alpha = F_\theta(q^\alpha)$ with respect to θ yields

$$\frac{d}{d\theta} q^\alpha = -\frac{\frac{\partial}{\partial \theta} F_\theta(q^\alpha)}{f_\theta(q^\alpha)}. \tag{8.2}$$

By Lemma 8.1, we can rewrite the above equation for $X(\theta) \sim F_\theta$ as

$$\frac{d}{d\theta} q^\alpha = \mathbb{E}[X'(\theta) | X(\theta) = q^\alpha].$$

For a given sequence $(X_\theta(i); 1 \leq i \leq m)$ denote the order statistic by $X_\theta(i:m)$, $1 \leq i \leq m$, i.e., we have $X_\theta(1:m) < X_\theta(2:m) < \cdots < X_\theta(m:m)$, where $(\cdot:m)$ is a bijection on $\{1, \ldots, m\}$, and observe that due to the continuity of F_θ there are almost surely no ties. Then, as shown in [113],

$$\lim_{m \to \infty} X_\theta(\lceil m\alpha \rceil : m) = q^\alpha \quad \text{a.s.},$$

where $\lceil m\alpha \rceil$ is the smallest integer greater than or equal to $m\alpha$. By continuity we arrive at

$$\lim_{m \to \infty} \mathbb{E}\left[X'(\theta) | X(\theta) = X_\theta(\lceil m\alpha \rceil : m)\right] = \frac{d}{d\theta} q^\alpha \quad \text{a.s.},$$

which shows that

$$\frac{\partial}{\partial \theta} X_\theta(\lceil m\alpha \rceil : m)$$

is an asymptotically unbiased estimator for $dq^\alpha / d\theta$.

※※※

For particular families of distributions, differentiability of $X(\theta)$ can be established directly without making use of Lemma 8.1. In fact, often we can compute the sample path derivatives without having explicit knowledge of the distribution, which establishes an interesting robustness property of IPA. For more details, see [63].

Definition 8.2. A parameter $\theta \in \Theta \subset \mathbb{R}$ of a family of probability distributions $\{F_\theta, \theta \in \Theta\}$ is called

- a *location parameter* if $F_\theta(x) = F_0(x - \theta)$, and
- a *scale parameter* if $F_\theta(x) = F_1(x/\theta)$, for $\theta \neq 0$.

Examples of location parameters are the mean of a normal distribution and the mean of the distribution of the random variable $X(\theta) = \theta + U$, where $U \sim U(-1, 1)$. Examples of scale parameters are the standard deviation of the normal distribution, the mean of the exponential distribution, and the mean of the random variable $X(\theta) = \theta U$, where $U \sim U(-1, 1)$. The following proposition can be used to verify assumption (i) in Theorem 8.1 without the need for using explicit representations of $X(\theta)$.

Proposition 8.1. *Let θ be a* location parameter *of the distribution function of $X(\theta)$. Then,*

$$\frac{d}{d\theta} X(\theta) = 1.$$

Let θ be a scale parameter *of distribution function of $X(\theta)$. Then,*

$$\frac{d}{d\theta} X(\theta) = X(1) = \frac{1}{\theta} X(\theta).$$

Proof. The proof follows from the fact that if θ is a location parameter, then $X(\theta) \stackrel{d}{=} \theta + X(0)$. If it is a scale parameter, then $X(\theta) = \theta X(1)$. □

We proceed now to discuss assumption (iii) in Theorem 8.1. We start with a technical definition.

Definition 8.3. A real-valued mapping f on $\Theta = (\theta_a, \theta_b) \subset \mathbb{R}$ is called *piecewise differentiable* on Θ if the set Θ_0 of locations of non-differentiability of f has no accumulation points in Θ. The set Θ_0 is called the non-differentiability set.

For a piecewise differentiable mapping f on Θ, we can split any closed subinterval $[\theta_0, \theta_1]$ of Θ up into a finite sequence of intervals (θ^i, θ^{i+1}), for $i \in I$, where I denotes the index set, such that f is differentiable on (θ^i, θ^{i+1}) and $[\theta_0, \theta_1] \subset \cup_i (\theta^i, \theta^{i+1})$. As the following lemma shows, we can establish the Lipschitz constant for continuous and piecewise differentiable mappings in a general form.

Lemma 8.2. *Let $\Theta \in \mathbb{R}$ be an open connected set. Let f be a real-valued mapping such that f is continuous throughout Θ and piecewise differentiable on Θ with non-differentiability set Θ_0. Then f is Lipschitz continuous on Θ with Lipschitz constant*

$$\sup \left\{ \left| \frac{d}{d\theta} f(\theta) \right| : \theta \in \Theta \setminus \Theta_0 \right\},$$

provided the above expression is finite.

Proof. Let $\theta_0, \theta_1 \in \Theta$, such that $\theta_0 < \theta_1$. By definition of the set Θ_0, we now label the points of Θ_0 that lie in $[\theta_0, \theta_1]$ as $\{\theta^i : i \in I\}$ in ascending order, with I the finite set of indices, where we assume the ease of presentation that $\theta_0, \theta_1 \in \Theta_0$. Then, writing $|f(\theta_0) - f(\theta_1)|$ as telescopic sum gives

$$|f(\theta_0) - f(\theta_1)| \le \sum_i |f(\theta^i) - f(\theta^{i+1})|,$$

where the sum is over all points θ^i that lie between θ_0 and θ_1. Since f is continuous throughout Θ and differentiable on the intervals (θ^i, θ^{i+1}), we may now apply the mean value theorem for the individuals intervals giving

$$|f(\theta^i) - f(\theta^{i+1})| \le |\theta^i - \theta^{i+1}| K_i,$$

where

$$K_i = \sup_{\theta \in (\theta^i, \theta^{i+1})} \left| \frac{d}{d\theta} f(\theta) \right|.$$

Hence,

$$|f(\theta_0) - f(\theta_1)| \le \sum_i |f(\theta^i) - f(\theta^{i+1})| \le \sum_i |\theta^i - \theta^{i+1}| \sup_i K_i = |\theta_0 - \theta_1| \sup_i K_i.$$

Noticing that the bound $\sup_i K_i$ is independent of the choice of θ_0 and θ_1, establishes the result. □

We now arrive at the following sufficient condition for assumption (iii) in Theorem 8.1 that is typically easily verifiable in applications.

Lemma 8.3. *Let $\Theta = (\theta_a, \theta_b) \subset \mathbb{R}$, with $-\infty < \theta_a < \theta_b < \infty$. Suppose that*

(i) *with probability one, the function $h(X(\theta))$ is continuous on Θ and piece-wise differentiable on Θ with non-differentiability set Θ_0 (where Θ_0 may depend on the realization),*
(ii) *there exists a random variable K such that $\mathbb{E}[K] < \infty$ and*

$$\sup \left\{ \left| \frac{d}{d\theta} h(X(\theta)) \right| : \theta \in \Theta \setminus \Theta_0 \right\} \le K,$$

then $h(X(\theta))$ is a.s. Lipschitz continuous on Θ with Lipschitz modulus K.

Proof. By Lemma 8.2, for $\theta_1 > \theta_0$, where $\theta_1, \theta_0 \in \Theta$, it holds almost surely that

$$|h(X(\theta_1)) - h(X_{\theta_0})| \le |\theta_1 - \theta_0| \sup_{\theta \in (\theta_a, \theta_b) \setminus \Theta_0} \left| \frac{d}{d\theta} h(X(\theta)) \right|.$$

The expression on the above right side serves almost surely as a Lipschitz modulus for $h(X(\theta))$ on Θ. Hence, any K such that

$$\sup \left\{ \left| \frac{d}{d\theta} h(X(\theta)) \right| : \theta \in \Theta \setminus \Theta_0 \right\} \le K$$

is a Lipschitz modulus as well and the proof follows from the assumption that $\mathbb{E}[K] < \infty$. □

By virtue of Lemma 8.3 we may apply the dominated convergence theorem (Theorem B.9) to mappings that have points of non-differentiability.

Example 8.3. Let $X(\theta)$ be exponentially distributed with mean $\theta > 0$. Using the representation put forward in (7.3), it follows that $X'(\theta) = X(\theta)/\theta$ as was developed in Example 7.2, which verifies condition (i) in Theorem 8.1. For any differentiable mapping h with derivative h',

$$\frac{d}{d\theta} h(X(\theta)) = \frac{1}{\theta} X(\theta) h'(X(\theta)).$$

GRADIENT ESTIMATION, FINITE HORIZON

Let $\hat{\Theta} = (\theta_l, \theta_r) \subset \Theta$ such that $\theta \in \hat{\Theta}$. Also by (7.3), it follows that $X(\theta)$ is monotone increasing in θ. Hence, for any differentiable mapping h with monotone increasing derivative

$$\sup_{\theta \in \hat{\Theta}} \frac{d}{d\theta} h(X(\theta)) \leq \frac{1}{\theta_l} X(\theta_r)\, h'(X(\theta_r)) \leq \frac{1}{\theta_l} X(\theta_r)\, |h'(X(\theta_r))| =: K. \tag{8.3}$$

Provided that $\mathbb{E}[X(\theta_r)\, |h'(X(\theta_r))|]$ is finite. Lemma 8.3 yields that condition (iii) in Theorem 8.1 is satisfied. Theorem 8.1 then implies that for any differentiable mapping with monotone derivative for which K in (8.3) is integrable, we can interchange derivative and expectation, that is,

$$\frac{d}{d\theta} \mathbb{E}[h(X(\theta))] = \frac{1}{\theta} \mathbb{E}[X(\theta)\, h'(X(\theta))].$$

Note that the condition on h can be simplified in some cases. Indeed, if h' is bounded, then $\mathbb{E}[X(\theta_r)] < \infty$ is sufficient for unbiasedness which is trivially satisfied.

We now turn to the mining example in Example 7.3. It is straightforward that h' in Example 7.3 is monotone increasing, and applying the above line of argument we see that the feedback in the example

$$Y_n = 1 - \frac{X(\theta_n)}{\theta_n} \left(\frac{1}{(1+X(\theta_n))^{1/2}} - \frac{X(\theta_n)}{2(1+X(\theta_n))^{3/2}} \right)$$

is an unbiased estimator for the desired derivative.   ✼✼✼

Example 8.4. We revisit Example 7.15 and establish unbiasedness of the IPA estimator for the sensitivity of the system's life to the mean lifetime of component 3. This example suffices to illustrate the case when other components are also included in the optimization.

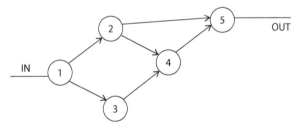

Use the Skorohod representation (see Example 7.2) to establish that T_3 has the same distribution as $-\theta \ln(1 - U(\omega))$ where $U(\omega) \sim U(0,1)$ and use this representation in what follows. Fix ω and rewrite the lifetime of the system as a function of θ as follows. To simplify notation, call $h(\theta, \omega) = L(T_1(\omega), T_2(\omega), T_3(\theta, \omega), T_4(\omega), T_5(\omega))$. Define

$$\bar{T}(\omega) \stackrel{\text{def}}{=} \min(T_1(\omega), T_4(\omega), T_5(\omega)).$$

Claim. We will next show that for fixed ω,

$$h(\theta, \omega) = \begin{cases} T_3(\theta, \omega) & \text{if } T_2(\omega) < T_3(\theta, \omega), \\ T_{i^*}(\omega)\ (i^* \neq 3) & \text{otherwise,} \end{cases}$$

$$= \begin{cases} \theta(-\ln(1-U(\omega))) & \text{if } T_2(\omega) < \theta(-\ln(1-U(\omega))) < \bar{T}(\omega), \\ T_{i^*}(\omega)\ (i^* \neq 3) & \text{otherwise,} \end{cases}$$

so that for each ω, h is a piecewise linear function of θ.

Proof of Claim: To simplify notation, drop the explicit dependency on the chosen value of ω. We distinguish the following cases:

(a) If $T_2 \geq \bar{T}$ then route R_2 breaks at time $\bar{T}$. Route R_3 breaks at time $\min(T_3(\theta), \bar{T})$. As for route R_1 it breaks at time $\bar{T}$ unless $\bar{T} = T_4$, in which case it breaks at $\min(T_1, T_2, T_5)$. So the life of the system is either $\bar{T}$, or L_1 which are both independent of θ. In this case, $i^* = \arg\max(\bar{T}, \min(T_1, T_2, T_5))$.

(b) If f $T_3(\theta) > \bar{T}$ then route R_3 breaks at time $\bar{T}$, so in this case the life of the system is also $\bar{T}$ unless $\bar{T} = T_4$. Therefore the life of the system is either $\bar{T}$, or L_1 which are both independent of θ. In this case, $i^* = \arg\max(\bar{T}, \min(T_1, T_2, T_5))$.

(c) Now suppose that $T_3(\theta) \leq T_2 < \bar{T}$. Route R_3 breaks at time $T_3(\theta)$ and routes R_1 and R_2 both break at time T_2, thus the life of the system is T_2.

(d) Finally, if $T_2 < T_3(\theta) < \bar{T}$ then routes R_1 and R_2 break at time T_2, and route R_3 breaks exactly at time $T_3(\theta) > T_2$, which is then the value of the system's life.

If $T_2(\omega) \geq \bar{T}(\omega)$ then $h(\theta, \omega) = T_{i^*}$ is independent of θ. Otherwise, if $T_2(\omega) < \bar{T}(\omega)$, define $\underline{\theta} = -T_2(\omega)/\ln(1 - U(\omega))$ and $\bar{\theta} = -\bar{T}(\omega)/\ln(1 - U(\omega))$. Then

$$h(\theta, \omega) = \begin{cases} T_2(\omega) & \text{if } \theta \leq \underline{\theta}, \\ \theta(-\ln(1 - U(\omega))) & \text{if } \underline{\theta} \leq \theta \leq \bar{\theta}, \\ T_{i^*}(\omega) & \text{if } \theta \geq \bar{\theta}, \end{cases}$$

and notice that this function is continuous and piecewise differentiable. By Lemma 8.3 the Lipschitz constant is bounded in absolute value by the largest slope, namely $-\ln(1 - U)$.

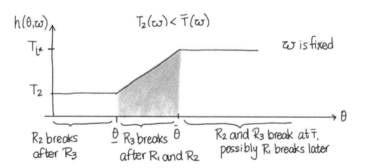

Because $-\ln(1 - U)$ is a random variable with distribution $\text{Exp}(1)$, it has bounded expectation, verifying condition (ii) in Theorem 8.1. Thus we can interchange derivative and expectation for h. The IPA estimator is the resulting derivative, expressed in terms of the original random variables $(T_i; i = 1, \ldots, 5)$, which is

$$\widehat{L}^{IPA}(\theta, \omega) = \begin{cases} \frac{T_3(\theta, \omega)}{\theta} & \text{if } T_2(\omega) < T_3(\theta, \omega) < \bar{T}(\omega) = \min(T_1(\omega), T_4(\omega), T_5(\omega)), \\ 0 & \text{otherwise.} \end{cases} \quad (8.4)$$

It follows from Theorem 8.2 that, although the derivative is a discontinuous random variable, it is unbiased for $dL(\theta)/d\theta$. It is important to realize that the expectation of $\widehat{L}^{IPA}(\theta)$ is independent of the "true" representation of the random variable $T_3(\theta)$ as a function of ω, because $T_3(\theta) \stackrel{d}{=} -\theta \ln(1 - U)$ is sufficient to establish the result. ✳✳✳

8.1.2 IPA for the Finite Horizon Problem

So far, we have considered the case of perturbing a single random variable. Many models that are of importance in applications are driven by a finite collection of random variables, and in the remainder of this section we explain how IPA can be used in these models. The following theorem states sufficient conditions for unbiasedness of the IPA estimator for static gradient estimation problem. To simplify the notation, we write $\frac{d}{d\theta}f(\theta_0)$ for the derivative of a differentiable mapping f at a point θ_0, i.e., we let

$$\left.\frac{d}{d\theta}f(\theta)\right|_{\theta=\theta_0} = \frac{d}{d\theta}f(\theta_0)$$

when this causes no confusion.

Theorem 8.2. *Let $X_i(\theta)$, $1 \leq i \leq N$, be (not necessarily statistically independent) random variables defined in an open connected set Θ with $\theta_0 \in \Theta$. Assume that*

(i) the sample path (or stochastic) derivative $dX_i(\theta)/d\theta$ exists with probability one at θ_0,
(ii) the cost function $L_N(X_1(\theta), \ldots, X_N(\theta))$ is differentiable at $(X_1(\theta_0), \ldots, X_N(\theta_0))$ with probability one, and
(iii) the cost function $L_N(X_1(\theta), \ldots, X_N(\theta))$ is almost surely Lipschitz continuous on Θ (and has thus by definition an integrable Lipschitz modulus).

Then the derivative at θ_0 satisfies

$$\frac{d}{d\theta}\mathbb{E}[L_N(X_1(\theta_0), \ldots, X_N(\theta_0))] = \mathbb{E}\left[\sum_{i=1}^{N} \frac{d}{d\theta}X_i(\theta_0) \frac{\partial}{\partial x_i} L_N(X_1(\theta_0), \ldots, X_N(\theta_0))\right].$$

Example 8.5. Let $\{X_n(\theta)\}$ be an iid sequence of exponentially distributed random variables with mean value θ. Moreover, for $N \geq 1$, let

$$T_\theta(N) = \sum_{k=1}^{N} X_k(\theta) = L_N(X_1(\theta), \ldots, X_N(\theta)).$$

By definition, the sequence $\{T_\theta(N)\}$ corresponds to the N-th arrival time of a Poisson process with rate $1/\theta$, for $\theta > 0$. It is easily seen that L_N satisfies the condition put forward in Theorem 8.2. In particular, it holds that $\partial L_N / \partial x_i = 1$, which yields

$$\frac{d}{d\theta}T_\theta(N) = \sum_{k=1}^{N} \frac{\partial}{\partial x_i} L_N(X_1(\theta), \ldots, X_N(\theta)) \frac{d}{d\theta}X_i(\theta) = \frac{1}{\theta}T_\theta(N)$$

for the IPA estimator. That the estimator is indeed unbiased can be verified here from the fact that $\mathbb{E}[T_\theta(N)] = N\theta$ and

$$\frac{d}{d\theta}\mathbb{E}[T_\theta(N)] = N = \frac{1}{\theta}\mathbb{E}[T_\theta(N)]$$

together with Example 7.2.

Let T be a fixed horizon and consider now the number of jumps of the Poisson process up to time T, which is given by

$$N_\theta(T) = \sum_{n=0}^{\infty} \mathbf{1}_{\{T_\theta(n) \leq T < T_\theta(n+1)\}} = \sup\{n : T_\theta(n) \leq T\}.$$

Since $N_\theta(T)$ is a piecewise constant mapping in θ it follows that

$$\frac{d}{d\theta} N_\theta(T) = 0$$

with probability one. Hence, the IPA estimator would yield the erroneous outcome 0, where it is well known that $\mathbb{E}[N_\theta(T)] = T/\theta$ and the true derivative is thus given by $-T/\theta^2$. To see why IPA fails for this performance function, note that $N_\theta(T)$ fails to be Lipschitz continuous in θ on a neighborhood of θ. While this example may (deceivingly) look like a finite horizon problem because T is deterministic, it is not because $N(T)$ is a random number. Thus, this problem belongs to the random horizon case. ✱✱✱

Example 8.6. We revisit the sojourn time example as introduced in Example 7.16 with notations and basic assumptions as detailed in this example. Customers arrive at a service station according to a renewal point process. The inter arrival times $\{A_n : n \in \mathbb{N}\}$ are iid, with $\mathbb{E}[A_n] < \infty$ and $\mathbb{P}(A_n = 0) = 0$. Customers are served in order of arrival, and consecutive service times are iid random variables $\{S_n(\theta) : n \in \mathbb{N}\}$. Interarrival times and service times are assumed to be mutually independent. The common distribution of the service times G_θ depends on parameter $\theta = \mathbb{E}[S_n(\theta)]$.

In addition we assume that θ is a scaling parameter of the service time distribution, which implies that condition (i) in Theorem 8.2 holds. Suppose that we are interested in solving the finite horizon gradient estimation problem with IPA for the average sojourn time over the first N customers in a single server queue:

$$L_N(X_1(\theta), \ldots, X_N(\theta)) = \frac{1}{N} \sum_{n=1}^{N} X_n(\theta). \tag{8.5}$$

Because L_N is an average, it suffices to show that each $X_n(\theta)$ is a.s. Lipschitz continuous. We use mathematical induction on (7.30), which can be rewritten as

$$X_{n+1}(\theta) = \begin{cases} S_{n+1}(\theta) & \text{if } X_n(\theta) < A_{n+1}, \\ S_{n+1}(\theta) + (X_n(\theta) - A_{n+1}) & \text{otherwise}. \end{cases} \tag{8.6}$$

Using the fact that θ is a scaling parameter, we obtain, for $n = 1$, that $X_1(\theta) = S_1(\theta) = \theta S_1(1)$ is almost surely Lipschitz continuous with modulus $l_1 = S_1(1)$. Indeed, it holds that

$$X_1(\theta + \delta) - X_1(\theta) = (\theta + \delta) S_1(1) - \theta S_1(1) = \delta S_1(1) = \delta l_1. \tag{8.7}$$

Having perturbed the service times, there is a "propagation" of the perturbation as illustrated in Figure 8.1. Indeed, the second client starts its service already delayed from the longer service time of the first customer. In addition, its own service time is also longer, so there are two contributions to the total increase in its sojourn time. Analogously, the n-th client will have an increase in sojourn time that accumulates the previous clients delays plus the increment in its own service time.

We now prove by induction that $X_n(\theta)$ is a.s. Lipschitz continuous with Lipschitz modulus

$$l_n = \sum_{k=1}^{n} S_k(1),$$

that is, we show that

$$|X_n(\theta + \delta) - X_n(\theta)| \leq |\delta| l_n. \tag{8.8}$$

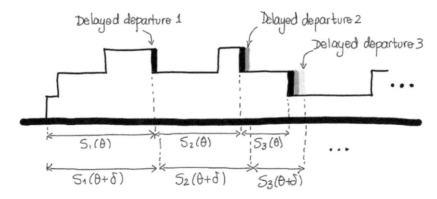

Figure 8.1. Visualization of the propagation of perturbation in the FCFS queue.

for $n \geq 1$. For $n = 1$, (8.8) follows from (8.7). Now suppose that (8.8) holds for n. We show that this implies that (8.8) holds for $n + 1$ as well. By (7.30) we have

$$X_{n+1}(\theta) = S_{n+1}(\theta) + \max(0, X_n(\theta) - A_{n+1}(\theta))$$
$$= S_{n+1}(\theta) + \max(A_{n+1}, X_n(\theta)) - A_{n+1}. \qquad (8.9)$$

Moreover, for $\delta > 0$, the system with mean service time $\theta + \delta$ is slower than the system under θ, and therefore $X_n(\theta + \delta) \geq X_n(\theta)$, for $n \geq 1$. We distinguish the following three cases:

(i) if $A_{n+1} < X_n(\theta)$, and therefore also $A_{n+1} < X_n(\theta + \delta)$, then

$$\max(A_{n+1}, X_n(\theta + \delta)) - \max(A_{n+1}, X_n(\theta)) = X_n(\theta + \delta) - X_n(\theta),$$

(ii) if $X_n(\theta) < A_{n+1}$ and $A_{n+1} < X_n(\theta + \delta)$, then

$$\max(A_{n+1}, X_n(\theta + \delta)) - \max(A_{n+1}, X_n(\theta)) = X_n(\theta + \delta) - A_{n+1}$$
$$\leq X_n(\theta + \delta) - X_n(\theta),$$

(iii) if $X_n(\theta) < X_n(\theta + \delta) \leq A_{n+1}$, then

$$\max(A_{n+1}, X_n(\theta + \delta)) - \max(A_{n+1}, X_n(\theta)) = 0$$
$$\leq X_n(\theta + \delta) - X_n(\theta).$$

This gives $X_{n+1}(\theta + \delta) - X_{n+1}(\theta) \leq S_{n+1}(\theta + \delta) - S_{n+1}(\theta) + X_n(\theta + \delta) - X_n(\theta)$, and, taking absolute values,

$$|X_{n+1}(\theta + \delta) - X_{n+1}(\theta)| \leq |S_{n+1}(\theta + \delta) - S_{n+1}(\theta)| + |X_n(\theta + \delta) - X_n(\theta)|.$$

Applying the induction hypothesis, we arrive at

$$|X_{n+1}(\theta + \delta) - X_{n+1}(\theta)| \leq \delta(S_{n+1}(1) + l_n) = \delta l_{n+1},$$

where we use the fact that as θ is a scaling parameter of the service times (i.e., $S_{n+1}(\theta + \delta) - S_{n+1}(\theta) = \delta S_{n+1}(1)$), which completes the induction argument. Since $\mathbb{E}[l_n] \leq n < \infty$, we have shown that L_N is a.s. Lipschitz continuous. The resulting IPA derivative is

$$\frac{d}{d\theta} L_N(X_1(\theta), \ldots, X_N(\theta)) = \frac{1}{N} \sum_{n=1}^{N} \frac{dX_n(\theta)}{d\theta}, \tag{8.10}$$

and following (8.9) it can be computed recursively using an auxiliary *derivative process*:

$$Z_{n+1}(\theta) \stackrel{\text{def}}{=} \frac{dX_{n+1}(\theta)}{d\theta} = \begin{cases} \frac{dS_{n+1}(\theta)}{d\theta} & \text{if } X_n(\theta) \leq A_{n+1}, \\ \frac{dS_{n+1}(\theta)}{d\theta} + Z_n(\theta) & \text{if } X_n(\theta) > A_{n+1}, \end{cases}$$

which can be calculated online, as the process $\{X_n(\theta)\}$ is being simulated or observed.

An alternative and popular computation procedure is provided now. Let $\alpha_n(\theta)$ be the index of the last customer prior to n that encounters the server idle, that is, $\alpha_n(\theta) = \max(k \leq n : X_{k-1}(\theta) < A_k)$, which is $\mathfrak{F}_n(\theta)$-measurable and therefore a stopping time. Then

$$X_n(\theta) = \sum_{i=\alpha_n(\theta)}^{n} S_i(\theta) - A_n, \tag{8.11}$$

and therefore we can write the derivative as

$$Z_n(\theta) = \frac{dX_n(\theta)}{d\theta} = \sum_{i=\alpha_n(\theta)}^{n} \frac{dS_i(\theta)}{d\theta},$$

(where the sum from n to n contains the one single term). This gives the final computation:

$$\frac{d}{d\theta} L_N(X_1(\theta), \ldots, X_N(\theta)) = \frac{1}{N} \sum_{n=1}^{N} Z_n(\theta) = \frac{1}{N} \sum_{n=1}^{N} \sum_{i=\alpha_n(\theta)}^{n} \frac{dS_i(\theta)}{d\theta} \stackrel{\text{d}}{=} \frac{1}{\theta N} \sum_{n=1}^{N} \sum_{i=\alpha_n(\theta)}^{n} S_i(\theta), \tag{8.12}$$

where the last equality follows because θ is a scaling parameter of the service time distribution.

We now argue that the conditions put forward in Theorem 8.2 are satisfied. The derivative of L_N is given in (8.12), which shows that (ii) is satisfied. In the above analysis, we used an induction argument for showing that L_N is Lipschitz continuous with integrable modulus, which shows that (iii) holds. Hence, (8.12) provides an unbiased gradient estimator. To see the formal equality of the derivative expression in the theorem and (8.12) and (8.10), respectively, note that $\partial L/\partial x_i = 1$. The pseudocode in Algorithm 8.1 uses f_θ to denote the density of the service times.

We finalize with an algorithmic analysis for the running time. While there are two nested **for** loops, which would yield a running time of order $O(N^2)$ in the worst case, a probabilistic analysis is more accurate here to find the growth rate of the expected running time. The total number of customers in one busy period is

$$\tau(\theta) = \min(n : A_n > X_{n-1}),$$

meaning that customer $\tau(\theta)$ finds the server idle and has no wait. The index α_n in (8.12) is the index of the first customer in the busy period of the current customer n, which implies that the total number of terms in the inner sum is stochastically bounded by $\tau(\theta)$. For θ such that the queue is stable, the stopping time $\tau(\theta)$ is finite w.p.1. Because

Algorithm 8.1 Finite horizon IPA for the average sojourn time using (8.12)

Initialize $X_1 = S_1 \sim f_\theta$, IPA $= 0, \alpha = 1$.
for $n = 1, \ldots, N$ **do**
    for $i = \alpha, \ldots, n$ **do**
        IPA $\leftarrow$ IPA $+ S_i$
    Generate $A_{n+1}, S_{n+1} \sim f_\theta$
    $\Delta = X_n - A_{n+1}$
    $X_{n+1} = \max(0, \Delta) + S_{n+1}$
    if $(\Delta < 0)$ **then** $\alpha = n + 1$
return IPA$/(N \theta)$.

consecutive cycles have the same distribution, this means that the expected running time is $\mathbb{E}[N\tau(\theta)] = O(N)$. ✳✳✳

Historical Note. We conclude our discussion of IPA by presenting some historical background. IPAs origins can be traced back to the work of Y. Ho and his coworkers in the late 1970s. Ho's research group studied the buffer allocation problem at the FIAT automobile company. More precisely, the question was how to optimally allocate a finite number of B buffer places to a serial transfer production line of N stations. For larger N and B evaluating all possible combinations becomes rapidly numerically infeasible. Comparing simulation traces for buffer allocations that differed by only changing the allocation of one buffer place from one machine to another, Ho realized that in most parts the simulation traces are identical, which gave rise to the idea to compute the effect of changing the allocation of one buffer place in buffer allocation alongside the simulation of the particular buffer allocation. This lead to [163], which is considered as the birth of sample-path perturbation analysis in engineering. It is worth noting that the buffer allocation problem is not related to sample-path differentiation as discrete perturbations are considered. However, asking the same question about the effect of a parameter change on the simulation trace, once applied to continuous parameters, quickly led to the development of IPA as tool for sample path differentiation. In the 1980s there was great interest in the study of systems that change their state only at discrete point in time and where a state transition is triggered by the occurrence of an "event" (see [160, 161, 162] for early references, and [63] for further development of event-driven IPA). These systems were called discrete event systems (DES) and discrete event dynamic systems (DEDS), respectively, and an event-driven formalism for perturbation analysis was introduced in [64]. Discrete event systems systems have been studied independently of IPA (see [14, 253, 307, 329] for some early references) and are still a topic of active research. The event-driven view on IPA required a thorough mathematical analysis of discrete-event computer simulations. Mathematically speaking, discrete-event Monte Carlo simulation models can be described as so-called *generalized semi-Markov process* (see [125, 278, 279, 307, 325]), and a sample-path differentiation theory for this type of (very general) models was developed. In particular, it could be shown that, under appropriate conditions, the time of the occurrence of events is differentiable if the so-called commuting condition holds, that is, if the state reached from a certain physical state is independent of the sequence in which the events occur, see [114, 115, 124, 144]. This conditions holds for many queueing systems that are of importance in practice. We refer to Chapter 12 in Part III for details. Monographs summarizing the early developments on IPA are [66, 114, 159]. For recent references, see [322, 331, 339, 340].

8.1.3 Smoothed Perturbation Analysis

The key limitation of the sample path approach is that the combined random variable $h(X(\theta))$ has to be a.s. Lipschitz continuous on Θ for some open neighborhood of the point of differentiation where this neighborhood may not depend on the sample path. For example, suppose that $X(\theta) \sim \text{Ber}(\theta)$ is a random variable with a Bernoulli distribution. If we use the Skorohod representation, then $X(\theta, U) \stackrel{d}{=} \mathbf{1}_{\{U \leq \theta\}}$, for U a uniform random variable. Here $\mathbb{E}[X(\theta)] = \mathbb{P}(U \leq \theta) = \theta$ so that the derivative satisfies $\frac{d}{d\theta}\mathbb{E}[X(\theta)] = 1$. However, for every U, $X(\theta, U)$ is piecewise constant with a jump at $U = \theta$, so that for every $\omega: U(\omega) \neq \theta$, $X'(\theta, U) = 0$. Because $\mathbb{P}(U(\omega) = \theta) = 0$, we have here that $X'(\theta, U) = 0$ a.s., and consequently $\mathbb{E}[X'(\theta, U)] = 0 \neq \frac{d}{d\theta}\mathbb{E}[X(\theta)] = 1$. Sometimes it is possible to overcome the obstacle of discontinuities using a conditioning (or "smoothing") approach, as illustrated in the following example.

Example 8.7. As an academic example, consider $X(\theta) \sim F_\theta$ where θ is a scale parameter, so that $X(\theta) \stackrel{d}{=} \theta Y$, where $Y \sim F_1$ is independent of θ. Suppose that we are interested in evaluating

$$\frac{d}{d\theta}\mathbb{P}(X(\theta) \leq \beta),$$

which renders $L(\theta, \omega) = \mathbf{1}_{\{X(\theta, \omega) \leq \beta\}}$ a step function. For this example, fixing ω yields $L(\theta, \omega) = 1$ for all $\theta \leq \beta/Y(\omega))$ and $L(\theta, \omega) = 0$ otherwise, yielding $L'(\theta, \omega) = 0$ for each ω. To smooth out the discontinuity in the derivative, we may use conditional expectations to partially integrate the jumps. There are three logical steps:

(i) For each θ, define the *critical event* by $\Omega^c\theta(\Delta) \stackrel{\text{def}}{=} \{\omega: X(\theta, \omega) \leq \beta < X(\theta + \Delta, \omega)\}$ and let $\Omega_\theta(\Delta) = \Omega \setminus \Omega_\theta^c(\Delta)$ be the complement of the critical event.

(ii) Express the expectation conditioning on $\Omega_\theta^c(\Delta)$:

$$\frac{d}{d\theta}\mathbb{E}[L(\theta)] = \lim_{\Delta \downarrow 0} \mathbb{E}\left[\overbrace{\frac{\mathbf{1}_{\{X(\theta+\Delta) \leq \beta\}} - \mathbf{1}_{\{X(\theta) \leq \beta\}}}{\Delta}}^{\Delta L(\theta)}\right] = \lim_{\Delta \downarrow 0} \mathbb{E}[\Delta L(\theta) \mid \Omega_\theta^c(\Delta)] \frac{\mathbb{P}(\Omega_\theta^c(\Delta))}{\Delta}$$

because $\mathbb{E}[\Delta L(\theta) \mid \Omega_\theta(\Delta)] = 0$ for all $\Delta > 0$. As $\Delta L(\theta) = -1$ on the critical event for all Δ, this yields

$$\frac{d}{d\theta}\mathbb{E}[L(\theta)] = -\frac{\mathbb{P}(\Omega_\theta^c(\Delta))}{\Delta}.$$

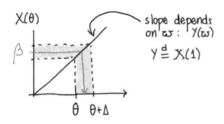
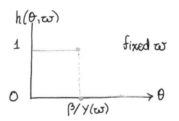

(iii) Calculate the *critical probability rate*:

$$\mathbb{P}(\Omega_\theta^c(\Delta)) = \mathbb{P}(\theta Y \leq \beta < (\theta + \Delta)Y) = \mathbb{P}\left(\frac{\beta}{\theta + \Delta} \leq Y < \frac{\beta}{\theta}\right) = F_1\left(\frac{\beta}{\theta}\right) - F_1\left(\frac{\beta}{\theta + \Delta}\right).$$

Using now that θ is a scale parameter of the distribution, it follows that $F_1(\beta/\theta) = F_\theta(\beta)$, which can be used to establish

$$\lim_{\Delta \downarrow 0} \frac{\mathbb{P}(\Omega_\theta^c(\Delta))}{\Delta} = \frac{\partial}{\partial \theta} F_\theta(\beta).$$

Naturally, for this academic example, the solution is straightforward from the problem statement because $F_\theta(\beta) = \mathbb{P}(X(\theta) \leq \beta)$; however, it illustrates the main three steps in the logical arguments used for more complicated problems. It is common to find that the critical probability rates depend on the density of the random variable. ✼✼✼

The above example shows how appropriate conditioning can yield an unbiased gradient estimator in case IPA fails. The basic general principle can be described as follows. Consider the problem of estimating the derivative w.r.t. θ of $L(\theta) \stackrel{\text{def}}{=} h(X(\theta))$ and assume that, for given $\theta_0 \in \Theta$ and for every ω the limits

$$\lim_{\Delta \downarrow 0} L(\theta_0 + \Delta) = L^+(\theta_0) \text{ and } \lim_{\Delta \uparrow 0} L(\theta_0 + \Delta) = L^-(\theta_0)$$

exist a.s. In case $L(\theta)$ is a.s. continuous at θ_0 we have $L^-(\theta_0) = L^+(\theta_0)$. Otherwise $\Delta L(\theta_0) = L^+(\theta_0) - L^-(\theta_0)$ expresses the height of the jump of $L(\theta)$ at θ_0; see Example 8.7.

For $\Delta \neq 0$, let $\Omega_{\theta_0}(\Delta) \subset \Omega$ be the event that $L(\theta)$ is differentiable at θ_0 and Lipschitz continuous on $[\theta_0 - |\Delta|, \theta_0 + |\Delta|]$. When $\Omega_{\theta_0}(\Delta)$ is a strict subset of Ω, we define the complement $\Omega_{\theta_0}^c(\Delta)$ as the *critical event*. This partition of Ω will help to deal with the expectation by conditioning on the critical event and "smoothing out" the discontinuities. Then

$$\frac{1}{\Delta}\mathbb{E}[L(\theta_0 + \Delta) - L(\theta_0)] = \frac{1}{\Delta}\mathbb{E}[L(\theta_0) - L(\theta_0)|\Omega_{\theta_0}(\Delta)]\mathbb{P}(\Omega_{\theta_0}(\Delta))$$
$$+ \mathbb{E}[L(\theta_0 + \Delta) - L(\theta_0)|\Omega_{\theta_0}^c(\Delta)]\frac{\mathbb{P}(\Omega_{\theta_0}^c(\Delta))}{\Delta}.$$

Typically, the probabilities on the above right-hand side are smooth and it holds that

$$\lim_{\Delta \to 0} \mathbb{P}(\Omega_{\theta_0}(\Delta)) \stackrel{\text{def}}{=} p_{\theta_0} \qquad (8.13)$$

and

$$\lim_{\Delta \to 0} \frac{1}{\Delta} \mathbb{P}(\Omega_{\theta_0}^c(\Delta)) \stackrel{\text{def}}{=} p'_{\theta_0}. \qquad (8.14)$$

Let $\Omega_{\theta_0} = \bigcap_{\Delta \neq 0} \Omega_{\theta_0}(\Delta)$. Provided that the limits in (8.13) and (8.14) exist, we obtain

$$\lim_{\Delta \to 0} \frac{1}{\Delta}\mathbb{E}[L(\theta_0 + \Delta) - L(\theta_0)] = \mathbb{E}\left[\frac{d}{d\theta}L(\theta_0)\bigg|\Omega_{\theta_0}\right] p_{\theta_0} + \mathbb{E}\left[L^+(\theta_0) - L^-(\theta_0)\big|\Omega_{\theta_0}^c\right] p'_{\theta_0},$$

where $\Omega_{\theta_0}^c$ is the complement of Ω_{θ_0}. The first part on the above right hand side is noticeably the IPA contribution to the derivative, and the second part is weighted by a probability rate of the jump size at θ_0. The above conditioning approach is known as *smoothed perturbation analysis*. While SPA has been successfully applied in many situations, it provides only ad hoc solutions as the conditioning depends on the structure of the sample path function and the type of dependence of $L(X(\theta))$ on θ. SPA was introduced in [133] and has been further developed in [46, 121, 132, 321]. For a monograph presenting a detailed account on SPA, we refer to [104]. Observe that in Example 8.7 we discussed an application of SPA where the IPA contribution was zero and $L^+(\theta_0) - L^-(\theta_0) = -1$ so that analysis reduced to the computation of the rate p'_{θ_0}.

For an elaborate example of SPA we refer to Section 10.3.1. Exercise 8.13 illustrates the application of conditioning to construct the SPA estimator for the reliability network of Example 8.4 when the goal is to estimate the sensitivity of a probability.

*8.1.4 An Indirect Approach to Establishing Unbiasedness

When presenting IPA we have chosen for a constructive approach. First we constructed the sample path derivatives, and then we asked whether they provide an unbiased estimator. This is the usual approach in the simulation literature. In this section we discuss a different approach to establishing the existence of a sample path derivative and the proof of unbiasedness. The approach is based on a basic result from measure theory for absolutely continuous mappings, and has been introduced in [117]. The resulting estimators are called "stochastic derivatives" in the applied mathematics literature, rather than IPA derivatives.

Definition 8.4. Let (X, d) be a metric space and let $I \subset \mathbb{R}$ be an interval. A mapping f from I to X is called *absolutely continuous* on I if for every $\epsilon > 0$, there exists $\delta > 0$ such that for every (finite or infinite) sequence of pairwise disjoint sub-intervals $[x_k, y_k]$ of I it holds

$$\sum_k |y_k - x_k| < \delta \implies \sum_k d(f(y_k), f(x_k)) < \epsilon.$$

If f is absolutely continuous, then f has a derivative almost everywhere, the derivative is Lebesgue integrable, and its integral is equal to the increment of f. The precise statement, known as fundamental theorem of integral calculus for Lebesgue integrals (see, e.g., [158]), states that if f is absolutely continuous, then the set of $x \in I$ such that f' fails to exist has Lebesgue measure zero, and it holds for $x, y \in I$ with $y > x$ that

$$f(y) - f(x) = \int_x^y f'(z) \lambda(dz),$$

where $\lambda(\cdot)$ denotes the Lebesgue measure. The usefulness of the concept of absolute continuity for gradient estimation stems from the fact that a Lipschitz-continuous function is absolutely continuous.

Let $X(\theta)$ be Lipschitz continuous on some finite interval Θ. In order to make use of arguments based on absolute continuity, we have to consider $X(\theta)$ as a random mapping on an extended sample space. More specifically, let, for all θ, $X(\theta)$ be defined on some underlying probability space $(\Omega, \mathfrak{F}, \mathbb{P})$. Let $\Theta = [\theta_a, \theta_b] \subset \mathbb{R}$. Equip $\mathbb{R}$ with the usual topology and equip Θ with its Borel field, denoted by $\mathcal{B}$. For the following consider $X(\theta)$ as random variable on the product space $(\Omega \times \Theta, \mathfrak{F} \otimes \mathcal{B}, \mathbb{P} \otimes \lambda)$, where $\mathfrak{F} \otimes \mathcal{B}$ denotes the product field of $\mathfrak{F}$ and $\mathcal{B}$ and $\mathbb{P} \otimes \lambda$ denotes the product probability measure of $\mathbb{P}$ and the Lebesgue measure. By absolute continuity of $X(\theta)$ it holds

$$X(\theta_b) - X(\theta_a) = \int_{\theta_a}^{\theta_b} X'(r) \lambda(dr),$$

and we conclude that $X(\theta)$ is differentiable (for Lebesgue, almost all θ in Θ). Note that since $X(\theta)$ is measurable as a mapping of θ, it follows that $X'(\theta)$ is measurable. Fubini's theorem now gives

$$\mathbb{E}[X(\theta_b)] - \mathbb{E}[X(\theta_a)] = \int_{\theta_a}^{\theta_b} \mathbb{E}[X'(r)] \lambda(dr),$$

from which we conclude that $\mathbb{E}[X(\theta)]$ is absolutely continuous and that $\mathbb{E}[X'(\theta)]$ is (for almost all θ) the derivative of $\mathbb{E}[X(\theta)]$ on Θ:

$$\frac{d}{d\theta}\mathbb{E}[X(\theta)] = \mathbb{E}[X'(\theta)]. \tag{8.15}$$

To summarize, the above approach is rather elegant in that interchanging differentiation and integration can be replaced by a simple application of Fubini's theorem. However, the fact that we only have Lebesgue almost everywhere differentiability on Θ makes the result in (8.15) somewhat esoteric, as for any given θ_0 we cannot say whether (8.15) holds or not.

8.2 DISTRIBUTIONAL APPROACH: BASIC RESULTS AND TECHNIQUES

In Chapter 7 we introduced the concept of *path wise* approach that leads to the IPA method and Section 8.1 developed the theoretical support for this method to be unbiased. The *distributional* approach considers the case where all the dependency on the parameter θ is is the distribution functions, or, more generally, in the measure. The distributional approach (also introduced in Chapter 7) leads to two different estimation methods: the score function (SF) method and the measure-valued differentiation method. We summarize here the main framework and basic results that will be used for both these methods. The following two sections provide the detailed analysis of SF and MVD, respectively.

Suppose that we wish to estimate the derivative of an expectation. We consider the expression

$$\frac{d}{d\theta}\mathbb{E}[h(X(\theta))] = \frac{d}{d\theta}\int_{\mathbb{R}} h(x) f_\theta(x)\,dx,$$

where $f_\theta(\cdot)$ is the probability (Lebesgue) density of the random variable $X(\theta)$. The method can also be used for discrete random variables, and in that case we seek

$$\frac{d}{d\theta}\mathbb{E}[h(X(\theta))] = \frac{d}{d\theta}\sum_{k\in\mathbb{N}} h(x_k)\, p_k(\theta),$$

where $p_k(\theta) = \mathbb{P}(X(\theta) = x_k)$ gives the distribution of the random variable $X(\theta)$. In this approach the dependency on θ is entirely in the measure.

The distributional approach is based on interchanging the derivative and expectation operators, focusing on the derivative of the θ-dependent *measure* inside the integral (or the summation).

The first model that we consider is that of Riemann densities, which leads to a simpler development of the main ideas.[1] Later in the chapter we extend the methods to general probability distributions. In Section 8.3.2, we discuss the general setting for SF involving Radon-Nikodym derivatives and absolute continuity of measures, and Section 8.4.3 presents the measure integral setting for MVD.

Definition 8.5. We call a mapping $f : \mathbb{R} \mapsto \mathbb{R}$ a Riemann density if $f(x) \geq 0$ for all $x \in \mathbb{R}$, and

$$\int_{\mathbb{R}} f(x)\,dx = 1,$$

---

[1] Integration via Rieman densities refers to classical integration, whereas Lebesgue densities require integration via the Lebesgue measure and densities then have to be dealt with as Radon-Nikodym derivates with respect to the Lebesgue measure.

see [22]. Moreover, for a real-valued mapping f, the set of all $x \in \mathbb{R}$, such that $f(x) \neq 0$ is called the *support* of f, denoted by $\mathrm{Supp}(f)$.

Definition 8.6. For any $v(x)$ such that $v(x) \geq 1$ for all $x \in \mathbb{R}^n$ the *weighted supremum norm*, or *v-norm* for short, of a real-valued mapping h is defined by

$$\|h\|_v = \sup_{\mathbb{R}^n} \frac{|h(x)|}{v(x)}.$$

We call v the *test function*.

Note that $\|h\|_v = c$ is equivalent to $|h(x)| \leq c v(x)$ for all $x \in \mathbb{R}^n$: a function h with finite v-norm can be bounded, up to a constant, by v. From this observation we notice that if h has a bounded v-norm then this implies that $h(x)$ is of order $O(v(x))$. The notion of the "Big O" to describe the growth rate of functions is very common in computer science. Here we will use the (equivalent) notion of v-norms, which leads to a simpler development of our formulas. If $\int v(x) f(x) dx < \infty$ for some Riemann density $f(x)$, then $\int |h(x)| f(x) dx < \infty$ for any Riemann integrable function h such that $\|h\|_v < \infty$.

A typical choice for v in applications is the exponential growth model:

$$v_\alpha(x_1, \ldots, x_n) = \alpha^{|x_1| + \ldots + |x_n|}, \quad \alpha \geq 1. \tag{8.16}$$

It is straightforward to verify that for the test function v in (8.16) and $\alpha > 1$, such that

$$\|x_1^p + \ldots + x_n^p\|_{v_\alpha} < \infty \quad \text{and} \quad \|x_1^p \times x_2^p \times \cdots \times x_n^p\|_{v_\alpha} < \infty, \tag{8.17}$$

for $p \in \mathbb{N}$, we have that $\int v_\alpha(x) f(x) dx < \infty$ implies finiteness of the integral (expectation) for sums and products of powers.

Theorem 8.3. *Let $\theta_0 \in \Theta$, with Θ an open set, and consider a family of (Riemann) densities $\{f_\theta(\cdot) : \theta \in \Theta\}$ with common support set S. If*

(i) *$f_\theta(x)$ is differentiable with respect to θ for all $x \in S$,*
(ii) *there exits a $k(x)$ such that for small $|\Delta| > 0$, $\quad \forall x \in S \quad |f_{\theta+\Delta}(x) - f_\theta(x)| \leq |\Delta| k(x)$, and there is a function $v \geq 1$ such that $\int_S v(x) k(x) dx < \infty$,*

then for all h with $\|h\|_v < \infty$,

$$\left. \frac{d}{d\theta} \int_S h(x) f_\theta(dx) \right|_{\theta = \theta_0} = \left. \int_S h(x) \frac{\partial}{\partial \theta} f_\theta(dx) \right|_{\theta = \theta_0}. \tag{8.18}$$

Proof. Condition (ii) means that the Lipschitz modulus of the density is upper bounded by $k(x)$ (independent of θ) and the weighted function $v(x) k(x)$ is integrable. Furthermore, from condition (i) the derivative of the densities are well defined. Therefore if $\|h\|_v < \infty$ (which implies $|h(x)| \leq v(x)$), then

$$\frac{1}{|\Delta|} \int_S |h(x)| \, |f_{\theta+\Delta}(x) - f_\theta(x)| \, dx \leq \int_S v(x) \, k(x) \, dx < \infty$$

and we can apply the dominated convergence theorem to interchange derivative and integration, which proves the claim. □

Condition (i) is straightforward to show in most applications. For condition (ii), following the mean value theorem, the function $k(x)$ in Theorem 8.3 can be chosen as

$$k(x) = \sup_{\theta \in \Theta_0} \left| \frac{\partial}{\partial \theta} f_\theta(x) \right|, \tag{8.19}$$

where Θ_0 is some open neighborhood of θ_0, the location of differentiation, provided that this is well defined. Verification of (8.19) will be carried out though examples in the following sections.

8.3 THE SCORE FUNCTION METHOD

In the following theorem we provide the precise statement for the SF estimator applied to Riemann densities. The theory for discrete distributions follows from the same line of argument and it is omitted here. Section 8.3.2 presents the general results. To simplify the presentation *we assume throughout this sections that "cost-functions" $h(\cdot)$ are Riemann integrable on bounded intervals.*

Theorem 8.4. *Let $\theta_0 \in \Theta$, with Θ an open set. Let f_θ, $\theta \in \Theta$, be a collection of (Riemann) densities with common support set S. If*

(i) *$f_\theta(x)$ is differentiable with respect to θ for all $x \in S$,*
(ii) *there exits a $k(x)$ such that for all $x \in S$ $|f_{\theta+\Delta}(x) - f_\theta(x)| \leq |\Delta| k(x)$, for small $|\Delta| > 0$, and there is a function $v \geq 1$ such that*

$$\int_S v(x) k(x) dx < \infty,$$

then it holds for all h with $\|h\|_v < \infty$, that

$$\left. \frac{d}{d\theta} \int_S h(x) f_\theta(dx) \right|_{\theta=\theta_0} = \int_S h(x) \mathsf{SF}(\theta_0, x) f_{\theta_0}(dx),$$

where

$$\mathsf{SF}(\theta_0, x) = \left. \frac{\partial}{\partial \theta} \log(f_\theta(x)) \right|_{\theta=\theta_0},$$

for all $x \in S$ and zero otherwise.

Proof. Note that for $\|h\|_v < \infty$, condition (ii) implies that we can interchange integration and differentiation for h; see the proof of Theorem 8.3 for details. By computation,

$$\lim_{\Delta \to 0} \frac{1}{\Delta} \left(\int_\mathbb{R} h(x) f_{\theta_0+\Delta}(dx) - \int_\mathbb{R} h(x) f_{\theta_0}(dx) \right) = \int_S h(x) \lim_{\Delta \to 0} \frac{1}{\Delta} \left(f_{\theta_0+\Delta}(x) - f_{\theta_0}(x) \right) (dx)$$

$$= \int_S h(x) \left. \frac{\partial}{\partial \theta} f_\theta(x) \right|_{\theta=\theta_0} dx$$

$$= \int_S h(x) \frac{\left. \frac{\partial}{\partial \theta} f_\theta(x) \right|_{\theta=\theta_0}}{f_{\theta_0}(x)} f_{\theta_0}(x) \, dx$$

$$= \int_\mathbb{R} h(x) \left. \frac{\partial}{\partial \theta} \log(f_\theta(x)) \right|_{\theta=\theta_0} dx,$$

where the third equality follows from the assumption that the densities have common support. □

The statement in Theorem 8.4 reads in random variable language as follows. Let $X(\theta)$ have density f_θ such that the collection $\{f_\theta : \theta \in \Theta\}$ satisfies the conditions in Theorem 8.4. It then holds that

$$\frac{d}{d\theta}\mathbb{E}[h(X(\theta))] = \mathbb{E}[h(X(\theta))\mathrm{SF}(\theta, X(\theta))].$$

The following example illustrates the application of Theorem 8.4 together with (8.19).

Example 8.8. Consider $\mathbb{E}[h(X(\theta))]$ for $X(\theta)$ exponential with mean θ. In the following we establish the conditions in Theorem 8.4. Condition (i) is satisfied, in particular

$$\frac{\partial}{\partial \theta} f_\theta(x) = \left(\frac{x}{\theta} - 1\right)\frac{1}{\theta^2}\exp(-x/\theta),\; x \geq 0,$$

where f_θ denotes the exponential density with mean θ. Given $\theta \in [\theta_0, \theta_1]$, it holds for all θ that

$$\left|\frac{\partial}{\partial \theta} f_\theta(x)\right| \leq \left(\frac{x}{\theta} + 1\right)\frac{1}{\theta^2}\exp(-x/\theta) \leq \frac{\theta_1}{\theta_0^2}\left(\frac{x}{\theta_0} + 1\right)\frac{1}{\theta_1}\exp(-x/\theta_1) =: k(x). \quad (8.20)$$

Condition (ii) in Theorem 8.4 can be expressed via a random variable representation as follows:

$$\int v(x)\, k(x)dx = \frac{\theta_1}{\theta_0^2}\int v(x)\left(\frac{x}{\theta_0} + 1\right)\frac{1}{\theta_1}e^{-x/\theta_1}\, dx = \left(\frac{\theta_1}{\theta_0^2}\right)\mathbb{E}\left[v(X(\theta_1))\left(\frac{X(\theta_1)}{\theta_0} + 1\right)\right] < \infty,$$

where $X(\theta_1)$ is an exponential random variable with mean θ_1. Since the exponential distribution has finite moment generating function, the above inequality holds for test function $v(x) = \alpha^x$, for some $\alpha \geq 1$. For any polynomially bounded h, i.e., h such that $\|h\|_v < \infty$, Theorem 8.4 thus yields

$$\frac{d}{d\theta}\mathbb{E}[h(X(\theta))] = \mathbb{E}\left[h(X(\theta))\frac{1}{\theta}\left(\frac{X(\theta)}{\theta} - 1\right)\right],$$

see Example 7.7. ✱✱✱

It is worth noticing that the statement of Theorem 8.4 helps to determine the *class of functions* $h(\cdot)$ for which the score function is unbiased, given the family of distributions. For this reason, in practice when identifying the growth function v we will seek a "worst" case of maximal growth: then the class of functions for which the score function is unbiased is larger and we do not have to repeat calculations when dealing with other problems or models that involve the same family of distributions.

Example 8.9. We revisit the normal distribution from Example 7.4. The normal distribution with mean μ and standard deviation σ, for $\sigma > 0$, has density

$$f(x; \mu, \sigma^2) = \frac{1}{\sigma\sqrt{2\pi}}e^{-\frac{1}{2}\frac{(x-\mu)^2}{\sigma^2}},\; x \in \mathbb{R}.$$

Denote by $\theta = \mu_0$ the point at which we want to take a derivative with respect to μ. For $\epsilon > 0$, choose $[\mu_0 - \epsilon, \mu_0 + \epsilon]$. Since μ is a shift parameter of $f(x; \mu, \sigma^2)$, we have for

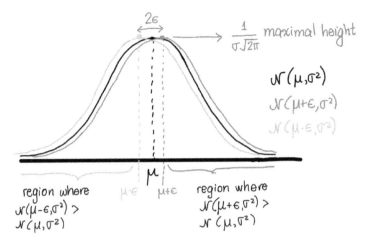

Figure 8.2. Visualization of inequalities.

$\mu \in [\mu_0 - \epsilon, \mu_0 + \epsilon]$

$$|f(x;\mu,\sigma^2) - f(x;\mu_0,\sigma^2)| \leq \begin{cases} f(x;\mu_0 - \epsilon, \sigma^2) & \text{if } x \leq \mu_0 - \epsilon, \\ \frac{1}{\sigma\sqrt{2\pi}} & \text{if } \mu_0 - \epsilon \leq x \leq \mu_0 + \epsilon, \\ f(x;\mu_0 + \epsilon, \sigma^2) & \text{if } \mu_0 + \epsilon \leq x, \end{cases}$$

which yields a bound $k(x)$ (see Figure 8.2).

Condition (ii) in Theorem 8.4 requires bounding the following integrals:

$$\int v(x) k(x) dx = \int_{-\infty}^{\mu_0 - \epsilon} v(x) f(x;\mu_0 - \epsilon, \sigma^2) dx + \frac{1}{\sigma\sqrt{2\pi}} \int_{\mu_0 - \epsilon}^{\mu_0 + \epsilon} v(x) dx$$
$$+ \int_{\mu_0 + \epsilon}^{\infty} v(x) f(x;\mu_0 + \epsilon, \sigma^2) dx.$$

Since the middle integral over a compact range is finite by definition, condition (ii) can be checked via

$$\int v(x) f(x;\mu_0 - \epsilon, \sigma^2) dx < \infty \quad \text{and} \quad \int v(x) f(x;\mu_0 + \epsilon, \sigma^2) dx < \infty.$$

As the normal distribution has a finite moment generating function, we see that condition (ii) holds for all polynomially bounded mappings h, i.e, for all h such that $\|h\|_v < \infty$, with $v(x) = \alpha^{|x|}$, for some $\alpha \geq 1$.

We now turn to the score function applied to the standard deviation. For μ fixed, note that the fact that σ is a scale parameter implies for $\sigma \leq \sigma_0 + \epsilon$, for $\epsilon > 0$, that there exist $M > 0$ such that

$$|f(x;\mu,\sigma^2) - f(x;\mu,(\sigma_0 + \epsilon)^2)| \leq \begin{cases} f(x;\mu,(\sigma_0 + \epsilon)^2)| & \text{if } x \leq -M + \mu \text{ and } M + \mu \leq x, \\ \frac{1}{(\sigma_0 + \epsilon)\sqrt{2\pi}} & \text{if } -M + \mu \leq x \leq M + \mu. \end{cases}$$

Following the line of argument for the mean μ, we see that condition (ii) for differentiation with respect to σ at σ_0 can be checked via

$$\int v(x) f(x; \mu, (\sigma_0 + \epsilon^2)) dx < \infty,$$

for all h such that $\|h\|_v < \infty$, where we let again $v(x) = \alpha^{|x|}$, for some $\alpha \geq 1$. ✻✻✻

Example 8.10. Let $f_\theta(x) = (1/\theta) \mathbf{1}_{\{[0,\theta]\}}$ denote the Riemann density of the uniform distribution. As the support of $f_\theta(x)$ depends on θ, and $f_\theta(x)$ fails to be differentiable at $\theta = x$, we cannot apply Theorem 8.4 to $f_\theta(x)$. ✻✻✻

Example 8.11. This is an example of what is known as the *optimal stopping* problem. Suppose you are selling a property. Offers come at random times. If you accept the offer then you sell the property. Otherwise you keep the property and wait for future offers. During the time that you keep the property, there are expenses incurred at a rate of c dollars per unit time. The amounts of the offers are iid positive random variables $\{Q_i\}$ with a bounded density $f(q)$, so that $\mathbb{E}[Q] < \infty$. We will assume that the intervals between offers are given by $\{T_i\}$, a sequence of iid random variables with unknown distribution.

You wish to determine a policy for stopping as follows. Let θ be your decision variable, and consider the rule:

$$X_\theta = \min\{n : Q_n \geq \theta\},$$

then you sell your property at this time, with a corresponding cost of

$$J(\theta) = \mathbb{E}\left[c \sum_{i=1}^{X_\theta} T_i - Q_{X_\theta}\right] \stackrel{\text{def}}{=} \mathbb{E}[c\, g(X_\theta)] - \mathbb{E}[Q_{X_\theta}],$$

where

$$g(X_\theta) = \sum_{i=1}^{X_\theta} T_i \quad \text{and} \quad \mathbb{E}[g(X_\theta)] = \mathbb{E}[X_\theta]\mathbb{E}[T_1].$$

We wish to find the optimal threshold policy that has best expected cost, that is,

$$\min_{\theta > 0} J(\theta). \tag{8.21}$$

This is a complex problem to solve, particularly when various distributions are not known analytically, and statistics or consecutive observations of the random variables must be used concurrently with the optimization. We are interested in estimating the derivative:

$$\frac{d}{d\theta} J(\theta) = c\, \mathbb{E}\left[\sum_{i=1}^{X_\theta} T_i\right] - \frac{d}{d\theta} \mathbb{E}[Q_{X_\theta}].$$

The problem can be greatly simplified by noticing that X_θ has a geometric distribution. To see this, let $\xi_n = \mathbf{1}_{\{Q_i \geq \theta\}}$, then $\{\xi_n\}$ are iid Bernoulli random variables with

$$p_\theta \stackrel{\text{def}}{=} \mathbb{P}(\xi = 1) = \int_\theta^\infty f(q)\, dq, \tag{8.22}$$

where $f(\cdot)$ is the density (assumed to exist) of the offer amount, and it holds that

$$p'_\theta = -f(\theta).$$

Then X_θ is the index of the first "success" in the sequence $\{\xi_n\}$, which has a geometric distribution. On the other hand, by construction, Q_{X_θ} has a density which is the conditional density $f(q)/p_\theta$, on $[\theta, \infty)$. That is, the distribution of Q_{X_θ} is independent of X_θ. This helps to solve the problem estimating separately the derivatives of $g(\theta)$ and of $\mathbb{E}[Q_{X_\theta}]$, which we now do.

The density function $f(q)$ is assumed to be known, so it may be possible to evaluate its expectation $\mathbb{E}[Q_{X_\theta}]$ as a function of θ and then take the derivative. However, for complicated distributions the integral may have to be calculated numerically, and derivatives may be difficult to evaluate. Alternatively, we can use

$$\frac{d}{d\theta}\mathbb{E}[Q_{X_\theta}] = \frac{d}{d\theta}\left(\frac{1}{p_\theta}\int_\theta^\infty q f(q)\, dq\right) = -\frac{p_\theta'}{p_\theta^2}\int_\theta^\infty q f(q)\, dq + \frac{1}{p_\theta}\theta f(\theta)$$

$$= \frac{1}{p_\theta}\left(\theta f(\theta) - p_\theta' \mathbb{E}[Q_{X_\theta}]\right)$$

$$= \frac{f(\theta)}{p_\theta}\left(\theta + \mathbb{E}[Q_{X_\theta}]\right),$$

so that $f(\theta)(\theta + Q_{X_\theta})/p_\theta$ is an unbiased estimator for the derivative. See Exercise 8.15 for an instance of this problem where Q_i has Weibull distribution. In that exercise, $p_\theta = 1 - F(\theta)$ is known analytically, but the moments of Q_{X_θ} are not available in closed form.

We now turn to the problem of estimating the derivative of $\mathbb{E}[g(X_\theta)]$. Conditioning on X_θ gives

$$\mathbb{E}[g(X_\theta)] = \sum_n \mathbb{E}[g(X_\theta)|X_\theta = n]\mathbb{P}(X_\theta = n).$$

Define

$$\hat{g}(n) \overset{\text{def}}{=} \mathbb{E}[g(X_\theta)|X_\theta = n] = n\mathbb{E}[T_1].$$

The geometric distribution is given by

$$\mathbb{P}(X_\theta = n) = p_\theta(n) \overset{\text{def}}{=} (1-p_\theta)^{n-1} p_\theta.$$

Taking the derivative with respect to θ gives

$$p_\theta'(n) = -f(\theta)\left((n-1)(1-p_\theta)^{n-2}p_\theta + (1-p_\theta)^{n-1}\right).$$

Choose neighborhood $(a,b) \subset [0,1]$ around the point of differentiation $\theta_0 \in (0,1)$. Note that by (8.22), p_θ is monotone decreasing in θ. Then

$$\sup_{\theta \in (a,b)} |p_\theta'(n)| \leq f^*\left((n-1)(1-p_b)^{n-2}p_a + (1-p_b)^{n-1}\right) =: k(n), \qquad (8.23)$$

where

$$f^* = \sup_{x \in (a,b)} |f(x)| < \infty,$$

and Lipschitz continuity of $p_\theta(n)$ follows from the mean value theorem, where $k(n)$ is the Lipschitz modulus. We now apply the discrete counterpart of Theorem 8.4 to the function $\hat{g}(n)$ and discrete density $p_\theta(n)$. For this, we let $v(n) = \alpha^n$, for $\alpha > 1$, which implies that

$$\|\hat{g}\|_v = \sup_n \frac{n\mathbb{E}[T_1]}{\alpha^n} < \infty.$$

It remains to show that $\sum_n v(n)k(n) < \infty$. To see this, use (8.23) to establish that

$$\sum_n v(n) f^* \left((n-1)(1-p_b)^{n-2} p_a + (1-p_b)^{n-1}\right)$$

$$= f^* \sum_n \left(\left(\frac{v(n) p_a}{(1-p_b) p_b}\right)(n-1)(1-p_b)^{n-1} p_b + \left(\frac{v(n)}{p_b}\right)(1-p_b)^{n-1} p_b\right)$$

$$= \frac{f^* p_a}{(1-p_b) p_b} \sum_n \left(v(n)(n-1)(1-p_b)^{n-1} p_b\right)$$

$$+ \frac{f^*}{p_b} \sum_n \left(v(n)(1-p_b)^{n-1} p_b\right).$$

As the geometric distribution has finite moments, we can conclude that the expressions on the above right-hand side are finite. It is straightforward to calculate the score function:

$$\mathrm{SF}(\theta, X_\theta) = \frac{\partial}{\partial \theta} \ln p_\theta(X_\theta) = \frac{\partial}{\partial \theta}\left((X_\theta - 1) \ln(1 - p_\theta) + \ln p_\theta\right) = -f'(\theta)\left(\frac{(X_\theta - 1)}{1 - p_\theta} + \frac{1}{p_\theta}\right).$$

To finalize, the unbiased derivative estimator using the score function is

$$J_{\mathrm{SF}}(\theta) = -f'(\theta)\left(\frac{(X_\theta - 1)}{1 - p_\theta} + \frac{1}{p_\theta}\right)\left(c \sum_{i=1}^{X_\theta} T_i\right) - \frac{f(\theta)}{p_\theta}(\theta + \mathbb{E}[Q_{X_\theta}]).$$

Hence, $Y_n = J_{\mathrm{SF}}(\theta_n)$ is an unbiased estimator for $J'(\theta)$ and we can use it to solve the problem in (8.21) via the stochastic approximation $\theta_{n+1} = \theta_n - \epsilon_n Y_n$, using the results from Chapter 4.

✲✲✲

8.3.1 SF for the Finite Horizon Problem

In the following we will discuss applying the SF method to vectors of independent random variables, or, equivalently, to products of measures. Let $f_{1,\theta}$ and $f_{2,\theta}$ be Riemann densities. Provided that $f_{1,\theta}(x)$ and $f_{2,\theta}(x)$ are differentiable with respect to θ at x it holds that

$$\frac{\partial}{\partial \theta}(f_{1,\theta}(x) f_{2,\theta}(x)) = \left(\frac{\partial}{\partial \theta} f_{1,\theta}(x)\right) f_{2,\theta}(x) + f_{1,\theta}(x) \frac{\partial}{\partial \theta} f_{2,\theta}(x)$$

$$= \left(\frac{\partial}{\partial \theta} \log(f_{1,\theta}(x)) + \frac{\partial}{\partial \theta} \log(f_{2,\theta}(x))\right) f_{1,\theta}(x) f_{2,\theta}(x).$$

Denoting the score function of $f_{1,\theta}(x)$ by $\mathrm{SF}_1(\theta, x)$ and the score function of $f_{2,\theta}(x)$ by $\mathrm{SF}_2(\theta, x)$, we arrive at following computational rule for the score function of the product of $f_{1,\theta}$ and $f_{2,\theta}$:

$$\frac{\partial}{\partial \theta} \log(f_{1,\theta}(x) f_{2,\theta}(x)) = \mathrm{SF}_1(\theta, x) + \mathrm{SF}_2(\theta, x). \tag{8.24}$$

In the light of (8.24), the extension of Theorem 8.4 to higher dimensional problems can be stated as follows, which provides the SF formula for the finite horizon gradient estimation problem.

Theorem 8.5. *Let Θ be an open set. Let $f_{i,\theta}$, $\theta \in \Theta$ and $1 \le i \le N$, be a collection of (Riemann) densities with common support set S, and consider test function $v : \mathbb{R}^N \mapsto \mathbb{R}$,* with $v \geq 1$. Assume that

(i) $f_{i,\theta}(x)$ are differentiable with respect to θ for all $x \in S$, for $1 \leq i \leq N$,
(ii) there exists a mapping k with

$$\left| \prod_{i=1}^{N} f_{i,\theta+\Delta}(x_i) - \prod_{i=1}^{N} f_{i,\theta}(x) \right| \leq |\Delta| k(x_1, \ldots, x_N), \quad (x_1, \ldots, x_N) \in S^N,$$

for all Δ such that $|\Delta| < \epsilon$, where ϵ is some small number independent of x, and

$$\int_{S^N} v(x_1, \ldots, x_N) k(x_1, \ldots, x_N) dx_1 \cdots dx_N < \infty.$$

Then it holds for all h with $\|h\|_v < \infty$ that

$$\frac{d}{d\theta} \int_{S^N} h(x_1, \ldots, x_N) \prod_{i=1}^{N} f_{i,\theta}(x_i) dx_1 \cdots dx_N$$

$$= \int_{S^N} h(x_1, \ldots, x_N) \sum_{j=1}^{N} \mathsf{SF}_j(\theta, x_j) \prod_{i=1}^{N} f_{i,\theta}(x_i) dx_1 \cdots dx_N,$$

where $\mathsf{SF}_i(\theta, x) = \frac{\partial}{\partial \theta} \log(f_{i,\theta}(x))$, $x \in S^N$, for $1 \leq i \leq n$.

Proof. Interchanging differentiation and integration follows from the same line of arguments as in the proof of Theorem 8.4. We now show how to rearrange the score function:

$$\frac{d}{d\theta} \int_{S^N} h(x_1, \ldots, x_N) \prod_{i=1}^{n} f_{i,\theta}(x_i) dx_1, \ldots, dx_N$$

$$= \int_{S^N} h(x_1, \ldots, x_N) \sum_{j=1}^{N} \left(\frac{\partial}{\partial \theta} f_{j,\theta}(x_j) \right) \prod_{i=1, i \neq j}^{N} f_{i,\theta}(x_i) dx_1, \ldots, dx_N$$

$$= \int_{S^N} h(x_1, \ldots, x_N) \left(\sum_{j=1}^{N} \mathsf{SF}_j(\theta, x_j) \right) \prod_{i=1}^{n} f_{i,\theta}(x_i) dx_1, \ldots, d_N,$$

which proves the claim. □

Example 8.12. We revisit Example 8.5 and use the notation defined therein. The $X_n(\theta)$'s have Riemann density $f_{1,\theta(x)} = f_{2,\theta(x)} = \exp(-x/\theta)/\theta$. Following Example 8.8, given $\theta \in [\theta_0, \theta_1]$, it holds for all θ that

$$\left| \frac{\partial}{\partial \theta} (f_{1,\theta}(x_1) f_{2,\theta}(x_2)) \right| \leq \frac{\theta_1^2}{\theta_0^2} \left(\frac{x_1 + x_2}{\theta_0} + 2 \right) \frac{1}{\theta_1} \exp(-x_1/\theta_1) \frac{1}{\theta_1} \exp(-x_2/\theta_1) =: k(x).$$

(8.25)

Condition (ii) in Theorem 8.5 can be expressed via a random variable representation as follows:

$$\int v(x) k(x) dx = \frac{\theta_1^2}{\theta_0^2} \mathbb{E}\left[v(X_{\theta_1}, Y_{\theta_1}) \left(\frac{X_{\theta_1} + Y_{\theta_1}}{\theta_0} + 2 \right) \right] < \infty,$$

where X_{θ_1} and Y_{θ_1} are independent exponential random variable with mean θ_1.

Since the exponential distribution has a finite moment generating function, the above inequality holds for

$$v(x_1, x_2) = \alpha^{x_1 + x_2},$$

for $\alpha \geq 1$. Moreover, for $h(x_1, x_2) = x_1 + x_2$, $x_1, x_2 \geq 0$, we have for $\alpha > 1$ that

$$||h||_v = \sup_{x_1, x_2 \geq 0} \frac{|h(x_1, x_2)|}{v(x_1, x_2)} < \infty$$

and we may apply the product rule of differentiation in Theorem 8.5 to h giving

$$\frac{d}{d\theta}\mathbb{E}[X_1(\theta) + X_2(\theta)] = \mathbb{E}\left[(X_1(\theta) + X_2(\theta))(\mathsf{SF}_1(\theta, X_1(\theta)) + \mathsf{SF}_2(\theta, X_2(\theta)))\right].$$

The extension to general n is straightforward and omitted here. To summarize, applying Theorem 8.5 to the finite sum of identically distributed exponentials gives

$$\frac{d}{d\theta}\mathbb{E}[T_\theta(n)] = \frac{d}{d\theta}\int \left(\sum_{i=1}^n x_i\right) \prod_{i=1}^n f_{i,\theta(x_i)}\, dx_1, \ldots, dx_n$$

$$= \int \left(\sum_{i=1}^n x_i\right) \sum_{j=1}^n \mathsf{SF}_j(\theta, x_j) \prod_{i=1}^n f_\theta(x_i)\, dx_1, \ldots, dx_n$$

$$= \mathbb{E}\left[T_\theta(n) \sum_{j=1}^n \mathsf{SF}_j(\theta, X_j(\theta))\right].$$

Like the IPA estimator, the SF estimator is thus unbiased. However, as we will show in Chapter 9, SF also yields an unbiased estimator for the Poisson counting process $N_\theta(T)$, for which IPA is biased (as shown in Example 8.5). ✲✲✲

The above example illustrates a proof technique for establishing condition (ii) in Theorem 8.5, which can be summarized in the following steps:

- The supremum over an interval $[\theta_0, \theta_1]$ of density f_θ is rewritten as a rescaled density η:

$$\sup_{\theta \in [\theta_0, \theta_1]} f_\theta(x) \leq c^f \eta_f(x).$$

In the exponential case we have for $\theta \in [\theta_0, \theta_1]$

$$\frac{1}{\theta}\exp(-x/\theta) \leq \frac{1}{\theta_0}\exp(-x/\theta_1) = \frac{\theta_1}{\theta_0}\frac{1}{\theta_1}\exp(-x/\theta_1), x \geq 0,$$

so that we may choose $c^f = \theta_1/\theta_0$, and $\eta_f(x) = f_{\theta_1}(x)$.
- The supremum over an interval $[\theta_0, \theta_1]$ of the derivative of density f_θ is rewritten as a rescaled density η_d times a cost function $z(x)$,

$$\sup_{\theta \in [\theta_0, \theta_1]} \left|\frac{d}{d\theta}f_\theta(x)\right| \leq z(x) c^d \eta_d(x).$$

In the exponential case we have, see (8.20),

$$\left|\frac{\partial}{\partial \theta}f_\theta(x)\right| \leq \frac{\theta_1}{\theta_0^2}\left(\frac{x}{\theta_0} + 1\right)\frac{1}{\theta_1}\exp(-x/\theta_1)$$

so that

$$c^d = \frac{\theta_1}{\theta_0^2}, z(x) = \frac{x}{\theta_0} + 1, \text{ and } \eta_d(x) = f_{\theta_1}(x).$$

- Using the mean value theorem, it then follows that $f_\theta(x)f_\theta(y)$ is Lipschitz with Lipschitz modulus

$$\sup_{\theta \in [\theta_0, \theta_1]} \left| \frac{\partial}{\partial \theta} \left(f_\theta(x) f_\theta(y) \right) \right| \leq z(x) c^d \eta_d(x) c^f \eta_f(y) + c^f \eta_f(x) z(y) c^d \eta_d(y)$$
$$= (z(x) c^d c^f + z(y) c^d c^f) \eta_d(x) \eta_f(y)$$
$$=: k(x,y).$$

- If the integrals $\int v(x) z(y) \eta(x) dx$ and $\int v(x) \eta(x) dx$ are finite, for $v(x) = \alpha^x, x \leq 0$ and $\alpha \geq 1$, and $\eta = \eta_d, \eta_f$. Then, by Fubini's theorem (where we omit the constants c^f and c^d for ease of notation)

$$\int \int \alpha^{x+y} (z(x) + z(y)) \eta_d(x) \eta_f(y) dx \, dy$$
$$= \int \int \alpha^{x+y} z(x) \eta_d(x) \eta_f(y) dx \, dy + \int \int \alpha^{x+y} z(y) \eta_d(x) \eta_f(y) dx \, dy$$
$$= \int \alpha^y \left(\int \alpha^x z(x) \eta_d(x) dx \right) \eta_f(y) \, dy + \int \alpha^x \left(\int \alpha^y z(y) \eta_f(y) dy \right) \eta_d(x) dx.$$

Hence, a sufficient conditions for (ii) in Theorem 8.5 is that for $\eta = \eta_f, \eta_d$

$$\int \alpha^x z(x) \eta(x) dx < \infty \text{ and } \int \alpha^x \eta(x) dx < \infty,$$

which holds in general if the cdf associated with η_f, η_d have a finite moment generating function.

The above scheme extends in a straightforward manner to n fold independent products. We call the above proof scheme the *product technique for the score function method*. The main technical property that allows us to conveniently use Fubini's theorem in the final step and to relate the analysis of the product to that of its elements, is *(i)* independence of the densities and the *(ii)* product form of v. The product technique for the score function method provides a blueprint for proofs. In concrete applications, however, adjustments may be required to fit the situation at hand. See, for example, the normal distribution, which, though it can be bounded in the same way as he exponential, leads to working with a more complex bound having three parts.

Example 8.13. Our final example considers again the sojourn time problem of Example 7.16 for a GI/G/1 queue. The function $L_N(X_1, \ldots, X_n)$ can be expressed as a function h depending only on the random variables $(A_1, \ldots, A_{N+1}; S_1(\theta), \ldots, S_N(\theta))$. It may be a complicated function containing recursive computations, but clearly it is a function of those variables only. Use the notation $g(\cdot)$ for the density of the inter arrival times A_i and $f_\theta(\cdot)$ for the density of the service times. Thus the expectation is of the form

$$\mathbb{E}[L_N(X_1, \ldots, X_N)] = \int h(a_1, \ldots, a_{N+1}; s_1, \ldots, s_N) \prod_{j=1}^{N+1} g(a_i)$$
$$\prod_{i=1}^{n} f_\theta(s_i) \, ds_1, \ldots, ds_n \, ; da_1, \ldots da_n.$$

In order not to use ambiguous notation, let $\text{SF}(\theta, x)$ be the score function for the service time distribution. As an illustration, in the particular case that the consecutive service times

$\{S_i(\theta)\}$ are iid exponential service times with mean θ, the SF estimator is given by

$$\hat{G}^{SF} = L_N(X_1, \ldots, X_n) \sum_{i=1}^{N} SF(\theta, S_i(\theta)) = \frac{1}{\theta} L_N(X_1, \ldots, X_n) \sum_{i=1}^{N} \left(\frac{X_i(\theta)}{\theta} - 1 \right). \quad (8.26)$$

Comparing with Example 8.6 it should be apparent that developing the stochastic derivative in IPA is highly dependent on the function L, but the same expression applies to any distribution for which θ is a scale parameter. In contrast, the formula for the SF gradient estimator is applicable for any bounded function L (it may even have discontinuities), but the score function depends on the service time distribution.

Unbiasedness of $\hat{G}^{SF}$ follows from the line of argument put forward in Example 8.12. Choose $[\theta_0, \theta_1] \subset (0, \infty)$. We now follow the product technique for the score function for establishing condition (ii) in Theorem 8.5.

In order to so, we note that the average sojourn time L_N can be bounded by the summing the interrival and service times

$$L_N = h(a_1, \ldots, a_{N+1}; s_1, \ldots, s_N) \le a_{N+1} + \sum_{i=1}^{N} (a_i + s_i)$$

and thus

$$\sup_{a_i, s_i} \frac{a_{N+1} + \sum_{i=1}^{N}(a_i + s_i)}{v(a_1, \ldots, a_{N+1}; s_1, \ldots, s_N)} < \infty,$$

for

$$v(a_1, \ldots, a_{N+1}; s_1, \ldots, s_N) = \alpha^{a_{N+1} + \sum_{i=1}^{N}(a_i + s_i)},$$

for $\alpha > 1$. Hence, $||h||_v < \infty$. Since the densities are independent, v is of product form, and the exponential has finite moment generating function, the product technique for the score function establishes condition (ii) in Theorem 8.5, and we obtain unbiasedness of the SF estimator.

Algorithm 8.2 Finite horizon SF for the average sojourn time using (8.26)

Initialize $X_1 = S_1 \sim f_\theta$, score $= 0$, $L = 0$
Define the Score Function $SF(\theta, x)$
for $n = 1, \ldots, N$ **do**
    $L \leftarrow L + X_n$
    score $\leftarrow$ score $+$ score(θ, S_n)
    Generate $A_{n+1} \sim g$, $S_{n+1} \sim f_\theta$
    $X_{n+1} = \max(0, X_n - A_{n+1}) + S_{n+1}$
return score $\times L/N$.

Algorithm 8.2 is straightforward to analyze, given that it only uses one **for** loop, so the running time is also linear in the number of customers, which in this case is both the expected and the worst case running time, of order $O(N)$. ✳✳✳

Historical Note. The score function in the context of simulation-based optimization method can be traced back to [255, 256]. Early references on the score function as discussed in this section are [266]. For further development of the SF methods see [116, 184, 270]. For a brief discussion of the use of the score function in statistics, we refer to Section 14.1.

*8.3.2 The General Setting: Absolute Continuity of Measures

In the development so far we used Riemann integrals, which restricted our analysis to pdfs $f_\theta(x)$ that are Riemann-integrable. Any bounded Riemann-integrable function is also Lebgesue-integrable, and it holds that

$$\int f(x)dx = \int f(x)\lambda(dx), \tag{8.27}$$

where $\lambda(\cdot)$ denotes the Lebesgue measure. The integral on the above right-hand side is called a *measure intergal*. The results obtained so far can thus be extended to measure integrals. In the following we will provide a more detailed disussion of this generalization.

Given a measure space $(\Omega, \mathfrak{F})$, such as, for example, $\mathbb{R}$ equipped with its Borel field, and a measure μ on $(\Omega, \mathfrak{F})$, one can define the (abstract) measure integral

$$\int \mathbf{1}_{\{A\}} \mu(dx) = \mu(A),$$

for any measurable set $A \in \mathfrak{F}$.

Example 8.14. In case that μ is constructed via a Riemann density like in (8.27), it holds for any measurable mapping h that

$$\int h(x)\mu(dx) = \int h(x)f(x)\lambda(dx),$$

provided the integrals exists, and f is called the Lebesgue density of μ. In particular, for $h = \mathbf{1}_{\{A\}}$ and A a measurable set, we have

$$\mu(A) = \int_A \mu(dx) = \int_A f(x)\lambda(dx).$$

※※※

We arrive at the following general definition of a density.

Definition 8.7. Let ν and μ be measures on a common measurable space $(\Omega, \mathfrak{F})$. A measurable mapping f is called a *μ-density of ν* if for all $A \in \mathfrak{F}$ it holds that

$$\nu(A) = \int_A f(x)\mu(dx).$$

The μ-density of ν is sometimes denoted by $\frac{d\nu}{d\mu}$ and it is also called the *Radon-Nikodym derivative*.

If the μ-density of a measre ν exists, then for any integrable function h and any $A \in \mathfrak{F}$

$$\int_A h(x)\,\nu(dx) = \int_A h(x)\frac{d\nu}{d\mu}(x)\,\mu(dx).$$

This transformation is called a "change of measure."

A sufficient condition for the Radon-Nikodym derivative to exist is that ν is absolutely continuous with respect to μ (written $\nu \ll \mu$), i.e., for any measurable set A, $\nu(A) = 0 \Longleftarrow \mu(A) = 0$. This is also expressed by saying that ν is *dominated* by μ. In the case that ν has Lebesgue density f_ν and μ has Lebesgue density f_μ, ν is absolutely continuous with respect to μ if $f_\mu(x) = 0$ implies $f_\nu(x) = 0$ for all x. If ν is *dominated* by μ and μ is also dominated by ν, then μ and ν are *equivalent*. If ν and μ are equivalent, then μ and ν a.s.

$$\frac{d\nu}{d\mu} = \left(\frac{d\mu}{d\nu}\right)^{-1}.$$

The advantage of the measure integral is that it allows to treat discrete measures and measures on a general state space within a unified framework as is illustrated in the following example.

Example 8.15. Let $X(\alpha)$ have Beta distribution with parameters $\alpha > 0, \beta > 0$, that is, the Lebesgue density of $X(\alpha)$ is

$$g_\alpha(x) = \frac{x^{\alpha-1}(1-x)^{\beta-1}}{B(\alpha, \beta)}, 0 \leq x \leq 1, \quad \text{where} \quad B(\alpha, \beta) = \frac{\Gamma(\alpha)\Gamma(\beta)}{\Gamma(\alpha+\beta)},$$

and $\Gamma(x) = \int_0^\infty y^{x-1} e^{-y} dy$ the Gamma function. Let $Y \sim g_1(x)$ be another Beta random variable. Then for any interval $A \subset [0, 1]$,

$$\mathbb{P}(X(\alpha) \in A) = \int_A g_\alpha(x)\, dx = \int_A \left(\frac{g_\alpha(x)}{g_1(x)}\right) g_1(x)\, dx.$$

Here we identify $\mu(dx)$ with the pdf $g_1(x)\, dx$, and $\nu_\alpha(dx)$ with $g_\alpha(x)\, dx$. The μ-density of ν_α can be obtained as the ratio of the densities

$$f_\alpha(x) = \frac{g_\alpha(x)}{g_1(x)} = \frac{x^{\alpha-1} B(1, \beta)}{B(\alpha, \beta)}.$$

Now suppose that $X(p)$ has a binomial distribution: $X \sim \text{Bin}(n, p)$ so

$$\mathbb{P}(X(p) = k) = \binom{n}{k} p^k (1-\theta)^{n-k},$$

and let $Y \sim \text{Bin}(n, 0.1)$. Then for any subset A of $\{1, \ldots, n\}$

$$\mathbb{P}(X(p) \in A) = \sum_{k \in A} \left(\frac{p^k(1-p)^{n-k}}{(0.1)^k(0.9)^{n-k}}\right) \binom{n}{k}(0.1)^k(0.9)^{n-k}.$$

Here, we identify ν_p as the $\text{Bin}(n, p)$ distribution and μ with the $\text{Bin}(n, 0.1)$ distribution, so the μ-density of ν_p can obtained from the μ-density of $\nu_{0.1}$ via ratio of the likelihoods

$$f_p(k) = \frac{p^k(1-p)^{n-k}}{(0.1)^k(0.9)^{n-k}}.$$

※※※

Absolute continuity of measures arises naturally in gradient estimation. To illustrate this, we give a proof of Theorem 8.4 for general measures. Assume that ν_θ is absolute continuous with respect to μ, for all $\theta \in \Theta$, where Θ is an open neighborhood of the point of differentiation θ_0, and denote the μ density of ν_θ by f_θ, i.e.,

$$f_\theta(x) = \frac{d\nu_\theta}{d\mu}(x); \quad \mu-a.s.$$

GRADIENT ESTIMATION, FINITE HORIZON

Assume that there is $k(x)$ such that $|f_{\theta+\Delta}(x) - f_\theta(x)| \leq k(x); \mu - a.s$, and, for a given, measurable cost function h is holds that

$$\int |h(x)| k(x) \mu(dx) < \infty,$$

then

$$\lim_{\Delta \to 0} \frac{1}{\Delta} \left(\int h(x) \nu_{\theta_0+\Delta}(dx) - \int h(x) \nu_{\theta_0}(dx) \right)$$

$$= \lim_{\Delta \to 0} \frac{1}{\Delta} \left(\int h(x) \frac{d\nu_{\theta_0+\Delta}}{d\mu}(x) \mu(dx) - \int h(x) \frac{d\nu_{\theta_0}}{d\mu}(x) \right) \mu(dx)$$

$$= \lim_{\Delta \to 0} \frac{1}{\Delta} \int h(x) \left(\frac{d\nu_{\theta_0+\Delta}}{d\mu}(x) - \frac{d\nu_{\theta_0}}{d\mu}(x) \right) \mu(dx)$$

$$= \int h(x) \lim_{\Delta \to 0} \frac{1}{\Delta} \left(f_{\theta_0+\Delta}(x) - f_{\theta_0}(x) \right) \mu(dx)$$

$$= \int h(x) \frac{\partial}{\partial \theta} f_{\theta_0}(x) \mu(dx)$$

$$= \int h(x) \frac{\frac{\partial}{\partial \theta} f_{\theta_0}(x)}{f_{\theta_0}(x)} \frac{d\nu_{\theta_0}}{d\mu}(x) \mu(dx)$$

$$= \int h(x) \text{SF}(\theta, x) \nu_{\theta_0}(dx).$$

In applications, the measures ν_θ, $\theta \in \Theta$, are typically equivalent measures and one chooses $\mu = \nu_{\theta_0}$.

8.4 MEASURE-VALUED DIFFERENTIATION

We provide in this section the technical foundations for measure-valued differentiation (MVD). Our starting point is the transfer of SF results to MVD expressions. In Section 8.4.1 we present basic technical results where we restrict ourself, like for SF, to Riemann densities. Then, in Section 8.4.2 we present the theory for the static gradient estimation problem. Finally, we discuss the extension to measure integrals in Section 8.4.3.

8.4.1 Basic Results and Techniques: The Score Function Revisited

We have already shown in Chapter 7 that in many cases the partial derivative of a pdf $f_\theta(x)$ with respect to θ can be written as a scaled difference between two pdfs. In light of Table 7.1, any theorem valid for the score function can be transferred to an MVD estimator. Indeed, if $f_\theta(x)$ is a pdf such that (7.21) holds for some cost function h, and if Theorem 8.4 applies to f_θ and h, then we have

$$\frac{d}{d\theta}\bigg|_{\theta=\theta_0} \int h(x) f_\theta(x) dx = \int h(x) \, \text{SF}(\theta_0, x) f_{\theta_0}(x) dx$$

$$= c_{\theta_0} \left(\int h(x) f_{\theta_0}^+(x) dx - \int h(x) f_{\theta_0}^-(x) dx \right),$$

where we make use of the fact that

$$\text{SF}(\theta_0, x) f_{\theta_0}(x) = c_{\theta_0} \left(f_{\theta_0}^+(x) - f_{\theta_0}^-(x) \right),$$

for all x. This observation allows us to freely switch from the SF representation of a gradient to a MVD representation. The choice of the representation depends on aspects such as practically of implementation and variance. Apart from these practical aspects, MVD provides a different view on gradient estimation in viewing the parameterized integral $\int h(x) f_\theta(x) dx$ in a slightly more abstract way, as detailed in Section 8.4.3 below.

We illustrate the MVD technique with the reliability problem out forward in Example 7.15.

Example 8.16. Revisit Example 7.15. In this example, we have five independent random variables $T_i \sim G_i$, which are independent of θ for $i \neq 3$, and $T_3(\theta) \sim \mathrm{Exp}(1/\theta)$. Let $T = (T_1, \ldots, T_5)$ and $h(t) = \mathbb{E}[L(T)|T_3 = t]$. Note that $h(t) \leq c + t$, where c is, for example, the sum of the expected values of the lifetime of components 1, 2, 4, and 5. Arguing like in Example 8.8, it follows that

$$\frac{d}{d\theta} \mathbb{E}[L(T)] = \int h(t) \frac{\partial}{\partial \theta} \left(\frac{1}{\theta} e^{-t/\theta} \right) dt, \tag{8.28}$$

where we use Theorem 8.4 for justifying the interchange of integration and differentiation. By inspection, we have

$$\frac{\partial}{\partial \theta} \left(\frac{1}{\theta} e^{-t/\theta} \right) = \frac{1}{\theta} \left(\frac{t}{\theta^2} e^{-t/\theta} - \frac{1}{\theta} e^{-t/\theta} \right),$$

which corresponds to a scaled difference between the Gamma$(2, 1/\theta)$ and the $\mathrm{Exp}(1/\theta)$ distribution. The corresponding MVD estimator is

$$\widehat{L}^{MVD} = \frac{1}{\theta} \Big(L(T_1, T_2, T_3(\theta) + X(\theta), T_4, T_5) - L(T_1, T_2, T_3(\theta), T_4, T_5) \Big), \tag{8.29}$$

with $X(\theta) \sim \mathrm{Exp}(1/\theta)$.

This is not the only way to represent this derivative. For example, instead of using $T_3 + X$, any other random variable with Gamma distribution will yield the same expectation for L. As well, any other decomposition for the derivative of the density as a scaled difference of densities will work as well.

Finally, comparing with expression (8.4), we note the following. The IPA formula established in (8.4) is unbiased when θ is a scale parameter of the distribution and does not require explicit knowledge of the distribution of T_3. In contrast, (8.29) is valid for $T_3 \sim \exp(1/\theta)$, and it requires modifying when $T_3(\theta)$ has a different distribution. However, the method also applies to other performance criteria. For example, if instead of the life of the system L we were interested in the risk measure $\mathbb{P}(L \leq \ell)$, for some given time ℓ, then for applying (8.29) we simply replace L by $\mathbf{1}_{\{L \leq \ell\}}$. In contrast, the corresponding IPA formula needs to be re-evaluated for different performance functions (actually, IPA is biased for the risk measure, as indicated in Exercise 8.12). ✼✼✼

Remark 8.1. In this monograph we distinguish between the *sample-path approach* via IPA and the *distributional approach* via SF and MVD. This distinction is less strict than it appears on the surface. Indeed, we show that distributional derivatives conditioned on a observation of the random variable itself can be expressed as IPA-like sample path derivatives. Recall that by Lemma 8.1 under appropriate smoothness, the IPA derivative of $X(\theta)$, denoted by

$X'(\theta)$, can be obtained from
$$-\frac{\frac{\partial}{\partial\theta}F_\theta(X(\theta))}{f_\theta(X(\theta))},$$
which can be written as
$$\mathbb{E}[X'(\theta)|X(\theta)] = -\frac{\frac{\partial}{\partial\theta}F_\theta(X(\theta))}{f_\theta(X(\theta))}.$$

Using the MVD expression
$$\frac{\partial}{\partial\theta}F_\theta(x) = c_\theta(F_\theta^+(x) - F_\theta^-(x)),$$
where $dF_\theta^\pm(x)\,dx = f_\theta^\pm(x)$, we obtain
$$\mathbb{E}[X'(\theta)|X(\theta)] = -\frac{c_\theta}{f_\theta(X(\theta))}(F_\theta^+(X(\theta)) - F_\theta^-(X(\theta))).$$

Alternatively, from $F_\theta(x) = \mathbb{E}[\mathbf{1}_{\{X(\theta)\leq x\}}]$ we obtain the SF representation
$$\mathbb{E}[X'(\theta)|X(\theta)] = -\frac{c_\theta}{f_\theta(X(\theta))}\mathbb{E}\left[\mathbf{1}_{\{\tilde{X}(\theta)\leq X(\theta)\}}\mathrm{SF}_\theta(\tilde{X}(\theta))\right],$$
where $\tilde{X}(\theta)$ is an iid copy of $X(\theta)$.

Example 8.17. This example illustrates the implementation of MVD for higher order derivatives for the special case of an exponential distribution. For $\alpha \in \{1, 2, \ldots\}$, let
$$f^e_{\alpha,\theta}(x) = \frac{x^{\alpha-1}}{(\alpha-1)!\,\theta^\alpha}e^{-\frac{x}{\theta}}$$
be the density of the distribution of the sum of α independent exponential random variables with mean θ, known as Gamma-$(\alpha,1/\theta)$ distribution and as α is intergervalued also as Erlang-$(\alpha,1/\theta)$ distribution.

The general formula for the n-th order derivative of the exponential distribution is as follows. We set
$$c_\theta^{(n)} = \frac{n!}{\theta^n},$$
for n even
$$f_\theta^{(n,+)}(x) = f^e_{\alpha+1,\theta}(x), \quad f_\theta^{(n,-)}(x) = f^e_{\alpha,\theta}(x)$$
and for n odd
$$f_\theta^{(n,+)}(x) = f^e_{\alpha,\theta}(x), \quad f_\theta^{(n,-)}(x) = f^e_{\alpha+1,\theta}(x),$$
then
$$\frac{d^n}{d\theta^n}\int_0^\infty h(x)\,f_\theta(x)\,dx = \frac{n!}{\theta^n}\left(\int_0^\infty h(x)\,f_\theta^{(n,+)}(x)\,dx - \int_0^\infty h(x)\,f_\theta^{(n,-)}(x)\,dx\right),$$
where interchanging integration and differentiation is justified for any polynomially bounded mapping h. Samples from the Gamma $(\alpha, 1/\theta)$ distribution can be obtained by summing α iid copies of exponentially distributed random variables with mean $1/\theta$. This leads to the following scheme for sampling an n-th order derivative of X_θ: let $\{X_\theta(k)\}$ be an iid sequence of exponentially distributed random variables with mean $1/\theta$, then, for any

polynomially bounded g, it holds that

$$\frac{d^n}{d\theta^n}\mathbb{E}[h(X_\theta(1))]$$
$$= (-1)^n \frac{n!}{\theta^n}\left(\mathbb{E}\left[h\left(\sum_{k=1}^{n+1} X_\theta(k)\right)\right] - \mathbb{E}\left[h\left(\sum_{k=1}^{n} X_\theta(k)\right)\right]\right)$$

and using common random numbers

$$\frac{d^n}{d\theta^n}\mathbb{E}[g(X_\theta(1))]$$
$$= (-1)^n \frac{n!}{\theta^n}\left(\mathbb{E}\left[h\left(\sum_{k=1}^{n+1} X_\theta(k)\right) - h\left(\sum_{k=1}^{n} X_\theta(k)\right)\right]\right).$$

Note that the above representation allows for a recursive estimation of higher-order derivatives: the $(n+1)$st derivative of $\mathbb{E}[h(X_\theta)]$ can be estimated from the same data as the n-th derivative and the additional drawing of one sample from an exponential distribution. In particular, taking h as the identity, one recovers

$$\frac{d^n}{d\theta^n}\mathbb{E}[X_\theta(1)] = (-1)^n \frac{n!}{\theta^n}\mathbb{E}[X_\theta(n+1)] = (-1)^n \frac{n!}{\theta^{n+1}} = \frac{d^n}{d\theta^n}\left(\frac{1}{\theta}\right).$$

✸✸✸

8.4.2 MVD for the Finite Horizon Problem

Before we present the main theorem establishing the conditions under which the MVD is unbiased, we recall the randomization of a summation, which will be used in one of statements of the theorem.

Example 8.18. Suppose that we are coding an algorithm that requires calculation of $\sum_{i}^{n}(a_i - b_i)$, where n may be very large and a_i, b_i require long running times to compute. Then the running time of the algorithm is $O(n)$. A common alternative in computer science is to *randomize* to achieve dimension reduction and only evaluate a fraction of the computations. In particular, certain randomizations turn the running time $O(1)$, albeit introducing errors that depend on the variance of the (random) computation.

Let a_i, b_i, $1 \le i \le n$, be some finite numbers. Then

$$\sum_{i=1}^{n}(a_i - b_i) = 2n\,\mathbb{E}[r_\sigma],$$

where σ is uniformly distributed on $\{1, \ldots, 2n\}$ and

$$r_i = \begin{cases} a_i & \text{if } i \le n, \\ -b_{i-n} & \text{if } i > n. \end{cases}$$

Indeed, it holds that

$$\sum_{i=1}^{n}(a_i - b_i) = 2n\left(\sum_{i=1}^{n} a_i \frac{1}{2n} - \sum_{i=n+1}^{2n} b_{i-n} \frac{1}{2n}\right)$$
$$= 2n\left(\sum_{i=1}^{n} a_i \mathbb{P}(\sigma = i) - \sum_{i=n+1}^{2n} b_{i-n}\mathbb{P}(\sigma = i)\right) = 2n\mathbb{E}[r_\sigma].$$

✸✸✸

GRADIENT ESTIMATION, FINITE HORIZON

Theorem 8.3, together with randomization introduced in Example 8.18, now gives the following MVD estimator for the finite horizon problem.

Theorem 8.6. *Let $\theta_0 \in \Theta$, with Θ an open set. Let $f_{i,\theta}$, $\theta \in \Theta$ and $1 \leq i \leq n$, be a collection of (Riemann) densities with common support set S. Assume that*

(i) $f_{i,\theta}(x)$ are differentiable with respect to θ for all $x \in S$, such that

$$\frac{\partial}{\partial \theta} f_{i,\theta}(x) = c_{i,\theta} \left(f_{i,\theta}^+(x) - f_{i,\theta}^-(x) \right),$$

with $c_{i,\theta}$ a finite constant and $f_{i,\theta}^\pm(x)$ densities with support S, for $1 \leq i \leq N$,
(ii) there exists a mapping k with

$$\left| \prod_{i=1}^N f_{i,\theta+\Delta}(x_i) - \prod_{i=1}^N f_{i,\theta}(x_i) \right| \leq |\Delta| k(x_1, \ldots, x_N), \quad (x_1, \ldots, x_N) \in S^N,$$

for all Δ such that $|\Delta| < \epsilon$, where ϵ is some small number independent of x, and there is a test function $v : \mathbb{R}^N \mapsto \mathbb{R}$, with $v \geq 1$ such that

$$\int_{S^N} v(x_1, \ldots, x_N) k(x_1, \ldots, x_N) dx_1 \cdots dx_N < \infty.$$

Then for all h with $\|h\|_v < \infty$

$$\frac{d}{d\theta} \int_{S^N} h(x_1, \ldots, x_N) \prod_{i=1}^N f_{i,\theta}(x_i) dx_1 \cdots dx_N$$

$$= \sum_{i=1}^N c_{i,\theta} \left(\int_{S^N} h(x_1, \ldots, x_N) \prod_{j=i+1}^N f_{j,\theta}(x_j) f_{i,\theta}^+(x_i) \prod_{j=1}^{i-1} f_{j,\theta}(x_j) dx_1 \cdots dx_N \right.$$

$$\left. - \int_{S^N} h(x_1, \ldots, x_N) \prod_{j=i+1}^N f_{j,\theta}(x_j) f_{i,\theta}^-(x_i) \prod_{j=1}^{i-1} f_{j,\theta}(x_j) dx_1 \cdots dx_N \right).$$

In addition, let σ be uniformly distributed on $\{1, \ldots, 2N\}$ independent of everything else, let $r(\sigma) = \mathbf{1}_{\{\sigma \leq N\}} - \mathbf{1}_{\{\sigma > N\}}$, and $\eta = \sigma(\mod N)$. Then the following expression holds:

$$\frac{d}{d\theta} \int_{S^N} h(x_1, \ldots, x_N) \prod_{i=1}^N f_{i,\theta}(x_i) dx_1 \cdots dx_N$$

$$= 2N \, \mathbb{E} \left[r(\sigma) \, c_{\eta,\theta} \int_{S^N} h(x_1, \ldots, x_N) \prod_{i=\eta+1}^N f_{i,\theta}(x_i) f_{\eta,\theta}^{[\sigma]}(x_\eta) \right.$$

$$\left. \prod_{i=1}^{\eta-1} f_{i,\theta}(x_i) \, dx_1 \cdots dx_N \right],$$

where the expected value is taken with respect to σ, and

$$f_{\eta,\theta}^{[\sigma]}(x) = \begin{cases} f_{\eta,\theta}^+(x) & \text{if } \sigma \leq N, \\ f_{\eta,\theta}^-(x) & \text{if } \sigma > N. \end{cases}$$

It is worth noting that the product technique for the score function can be applied for establishing condition (ii) in Theorem 8.6. The second representation in Theorem 8.6 is called the *randomized MVD* and was first introduced in [47]. It allows the code to run with one observation of either the "plus" or the "minus" random variable replacing only one of the original random variables in the cost-function $h(x_1, \ldots, x_n)$. Although simple to implement, this version may introduce large variance.

In terms of random variables, the full version estimator can be expressed as follows: we define the i-th *phantom* plus (minus) process by

$$\xi_i(\theta)^\pm = (X_1(\theta), \ldots, X_{i-1}(\theta), X_i^\pm(\theta), \tilde{X}_{i+1}^\pm(\theta, i), \ldots, \tilde{X}_n^\pm(\theta, i)),$$

where the distribution of $\tilde{X}_k^\pm(\theta, i)$, $k > i$ is either the same as $X_k(\theta)$ (if the original variables are independent), or it satisfies the original Markovian dynamics starting at $k = i$ from the state $X_i^\pm(\theta)$.

Thus the estimator can be written as

$$\frac{d}{d\theta}\mathbb{E}[h(X_1(\theta), \ldots, X_n(\theta)))] = \mathbb{E}\left[\sum_{i=1}^n c_{i,\theta}\left(h(\xi_i^+(\theta)) - h(\xi_i^-(\theta))\right)\right].$$

Randomization can be applied to decrease the number of parallel phantom processes. However, it comes at the price of a higher variance of the estimator. Randomization is readily extended to the case of randomly picking k, with $k \leq N$, points of differentiation. Balancing memory requirement and computational burden with variance of the estimator, one can estimate the optimal value of k.

We note that we may alternatively only randomize over the position of the perturbation while always keeping a positive and negative part. To see this, let the conditions of Theorem 8.6 be satisfied and let $X^\pm(\theta)$ have pdf $f_{1,\theta}^\pm$ and let $Y^\pm(\theta)$ have pdf $f_{2,\theta}^\pm$. Then the full representation in Theorem 8.6 can be expressed as

$$\frac{d}{d\theta}\mathbb{E}[h(X(\theta), Y(\theta))] = \mathbb{E}\left[c_{1,\theta}\, h(X(\theta)^+, Y(\theta)) + c_{2,\theta}\, h(X(\theta), Y(\theta)^+)\right]$$
$$- \mathbb{E}\left[c_{1,\theta} h(X(\theta)^-, Y(\theta)) + c_{2,\theta} h(X(\theta), Y(\theta)^-)\right].$$

We may now introduce the random vectors $\xi^\pm(\theta)$ such that

$$\xi^+(\theta) = \begin{cases} (X(\theta)^+, Y(\theta)) & \text{with probability } \frac{c_{1,\theta}}{c_{1,\theta}+c_{2,\theta}}, \\ (X(\theta), Y(\theta)^+) & \text{with probability } \frac{c_{2,\theta}}{c_{1,\theta}+c_{2,\theta}}, \end{cases}$$

and

$$\xi^-(\theta) = \begin{cases} (X(\theta)^-, Y(\theta)) & \text{with probability } \frac{c_{1,\theta}}{c_{1,\theta}+c_{2,\theta}}, \\ (X(\theta), Y(\theta)^-) & \text{with probability } \frac{c_{2,\theta}}{c_{1,\theta}+c_{2,\theta}}. \end{cases}$$

With this notation we arrive at

$$\frac{d}{d\theta}\mathbb{E}[h(X(\theta), Y(\theta))] = (c_{1,\theta} + c_{2,\theta})\, \mathbb{E}\left[h(\xi^+(\theta)) - h(\xi^-(\theta))\right].$$

The full expression of the MVD in Theorem 8.6 involves a sum, where each term i has the i-th variable replaced by the "plus" and "minus" random variables to calculate the difference. At first glance it would seem that MVD requires n parallel simulations, each one with two separate calculations. However, this is not always necessary: much effort has been dedicated to express the full sum (including all terms) as a "single-run" estimator without elaborating on randomization. Such algorithms typically result in much less memory and running time requirements and typically present variance reduced by $O(1/n)$ compared to the randomized version (which might be much faster to implement in preliminary experiments). This is explained in the following example.

Example 8.19. Refer to Example 8.13 of a $GI/G/1$ queue. Customers arrive at a service station according to a renewal point process. The inter arrival times $\{A_n : n \in \mathbb{N}\}$ are iid, with $\mathbb{E}[A_n] < \infty$ and $\mathbb{P}(A_n = 0) = 0$. Customers are served in order of arrival, and consecutive service times are iid random variables $\{S_n(\theta) : n \in \mathbb{N}\}$. Interarrival times and service times are assumed to be mutually independent. The common distribution of the service times G_θ depends on parameter $\theta = \mathbb{E}[S_n(\theta)]$.

We now apply Theorem 8.6 to our sojourn time example; for details on establishing the conditions of the theorem we refer to Example 8.13, where we have detailed the SF version of the estimator.

Applying the first statement of Theorem 8.6, leads to the following estimator for the N-th sojourn time. For $i \leq N$, let $\{X_n^\pm(\theta; i), n = 1, \ldots, N\}$ be defined as follows. Up to time i, $X_n^\pm(\theta; i)$ behaves just like $X_n(\theta)$, i.e., following (7.30),

$$X_{n+1}^\pm(\theta; i) = \max(0, X_n^\pm(\theta; i) - A_{n+1}) + S_{n+1}(\theta),$$

for $n \leq i$. At the transition from $X_{i-1}^\pm(\theta; i)$ to $X_i^\pm(\theta; i)$ service time is $S_i^\pm(\theta)$, defined from the MVD derivative of this distribution:

$$X_i^\pm(\theta; i) = \max(0, X_{i-1}^\pm(\theta; i) - A_i) + (S_i^\pm(\theta)),$$

For example, in the case that the service times are exponential, then $S_i^+(\theta) = S_i(\theta) + Y_i(\theta)$, where $\{Y_i(\theta)\}$ are iid exponential random variables with mean θ and independent of everything else. The "minus" process will be the same as the nominal, that is, $S_i^-(\theta) = S_i(\theta)$. For $n > i$, the transitions of $X_n^\pm(\theta; i)$ follow the standard update formula

$$X_{n+1}^\pm(\theta; i) = \max(0, X_n^\pm(\theta; i) - A_{n+1}) + S_{n+1}(\theta).$$

Consider a cost function that depends only on the N-th sojourn time $h(X_N)$. We wish to estimate the derivative w.r.t. θ of $\mathbb{E}[h(X_N)]$. Then, by Theorem 8.6,

$$\frac{d}{d\theta}\mathbb{E}[h(X_N(\theta))] = \sum_{i=1}^{N} c_{i,\theta} \mathbb{E}\left[h(X_N^+(\theta; i)) - h(X_N^-(\theta; i))\right]. \quad (8.30)$$

When service times are iid, the normalizing factor $c_{i,\theta}$ is independent of i and we write c_θ. As illustration, in the particular case of exponential service times $c_{i,\theta} = 1/\theta$ the above expression simplifies to

$$\mathbb{E}\left[\sum_{i=1}^{N} \frac{1}{\theta} h(X_N^+(\theta; i))\right] - \frac{N}{\theta}\mathbb{E}[h(X_N(\theta))].$$

Using randomization, we select σ out of $\{1, \ldots 2N\}$ uniformly and for $\sigma \leq N$, we generate the positive version of $\{X_i^+(\theta; i)\}$ perturbed at transition $i = \sigma$, and for $\sigma > N$ we generate the negative version of $\{X_n^-(\theta; i)\}$ perturbed at transition $i = \sigma$, which leads to the following single run estimator,

$$\frac{d}{d\theta}\mathbb{E}[h(X_N(\theta))] = \frac{2N}{\theta}\mathbb{E}\left[r(\sigma)h(X_N^s(\theta;\sigma \bmod n)\right], \quad (8.31)$$

where σ is uniformly distributed on $\{1,\ldots,2N\}$, $s = $ "+", for $\sigma \leq N$, and $s = $ "-", for $\sigma > N$, and $r(\sigma) = \mathbf{1}_{\{\sigma \leq N\}} - \mathbf{1}_{\{\sigma > N\}}$. It is worth noting that the above single process version has larger variance than the estimator in (8.30) but comes at the ease of having to only perturb once for a single sample path. Algorithm 8.3 provides the pseudo-code for simulating this queue model and generating a sampled process $\{\xi_n(\theta); 1 \leq n \leq N\}$, which satisfies

$$\mathbb{E}\left[\frac{2N}{\theta}h(\xi_N(\theta))\right] = \frac{d}{d\theta}\mathbb{E}[h(X_N(\theta))].$$

Algorithm 8.3 Randomized MVD for the N-th sojourn time using (8.31)

Read cost function h, θ, and density of the service time f_θ with MVD $(c_\theta, f_\theta^+, f_\theta^-)$.
Initialize: $X_0 = 0$.
for $n = 1,\ldots,N$ **do**
  Generate A_n and $S_n \sim f_\theta$
Generate a uniform random number $\sigma \in \{1,\ldots,N\}$
Generate $U \sim U(0,1)$,
if $(U < 0.5)$ **then**
  $r = +1$
  Generate $Y \sim f_\theta^+$
else
  $r = -1$
  Generate $Y \sim f_\theta^-$
for $n = 1,\ldots,\sigma-1$ **do**
  $X_n = \max(0, X_{n-1} - A_n) + S_n$
$X_\sigma = \max(0, X_{\sigma-1} - A_\sigma) + S_\sigma + Y$
for $n = \sigma+1,\ldots,N$ **do**
  $X_n = \max(0, X_{n-1} - A_n) + S_n$
return $r(2N/\theta)h_N(X)$

We finish the example considering the problem formulation in Example 7.16, where the cost function h is the average sojourn time of the first N customers, namely,

$$h_N(X(\theta)) = L_N(X_1(\theta),\ldots,X_N(\theta)) = \frac{1}{N}\sum_{n=1}^{N} X_n(\theta),$$

so that this function depends on all the sojourn times and not just on the last one. Using the definitions of the phantom processes $\{X_n^\pm(\theta;i)\}$, the MVD estimation satisfies

$$\frac{1}{\theta}\mathbb{E}\left[\sum_{i=1}^{N}\left(h_N(X^+(\theta;i)) - h_N(X^-(\theta;i))\right)\right] = \frac{d}{d\theta}\mathbb{E}[h_N(X(\theta))].$$

The full MVD estimator is given by

$$\hat{G}^{(\text{MVD})} = \frac{c_\theta}{N}\sum_{i=1}^{N}\sum_{n=1}^{N}\left(X_n^+(\theta;i) - X_n^-(\theta;i)\right).$$

At first glance, there are two sums in the above expression, one over the index of the "phantom" processes (i) and another one over the customer numbers (n). Ordinarily this would entail $O(N^2)$ calculations. Memory requirements are just as demanding, because each phantom process must keep N values. However, this is not the way we should program the code. Instead we use the *difference process*; see [155, 316, 310, 306]. For this we define the i-th *phantom difference process* by

$$d_n(\theta; i) = X_n^+(\theta; i) - X_n^-(\theta; i). \tag{8.32}$$

In the particular case of the exponential service times the minus processes are the same as the nominal process, but our general formulas and code hold for other distributions, using the appropriate "plus" and "minus" distributions from Table 7.1.

Given the sequence $\xi = (A_k; S_{k+1}; k = 0, \ldots, N)$, let $(Y_i; i = 1, \ldots, N)$ be iid random variables defined via the initial differences:

$$Y_i \stackrel{\text{def}}{=} X_i^+(i) - X_i^-(i),$$

which in the example of exponential service distribution correspond to iid exponential random variables with mean θ, independent of $\{X_n\}$.

We claim that the i-th phantom difference process satisfies the recurrence:

$$d_n(\theta; i) = \begin{cases} 0 & \text{if } n < i, \\ Y_i & \text{if } n = i, \\ \max(0, d_{n-1}(\theta; i) - (A_n - X_{n-1}^-(\theta))_+) & \text{if } n > i. \end{cases} \tag{8.33}$$

To see this, we assume without loss of generality that $Y_i > 0$ for the argument that follows (otherwise the argument can be easily adjusted reversing the sign of the difference process).

To prove the claim, it is easier to refer to Figure 8.3. Fix an arbitrary index i (in the figure, $i = 2$). From the definition of the i-th phantom process $X_n(\theta; i)$ the first and second case follow: $d_i(\theta; i) = Y_i$, and we say that the i-th phantom "is born at step i." The figure illustrates the case $i = 2$. Because the i-th customer has a longer service time (by Y_i) this delays the other customers coming after her/him/it (shown in gray in the figure). Let $\Delta_n(i) = A_n - X_{n-1}^-(\theta)$. This quantity determines if the arrival of the n-th customer in the baseline "minus" process is after the departure time of customer $n - 1$, or not, and it is at the basis of the logical reasoning. If $\Delta_n(i) < 0$ then the new arrival finds customer $n - 1$ in the queue and it must wait. The delay in service at the phantom queue of the "plus" process "pushes" the new customer to wait the same extra amount of time, and in this case $d_n(\theta; i) = d_{n-1}(\theta; i)$: this is illustrated in the figure when $n = 3$: $d_3(\theta; 2) = d_2(\theta; 2)$. Now suppose that $\Delta_n(i) > 0$, so customer n arrives after $n - 1$ has departed the queue. There are two situations here: one when the arrival A_n happens before customer $n - 1$ has departed in the phantom queue (because it was delayed) or after. In the first case the new customer will not be delayed as much as previous ones, and the reduction in the delay is exactly the amount $\Delta_n(i)$. That is, when $\Delta_n(i) > 0$,

$$d_n(\theta; i) = d_{n-1}(\theta; i) - \Delta_n(i),$$

provided that this quantity is positive. In Figure 8.3 this happens for $n = 5$ and the (reduced) delay is shown in hashed lines. Specifically, the illustration shows that $d_5(\theta; i) = d_4(\theta; i) - (A_5 - X_4(\theta))$. Finally, if this quantity is negative it means that customer n finds an empty queue in both the nominal and the phantom queue, thus we say that the i-th phantom "dies"

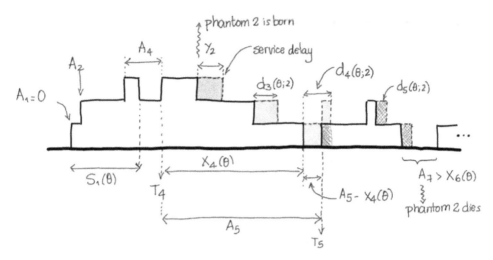

Figure 8.3. Process $\{X_n^-(\theta)\}$ (shown in solid black line) and corresponding phantom "plus" process $\{X_n^+(\theta; i)\}$ for $i = 2$ (dashed line). The inter arrival times are $\{A_n\}$. The arrival epoch of customer n is T_n. Service times are $\{S_n(\theta)\}$. The initial delay on the second phantom process ($i = 2$) is initialized at $d_2(\theta; 2) = Y_2$. This means that the service completion of this customer in the phantom queue is delayed by this amount. Consequently, the following customer starts service later and is also delayed. The delays propagate to future customers until the phantom queue is empty.

at this step (in the figure, the second phantom process dies at step $n = 7$). We denote by $\tau_i(\theta)$ the lifetime of the i-th phantom difference: it is born at step i and it dies at step $i + \tau_i(\theta)$. Specifically,

$$\tau_i(\theta) \stackrel{\text{def}}{=} \min(n: d_n(i; \theta) \leq 0). \tag{8.34}$$

Once $d_n(\theta; i) = 0$ it follows that $d_k(\theta; i) = 0$ for all $k \geq n$. In terms of Markov chains, this is known as "coupling" of two Markov chains. This establishes (8.33).

The case of general service distribution is almost exactly the same, except that we need to implement the corresponding random variables for $X^\pm(\theta; i)$. Accordingly, (8.33) must be modified for other models than exponential, and it depends on whether the initial difference is positive or negative. Algorithm 8.4 uses a list of "living phantoms" that updates to reflect the fact that different processes will have an a.s. finite lifetime. In this pseudocode we assume that the initial difference is positive, so (8.33) holds.

Looking at Algorithm 8.4 it may still seem that the double sum (in the nested **for** loops) is of order $O(N^2)$, corresponding to $\sum_{i=1}^{N} \sum_{j=1}^{i} c$, where c is the running time of the basic operations in the code (constant time). However, this is not a tight bound, but rather the "worst" case. Because each phantom process follows the same dynamics as the nominal queue (except for its initial value) and the queue is stable, then the phantom difference processes are supermartingales (see Section B.6 in Appendix B) and therefore they die with probability 1. Their lifetimes $\{\tau_i(\theta)\}$ are iid and finite a.s. Thus, an expected running time analysis yields a *linear* running time of the MVD when implemented in this manner, because the expected running time is $\sum_{i=1}^{N} \mathbb{E}[\tau_i(\theta)] = O(N)$. Compared to Example 8.6 and Example 8.13 it should be apparent that these algorithms have the same order of running time. Notice, however, that SF and MVD can be applied to other models for the service distribution or for the cost function where IPA may not be applicable. ✳✳✳

Algorithm 8.4 Full MVD–Single run general service distribution

Initialize MVD $= 0$, $X_0 = 0$, $\mathcal{L} = \emptyset$.
for $n = 1, \ldots, N$ **do**
　　Generate A_n, $S_n \sim f_\theta$ and $Y_n^\pm \sim f_\theta^\pm$
Initialize nominal sojourn time: $X_1 = 0$ and estimator MVD $= 0$.
for $i = 1, \ldots N$ **do**
　　$d_i(i) = Y_i^+ - Y_i^-$ (birth of i-th phantom difference process)
　　$X_i^\pm(i) = Y_i^\pm + \max(0, X_{i-1} - A_i)$
　　$\Delta(i) = A_i - X_{i-1}^-$
　　$X_i = \max(0, X_{i-1} - \Delta(i)) + S_i$
　　APPEND(i) to list $\mathcal{L}$
　　for $j = 1 \in \mathcal{L}$ **do**
　　　　$d_i(j) = \max(0, \max(0, d_i(j) - \Delta(i)))$
　　　　if $d_i(j) = 0$ **then** REMOVE(j) from list $\mathcal{L}$ (death of the j-th phantom difference process)
　　　　MVD $\leftarrow$ MVD $+ d_i(j)$
return MVD(c_θ/N).

Historical Note. Weak differentiation in the context of optimization was introduced by Pflug in [241], and has been further developed by Pflug in [242, 243, 244]. For a monograph summarizing these early results on weak differentiation, see [245]. The concept of weak differentiation has been further developed into that of MVD in a sequence of articles; see [153, 154]. As illustrated in Example 8.19, the number of parallel phantoms, i.e., the "+" and "−" may become prohibitively large when N is very large, and this may create serious memory problems with the computer, slowing down the computation. For many cases, phantom processes can be merged in the simulation simply by managing a list and erasing from the list those phantoms for which the "plus" and "minus" processes attain the same state value, because in this case the two Markov chains achieve coupling and their future trajectories are identical (straightforward to implement). An analysis of the number of parallel phantoms needed at each time in the simulation is provided in [190]. This type of analysis generalizes the standard method of probabilistic analysis of algorithms. To address the memory problem, one can use the randomized estimator; in words, only two phantoms are generated at a randomly chosen state. In general, one can fix a priori the maximal number of phantoms that can be simulated in parallel, say M, and then randomly generate the phantoms only once per M steps.

Phantom processes were introduced in [305] in order to create more efficient code. In [47] and other early work in [308] the focus was on estimating derivatives with respect to the rate of a point process and for those examples, the "phantom" systems start deleting customers. For this phenomenon the name *disappearing phantom* was coined. In [306] another implementation is studied, where the "phantom" system starts with swapping the actions, and we call them the *swapping phantoms*. The work in [190] considers a more complex MDP formulation where it is not possible to simulate the parallel phantom systems. An alternative approach is implemented, where we use a "cut and paste" argument. When a phantom system starts at a state/action pair (x, a) different from the nominal system (assumed to be observed in real time), the code "freezes" the computation until the nominal process hits this state. Imagining that the phantom system evolves from that first step as the nominal one, the difference process is zero until the nominal has completed N steps, at which point the phantom system still needs to calculate the contribution of the remaining steps. This is done via conditional expectations, for variance reduction and speed of algorithms. We call these the *frozen phantoms*. Phantom estimators appear in the literature also under the name of rare perturbation analysis (RPA) [65, 306, 308, 312, 314, 315, 316]. It is worth noting that when dealing with certain jumps or discontinuities due to threshold-like control variables, SPA formulas also require computation on parallel systems, which they perform via "off-line" computation [104, 311].

With the advent of phantom estimators comes the search for techniques for reducing the computational burden. For discrete-state processes, a favorable approach is to construct the perturbed path (read "phantom") from the nominal path by a "cut-and-paste" approach. An early reference on the cut-and-paste idea for IPA estimation is [164] and for a thorough treatment we refer to the seminal work [60]. Applying these methods to MVD was not done until [190]. Another approach of reading the phantom information from the nominal sample path is by applying an importance sampling approach, see [148] for the general state space, and [204] for an efficient algorithm for constructing a good dominating measure for discrete-state space models. For continuous state-space processes, much research has been done on constructing coupling schemes. The relation between coupling schemes and sample-path perturbation analysis has been explained in [45]. A detailed analysis of coupling techniques useful in gradient estimation can be found in [83]. For Markov processes with general state space, we refer to [154, 306] for a discussion on couplings of phantoms.

*8.4.3 The Weak Differentiation Approach: Measure Integrals

Let $X \in S$ have distribution μ on some measurable space $(S, \mathcal{S})$. The expected value of $h(X)$ reads

$$\mathbb{E}[h(X)] = \int_S h(x)\, \mu(dx)$$

and it can be interpreted as a bilinear mapping $\langle \cdot, \cdot \rangle : (h, \mu) \mapsto \mathbb{E}[f(X)]$. Suppose that μ depends on some parameter θ and write μ_θ and X_θ, respectively. Properties of μ_θ such as continuity or, as we will see later on, differentiability can be introduced via families of test functions. For example, the sequence of measures $\{\mu_{\theta_n}\}$ is said to be weakly convergent toward a measure μ_θ if for any $\{\theta_n\}$ such that $\theta_n \to \theta$ as n tends to ∞ it holds that

$$\lim_{n \to \infty} \langle h, \mu_{\theta_n} \rangle = \langle h, \mu_\theta \rangle \quad \forall h \in C^b(S), \tag{8.35}$$

where $C^b(S)$ is the set of continuous and bounded mappings from S to $\mathbb{R}$. Note that weak convergence is also called weak continuity. A natural question is why not simply define continuity of measures via set-wise continuity $\lim_{n \to \infty} \mu_{\theta_n}(A) = \mu_\theta(A)$ for all $A \in \mathcal{S}$? The reason is that this defintion is too restrictive as it makes too few sequences convergent. Indeed, (8.35) may hold for a sequence $\{\mu_{\theta_n}\}$ that does not satisfy set-wise continuity. To see this, let μ_{θ_n} denote the uniform distribution on $[1 - \theta_n, 1]$ and consider convergence of μ_{θ_n} as $\theta_n \to 1$. Then

$$\lim_{n \to \infty} \langle h, \mu_{\theta_n} \rangle = \frac{1}{1 - \theta_n} \int_{1-\theta_n}^1 h(x)\, dx = h(1) = \int_S h(x)\, \delta_1(dx),$$

where δ is the Dirac measure with unit measure in point 1. Hence, μ_{θ_n} converges weakly toward δ_1. Set-wise convergence does not hold. Indeed, let $A = \{1\}$, i.e., a singleton. Then

$$\lim_{n \to \infty} \mu_{\theta_n}(A) = 0 \neq 1 = \delta_1(1).$$

GRADIENT ESTIMATION, FINITE HORIZON

Apart from this, admittedly somewhat technical argument, the definition in (8.35) has the advantage to capture the core application of continuity of measures, namely that of continuity of expected values for a pre defined class of functions.

Let F be the cdf of X and let μ denote the measure of X, then

$$F(x) = \mathbb{P}(X \leq x) = \mu\Big(\{s \in S : X(s) \leq x\}\Big),$$

for all x. Denoting the probability density function of X by f_X, we have

$$\mathbb{E}[h(X)] = \int h(x) f_X(x)\, dx = \int h(x) \mu(dx),$$

where we assume that the expression on the above left hand side exists. A subtle but important point in the relation between measures and cdfs is that (point-wise) convergence of the sequence of cdfs F_n toward a cdf F only characterizes F on continuity points of F. In words, if mass is shifted toward a single point when taking the limit, then the limit of F_n bears no information on F at that particular point.

Example 8.20. Let U_θ be uniformly distributed on $[0, \theta]$. Letting $U_\theta = \theta U$, with U uniformly distributed on $[0, 1]$, the sample path derivative becomes

$$\frac{d}{d\theta} U_\theta = U = \frac{1}{\theta} U_\theta.$$

For the distributional approach, note that U_θ has density

$$f_\theta(x) = \frac{1}{\theta} \mathbf{1}_{\{x \in [0,\theta]\}},$$

which fails to be differentiable with respect to θ, and consequently SF doesn't apply; see also Example 8.10. For MVD, note that for any continuous mapping h basic analysis yields

$$\frac{d}{d\theta} \int h(x) f_\theta(x)\, dx = \frac{d}{d\theta}\left(\frac{1}{\theta} \int_0^\theta h(x)\, dx\right) = -\frac{1}{\theta^2} \int_0^\theta h(x)\, dx + \frac{1}{\theta} h(\theta)$$

$$= \frac{1}{\theta}\left(h(\theta) - \frac{1}{\theta} \int_0^\theta h(x)\, dx\right)$$

$$= \frac{1}{\theta}\left(h(\theta) - \int h(x) f_\theta(x)\, dx\right).$$

Hence, we obtain for the uniform distribution the estimators

$$\psi^{\text{IPA}}(h, U_\theta, \theta) = \frac{U_\theta}{\theta} h'(U_\theta),$$

$$\psi^{\text{MVD}}(h, U_\theta, \theta) = \frac{1}{\theta}\left(h(\theta) - h(U_\theta)\right).$$

In summary, for this example

$$\text{IPA} \quad \frac{1}{\theta} \mathbb{E}\big[U_\theta h'(U_\theta)\big] = \frac{d}{d\theta} \mathbb{E}\big[h(U_\theta)\big] = \begin{cases} \text{not applicable} & \text{SF}, \\ \frac{1}{\theta} \mathbb{E}\big[h(\theta) - h(U_\theta)\big] & \text{MVD}. \end{cases}$$

Table 8.1. Measure-valued differentiability of common distributions.

μ_θ	c_θ	μ_θ^+	μ_θ^-
Pareto(θ, β) type I	β/θ	Pareto(θ, β) type I	Dirac(θ)
Uniform$(0, \theta)$	$1/\theta$	Dirac(θ)	Uniform$(0, \theta)$

Note that for IPA to be unbiased h has to differentiable, whereas for the weak differentiation approach integrability of h is sufficient. ✱✱✱

The above example illustrates the problem raised by having convergence only at points of continuity. All instances of F_n were continuous distributions and taking the limit as n tends to ∞ the mass shifted toward the single point $\{1\}$, which results in a non continuous limit F.

8.4.3.1 Measure-Valued Differentiation: General Theory

For $\theta \in \Theta \subset \mathbb{R}$, let μ_θ denote a probability measure, and denote the set of absolutely μ_θ integrable mappings for any $\theta \in \Theta$ by $L^1(\mu_\theta, \Theta)$, i.e.,

$$h \in L^1(\mu_\theta, \Theta) \quad \Leftrightarrow \quad \forall \theta \in \Theta: \quad \int |h(x)| \mu_\theta(dx) < \infty.$$

A measure is called a *signed measure* if it assigns negative mass to certain measurable sets.

Definition 8.8. Let $\mathcal{D} \subset L^1(\mu_\theta, \Theta)$. The probability measure μ_θ is called $\mathcal{D}$-*differentiable* if a signed measure μ'_θ exists such that for all $h \in \mathcal{D}$

$$\lim_{\Delta \to 0} \frac{1}{\Delta} \left(\int h(s) \mu_{\theta+\Delta}(ds) - \int h(s) \mu_\theta(ds) \right) = \int h(s) \mu'_\theta(ds).$$

Let c_θ be a constant and μ_θ^+ and μ_θ^- two probability measures such that

$$\int h(s) \mu'_\theta(ds) = c_\theta \left(\int h(s) \mu_\theta^+(ds) - \int h(s) \mu_\theta^-(ds) \right),$$

then $(c_\theta, \mu_\theta^+, \mu_\theta^-)$ is called a $\mathcal{D}$-derivative of μ_θ.

The fact that μ'_θ is a signed measure allows one to write it as the difference between two positive measures. This fact is known as *Hahn-Jordan decomposition*. In the case that μ_θ has a differentiable density, the Hahn-Jordan decomposition can be constructed explicitly. While the Hahn-Jordan decomposition typically leads to an estimator with small variance, Pflug provides in Example 4.21 in [245] an instance where the Hahn-Jordan decomposition has larger variance than an alternative representation.

To understand the power of the concept of weak differentiation it is important to realize that differentiability of f_θ is not essential for this approach; see Example 8.20. Table 8.1 provides extensions of Table 7.1 to measures that cannot be analyzed via their densities. Here, Pareto(θ, β) type I denotes the Pareto type I distribution cdf $1 - (\theta/x)^\beta$, $x \geq \theta$.

8.4.3.2 Weak Differentiation and Banach Spaces

For putting the basic problem of gradient estimation into a broader perspective, we start off by brining structure to the cost functions we may want to analyze. In applications, one usually has a certain class of mappings in mind. For example, one might be interested in continuous mappings only, denoted by $\mathcal{C}$, or in the much wider class of measurable mappings, denoted by $\mathcal{B}$. In general, let $\mathcal{D}$ be set of cost functions such that

$$\left.\frac{d}{d\theta}\right|_{\theta=\theta_0} \int h(x) f_\theta(x) dx = c_{\theta_0} \left(\int h(x) f_{\theta_0}^+(x) dx - \int h(x) f_{\theta_0}^-(x) dx \right) \qquad (8.36)$$

holds for all $h \in \mathcal{D}$, implying that all integrals are finite and well-defined. Then, we call $(c_\theta, f_\theta^+, f_\theta^-)$ a $\mathcal{D}$-derivative of f_θ.

Let $\mathcal{H} \subset L^1(\mu_\theta, \Theta)$ denote the range of mappings one is interested in from the outset. To characterize analytical properties of $\mathcal{H}$ it is helpful to choose a mapping $v \geq 0$, such that $\int v(x) f_\theta(x) dx$ is finite, and consider the set

$$\mathcal{D}_v := \mathcal{D}_v(\mathcal{H}) = \{h \in \mathcal{H} : |h(x)| \leq cv(x) \text{ for all } x \text{ and some finite constant } c\}.$$

The set $\mathcal{H}$ is called the *base set* (usually $\mathcal{C}$ or $\mathcal{B}$). Notice that the subset $\mathcal{D}_v$ corresponds to all functions $h \in \mathcal{H}$ with growth limited by v, that is, functions that are $O(v(x))$. Put differently, $\mathcal{D}_v$ is the set of all $h \in \mathcal{H}$ with finite v_p-norm (see Definition 8.6 of $\|\cdot\|_v$).

As we have seen in the previous section, a choice for v that is useful in analyzing the static problem is $v_\alpha(x) = \alpha^{|x|}$, for $\alpha \geq 1$; see (8.16) and the examples in (8.17). Then, for example, the exponential distribution is weakly differentiable with respect to the set $\mathcal{B}_{v_\alpha} \stackrel{\text{def}}{=} \mathcal{D}(\mathcal{B})_{v_\alpha}$, which stems from the fact that (8.36) holds for v_α. Furthermore, note that $\mathcal{C}_{v_0} := \mathcal{D}(\mathcal{C})_{v_0}$ is the set of continuous and bounded mappings.

The following result shows that under appropriate conditions, weak differentiability implies norm Lipschitz continuity, which is a property that helps by simplifying the technical analysis as well as illustrating the weak differentiation behaves "almost" as strong (i.e., norm-wise) differentiability. Recall that $(\mathcal{D}_v, \|\cdot\|_v)$ is called a *Banach space*, if the point limit of sequences from $\mathcal{D}_v$ in v-norm sense belong to $\mathcal{D}_v$ as well.

Lemma 8.4. *Let f_θ be $\mathcal{D}$-differentiable, such that $\mathcal{D}$ equipped with norm $\|\cdot\|_v$ becomes a Banach space. Then a finite constant M exists such that for all Δ such that $\theta + \Delta \in \Theta$ it holds that*

$$\|f_{\theta+\Delta} - f_\theta\|_v \leq |\Delta| M.$$

In words, $\mathcal{D}$-differentiability implies Lipschitz continuity.

Proof. Using the shorthand notation $\langle h, \mu_\theta \rangle$ for the μ_θ integral of h, it holds, under the assumption in the lemma, that the sequence

$$\frac{1}{\Delta}(\langle h, \mu_{\theta+\Delta} \rangle - \langle h, \mu_\theta \rangle)$$

converges for any h in $\mathcal{D}$ as Δ tends to zero. Hence,

$$\sup_{\Delta \neq 0} \left| \frac{1}{\Delta}(\langle h, \mu_{\theta+\Delta} \rangle - \langle h, \mu_\theta \rangle) \right| < \infty,$$

for any $h \in \mathcal{D}$. The Banach-Steinhaus theorem then implies that the above set is also bounded in norm-sense, i.e.,

$$\sup_{\Delta \neq 0} \left\| \frac{1}{\Delta} (\mu_{\theta+\Delta} - \mu_\theta) \right\|_v \stackrel{\text{def}}{=} M < \infty,$$

which proves the claim. □

The set $L^1(\mu_\theta, \Theta)$ is the maximal set of mappings h having the property that the integral $\int h(x)\mu_\theta(dx)$ is well defined for any θ. In practice, the set of mappings that are feasible for differentiation is smaller than $L^1(\mu_\theta, \Theta)$. Indeed, recall that for the uniform distribution $\int h(x)\mu_\theta(dx)$ exist for any measurable a.s finite mapping h, whereas the integral is only differentiable when h is continuous; see Example 8.20.

Given that μ_θ and ν_θ are weakly differentiable, does it then hold that the product probability measure $\mu_\theta \times \nu_\theta$ is weakly differentiable as well? In case that μ_θ and ν_θ have differentiable densities, this question can be answered along the lines detailed in the proof of Theorem 8.5. In the general case the proof becomes more elaborate. However, if the functional space $(\mathcal{D}_v, \|\cdot\|_v)$ is a Banach space, then weak differentiability of product measures can be established in a mathematically comprehensive way. The supporting theory for this is provided in the next section.

8.4.3.3 Differentiability of Products of Measures

An interesting feature of MVD is that $\mathcal{D}$-differentiability of products of probability measures can be deduced from that of the elements of the products without further assumptions, provided that the set $\mathcal{D}$ is well-chosen. For the analysis we require an extension of the v-norm defined in the previous section to measures.

Definition 8.9. Let μ be a measure on $\mathbb{R}$, and let $v(x) \geq 1$ for all x in the support of μ. The *weighted supremum norm*, or *v-norm* for short, is defined as

$$\|\mu\|_v = \int v(x)|\mu|(dx).$$

Note that $|\mu|(\cdot)$ denotes the absolute measure. In case that μ is a positive measure, such as a probability measure, it holds that $|\mu| = \mu$, in case that μ is a signed measure we consider the Hahn-Jordan decomposition of μ and let

$$\int v(x)|\mu|(dx) = \int v(x)\mu^+(dx) + \int v(x)\mu^-(dx).$$

We now turn to the proof of the product rule for differentiation. It establishes that the "usual" chain rule for functions also applies to measure differentiation. Consider a set $\mathcal{T}$ of real-valued mappings on $\mathbb{R}^2$. Furthermore, let μ and ν be (probability) measures on $\mathbb{R}$, and denote the product measure on $\mathbb{R}^2$ by $\mu \otimes \nu$. Note that if μ has Lebesgue density f and if ν has Lebesgue density g, then $\mu \otimes \nu$ has Lebesgue density $f(x)g(y)$ for $(x,y)^\top \in \mathbb{R}^2$. The product rule will have to answer under what conditions $\mu \otimes \nu$ is differentiable, provided μ and ν are. In this section, differentiability is defined in the weak sense and thus relative to the set of test functions, it may happen that μ is $\mathcal{D}^\mu$ differentiable and ν is $\mathcal{D}^\nu$-differentiable with $\mathcal{D}^\mu \neq \mathcal{D}^\nu$. The set of mappings the product measure can differentiate will thus be a subset of the mappings $h \in \mathcal{T}$ so that $h(\cdot, y) \in \mathcal{D}^\mu$ for all y and $h(x, \cdot) \in \mathcal{D}^\nu$ for all x. As it turns out the v-norm is a nice tool for handling this situation. While the product rule can be established in a more general setting (see [151]) we will prove the statement in a simpler form that is sufficient for the current text. For this, we write $f \otimes g$ for the product of real-valued mappings f and g, i.e., $(f \otimes g)(x, y) = f(x)g(y)$, for $x, y \in \mathbb{R}$. The following product rule of weak differentiation can be obtained, where we use the fact that $(\mathcal{D}, ||\cdot||_v)$ is a Banach space, if $||h||_v < \infty$ for any $h \in \mathcal{D}$. For Banach spaces $(\mathcal{D}^\mu, ||\cdot||_v)$ and $(\mathcal{D}^\nu, ||\cdot||_w)$, we call $(\mathcal{D}, ||\cdot||_{v \otimes w})$ the corresponding product space where $h \in \mathcal{D}$ if and only if $h(\cdot, y) \in \mathcal{D}^\mu$ for all y and $h(x, \cdot) \in \mathcal{D}^\nu$ for all x.

Theorem 8.7. *Let $(\mathcal{D}^\mu, ||\cdot||_v)$ and $(\mathcal{D}^\nu, ||\cdot||_w)$ be Banach spaces on $\mathbb{R}$, with corresponding product space $(\mathcal{D}, ||\cdot||_{v \otimes w})$.*

If μ_θ is $\mathcal{D}^\mu$-differentiable and ν_θ is $\mathcal{D}^\nu$-differentiable, then the product measure $\mu_\theta \otimes \nu_\theta$ is $\mathcal{D}$-differentiable, that is, for any $h \in \mathcal{D}$ it holds that

$$\frac{d}{d\theta} \int h(x,y) \mu_\theta(dx) \nu_\theta(dy) = \int h(x,y) \mu'_\theta(dx) \nu_\theta(dy) + \int h(x,y) \mu_\theta(dx) \nu'_\theta(dy).$$

Moreover, if in addition, μ_θ has $\mathcal{D}^\mu$-derivative $(c_\theta^\mu, \mu_\theta^+, \mu_\theta^-)$ and that ν_θ has $\mathcal{D}^\nu$-derivative $(c_\theta^\nu, \nu_\theta^+, \nu_\theta^-)$. Then it holds that for any $h \in \mathcal{D}$ that

$$\frac{d}{d\theta} \int h(x,y) \mu_\theta(dx) \nu_\theta(dy)$$
$$= (c_\theta^\mu + c_\theta^\nu) \left(\frac{c_\theta^\mu}{c_\theta^\mu + c_\theta^\nu} \int h(x,y) \mu_\theta^+(dx) \nu_\theta(dy) + \frac{c_\theta^\nu}{c_\theta^\mu + c_\theta^\nu} \int h(x,y) \mu_\theta(dx) \nu_\theta^+(dy) \right.$$
$$\left. - \frac{c_\theta^\mu}{c_\theta^\mu + c_\theta^\nu} \int h(x,y) \mu_\theta^-(dx) \nu_\theta(dy) + \frac{c_\theta^\nu}{c_\theta^\mu + c_\theta^\nu} \int h(x,y) \mu_\theta(dx) \nu_\theta^-(dy) \right).$$

Proof. For Δ such that $\theta + \Delta \in \Theta$, set

$$\bar{\mu}_\Delta = \frac{\mu_{\theta+\Delta} - \mu_\theta}{\Delta} - \mu'_\theta; \quad \bar{\nu}_\Delta = \frac{\nu_{\theta+\Delta} - \nu_\theta}{\Delta} - \nu'_\theta.$$

To simplify notation, we write $\mu_n \stackrel{\mathcal{H}}{\Longrightarrow} \nu$ for

$$\lim_{n \to \infty} \langle h, \mu_n \rangle = \langle h, \nu \rangle, \quad \text{for all } h \in \mathcal{H}.$$

By hypothesis, $\bar{\mu}_\Delta \stackrel{\mathcal{D}^\mu}{\Longrightarrow} \varnothing$ and $\bar{\nu}_\Delta \stackrel{\mathcal{D}^\nu}{\Longrightarrow} \varnothing$, for $\Delta \to 0$, where $\varnothing$ denotes the null measure. Simple algebra shows that the proof of the claim follows from

$$\Delta(\bar{\mu}_\Delta + \mu'_\theta) \times (\bar{\nu}_\Delta + \nu'_\theta) + \mu_\theta \times \bar{\nu}_\Delta + \bar{\mu}_\Delta \times \nu_\theta \stackrel{\mathcal{D}}{\Longrightarrow} \varnothing, \quad (8.37)$$

for $\Delta \to 0$. Hence, to conclude the proof, we show that each term on the left side of (8.37) converges weakly to null measure $\varnothing$.

Since $\bar{\mu}_\Delta + \mu'_\theta \stackrel{\mathcal{D}^\mu}{\Longrightarrow} \mu'_\theta$ and $\bar{\nu}_\Delta + \nu'_\theta \stackrel{\mathcal{D}^\nu}{\Longrightarrow} \nu'_\theta$, applying Lemma 8.4 yields

$$\sup_{\Delta \in V \setminus \{0\}} ||\bar{\mu}_\Delta + \mu'_\theta||_v < \infty \quad \text{and} \quad \sup_{\Delta \in V \setminus \{0\}} ||\bar{\nu}_\Delta + \nu'_\theta||_w < \infty,$$

for any compact neighborhood V of 0. By simple algebra,

$$\left| \Delta \int h(s,t)((\bar{\mu}_\Delta + \mu'_\theta) \times (\bar{\nu}_\Delta + \nu'_\theta))(ds, dt) \right|$$
$$\leq |\Delta| \left| \int \frac{h(s,t)}{v(s)w(t)} ((v(s)|\bar{\mu}_\Delta + \mu'_\theta|) \times (w(t)|\bar{\nu}_\Delta + \nu'_\theta|))(ds, dt) \right|$$

$$\leq |\Delta| \int \|h\|_{v\otimes w} \left((v(s)|(\bar{\mu}_\Delta + \mu'_\theta|) \times (w(t)|\bar{v}_\Delta + v'_\theta|) \right) (ds, dt)$$

$$\leq |\Delta| \cdot \|h\|_{v\otimes u} \cdot \|\bar{\mu}_\Delta + \mu'_\theta\|_v \cdot \|\bar{v}_\Delta + v'_\theta\|_w.$$

Letting $\Delta \to 0$ in the above inequality it follows that the first term in (8.37) converges weakly to $\varnothing$.

The second and the third terms in (8.37) are symmetric so they can be treated similarly. For instance, for the second term in (8.37) note that

$$\int h(s,t)(\mu_\theta \times \bar{v}_\Delta)(ds, dt) = \int \int h(s,t)\mu_\theta(ds)\, \bar{v}_\Delta(dt) = \int H_\theta(h,t)\bar{v}_\Delta(dt),$$

where $H_\theta(h,t) = \int h(s,t)\mu_\theta(ds)$ for all t and for all h. Therefore,

$$\forall t \in T: \frac{|H_\theta(h,t)|}{w(t)} \leq \frac{\|h(\cdot,t)\|_v}{w(t)} \|\mu_\theta\|_v \leq \|h\|_{v\otimes w} \|\mu_\theta\|_v,$$

where the second inequality follows from

$$\forall s \in S, t \in T: |h(s,t)| \leq \|h\|_{v\otimes w} v(s) w(t).$$

Consequently, $H_\theta(h,\cdot) \in \mathcal{D}^v$, for $h \in (\mathcal{D}, \|\cdot\|_{v\otimes w})$. We have assumed that v_θ is $\mathcal{D}^v$-differentiable, which yields $\bar{v}_\Delta \overset{\mathcal{D}^v}{\Longrightarrow} \varnothing$. Hence,

$$\lim_{\Delta \to 0} \int H_\theta(h,t)\bar{v}_\Delta(dt) \to 0,$$

which shows that the second term in (8.37) converges weakly to $\varnothing$. This concludes the proof. □

Remark 8.2. One of the virtues of the v-norm is that one can show that if $\mathcal{D}$ is either the set of continuous or measurable mappings from $\mathbb{R}^2$ to $\mathbb{R}$ with finite $v \otimes v$ norm, then $(\mathcal{D}, \|\cdot\|_{v\otimes v})$ is the product space corresponding to $(\hat{\mathcal{D}}, \|\cdot\|_v)$, with $\hat{\mathcal{D}}$ being the set of continuous or measurable mappings from $\mathbb{R}$ to $\mathbb{R}$ with finite v norm, see [151] for a proof. The introduction of the product space $(\mathcal{D}, \|\cdot\|_{v\otimes v})$ is the counterpart of product technique for the score function method.

8.5 PRACTICAL CONSIDERATIONS

IPA and SF gradient estimators can be evaluated along the trajectory, as the system is observed or simulated, and are called *on-line* or *single-run* estimators. The advantage of such constructions is practical: if the estimation is implemented in a real time system for control or supervisory purposes, then it is not desirable to wait until the end of the horizon N (or the end of a cycle) to evaluate the gradient. Instead, a single-run estimator provides a formula that can be coded in terms of partial computations that use only the past history of the observations of the queue occupancy and/or waiting times.

It is worth noting that studying the effect of a finite perturbation, as required by MVD, does not necessarily lead to an off-line parallel computation. As typically the natural filtration of the nominal process cannot provide enough information to build the derivative, it must be augmented. However, often we can find a *minimal* augmentation of the natural filtration of the process to build a version of the parallel processes. This version of the derivative process should be as close as possible to a single-run adapted process. A favorable approach is to construct the perturbed path (read "phantom") from the nominal path by a "cut-and-paste" approach. An early reference on this is [164] and for a thorough treatment we refer to the seminal work [60] as well as [164]. Another approach of reading the phantom information from the nominal sample path is by applying an importance sampling approach; see [148] for the general state space and [204] for an efficient algorithm for constructing a good dominating measure for discrete-state space models. In the best of cases, a single–run MVD estimator can be constructed, combining low variance with generality in use, and requiring a minimal amount of added information in order to build a derivative process, see [147, 155, 190, 306, 308, 312, 316]. As proved theoretically in [156] for the special case of the normal distribution, MVD estimators may have a lower variance than IPA estimators.

8.6 EXERCISES

Exercise 8.1. Refer to the queuing model of Example 7.16 and Example 8.6 . Let $J(\theta) = \mathbb{E}[L(X_1(\theta),\ldots,X_N(\theta))]$, where $L(X_1,\ldots,X_N)$ is such that $\frac{\partial L}{\partial x_i}(X_1,\ldots,X_N)$ is continuous and uniformly bounded in absolute value by a constant l for all i. Show that if θ is a scale parameter of the service distribution satisfying $\mathbb{E}[S_1(\theta)/\theta] < \infty$, then the IPA derivative is unbiased for $\nabla_\theta J(\theta)$.

Exercise 8.2. Refer to Example 7.16 and Example 8.6. Assume that θ is a location parameter of the service times in a FCFS GI/G/1 queue. Specifically, assume that the service times have a representation of the form $S_n(\theta) \stackrel{d}{=} \theta + F_0^{-1}(U_n)$, where $\{U_n\}$ is a sequence of iid uniform random variables $U(0,1)$. For the queueing model, we consider only positive values $\theta > 0$:

(a) Specify under which assumptions for $F_0(x)$ is $X_n(\theta)$ a.s. Lipschitz continuous for all $n \leq N$.
(b) Calculate the derivative of the sample average

$$L(X_1(\theta),\ldots,X_N(\theta)) = \frac{1}{N}\sum_{n=1}^{N} X_n(\theta).$$

(c) Under the assumptions in (a), prove that the IPA estimator of $\mathbb{E}[L]$ is unbiased.

Exercise 8.3. We revisit the sojourn time example as introduced in Example 7.16 and Example 8.6. with notations and basic assumptions as detailed in these examples. Provided that the service times $S_n(\theta)$ are almost surely Lipschitz continuous on some fixed interval Θ, show that

$$\frac{d}{d\theta}\mathbb{E}\left[\frac{1}{N}\sum_{n=1}^{N} X_n(\theta)\right] = \mathbb{E}\left[\frac{1}{N}\sum_{n=1}^{N} \frac{dX_n(\theta)}{d\theta}\right],$$

for almost all $\theta \in \Theta$.

Exercise 8.4. Show that if ν_θ has a μ-density f_θ for all $\theta \in \Theta \subset \mathbb{R}$, then ν_θ has ν_{θ_0}-density f_θ/f_{θ_0} for any $\theta_0 \in \Theta$, provided that the support of f_θ is a subset of that of f_{θ_0} for all $\theta \in \Theta$.

Exercise 8.5. Let $\mathsf{SF}(\theta, X(\theta))$ be the score function of a random variable $X(\theta)$ with pdf $f_\theta(x)$. Prove

$$\mathbb{E}\left[\mathsf{SF}(\theta, X(\theta))\right] = 0.$$

Exercise 8.6. Let f denote a density of random variable X with support $(-\infty, \infty)$. Show that
$$f(x) = \mathbb{E}\left[\mathbf{1}_{\{X \leq x\}} \frac{f'(X)}{f(X)}\right].$$
This result shows that the value of the density at a point can be obtained via the score function.

Exercise 8.7. Revisit Example 8.2. Establish the corresponding SF and MVD quantile sensitivity estimators.

Exercise 8.8. Let $\{X_n\}$ be a finite Markov chain with continuously differentiable transition probabilities $P_{i,j}(\theta)$ and known initial state $X_0 = x_0$.

(a) Show that the score function for a horizon N is
$$\mathrm{SF}_N(\theta; X_1, \ldots, X_N) = \sum_{n=1}^{N} \frac{d}{d\theta} \log P_{X_{n-1}, X_n}(\theta).$$

(b) Let $c: S \to \mathbb{R}$ be a bounded function and consider the cost function:
$$C(\theta) = \mathbb{E}\left[\frac{1}{N}\sum_{n=1}^{N} c(X_n)\right]. \tag{8.38}$$

Show that
$$\frac{d}{d\theta}C(\theta) = \mathbb{E}\left[\mathrm{SF}_N(\theta; X_1, \ldots, X_N) \frac{1}{N}\sum_{n=1}^{N} c(X_n)\right]$$
$$= \mathbb{E}\left[\frac{1}{N}\sum_{n=1}^{N} c(X_n) \mathrm{SF}_n(\theta; X_1, \ldots, X_n)\right],$$
provided that the integrands above have finite expectation.

(c) Argue by inspection that the second estimator, namely
$$\frac{1}{N}\sum_{n=1}^{N} c(X_n) \mathrm{SF}_n(\theta; X_1, \ldots, X_n)$$
must necessarily have smaller variance than the one using $\mathrm{SF}_N(\theta; X_1, \ldots, X_N)$. This is a usual approach for variance reduction whenever applying SF to Markov chains.

Exercise 8.9. Refer to Exercise 7.7 on the optimal strategic investment for the mining model. Show that the IPA, the SF, and the MVD methods provide unbiased estimators for the derivative.

Exercise 8.10. Consider the reliability model of Examples 7.15, 8.4, and 8.16, illustrated in Figure 8.4. The life of the system is
$$L(\theta) = \max\left(\min(T_1, T_2, T_5), \min(T_1, T_2, T_4, T_5), \min(T_1, T_3(\theta), T_4, T_5)\right),$$
where the components' lives are five independent random variables: $T_i \sim G_i, i \neq 3$ (independent of θ) and $T_3(\theta) \sim$ Exponential $(1/\theta)$.

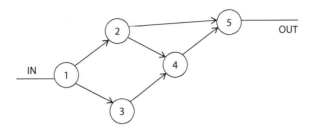

Figure 8.4. Reliability network.

(a) Calculate the SF estimator for the derivative of $\mathbb{E}[L(\theta)]$ and show that it is unbiased.
(b) Program a simulation to generate independent samples of $L(\theta)$ and of the three derivative estimators: the IPA (given in Example 8.4), the SF, and the MVD (given in Example 8.16). For the same amount of samples, compute a table to compare variances and CPU times.

Exercise 8.11. Refer to the reliability model of Examples 7.15, 8.4, and 8.16, illustrated in Figure 8.4. Consider now the model where the lifetimes have a Weibull distribution with parameters (λ_i, k). In particular, for component 3 we have $\mathbb{P}(T_3 \leq x) = G_3(x) = 1 - e^{-(x/\theta)^k}$, so that now $\mathbb{E}(T_3) = \theta\, \Gamma(1 + 1/k)$.

(a) Find the representation $T_3 = G_3^{-1}(U)$ and show that the IPA estimator

$$\widehat{L}^{IPA}(\theta, \omega) = \begin{cases} \frac{T_3(\theta,\omega)}{\theta} & \text{if } T_2(\omega) < T_3(\theta,\omega) < \bar{T}(\omega) = \min(T_1(\omega), T_4(\omega), T_5(\omega)), \\ 0 & \text{otherwise,} \end{cases}$$
(8.39)

is valid for this case. For what family of distributions G_3 will the IPA estimator remain unchanged?
(b) Explain how the SF and MVD estimators of Exercise 8.10 should be modified for this problem.

Exercise 8.12. Consider the reliability model of Examples 7.15, 8.4, and 8.16, illustrated in Figure 8.4. Suppose now that we are interested in estimating the sensitivity:

$$\frac{d}{d\theta}\mathbb{P}(L > \bar{l}),$$
(8.40)

for some specific bound $\bar{l}$, and assume that the lifetime distributions are exponential: $G_i \sim$ Exponential $(1/\lambda_i)$, with $\lambda_3 \theta = 1$.

(a) Use the same representation as in Example 6.14 and show that IPA is biased. Explain why.
(b) Explain how the SF and MVD estimators of Exercise 8.10 should be modified for this problem.

Exercise 8.13. We now present the reliability problem related to a network of logically interconnected components as illustrated in Figure 8.4. For such systems, maintenance rules must be specified in order to repair or replace failed components. Suppose that the system has been operating for some time, maybe with some components having been replaced already, so not all of them have the same age. If one looks at such systems at any arbitrary time, there is a probability $p_i \in (0, 1)$ that one finds component i working. Let X_i be the indicator that component i is working. We assume that $\{X_i\}$ are independent, Bernoulli(p_i) random variables defined on a common probability space $(\Omega, \mathfrak{F}, \mathbb{P})$. The *reliability* of the system is defined as the probability that the system works, that is, $\mathbb{P}(\phi(X_1,\ldots,X_5)=1)$, where the reliability function $\phi(\cdot)$ is the indicator that the system works. Suppose that $\theta = p_3$ is the variable of interest, and that we wish to estimate the sensitivity:

$$\frac{d}{d\theta}\mathbb{P}[\phi(X_1,\ldots,X_5)=1] = \frac{d}{d\theta}\mathbb{E}[\phi(X_1,\ldots,X_5)]. \tag{8.41}$$

(a) Show that the reliability ϕ of Figure 8.4 can be expressed as

$$\phi(X_1,\ldots,X_n) = X_1 X_5 \max(X_2, X_3 X_4).$$

(b) Consider the representation $X_3 = \mathbf{1}_{\{U \le \theta\}}$, and show that IPA is biased for (8.41).
(c) Calculate the SF estimator for (8.41) and show that it is unbiased for (8.41).
(d) Calculate the MVD estimator for (8.41) and show that it is unbiased for (8.41).
(e) Use the properties of conditional expectation, namely $\mathbb{E}[\phi] = \mathbb{E}[\mathbb{E}(\phi \mid \mathcal{G})]$ for any $\mathcal{G} \subset \mathfrak{F}$, and show that $\mathbb{E}[\phi \mid X_1, X_2, X_4, X_5]$ is now absolutely continuous in θ. Find the IPA for this new representation and compare with the MVD estimator found in (d). Conditioning can, in general, "smooth out" discontinuities of functions, allowing for an interchange between derivative and expectation, and the method is known as smoothed perturbation analysis.

Exercise 8.14. Revisit Example 8.5. Derive the MVD estimator for the Poisson counting process and establish unbiasedness of the estimator.

Exercise 8.15. Refer to Example 8.11. Your boss Mr. Casas has mentioned in a meeting that his wife is expecting a baby and they are moving out of Manhattan. He has therefore put his one bedroom Upper West Side condo (off of Central Park) out for sale, and he needs to sell before the baby is born. His real estate agent Karen has figured out a strategy: she recommends accepting the first offer that comes above the mean value for similar properties in the area. After the meeting, you go home and ponder this because it reminds you of a similar problem in your class of stochastic optimization: "Could I do better for my boss?" you think. "Hmm, may be I can have that promotion earlier." And so you start working on the problem. Karen has accepted your offer for lunch and has given you a lot of information. Offers come according to a renewal process at a rate $\lambda = 0.2$. Through exhaustive statistical analyses, her firm has estimated that such properties may expect offers of Q millions of dollars, where Q has a Weibull(1,2) distribution, that is,

$$\mathbb{P}(Q \le x) = 1 - e^{-x^2}.$$

"So there," she claims very pleased. "He should be expecting at least \$2 million for his condo." "Should he, now?" you think. You do a quick Google search and find out that the *median* of this distribution is only 0.8325; this distribution is skewed to the left. That is, *most* offers are well below the mean.

You remember that the success probability is $p_\theta = \mathbb{P}(Q > \theta) = e^{-\theta^2}$, for $\theta \ge 0$ and $p_\theta = 1$ otherwise, which is 0.0183 when $\theta = 2$ as Karen suggests. Let X_θ be the first success in consecutive offers, that is, the offer that is accepted. Mathematically: $X_\theta = \min\{n: Q_n \ge \theta\}$, then X_θ has a geometric distribution with parameter p_θ and mean $1/p_\theta \approx 55$ for $\theta = 2$. Because offers are expected to come every five days in the mean, doing a quick analysis you calculate your boss' expected time until selling at about 275 days, over nine months.

GRADIENT ESTIMATION, FINITE HORIZON

You get back home and call Mr. Casas to get some other information. He tells you that the property incurs an expense of $300 per day during the time it is on sale (maintenance, taxes, insurance, management costs, advertising, etc.) With this information, you set out to find your boss a better strategy, using SA for optimization. Make the assumption that for any $n \in \mathbb{N}$,

$$\mathbb{E}\left[\left(\sum_{i=1}^{n} T_i\right)^r\right] = O(n^r).$$

(a) Consider the optimization problem:

$$\min_{\theta} J(\theta) \stackrel{\text{def}}{=} \mathbb{E}\left[c \sum_{i=1}^{X_\theta} T_i - Q_{X_\theta}\right],$$

where $\{T_i\}$ is a sequence of iid random variables with mean $1/\lambda = 5$ days, and $c = 3 \times 10^{-4}$ (in million dollars) is the cost per day. For $n \in \mathbb{N}$, let $\{Q_1^{(n)}, \ldots Q_{X_\theta^{(n)}}^{(n)}\}$ be a sample for the Weibull distribution, stopped at $X_{\theta_n}^{(n)}$. Let $f(q)$ be the density of the offer amounts $\{Q_n\}$. Show that the algorithm

$$\theta_{n+1} = \theta_n + \epsilon \left(p'_{\theta_n} \left(\frac{(X_{\theta_n}^{(n)} - 1)}{1 - p_{\theta_n}} + \frac{1}{p_{\theta_n}} \right) c \sum_{i=1}^{X_{\theta_n}^{(n)}} T_i + \frac{\theta f(\theta) - p'_\theta Q_{X_\theta^{(n)}}^{(n)}}{p_\theta} \right) \quad (8.42)$$

converges to the optimal solution. Specifically, show that the interpolated processes $\vartheta^{\epsilon}(\cdot)$ converge in distribution as $\epsilon \to 0$ to the solution of the ODE:

$$\frac{d}{dt}\vartheta(t) = -\frac{\partial}{\partial \theta} J(\theta).$$

Specify which theorems you are using, and prove the corresponding assumptions. (You may assume that for all $n \in \mathbb{N}$, $\theta_n \geq 0.0001 > 0$ in your proof; Mr. Casas would not be selling his property for less than $100!).

(b) Let Y_θ be a random variable with distribution $\mathbb{P}(Y_\theta = k) = k \, p_\theta^2 \, (1 - p_\theta)^{k-1}$. Show that the following estimator

$$\hat{J}'_{MVD}(\theta) = 2\theta c \left(\sum_{i=1}^{Y_\theta+1} T_i - \sum_{i=1}^{X_\theta} T_i \right) - 2\theta^2 - 2\theta \, Q_{X_\theta^{(n)}}^{(n)} \quad (8.43)$$

is an MVD estimator for $J'(\theta)$, and argue that it is unbiased.

(c) Instead of of SF estimator in (8.42), consider using the stochastic approximation with the MVD estimator (8.43). Show that this algorithm also converges as $\epsilon \to 0$.

(d) Program the SA algorithms using the SF and the MVD methods. Use $T_i \sim U(2, 8)$ uniform random variables. Compare results and discuss the following:

- How much more can Mr. Casas expect to gain from your work?
- If Mr. Casas uses your estimated optimal threshold θ^*, what is the expected time to sell his property?
- How robust is our program so that you can apply it to other clients and other situations?
- If you were to ensure the client that the probability that the property sells within four months is at least 90%, how would you modify your algorithm to incorporate the constraint?

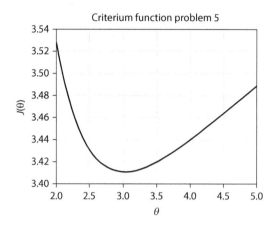

Figure 8.5. The criterium function of Exercise 8.17.

Exercise 8.16.

(a) Are the following functions Lipschitz continuous?
   (i) $y(\theta) = \theta e^{\theta}$ on $(1, 2)$.
   (ii) $y(\theta) = |\theta|$ on $(-1, 1)$.
   (iii) $y(\theta) = 1/\theta$ on $(0, 1)$.

(b) Are the following variables almost surely Lipschitz continuous?
   (i) $Y(\theta) = \theta^2 U$ for $\theta \in (1, 2)$ and U uniform on $(0, 1)$.
   (ii) $Y(\theta) = \frac{\theta^2}{U - 0.5}$ on $(1, 2)$ and U uniform on $(0, 1)$.

Exercise 8.17. Let $X_1(\theta)$ and $X_2(\theta)$ be independent and both with the Weibull$(\theta, 1)$ distribution. Define the performance function $h(x_1, x_2) = (x_1 + x_2)^2$, and the cost function $J(\theta) = \mathbb{E}[h(X_1(\theta), X_2(\theta))]$. Consider the optimization problem $\min_{\theta > 0} J(\theta)$. Figure 8.5 shows that there is a local minimum on $[2, 5]$.

(a) Describe the SA algorithm for solving the minimization problem. Specifically, give the recursive formula.
(b) Derive a single-run (or single sample) unbiased estimator $\psi(\theta)$ of $J'(\theta)$, which is based on the score function.
(c) Show the interchange conditions of the SF method in any $\theta \in [2, 5]$.

Exercise 8.18. Train A arrives at the central station at time $X(\theta)$ and it is next scheduled for departure at time s. The delay is $\max(0, X(\theta) - s) = (X(\theta) - s)_+$. The parameter θ governs the distribution of $X(\theta)$ and can in principle be adjusted (speed of train is one example; make and model, or driver's experience are other examples), perhaps satisfying some constraints. In order to carry out an optimization procedure it is first necessary to estimate the *sensitivity* or derivative of the expected delay with respect to θ. In other words, we wish to estimate

$$\frac{d}{d\theta} \mathbb{E}[h(X(\theta))]; \quad h(x) = \max(0, x - s).$$

For this exercise consider $X(\theta) \sim \text{Erlang}(3, 1/\theta)$, with density $f_\theta(x) = (x^2/2\theta^3) e^{-x/\theta}$ for $x > 0$.

(a) Calculate the score function for this distribution.
(b) Verify the conditions of Theorem 8.3. What family of functions $v(\cdot)$ can be used? Is h in the family?
(c) Give the simulation algorithm for generating n replications of $D^{SF}(\theta)$. From these n replications, give the sample average, the sample variance, and provide the formula for estimating a confidence interval at confidence level α.

Exercise 8.19. Consider $L(\theta) = \mathbb{E}[h(X(\theta))]$, where $h(X(\theta)) = \max(X(\theta) - s, 0)$, for some given $s > 0$, and where $X(\theta)$ has an Erlang$(3, 1/\theta)$ distribution for $\theta > 0$ with pdf

$$\frac{1}{2\theta^3} x^2 e^{-x/\theta}, x \geq 0.$$

(a) Derive the MVD estimator $D^{MVD}(\theta)$ of $L'(\theta)$.
   (*Hint:* it is a difference of two estimators involving Erlang-$(4, 1/\theta)$ and Erlang-$(3, 1/\theta)$ distributions.)
(b) Give the simulation algorithm for generating n replications of $D^{MVD}(\theta)$, including how you generate from the Erlang$(4, 1/\theta)$ and Erlang$(3, 1/\theta)$ distributions. Make sure to exploit CRN as much as possible. From these n replications, give the sample average, the sample variance, and provide the formula for estimating a confidence interval at confidence level α.
(c) Give the expression for a randomized MVD estimator $D^{MVDrand}(\theta)$ that involves a single estimator (instead of the difference in (b)). Show that $\mathbb{E}[D^{MVDrand}(\theta)] = \mathbb{E}[D^{MVD}(\theta)]$.

Exercise 8.20. Let h be globally Lipschitz continuous with Lipschitz constant L, and let $X(\theta)$ have finite second moment for all θ. Show that if θ is a scale parameter of the cdf of $X(\theta)$, then the variance of the IPA estimator for $\mathbb{E}[h(X(\theta))]$ is bounded by

$$\frac{L^2}{\theta^2} \text{Var}(X(\theta)).$$

Chapter Nine

Gradient Estimation, Markovian Dynamics

This chapter presents the development of gradient estimation algorithms for the case when the target field involves the gradient of a long-term, random horizon, or stationary average of the observed process. In parallel to Chapter 5, such gradient estimators can be used to drive stochastic approximations for optimization and learning.

9.1 THE INFINITE HORIZON PROBLEM

To consider the infinite horizon problem (see Definition 7.2) where the cost (or constraint) function is a long-term average of the form:

$$\lim_{N \to \infty} \frac{1}{N} \sum_{n=1}^{N} \mathbb{E}[L(\xi_n(\theta), \theta)], \tag{9.1}$$

where $L(x, \theta)$ represents some cost function for state x at θ. The long-run behavior captures *infinite horizon* models, where an asymptotic cost rate is the performance (or constraint) of interest. Following Definition 7.2, we assume throughout this chapter that the underlying process $\{\xi_n(\theta)\}$ is a Markov process, for $\theta \in \Theta$. If the Markov process is ergodic then the long-term average is also the stationary average (see Section C.1 in Appendix C). The running sample average for finite N is a *consistent* estimator of the infinite horizon and methods for estimating confidence intervals for the infinite horizon average are discussed in Appendix D.

In this section we extend the gradient estimation results for the finite horizon problem developed in Chapter 8, where we have already provided conditions for the unbiasedness of the finite horizon problem:

$$\frac{d}{d\theta} \frac{1}{N} \sum_{n=1}^{N} \mathbb{E}[L(\xi_n(\theta), \theta)] = \frac{1}{N} \sum_{n=1}^{N} \mathbb{E}[\phi(\xi_n(\theta), \theta)], N \geq 1, \tag{9.2}$$

where ϕ may represent an IPA, SF, or MVD estimator. It is rather natural to ask whether we can apply these results and take the limit as $N \to \infty$ for obtaining a consistent estimator for $dJ(\theta)/d\theta$. The answer is, "not so fast." One has to show that the following is true:

$$\lim_{N \to \infty} \frac{d}{d\theta} \mathbb{E}\left[\frac{1}{N} \sum_{n=1}^{N} L(\xi_n(\theta), \theta)\right] = \frac{d}{d\theta} \lim_{N \to \infty} \mathbb{E}\left[\frac{1}{N} \sum_{n=1}^{N} L(\xi_n(\theta), \theta)\right]. \tag{9.3}$$

To establish (9.3), we will impose the quite natural condition that the underlying processes $\{\xi_n(\theta) : n \geq 0\}$, for $\theta \in \Theta$, are ergodic Markov chains, so that it holds

that
$$\lim_{N\to\infty} \frac{1}{N} \sum_{n=1}^{N} \mathbb{E}[L(\xi_n(\theta), \theta)] = \lim_{N\to\infty} \frac{1}{N} \sum_{n=1}^{N} L(\xi_n(\theta), \theta) \quad \text{a.s.,}$$

and

$$\lim_{N\to\infty} \frac{d}{d\theta} \frac{1}{N} \sum_{n=1}^{N} \mathbb{E}[L(\xi_n(\theta), \theta)] = \lim_{N\to\infty} \frac{1}{N} \sum_{n=1}^{N} \mathbb{E}[\phi(\xi_n(\theta), \theta)]$$

exist. Although seemingly straightforward, proving (9.3) has been the subject of numerous studies. The first result was developed in [305] and published in [313]. We now state without proof a special case of Theorem 4 in [313] that is particularly useful for application of stochastic approximation for the dynamic model of Chapter 5. Elaborating on norm arguments in combination with an operator approach, a statement equivalent to that presented in the theorem below is established with explicit proof in Theorem 13.4.

Theorem 9.1. *Let $\Theta \subset \mathbb{R}$ be a compact set and for each $\theta \in \Theta$, let $\{\xi_n(\theta)\}$ be a positive Harris-recurrent ergodic Markov chain. Let $c(\xi_n(\theta), \theta)$ denote an instantaneous cost. Assume*

(i) *For each θ the process has a unique invariant measure $\pi_\theta(\cdot)$ and the set $\{\pi_\theta; \theta \in \Theta\}$ is tight. For each θ the cost function c is π_θ-integrable.*

(ii) *For all N, the finite step transition is differentiable: there is a function g such that*

$$\frac{d}{d\theta}\mathbb{E}[c(\xi_N(\theta), \theta) \mid \xi_0(\theta) = x] = \mathbb{E}[g(\xi_N(\theta), \theta) \mid \xi_0(\theta) = x].$$

(iii) *For each finite N, and there is a constant K independent of x such that for $\theta \in \Theta$*

$$-KN \leq \mathbb{E}\left[\sum_{n=1}^{N} g(\xi_n(\theta), \theta) \mid \xi_0(\theta) = x\right] \leq KN.$$

(iv) *For any π_θ-integrable function f the partial sums*

$$\sum_{n=1}^{N} \left(\mathbb{E}[f(\xi_n(\theta)) \mid \xi_0(\theta) = x] - \int \pi_\theta(dy) f(y) \right)$$

are uniformly bounded in (x, N) and converge point-wise as $N \to \infty$.

Then

$$\lim_{N\to\infty} \mathbb{E}[g(\xi_N(\theta), \theta) \mid \xi_0(\theta)] = \frac{d}{d\theta} \int c(y, \theta) \pi_\theta(dy) \quad \text{a.s.}$$

The fact that this result implies (9.3) follows using a Cesàro sum (see Section A.10 in Appendix A):

$$\lim_{N\to\infty} \frac{1}{N} \sum_{n=1}^{N} \mathbb{E}[g(\xi_n(\theta), \theta) \mid \xi_0(\theta)] = \lim_{N\to\infty} \mathbb{E}[g(\xi_N(\theta), \theta) \mid \xi_0(\theta)] \quad \text{w.p.1.}$$

Assumption (i) is on the behavior of the family of limiting measures as a parametrized by θ. As the following examples show, if $\xi_n(\theta)$ is a regenerative process, then π_θ is well defined, and tightness is typically satisfied when Θ is bounded. Note that $\xi_n(\theta)$ usually is given by the nominal process supplemented with additional information needed to be able to evaluate the gradient. Assumptions (ii) and (iii) are the same type of assumptions that we have used to develop the finite horizon gradient estimators for Markov chains, and we can elaborate on the results put forward in Chapter 8. Assumption (iv) is implied when the chain is geometrically ergodic; see Appendix C.

Remark 9.1. An alternative approach to establishing (9.3) can be developed provided that

$$f_N(\theta) = \frac{1}{N} \sum_{n=1}^{N} \mathbb{E}[L(\xi_n(\theta), \theta)]$$

is convex. Indeed, let us assume that $\{f_N(\theta)\}$ is a sequence of functions on Θ that converge point-wise for N to ∞ for all $\theta \in \Theta$ to some limiting function f. If $f_N(\theta)$ is convex on Θ, for every N, and f is differentiable at θ_0 with θ_0 an interior point of Θ, then

$$\lim_{N \to \infty} \frac{d^- f_N(\theta_0)}{d\theta} = \lim_{N \to \infty} \frac{d^+ f_N(\theta_0)}{d\theta} = \frac{df(\theta_0)}{d\theta},$$

where $d^\pm f_N(\theta_0)/d\theta$ denotes the one sided derivative of f_N at θ_0 and $df(\theta_0)/d\theta$ the derivative of f with respect to θ at θ_0. If, in addition, $\{f_N\}$ are differentiable at θ_0, then

$$\lim_{N \to \infty} \frac{df_N(\theta_0)}{d\theta} = \frac{df(\theta_0)}{d\theta}.$$

For a proof see Theorem 25.7 in [261]. Analysis of gradient estimation based on this convexity result has been developed in [126] and [165]. We refer to [114, 122, 123, 166] for details and examples.

We now present three examples, all of which consider a $GI/G/1$ queue with interarrival times $\{A_n\}$, service times $\{S_n(\theta)\}$ with scaling parameter θ, and first-come, first-served (FCFS) discipline. For this model the sojourn times satisfy the Lindley equation (8.6):

$$X_{n+1}(\theta) = S_{n+1}(\theta) + (X_n(\theta) - A_{n+1})_+.$$

We have already discussed that if $0 \le \theta < \mathbb{E}[A_1]$, then the queue is stable and the sequence of sojourn times $\{X_n(\theta)\}$ is a regenerative process. Let Θ be a compact set with $\sup_\Theta(\theta) \stackrel{\text{def}}{=} \bar{\theta} < \mathbb{E}[A_1]$. Then $\tau(\theta) = \min(n > 1 : A_n > X_{n-1})$ is finite w.p.1 for all $\theta \in \Theta$. The problem is now the estimation of the infinite horizon derivative:

$$\frac{d}{d\theta} \mathbb{E}\left[\lim_{N \to \infty} \frac{1}{N} \sum_{n=1}^{N} X_n(\theta) \right].$$

Example 9.1 presents the extension of the IPA estimator from the finite to the infinite horizon problem. Example 9.2 presents the case for the SF estimator, and finally Example 9.3 does the analysis for the MVD estimator.

Example 9.1. Equation (8.12) in Example 8.6 is an unbiased estimator of the finite horizon average sojourn time. We now show how Theorem 9.1 can be applied to the infinite horizon version of this estimator. The finite horizon derivative estimator in (8.12) is

$$\hat{G}^{(\text{IPA})} = \frac{d}{d\theta} \frac{1}{N} \sum_{n=1}^{N} X_n(\theta) = \frac{1}{N} \sum_{n=1}^{N} Z_n(\theta) = \frac{1}{N\theta} \sum_{n=1}^{N} \sum_{i=\alpha_n(\theta)}^{n} S_i(\theta),$$

where $\alpha_n(\theta)$ is the index of the first customer in the busy period where customer n belongs. To write the example in the notation of Theorem 9.1, we introduce the underlying process $\xi_n(\theta) = (X_n(\theta), Z_n(\theta))$ so that here $c(\xi_n(\theta), \theta) = X_n(\theta)$ and $g(\xi_n(\theta), \theta) = Z_n(\theta) = \sum_{i=\alpha_n(\theta)}^{n} S_i/\theta$. Notice that for this problem, the regeneration points of $X_n(\theta)$ are the same as the regeneration points of the derivative process $Z_n(\theta)$, where they both reset at zero.

Conditions (i), (ii), and (iv) are satisfied under the model assumption that $\theta \in \Theta$ is a scaling parameter of the service distribution because the process is positive Harris recurrent with geometric convergence rate to the stationary measure. Condition (iii) follows from the analysis of the algorithm for (8.12). The expected number of terms in the sum describing $Z_1(\theta)$ is bounded by

$$\mathbb{E}\left[\sum_{i=1}^{\tau(\theta)} S_i\right] = K(\theta),$$

where $K(\theta)$ is the expected value of the duration of a busy cycle (in time units). This length is finite w.p.1 for stable queues, so $\sup_\Theta(K(\theta)) < \infty$ and condition (iii) holds. Therefore, Theorem 9.1 implies that

$$\frac{d}{d\theta}\mathbb{E}\left[\lim_{N\to\infty}\frac{1}{N}\sum_{n=1}^{N} X_n(\theta)\right] = \lim_{N\to\infty}\mathbb{E}\left[\frac{1}{\theta}\sum_{n=1}^{N}\sum_{i=\alpha_n(\theta)}^{n} S_i(\theta)\right]. \quad (9.4)$$

For further readings on IPA results for the infinite horizon optimization problem for the queuing system we refer the reader to [57, 122, 166, 298, 335]. ✸✸✸

Example 9.2. In Example 8.13 we calculated the SF derivative estimator for the finite horizon average sojourn times in a GI/G/1 queue, exemplifying with exponential service times. From (8.26) we have for this problem

$$\hat{G}_1^{(\text{SF})} = \left(\frac{1}{N}\sum_{n=1}^{N} X_n(\theta)\right)\left(\sum_{n=1}^{N} \text{SF}(\theta, S_n(\theta))\right),$$

where $\text{SF}(\theta, S_n(\theta))$ is the score function associated with the service distribution. At first sight it is not apparent that condition (iii) is satisfied. However, it is possible to use a variance reduction method to rewrite the estimator resulting in a more efficient one (see Exercise 8.8):

$$\hat{G}_2^{(\text{SF})} = \frac{1}{N}\sum_{n=1}^{N} X_n(\theta)\,\text{SF}_n(\theta, S_1(\theta), \ldots, S_n(\theta)),$$

where the (partial) score function is given by $\text{SF}_n(\theta, s_1, \ldots, s_n) = \sum_{i=1}^{n} \text{SF}(\theta, s_i)$.

Let $\alpha_n(\theta)$ be the index of the customer that initiates the busy period where customer n belongs, as we did in the previous example. We now use the observation that if $i < \alpha_n(\theta)$ then customers i and n belong to different and separate busy periods, which implies that X_n is statistically independent of $S_i(\theta)$ for $i < \alpha_n$. Then using that $\mathbb{E}[\text{SF}(\theta, S_i(\theta))] = 0$, we obtain a further simplified expression for the SF:

$$\hat{G}_3^{(\text{SF})} = \frac{1}{N}\sum_{n=1}^{N} X_n(\theta) \sum_{i=\alpha_n(\theta)}^{n} \text{SF}(\theta, S_i(\theta)).$$

Call $Z_n = \sum_{i=\alpha(\theta)}^{n} \text{SF}(\theta, S_i(\theta))$. Using this notation we define the underlying process as $\xi_n(\theta) = (X_n(\theta), Z_n(\theta))$, so that $c(\xi_n(\theta), \theta) = X_n(\theta)$ and $g(\xi_n(\theta), \theta) = X_n(\theta) Z_n(\theta)$.

Condition (iii) now requires that $\mathbb{E}[X_n(\theta)\,Z_n(\theta)]$ be bounded. A common approach to prove this claim is the regenerative approach that uses the fact that both X_n and Z_n have the same regeneration points and $X_n(\theta)Z_n(\theta)$ is zero at the regeneration points. Under this condition, Theorem 9.1 implies that

$$\frac{d}{d\theta}\mathbb{E}\left[\lim_{N\to\infty}\frac{1}{N}\sum_{n=1}^{N}X_n(\theta)\right] = \lim_{N\to\infty}\mathbb{E}\left[\frac{1}{N}\sum_{n=1}^{N}X_n(\theta)\sum_{i=\alpha_n(\theta)}^{n}\mathrm{SF}(\theta,S_i(\theta))\right].$$

※※※

Example 9.3. For the finite horizon average sojourn time in a GI/G/1 queue Example 8.19 established the MVD derivative estimator, showing that

$$\hat{G}^{(\mathrm{MVD})} = \frac{c_\theta}{N}\sum_{i=1}^{N}\sum_{n=1}^{N}(X_n^+(\theta;i) - X_n^-(\theta;i)) = \frac{c_\theta}{N}\sum_{i=1}^{N}\sum_{n=i}^{(i+\tau_i(\theta))\wedge N}d_n(\theta;i)$$

is an unbiased estimator of the finite horizon derivative:

$$\mathbb{E}[\hat{G}^{(\mathrm{MVD})}] = \frac{d}{d\theta}\mathbb{E}\left[\frac{1}{N}\sum_{n=1}^{N}X_n(\theta)\right].$$

We recall here that (assuming that $X_i^+(\theta) > X_i^-(\theta)$) the difference processes (8.33) for general service distribution satisfy:

$$d_n(\theta;i) = \begin{cases} 0 & \text{if } n < i, \\ X_i^+(\theta;i) - X_i^-(\theta;i) & \text{if } n = i, \\ \max(0, d_{n-1}(\theta;i) - (A_n - X_{n-1}^-(\theta))_+) & \text{if } n > i. \end{cases}$$

If the initial difference is negative, the argument can be modified in a straightforward manner and details are omitted here. These processes are super-martingales with a.s. finite lifetime: $\tau_i(\theta) = \min(n\colon d_n(\theta;i)=0)$. Define the cumulative sum

$$Z_i(\theta) = \sum_{n=i}^{i+\tau_i(\theta)} d_n(\theta;i),$$

where $\tau(i)$ is the random lifetime of the i-th phantom difference process $\{d_n(\theta;i);n\geq i\}$. Because the difference processes are non increasing w.p.1 we can bound

$$Z_i(\theta) \leq \sum_{n=i}^{i+\tau_i(\theta)} d_i(\theta,i) = \tau_i(\theta)\,d_i(\theta;i),$$

with $d_i(\theta;i) = Y_i^+ - Y_i^-$. This random variable depends on the service distribution (see Table 7.1 and Table 8.1) and for most common distributions it is an integrable random variable, so its expectation is finite. The running time analysis of Algorithm 8.4 in Example 8.19 also shows that $\mathbb{E}[\tau_i(\theta)]$ is bounded for all $\theta\in\Theta$, so condition (iii) holds for the estimation when $\mathrm{Cov}(d_i(\theta;i),\tau_i(\theta)) < \infty$.

The problem with this definition of $Z_n(\theta)$ is that it "depends on the future," that is, it is not measurable w.r.t. the history of the process. We now provide an alternative formulation of the MVD estimator that is appropriate for programming the SA procedure of Chapter

5. Let $\mathfrak{F}_n = \sigma(X_k(\theta), X_k^+(\theta), X_k^-(\theta); k \leq n)$ be the history of the (augmented) process up to time n. Let $\mathcal{L}_n(\theta)$ be the set of "living phantoms" at time n. Specifically

$$\mathcal{L}_n(\theta) = \{j \leq n : j + \tau_j(\theta) > n\} \in \mathfrak{F}_n,$$

where the measurability statement follows from the fact that $d_n(\theta; j) > 0 \iff j + \tau_j(\theta) > n$, and this is an event in $\mathfrak{F}_n$. We now rewrite the double sum as

$$\hat{G}^{(\text{MVD})} = \frac{c_\theta}{N} \sum_{n=1}^N \sum_{i \in \mathcal{L}_n(\theta)} d_n(\theta; i).$$

Call now

$$Z_n(\theta) \stackrel{\text{def}}{=} \sum_{i \in \mathcal{L}_n(\theta)} d_n(\theta; i).$$

We can state the model in the terminology of Theorem 9.1. The underlying process for the derivative estimation is now defined by

$$\xi_n(\theta) = (X_n(\theta), \mathcal{L}_n(\theta), (X_n^\pm(\theta; i), i \in \mathcal{L}_n))$$

and it is a Markov chain adapted to the filtration $\{\mathfrak{F}_n\}$ of the augmented process. Note that this is exactly the information that we need to keep at each iteration in order to program the algorithm. Use $c(\xi_i(\theta), \theta) = X_i(\theta)$ and $g(\xi_i(\theta), \theta) = c_\theta Z_i(\theta)$. In order to verify the conditions of Theorem 9.1 one uses regeneration. For example, when the inter arrivals have an unbounded support, then there is a positive probability that all living phantoms die at the next step (when the next inter arrival time is very large). Conditions (i), (ii), and (iv) follow from the ergodicity of the original queue process, together with the regeneration of $Z_n(\theta)$. Only condition (iii) needs to be verified. In Example 13.10 in Chapter 13, we present an alternative approach to this result.

Under this condition, Theorem 9.1 implies that

$$\frac{d}{d\theta}\mathbb{E}\left[\lim_{N\to\infty} \frac{1}{N} \sum_{n=1}^N X_n(\theta)\right] = \lim_{N\to\infty} \mathbb{E}\left[\frac{1}{\theta N} \sum_{i=1}^N \sum_{n=1}^N (X_n^+(\theta; i) - X_n^-(\theta; i))\right]$$

$$= \lim_{N\to\infty} \mathbb{E}\left[\frac{1}{N} \sum_{n=1}^N \left(\frac{1}{\theta} \sum_{i \in \mathcal{L}_n(\theta)} d_n(\theta; i)\right)\right].$$

※※※

Remark 9.2. Gradient estimation for the infinite horizon can be implemented into stochastic approximations when the problem follows the dynamic model of Chapter 5, see Definition 5.1. To better understand the delicacies of the stationary setting, we discuss in the following a problem for which the target vector field is the derivative of the stationary sojourn time in a G/G/1 queue.

Let $X_n(\theta)$ be the n-th sojourn time in a G/G/1 queue (for notation and deduction we refer to Example 8.6), then

$$\phi(\xi_n(\theta), \theta) = Z_n(\theta) = \begin{cases} \frac{dS_n(\theta)}{d\theta} & \text{if } X_{n-1}(\theta) \leq A_n, \\ \frac{dS_n(\theta)}{d\theta} + Z_{n-1}(\theta) & \text{if } X_{n-1}(\theta) > A_n, \end{cases} \quad (9.5)$$

with $\xi_n(\theta) = (A_n, S_n(\theta), X_n(\theta), Z_n(\theta))$. In words, the underlying process $\xi_n(\theta)$ comprises the n-th inter arrival, service and sojourn times as well as the value of the auxiliary derivative process. Denoting the stationary distribution (existence assumed) of this Markov process by $\hat{\pi}_\theta$, (5.6) requires that

$$G(\theta) = \int g(\xi,\theta)\hat{\pi}_\theta(d\xi) = \int \left(\int \phi(x,\theta) P_\theta(\xi, dx) \right) \hat{\pi}_\theta(d\xi)$$
$$= \int \mathbb{E}[\phi(X,\theta)|\xi]\hat{\pi}_\theta(d\xi), \tag{9.6}$$

with X distributed according to $P_\theta(\xi,\cdot)$. Note that (9.6) cannot yet be made operational for optimization (via Theorem 5.3 or Theorem 5.4) as we do not know whether

(i) the expression on the righthand side of (9.6) yields the correct target field, that is, we still have to show that

$$G(\theta) = \frac{d}{d\theta} \mathbb{E}_{\pi_\theta}[X(\theta)],$$

where $X(\theta)$ is a sample of the stationary wait following distribution π_θ, and that
(ii) the underlying process $\xi_n(\theta_n)$ is ergodic.

The examples presented in Chapter 5 were less challenging, because the target field was a simple loss function and could be directly sampled from the nominal process itself (no auxiliary process was needed), which made (i) trivial. The main work in Chapter 5 was therefore showing (ii), that is, that the long-run average of the sojourn times converges uniformly on some θ interval toward their stationary distribution, which is required for (5.6). Example 9.1 shows how (i) and (ii) can be established by means of Theorem 9.1 for the case of θ being a scale parameter of distribution of the service times. Specifically, (9.4) shows that (i) holds for

$$\phi(\xi_n(\theta),\theta) = Z_n(\theta),$$

where $Z_n(\theta)$ is obtained via $X_{n-1}(\theta_{n-1})$ and $Z_{n-1}(\theta_{n-1})$ (available thought $\xi_{n-1}(\theta_{n-1})$) following recursion (9.5), and thus (9.6) holds.

It is worth noting that $\phi(\cdot,\theta)$ in (9.5) is noticeable the IPA derivative of a single sojourn time conditioned on the previous sojourn time together with the derivative process being stationary value. It is important to note here that even though sampling of ξ is practically impossible, if follows from (9.4) that we may work with the finite-horizon IPA estimator $\hat{G}^{(\text{IPA})}$ instead, for which we have neat evaluation formulas. But doesn't (9.4) suggest that we have to take the limit in N for $\hat{G}^{(\text{IPA})}$ to become an unbiased estimator for $d\mathbb{E}_{\pi_\theta}[X(\theta)]/d\theta$? It is one of the marvels of stochastic approximation that the answer to this question is "no!" (just as is the case in Chapter 5). To see this, revisit, say, Theorem 5.4). Condition (a3) therein only requires that the gradient estimator is asymptotically unbiased. Thus, it is sufficient to use $\hat{G}^{(\text{IPA})}$ for any N as feedback Y_n in a stochastic approximation. To summarize, while $\hat{G}^{(\text{IPA})}$ is only asymptotically unbiased for $d\mathbb{E}_{\pi_\theta}[X(\theta)]/d\theta$, SA still traces the correct target ODE (namely that given by $d\mathbb{E}_{\pi_\theta}[X(\theta)]/d\theta$, independent of the choice of N. The choice of N is thus driven by considerations of numerical stability and variance reduction only.

A guided application on infinite horizon gradient estimation in stochastic approximation is provided in Exercise 9.3, where detailed steps to implement the optimization are given. Moreover, Section 10.1 details how Theorem 9.1 can be used to implement gradient estimation into a stochastic approximation for optimization and learning for the infinite horizon problem.

9.2 THE RANDOM HORIZON PROBLEM

In this section we present gradient estimators for the performance over a random horizon, see Definition 7.3. In Chapter 8 we already provided conditions such that for the finite horizon problem

$$\frac{d}{d\theta}\mathbb{E}[L(\xi_1(\theta),\ldots,\xi_N(\theta),\theta)] = \mathbb{E}[\phi(\xi_1(\theta),\ldots,\xi_N(\theta),\theta)], N \geq 1,$$

where ϕ may represent an IPA, SF, or MVD estimator.

It is rather natural to ask whether we can apply these results and and simply replace N by a random stopping time τ_θ. This section addresses the answer to this question.

9.2.1 IPA Results

The IPA methodology has been well developed for particular stopping times in the setting of discrete event systems, and we refer to Chapter 12 for details on the IPA theory for discrete event systems. Generally speaking, however, the random horizon problems are a big challenge to IPA due to the discrete nature of the (often integer-valued) time horizon which causes non-differentiability of the sample function for a random horizon problem.

It is worth noting that an approach based on joint regeneration of sample derivatives and the basic process does show that IPA can successfully work in connection with random horizons given by cycles of a regenerative process. A very elegant way of establishing IPA estimators via regeneration is by means of almost sure absolute continuity (see Section 8.1.4 for finite horizon results) and has been first used by Glasserman in [120], and further developed in [117, 118]. For applications to waiting times in multi server queues, tandem queues and acyclic fork-join networks, we refer to [117]. These results, however, do not imply the IPA is applicable for gradient estimation over random horizon costs, see [122, 142, 298] for more details.

9.2.2 The Distributional Approach

For $\theta \in \Theta$, let $\{X_i(\theta) : i \in \mathbb{N}\}$ be a family stochastic processes defined on a common probability space, such that $X_i(\theta)$ is a measurable mapping on some measurable space $(S, \mathcal{S})$. Let $\tau_\alpha(\theta)$ denote the first entrance time of $X_n(\theta)$ into the set α, i.e.,

$$\tau_\alpha(\theta) = \inf\{n > 0 : X_n(\theta) \in \alpha\},$$

where we set $\tau_\alpha(\theta) = \infty$ if the set on the above right-hand side is empty. Note that the event $\{\tau_\alpha(\theta) = n\}$ is equivalent to the event $\{X_i(\theta) \notin \alpha, 1 \leq i < n; X_n(\theta) \in \alpha\}$; so that $\tau_\alpha(\theta)$ is a stopping time adapted to the natural filtration of $\{X_n(\theta) : n \geq 1\}$ (see Definition B.9) in Appendix B. Given a measurable function $L_n : S^n \to \mathbb{R}$, for $n \in \mathbb{N}$, the random horizon gradient estimation problem is to find an unbiased estimator for

$$\frac{d}{d\theta}\mathbb{E}[L_{\tau_{\theta_\alpha}}(X_1(\theta),\ldots,X_{\tau_{\theta_\alpha}}(\theta))],$$

provided the expected values exists.

The SF and MVD formulas for the static gradient estimation problem are extended to the random horizon problem in the following way. Denote by $f_\theta(x_1,\ldots,x_n)$ the joint density of $(X_1(\theta),\ldots,X_n(\theta))$, and for a fixed $\theta_0 \in \Theta$ introduce the likelihood ratio, or "change of measure" mapping

$$\ell_n(\theta_0, \theta; x_1, \ldots, x_n) = \frac{f_\theta(x_1, \ldots, x_n)}{f_{\theta_0}(x_1, \ldots, x_n)}, \quad (9.7)$$

provided that $f_{\theta_0}(x_1, \ldots, x_n) \neq 0$ and set it to zero otherwise, see Theorem B.2 in Appendix B. If $f_{\theta_0}(x_1, \ldots, x_n) \neq 0$ everywhere, or, more generally, if the distribution of $(X_1(\theta), \ldots, X_n(\theta))$ is absolutely continuous with respect to the distribution of $(X_1(\theta_0), \ldots, X_n(\theta_0))$ for fixed θ_0, then

$$\mathbb{E}[L_n(X_1(\theta), \ldots, X_n(\theta))] = \int_{\mathbb{R}^n} L_n(x_1, \ldots, x_n) f_\theta(x_1, \ldots, x_n) dx_1 \ldots dx_n$$

$$= \int_{\mathbb{R}^n} L_n(x_1, \ldots, x_n) \ell_n(\theta_0, \theta; x_1, \ldots, x_n)$$

$$f_{\theta_0}(x_1, \ldots, x_n) dx_1 \ldots dx_n$$

$$= \mathbb{E}[L_n(X_1(\theta_0), \ldots, X_n(\theta_0)) \ell_n(\theta_0, \theta; X_1(\theta_0), \ldots, X_n(\theta_0))]. \quad (9.8)$$

This expression can be directly translated to a random horizon experiment. This fact allows to apply the change of measure to the event $\{\tau_\alpha(\theta) = n\}$. We obtain the following:

$$\mathbb{E}[L_{\tau_\alpha(\theta)}(X_1(\theta), \ldots, X_{\tau_\alpha(\theta)}(\theta))]$$

$$= \sum_{n=1}^{\infty} \mathbb{E}[L_n(X_1(\theta), \ldots, X_n(\theta)) 1_{\tau_\alpha(\theta)=n}]$$

$$= \sum_{n=1}^{\infty} \mathbb{E}[L_n(X_1(\theta_0), \ldots, X_n(\theta_0)) 1_{\tau_\alpha(\theta_0)=n} \ell_n(\theta_0, \theta; X_1(\theta_0), \ldots, X_n(\theta_0))]$$

$$= \mathbb{E}[L_{\tau_\alpha(\theta_0)}(X_1(\theta_0), \ldots, X_{\tau_\alpha(\theta_0)}(\theta_0)) \ell_{\tau_\alpha(\theta_0)}(\theta_0, \theta, X_1(\theta_0), \ldots, X_{\tau_\alpha(\theta_0)}(\theta_0))], \quad (9.9)$$

where the first and the last equality follow from the monotone convergence theorem where one applies the argument first to the positive part and then to the negative part of L_n and $L_n \ell_n$, respectively. Taking derivative yields

$$\frac{d}{d\theta} \mathbb{E}[L_{\tau_\alpha(\theta)}(X_1(\theta), \ldots, X_{\tau_\alpha(\theta)}(\theta))]$$

$$= \mathbb{E}\left[L_{\tau_\alpha(\theta_0)}(X_1(\theta_0), \ldots, X_{\tau_\alpha(\theta_0)}(\theta_0)) \right.$$

$$\left. \frac{d}{d\theta} \ell_{\tau_\alpha(\theta_0)}(\theta_0, \theta; X_1(\theta_0), \ldots, X_{\tau_\alpha(\theta_0)}(\theta_0)) \right], \quad (9.10)$$

provided that interchanging expectation and differentiation is justified and that $\ell_n(\theta_0, \theta; x_1, \ldots, x_n)$ is a differentiable mapping of θ. The derivative of the likelihood ratio ℓ is noticeably the score function for the entire path $X_1(\theta_0), \ldots, X_{\tau_\alpha(\theta_0)}(\theta_0)$. We summarize in the following theorem, which was first established in [155].

Theorem 9.2. *For $\theta \in \Theta$, let $\{X_i(\theta) : i \in \mathbb{N}\}$ be a family stochastic processes and define*

$$\tau_\alpha(\theta) = \inf\{n > 0 : X_n(\theta) \in \alpha\},$$

Let Θ be an open set. Assume that for any integer n the likelihood ratio in (9.7) $\ell_n(\theta_0, \theta; x_1, \ldots, x_n)$ is a differentiable mapping of θ and that (9.8) holds for all $n \geq 1$.

Consider any any measurable mapping $L_n: S^n \to \mathbb{R}$. *If*

$$\frac{d}{d\theta}\mathbb{E}[L_n(X_1(\theta),\ldots,X_n(\theta))]$$
$$= \mathbb{E}\left[L_n(X_1(\theta_0),\ldots,X_n(\theta_0))\frac{d}{d\theta}\ell_n(\theta_0,\theta;X_1(\theta_0),\ldots,X_n(\theta_0))\right], n \geq 1,$$

then

$$\frac{d}{d\theta}\mathbb{E}[L_{\tau_\alpha(\theta)}(X_1(\theta),\ldots,X_{\tau_\alpha(\theta)}(\theta))]$$
$$= \mathbb{E}\left[L_{\tau_\alpha(\theta)}(X_1(\theta),\ldots,X_{\tau_\alpha(\theta)}(\theta))\frac{d}{d\theta}\ell_{\tau_\alpha(\theta)}(\theta,\theta;X_1(\theta),\ldots,X_{\tau_\alpha(\theta)}(\theta))\right].$$

Remark 9.3. With the setting in Theorem 9.2, we may use

$$L_n(X_1(\theta),\ldots,X_n(\theta))\frac{d}{d\theta}\ell_n(\theta,\theta;X_1(\theta),\ldots,X_n(\theta))\mathbf{1}_{\{n \leq \tau_\alpha(\theta)\}}$$

as biased gradient estimator that allows for controlling the numerical budget for evaluating the estimator. By construction, the bias tends to zero as $n \to \infty$.

The derivative expression put forward in Theorem 9.2 takes a nice form for (general state-space) Markov chains. To see this, let $\{X_n(\theta)\}$ be a Markov process with transition density $h_\theta(x_1, x_2)$. Then the density of realization $x_1, \ldots, x_n$ is given by a product form:

$$f_\theta(x_1,\ldots,x_n) = \prod_{k=1}^{n-1} h_\theta(x_k, x_{k+1}).$$

Assuming that $h_\theta(x_1, x_2)$ is differentiable with respect to θ, it follows that

$$\frac{d}{d\theta}\bigg|_{\theta=\theta_0}\ell_n(\theta_0,\theta;x_1,\ldots,x_n) = \frac{d}{d\theta}\bigg|_{\theta=\theta_0}\frac{f_\theta(x_1,\ldots,x_n)}{f_{\theta_0}(x_1,\ldots,x_n)}$$
$$= \sum_{k=1}^{n-1}\frac{\frac{d}{d\theta}\big|_{\theta=\theta_0}h_\theta(x_k,x_{k+1})}{h_{\theta_0}(x_k,x_{k+1})}$$
$$= \sum_{k=1}^{n-1}\mathsf{SF}(\theta_0,x_k,x_{k+1}), \tag{9.11}$$

where

$$\mathsf{SF}(\theta,i,j) = \frac{\frac{d}{d\theta}h_\theta(i,j)}{h_\theta(i,j)}, \quad i,j \geq 0,$$

is the score function of the one-step transition probabilities. Provided that the conditions in Theorem 9.2 are satisfied, one obtains the following versions of the SF estimator

$$\frac{d}{d\theta}\mathbb{E}\left[\sum_{n=1}^{\tau_\alpha(\theta)}X_n(\theta)\right] = \mathbb{E}\left[\left(\sum_{n=1}^{\tau_\alpha(\theta)}X_n(\theta)\right)\sum_{k=1}^{\tau_\alpha(\theta)-1}\mathsf{SF}(\theta,X_k(\theta),X_{k+1}(\theta))\right]$$
$$= \mathbb{E}\left[\sum_{n=1}^{\tau_\alpha(\theta)}X_n(\theta)\sum_{k=1}^{n-1}\mathsf{SF}(\theta,X_k(\theta),X_{k+1}(\theta))\right],$$

where the last equality follows from (9.9).

We illustrate the application of the score function with an example related to the queue-length process of the M/M/1 queue. For more examples from varying application domains, we refer to [154].

Example 9.4. Consider the embedded queue-length process $\{X_n(\theta)\}$ of the M/M/1 queue, with mean inter arrival time $1/\lambda$ and mean service time $1/\theta$, where we assume $\lambda < \theta$ for stability. Note that $\{X_n(\theta)\}$ is a random walk on $\mathbb{N}$. With the notation introduced in Section 9.3.1 we have for $i > 0$

$$\mathsf{SF}(\theta, i, i+1) = -(\lambda + \theta), \quad \mathsf{SF}(\theta, i, i-1) = \frac{\lambda}{\theta}(\lambda + \theta), \tag{9.12}$$

and zero otherwise. Suppose that we are interested in the sensitivity of the first return time to the empty state with respect to θ. Hence, we start the queue empty, i.e., $X_1(\theta) = 0$, and let $\alpha = \{0\}$. In order to return to the empty state, the queue has to have equally many arrivals as departures. Hence, if the queue returns at $\tau_\alpha(\theta)$ back to 0, any such path has in total $(\tau_\alpha(\theta) - 1)/2$ arrivals and $(\tau_\alpha(\theta) - 1)/2$ departures. Together with (9.11) and (9.12), this gives

$$\left.\frac{d}{d\theta}\right|_{\theta=\theta_0} \ell_{\tau_\alpha(\theta_0)}(\theta_0, \theta; X_1(\theta_0), \ldots, X_{\tau_\alpha(\theta_0)}(\theta_0)) = \sum_{k=1}^{\tau_\alpha(\theta_0)-1} \mathsf{SF}(\theta_0, X_k, X_{k+1})$$

$$= \frac{\tau_\alpha(\theta_0) - 1}{2}\left(1 - \frac{\lambda}{\theta_0}\right)(\lambda + \theta_0).$$

By Theorem 9.2, the sensitivity of the expected mean return time at $\theta_0 = \theta$ is given by

$$\frac{d}{d\theta}\mathbb{E}[\tau_\alpha(\theta)] = \mathbb{E}\left[\tau_\alpha(\theta) \sum_{k=1}^{\tau_\alpha(\theta)-1} \mathsf{SF}(\theta, X_k, X_{k+1})\right]$$

$$= \frac{1}{2\theta}\left(\theta^2 - \lambda^2\right)\mathbb{E}\left[\tau_\alpha(\theta)(\tau_\alpha(\theta) - 1)\right].$$

✳✳✳

If $\{X_n(\theta)\}$ is a regenerative process and α a set such $\{X_n(\theta)\}$ regenerates upon entering α, then Theorem 9.2 can be used for applying the SF method to the long-run problem, see Section 7.3.2. The above analysis requires α to be independent of θ. For general state-space Markov chains it is also of interest to let α depend on θ as well. For example, in a queueing model, if $\alpha = (\theta, \infty)$, then $\tau_{(\theta, \infty)}$ is the number of transitions until the sojourn time of a customer exceeds θ, which is an important criterium in quality control. The analysis of this case is significantly more complex and we refer to [152, 239] for details.

Remark 9.4. Theorem 9.2 requires that the score function be computed up to $\tau(\theta)$. Note that this is also the case when h only depends on $X_1(\theta)$ up to $X_k(\theta)$ for $k < \tau(\theta)$.

Example 9.5. Drones are to be deployed in an emergency situation for a rescue mission. Each drone has a limited battery capacity, so the energy is mostly used for flying and collecting images, sound, and for other sensors (temperature, etc). These actions create a stream of "tasks" that must be performed. Specifically, speech and image processing, plus data analysis of the collected information consume significant amounts of energy that the drones cannot provide. Thus a cloud server is used to perform the tasks that are generated by the drones' activities. However, the server must rotate to address the accumulated tasks of the drones one at a time, according to some polling policy.

For a single drone, this means that tasks are produced according to a counting process $\{N_\theta(t)\}$ at rate $\theta > 0$. At a random time S the server will empty the task queue to perform the tasks locally. The accumulation of tasks at the drone can be described as a bulk server queue. Each task (e.g., taking a video, or a single image) requires a random amount of work (energy) from the server, denoted W_i. Additionally, tasks have deadlines: there is a penalty for the tasks that cannot be completed δ seconds after being collected.

Let $\{T_i(\theta)\}$ be the interarrival times of the counting process $N_\theta(\cdot)$ with corresponding arrival times $A_i(\theta) = \sum_{k=1}^{i} T_i(\theta)$. As a simplified problem, consider finding the optimal rate θ: the profit is modeled as a non decreasing function $P(\theta)$ (usually of a logarithmic form, following economic theory) and the cost of computation is $C(\theta)$, which is also increasing. In addition, impose the constraint that the probability of not meeting the deadline is bounded by a tolerance α. This corresponds to minimizing $J(\theta)$ subject to $g(\theta) \leq 0$ on $\theta \in (0, \infty)$, where

$$J(\theta) = C(\theta) - \kappa P(\theta); \quad C(\theta) \stackrel{\text{def}}{=} \mathbb{E}\left[\sum_{i=1}^{N_\theta(S)} W_i\right]$$

$$g(\theta) \stackrel{\text{def}}{=} \mathbb{E}\left[\sum_{i=1}^{N_\theta(S)} \mathbf{1}_{\{W_i + S - A_i(\theta) > \delta\}}\right] - \alpha.$$

For this problem, it is possible to extend the finite horizon score function to a random horizon one, the details are left to the reader in Exercise 9.2. With this formulation, it is possible to estimate both the derivative of $J(\theta)$ as well as that of the constraint function $g(\theta)$. Algorithms such as the penalty function (Algorithm 1.2) can then be implemented to solve the problem. ✯✯✯

Score function estimators can be transformed into MVD estimators as explained in Section 8.4.1. Indeed, adapting the line of argument put forward in (9.11) for MVD gives

$$\frac{d}{d\theta}\ell_n(\theta_0, \theta; x_1, \ldots, x_n)$$

$$= c_\theta \left(\left(\frac{\frac{d}{d\theta} f_\theta^+(x_1; x_2)}{f_{\theta_0}(x_1; x_{i2})} - \frac{\frac{d}{d\theta} f_\theta^-(x_1; x_2)}{f_{\theta_0}(x_1; x_2)} \right) \ell_{n-1}(\theta_0, \theta; x_2, \ldots, x_n) \right.$$

$$+ \sum_{i=2}^{n-2} \ell_i(\theta_0, \theta; x_1, \ldots, x_i) \left(\frac{\frac{d}{d\theta} f_\theta^+(x_i; x_{i+1})}{f_{\theta_0}(x_i; x_{i+1})} - \frac{\frac{d}{d\theta} f_\theta^-(x_i; x_{i+1})}{f_{\theta_0}(x_i; x_{i+1})} \right)$$

$$\ell_{n-i}(\theta_0, \theta; x_{i+1}, \ldots, x_n)$$

$$\left. + \ell_{n-1}(\theta_0, \theta; x_1, \ldots, x_{n-1}) \left(\frac{\frac{d}{d\theta} f_\theta^+(x_{n-1}; x_n)}{f_{\theta_0}(x_{n-1}; x_n)} - \frac{\frac{d}{d\theta} f_\theta^-(x_{n-1}; x_n)}{f_{\theta_0}(x_{n-1}; x_n)} \right) \right).$$

To obtain the expression for the derivative estimator, we replace the score function in (9.9) with an MVD split. The i-th phantom processes are defined as usual: they correspond to the nominal process up to the transition between steps $i-1$ and i: this transition is sampled from $f_\theta^\pm$, respectively. Given the i-th phantom processes $X_k^\pm(\theta, i)$, we define the corresponding stopping times:

$$\tau^\pm(\theta, i) = \min(n: X_n^\pm(\theta, i) \in \alpha).$$

When exchanging the order of the outer sum over the horizon n with the expectation, we have to note that the phantoms run until they hit the set α; for details and a full proof, we refer to the taboo kernel approach in Section 13.2. The resulting MVD version of (9.10) becomes

$$\frac{d}{d\theta}\mathbb{E}[L_{\tau_\alpha(\theta)}(X_1(\theta),\ldots,X_{\tau_\alpha(\theta)}(\theta))]$$

$$= c_\theta \mathbb{E}\left[\sum_{i=1}^{\tau_\alpha(\theta)} \left(L_{\tau_\alpha^+(\theta,i)}(X_1^+(\theta,i),\ldots,X_{\tau_\alpha^+(\theta,i)}^+(\theta,i))\right.\right.$$

$$\left.\left. - L_{\tau_\alpha^-(\theta,i)}(X_1^-(\theta,i),\ldots,X_{\tau_\alpha^-(\theta,i)}^-(\theta,i))\right)\right].$$

Example 9.6. Estimating derivatives of ruin probabilities first appeared in [308, 316] and was later developed in [154] using the MVD estimator. We present here a simplified version of the problem. Following the canonical model in risk theory (see [106]) an insurance company has a constant rate of income of c euros per unit time coming from the premium payments of its clients. At random times $\{A_k\}$ claims are made and the corresponding repayments are honored to the clients. Let $N(\cdot)$ be the counting process of claim's repayments. Then the surplus process satires:

$$U(t) = u + ct - \sum_{k=1}^{N(t)} Y_k, \qquad (9.13)$$

where u is the initial endowment, and successive claims are paid the amounts $\{Y_k\}$. The *ruin probability* is defined as follows. Let

$$\eta = \min(n : U(A_n) < 0)$$

denote the index of the first claim that "ruins" the company: at this moment the payment in question needs more money than the current surplus. Correspondingly we call $\tau = A_\eta$ the time of ruin. Then the ruin probability is

$$\phi(\lambda) \stackrel{\text{def}}{=} \mathbb{P}(\tau < \infty).$$

The goal is to estimate the sensitivity of ϕ with respect to the incoming claims rate λ, that is, $d\phi/d\lambda$. The canonical model in risk theory assumes that $N(\cdot)$ is a homogeneous Poisson process with rate λ, and $\{Y_k\}$ are assumed to be iid with a general distribution with density g (sometimes assumed to have a finite moment generating function). Much effort has been dedicated to calculations of ϕ both with analytical formulas as well as with Monte Carlo simulation (e.g., [10, 12, 55, 92]). It is known that if $c < \lambda \mathbb{E}[Y_k]$ then $\phi(\lambda) = 1$, which leads to the fact that insurance companies adjust their premiums to satisfy $c > \lambda \mathbb{E}[Y_k]$. But, under this realistic assumption, simulation is notoriously difficult: if $\phi(\lambda) < 1$ then when simulating the process $U(\cdot)$ there must be trajectories where ruin will never occur. Thus, it is not straightforward to know when to stop a simulation and "declare" that ruin did not occur for that particular trajectory. In simulation theory, this is is an example of *rare event estimation*, for which the technique of change of measure, called *important sampling*, is the most adequate model [264].

The idea is to simulate another surplus process where $N(\cdot)$ has a different rate $\tilde\lambda$ and where the distribution of Y_k has density $\tilde g$ instead of the actual density. Then, application of the Radon-Nikodym theorem (Theorem B.2) allows to calculate expectations with respect to the original measure, using observations from the simulated process. The likelihood ratio is

$$\ell(t_1, y_1; \ldots; t_\nu, y_\nu) = \prod_{k=1}^{\nu} \left(\frac{\lambda e^{-\lambda t_k}}{\tilde{\lambda} e^{-\tilde{\lambda} t_k}} \right) \prod_{k=1}^{\nu} \frac{g(y_k)}{\tilde{g}(y_k)}.$$

For a given constant R, use the exponential tilting [264] known in risk theory as the Lundberg transformation:

$$\tilde{g}(y) = \frac{e^{Ry} g(y)}{M_Y(R)},$$

where M_Y is the moment generating function of g. The likelihood ratio is then

$$\ell(t_1, y_1; \ldots; t_\nu, y_\nu) = K(\lambda)^\nu \exp\left(\sum_{k=1}^{\nu} \left((\lambda - \tilde{\lambda}) t_k - R y_k \right) \right); \quad K(\lambda) \stackrel{\text{def}}{=} \frac{\lambda M_Y(R)}{\tilde{\lambda}}.$$

Because $\phi(\lambda) = \mathbb{E}[\mathbf{1}_{\{\tau < \infty\}}]$, if ruin is certain under the new of measure, then estimation of this probability is done using Theorem B.2:

$$\phi(\lambda) = \tilde{\mathbb{E}}[\ell(T_1, Y_1; \ldots; T_\nu, Y_\nu)],$$

where $\{T_i\}$ follow a Poisson process with rate $\tilde{\lambda}$ and $\{Y_k\}$ are iid with density $\tilde{g}$. Finally, using $\tilde{\lambda} = \lambda + cR$ for the change of measure, the likelihood simplifies to

$$\ell(T_1, Y_1; \ldots; T_\nu, Y_\nu) = K(\lambda)^\nu \exp(R(c\tau - S(\tau))); \quad S(\tau) \stackrel{\text{def}}{=} \sum_{i=1}^{\nu} Y_i,$$

where we use the fact that $\tau = A_\nu = \sum_{k=1}^{\nu} T_k$ is the epoch of ruin. In order to estimate the derivative we first use the chain rule of differentiation, noting that $K'(\lambda) = K(\lambda)(1/\lambda - 1/\tilde{\lambda})$, to obtain

$$\frac{d\phi(\lambda)}{d\lambda} = \mathbb{E}\left[\nu K(\lambda)^\nu \left(\frac{1}{\lambda} - \frac{1}{\tilde{\lambda}} \right) e^{R(c\tau - S(\tau))} \right] + \mathbb{E}\left[K(\lambda)^\nu \frac{d}{d\tilde{\lambda}} \mathbb{E}[e^{R(c\tau - S(\tau))} \mid \nu] \right].$$

The MVD estimator is now used for the second term. Because the inter arrival times have exponential distribution with intensity parameter $\tilde{\lambda}$, using Table 7.1 the MVD will split at each step i into a plus process that has the same distribution as the original "nominal" process, and a minus process where the i-th interarrival time has a Gamma$(2, \tilde{\lambda})$ distribution. For each $i = 1, \ldots, \nu$ define the sequences

$$A_k^-(i) = \begin{cases} A_k, & \text{if } k < i, \\ A_i + X_i; \quad X_i \sim \text{Exp}(\lambda), & \text{if } k = 1, \\ A_{k-1}^-(i) + T_k, & \text{if } k > i, \end{cases}$$

that will drive the corresponding phantom surplus processes $U^-(i; t)$ with corresponding ruin times

$$\nu^-(i) = \min(n: u + c A_n^-(i) - S_n < 0); \quad \tau^-(i) = \inf(t: U^+(t) < 0).$$

Using this approach, we obtain the MVD estimator

$$\nu K(\lambda)^\nu \left(\frac{1}{\lambda} \right) e^{R(c\tau - S(\tau))} - \frac{K(\lambda)^\nu}{\tilde{\lambda}} \sum_{i=1}^{\nu} e^{R(c\tau^-(i) - S(\tau^-(i)))}. \qquad (9.14)$$

Finally, we point out that when R is chosen as the adjustment coefficient, then $K(\lambda) = 1$, which not only simplifies computation for estimating the derivative, but it also ensures variance reduction for estimating $\phi(\lambda)$.

An efficient way to estimate the difference process here proceeds as for the RPA phantom differences for general point processes introduced in [47]. Specifically, we notice that the above "minus" process has the same distribution as the following process: claims arrive at times $\{T_k\}$, but the i-th claim "disappears" in the i-th phantom process, so that it is replaced by a null claim. With this, the phantom process can be represented by the underlying random sequence:

$$(T_k^-(i), Y_k^-(i)) = \begin{cases} (T_k, Y_k), & \text{if } k < i, \\ (T_i, 0), & \text{if } k = i, \\ (T_k, Y_k), & \text{if } k > i. \end{cases}$$

This has the advantage of using common random variables to couple the nominal and phantom processes, making computation more efficient. Indeed, with this representation it follows that $\nu \leq \nu^-(i)$ w.p.1, because in each phantom process one of the claims is nullified, but all arrivals are synchronized. The derivative processes at claim times in (9.14) are

$$\phi_n \stackrel{\text{def}}{=} e^R \exp \sum_{k=1}^n (cT_k - Y_k),$$

$$\phi_n^-(i) \stackrel{\text{def}}{=} e^R \exp \sum_{k=1}^n (cT_k^-(i) - Y_k^-(i)).$$

Computation of the derivative can be efficiently implemented recursively noting that for $n \leq \nu$,

$$\phi_{n+1} = \phi_n e^{cT_{n+1} - Y_{n+1}},$$

$$\phi_{n+1}^-(i) = \phi_{n+1} e^{Y_i}; i \leq n.$$

For $n \geq \nu$ only the phantom processes survive, satisfying $\phi_{n+1}^-(i) = \phi_n^-(i) e^{cT_{n+1} - Y_{n+1}}$, for $n \leq \nu^-(i)$. Algorithm 9.1 shows the pseudocode of the MVD algorithm for derivative estimation of the ruin probability. It requires λ, g as inputs in order to construct the corresponding $\tilde{\lambda}$ and $\tilde{g}$ (not shown). ✹✹✹

Section 10.3 shows how Theorem 9.2 can be used in the context of stochastic approximatiom for optimization and learning.

9.3 THE STATIONARY PROBLEM

In this section we present the main ideas concerning estimation of the derivative of a stationary average, see Definition 7.5. The general theory and main results are shown in Section 13.2, but for a simpler presentation, we consider in this section an ergodic Markov chain $\{X_n(\theta)\}$ on a finite state space S; for basic definitions we refer to Section C.3 in Appendix C. There is a unique stationary measure π_θ for such processes, satisfying the equilibrium equation

$$\pi_\theta P_\theta = \pi_\theta,$$

and it is also the (unique) limit measure $\lim_{n \to \infty} P_\theta(i,j) = \pi_\theta(j)$, for all $i, j \in S$. In matrix notation, it follows that $P^n \to \Pi_\theta$, called the *ergodic projector*, and it has all rows equal to π_θ.

Algorithm 9.1 Full (parallel) MVD for the ruin probability

Initialize $n = 0$, $\phi = 1$, $U = u$, $\mathcal{L} = \emptyset$.
while $(U > 0)$ **do**
   Generate $T \sim \text{Exp}(\tilde{\lambda})$
   Generate $Y \sim \tilde{g}$
   $n \leftarrow n + 1$
   $\phi \leftarrow \phi e^{cT-Y}$
   factor $= e^Y$, tmp $= Y$
   $U \leftarrow U + (cT - Y)$
   for $(i = 1, \ldots n)$ **do**
     $\phi_{\text{phan}}(i, n) = \phi \times$ factor (i)
MVD $= n e^R \phi / \lambda$
$k = n$ (this is the ruin time for the nominal)
for $(i = 1 \ldots, n)$ **do**
   $U_{\text{phan}}(i) = U + \text{tmp}(i)$
   if $(U_{\text{phan}}(i) > 0)$ **then** APPEND(i) to list $\mathcal{L}$
while $\mathcal{L}$ not empty **do**
   Generate $T \sim \text{Exp}(\tilde{\lambda})$
   Generate $Y \sim \tilde{g}$
   $k \leftarrow k + 1$
   $C1 = cT - Y$
   $C2 = e^{C1}$
   for $(i \in \mathcal{L})$ **do**
     $U_{\text{phan}}(i) \leftarrow U_{\text{phan}}(i) + C1$
     $\phi_{\text{phan}}(i) \leftarrow \phi_{\text{phan}}(i) C2$
     if $(U_{\text{phan}}(i) < 0)$ **then**
       REMOVE(i) from list $\mathcal{L}$
       MVD $\leftarrow$ sc MVD$-\phi_{\text{phan}}(i)/\tilde{\lambda}$
return MVD

If there is a way to calculate directly π_θ then this problem belongs to the static or finite horizon class of problems. However, in most practical applications only the transition kernel P_θ is known, so we assume here that π_θ is not known.

Assuming that π_θ and P_θ are element-wise differentiable with respect to θ, differentiating the equilibrium equation yields

$$\pi'_\theta P_\theta + \pi_\theta P'_\theta = \pi'_\theta,$$

where $(\cdot)'$ denotes differentiation with respect to θ. Hence,

$$\pi'_\theta (I - P_\theta) = \pi_\theta P'_\theta, \tag{9.15}$$

where I denotes the identity matrix of appropriate size. If $(I - P_\theta)$ were invertible, the resulting expression for the derivative would be $\pi'_\theta = \pi_\theta P'_\theta (I - P_\theta)^{-1}$. Unfortunately, $I - P_\theta$ fails to be invertible for Markov chains. This can be argued noting that invertibility would lead to $(I - P_\theta)^{-1} = \sum_n P_\theta^n$, but $P^n \to \Pi_\theta$ so this series is not summable.

To overcome the obstacle that $(I - P_\theta)$ is not invertible, (9.15) is modified as follows.

For any distribution μ it holds that $\mu \Pi_\theta = \pi_\theta$. Thus $\pi_{\theta+\delta} \Pi_\theta = \pi_\theta$ and it is straightforward to verify that $\pi'_\theta \Pi_\theta = 0$. Hence, adding $\pi'_\theta \Pi_\theta$ on both sides of (9.15) yields

$$\pi'_\theta (I - P_\theta + \Pi_\theta) = \pi_\theta P'_\theta.$$

It is shown in Chapter 13 that the matrix $(I - P_\theta + \Pi_\theta)$ is invertible, which gives

$$\pi'_\theta = \pi_\theta P'_\theta (I - P_\theta + \Pi_\theta)^{-1}$$

$$= \pi_\theta P'_\theta \sum_{n=0}^{\infty} (P_\theta - \Pi_\theta)^n,$$

and it follows from Theorem 13.3 that the expression on the above right-hand side is equal to

$$\pi'_\theta = \pi_\theta P'_\theta \sum_{n=0}^{\infty} (P_\theta^n - \Pi_\theta) = \pi_\theta \sum_{n=0}^{\infty} P'_\theta (P_\theta^n - \Pi_\theta) = \pi_\theta \sum_{n=0}^{\infty} P'_\theta P_\theta^n. \qquad (9.16)$$

While in earlier sections of Part II differentiation expressions are almost directly translatable into statistical estimators, we are now confronted with closed-form expression for derivatives that have no simple interpretation in terms of simulation or on-line experiments. Therefore, we will conclude this section by providing a simulation scheme for the derivative of $\pi_\theta f$, for a measurable function f, based on (9.16).

The expression for π'_θ in (9.16) is interpreted as follows: the initial state $X_\theta(0) \sim \pi_\theta$, and then the process evolves according to a Markov kernel dictated by the right hand side of (9.16). Specifically, using the MVD representation $(c_\theta, P_\theta^+, P_\theta^-)$ for P'_θ, for any measurable function f, given initial state x

$$\sum_{n=0}^{\infty} (P'_\theta P_\theta^n f)(x) = \mathbb{E}\left[c_\theta(x) \sum_{n=0}^{\infty} f(X_n^+(\theta)) - f(X_n^-(\theta)) \Big| X_0^+(\theta) = X_0^-(\theta) = x\right],$$

where the transitions of the processes $X_n^\pm(\theta)$, for $n \geq 1$, are governed by P_θ. The initial transition from $x = X_0^+(\theta)$ to $X_1^+(\theta)$ is generated from P_θ^+ and the initial transition from $x = X_0^-(\theta)$ to $X_1^-(\theta)$ is generated from P_θ^-.

By the strong Markov property it follows that once $X_n^+(\theta)$ and $X_n^-(\theta)$ hit the same state for some n, we can merge the paths, i.e., $X_k^+(\theta) = X_k^-(\theta)$ for all $k \geq n$. Let

$$\eta_\theta^\pm(x) = \min(k > 1: X_n^+(\theta, x) = X_n^-(\theta, x).$$

Then the above estimator is equivalent to

$$\sum_{n=0}^{\infty} (P'_\theta P_\theta^n f)(x) = \mathbb{E}\left[c_\theta(x) \sum_{n=0}^{\eta_\theta^\pm(x)} f(X_n^+(\theta, x)) - f(X_n^-(\theta, x))\right]. \qquad (9.17)$$

It now remains to establish how can we implement the initial condition, because $X_0^\pm(\theta)$ should have the stationary distribution π_0.

Because $\{X_n(\theta)\}$ is an ergodic Markov chain on a finite state space, it is also a regenerative process. Call $\nu(\theta)$ the cycle length for the regeneration point, assumed to be $X_n(\theta) = 0$. For this process, the stationary average is equivalent to the cycle average (see Theorem C.2 in Appendix C):

$$\pi_\theta f = \frac{\mathbb{E}\left[\sum_{i=0}^{\nu_\theta - 1} f(X_i(\theta)) \Big| X_0(\theta) = 0\right]}{\mathbb{E}[\nu_\theta]},$$

see Section 7.3. Using randomization, let $\sigma(n) \sim U\{0, \ldots, n\}$ and define

$$X^\pi(\theta) \stackrel{\text{def}}{=} X_k(\theta), \quad \text{for } k = \sigma(\nu_\theta). \qquad (9.18)$$

Then
$$\pi_\theta f = \frac{\mathbb{E}\left[\sum_{i=0}^{\nu_\theta-1} f(X_i(\theta)) \big| X_0(\theta) = 0\right]}{\mathbb{E}[\nu_\theta]} = \frac{\mathbb{E}[\nu_\theta f(X^\pi(\theta))]}{\mathbb{E}[\nu_\theta]}.$$

Putting the results together yields the following simulation scheme for estimating the derivative of the stationary a average of a function f. First, simulate one cycle and sample $X^\pi(\theta)$ as constructed in (9.18). Then use (9.17) with initial state $X_0^\pm(\theta) = X^\pi(\theta)$. This yields

$$\hat{D}(\theta) \stackrel{\text{def}}{=} \frac{\nu_\theta\, c_\theta(X^\pi(\theta))}{\mathbb{E}[\nu_\theta]} \sum_{n=0}^{\eta_\theta^\pm(X^\pi(\theta))} \left(f(X_n^+(\theta, X^\pi(\theta))) - f(X_n^-(\theta, X^\pi(\theta)))\right) \quad (9.19)$$

and provides an unbiased estimator: $\mathbb{E}[D(\theta)] = (\pi_\theta f)'$. Naturally, there is a problem with the denominator $\mathbb{E}[\nu_\theta]$ that requires estimation as well. Although theoretically sound, this approach is not practical for simulation, because it requires at least two simulations to build the estimator. However, it is possible to implement this formula to achieve the desired result.

Example 9.7. For completeness, we look again at the queueing system of Example 8.19. It represents a $GI/G/1$ queue with interarrival times $\{A_n\}$, service times $\{S_n(\theta)\}$ with scaling parameter $\theta > 0$, and FCFS discipline. For this model the sojourn times satisfy the Lindley equation (8.6). Assuming stability condition $\mathbb{E}[A_n] > \mathbb{E}[S_n]$, the sojourn time process $\{X_n(\theta)\}$ is an ergodic Markov chain with unique stationary measure π_θ. Furthermore, the infinite horizon average is the same as the stationary average:

$$\lim_{N \to \infty} \mathbb{E}\left[\frac{1}{N} \sum_{n=1}^{N} X_n(\theta)\right] = \int_{\mathbb{R}^+} x\, \pi_\theta(dx).$$

To implement (9.19) a preliminary simulation of a single cycle $(X_1(\theta), \ldots X_{\nu_\theta}(\theta))$ is used to choose uniformly the initial state x. Next, only two processes are required: a "plus" and a "minus" process. Because the dynamics of the process follow the Lindley recursion (8.6) the difference process can be calculated exactly as we have shown in Example 8.19.

Algorithm 9.2 shows the implementation of (9.19) to the queueing example. Here, we assume that $S_1^+(\theta) > S_1^-(\theta)$, so the difference process follows the same dynamics as in (8.33). It should be apparent that while there seems to be savings in computational time because there is only one split at the initial stage into the "plus" and "minus" processes, there is a problem for estimating the denominator $\mathbb{E}[\nu_\theta]$, which may require a large number of independent simulations. ✹✹✹

Alternatively, an SF estimator can be derived. For this we note

$$\sum_{n=0}^{\infty} (P_\theta^n f - \Pi_\theta f)(x) = \sum_{n=0}^{\infty} \left((P_\theta^n f)(x) - \pi_\theta f\right),$$

where assume that the stationary cost $\pi_\theta f$ is known. Assuming that P_θ has a point differentiable (Rieman) conditional density, so that

$$\int h(y) P_\theta(x, dy) = \int h(y) f_\theta(x, y) dy, \quad (9.20)$$

we can then introduce the score function

$$\mathrm{SF}_\theta(x, y) = \frac{\partial}{\partial \theta} \log f_\theta(x, y),$$

Algorithm 9.2 Stationary MVD general service distribution

Initialize MVD = 0.
Simulate one cycle $(X_n(\theta); n = 1, \ldots \nu(\theta))$
$x \sim U\{X_1(\theta), \ldots, X_{\nu(\theta)}(\theta)\}$
$X_1^+(\theta) = S_1^+(\theta) + (x - A_1)_+$
$X_1^-(\theta) = S_1^-(\theta) + (x - A_1)_+$
$d = S_1^+(\theta) - S_1^-(\theta)$
while $(d > 0)$ **do**
   Generate A, S
   $X^\pm \leftarrow S + \max(0, X^\pm - A)$
   $d \leftarrow \max(0, d - (A - X_n^-(\theta))$
   MVD$\leftarrow$ MVD $+d$
MVD $\leftarrow$ MVD$\times(\nu(\theta)\, c_\theta(x))$.
for $(j = 1 \rightarrow$ NumSim$)$ **do**
   Simulate j-th cycle $(X_n(\theta; j); n = 1, \ldots, \nu(\theta; j))$
$\bar{\nu}$ = Average$(\nu(\theta; j))$.
return MVD$/\bar{\nu}$
---

which leads to the following unbiased SF estimator for $(\pi_\theta f)'$

$$\frac{\eta_\theta}{\mathbb{E}[\eta_\theta]} \sum_{n=0}^{\infty} \mathrm{SF}_\theta(X^\pi(\theta), X_1(\theta))(f(X_n(\theta)) - \pi_\theta f).$$

For ergodic chains, $\mathbb{E}[f(X_n(\theta))]$ converges towards $\pi_\theta f$ and the infinite sum can be approximated by replacing ∞ with some large finite upper bound of the sum. Hence, the SF estimator is only asymptotically unbiased in case $\pi_\theta f$ is not known. Moreover, finding a closed-form expression for $f_\theta(x, y)$ in (9.20) is a nontrivial task even for simple systems, such as the Markov chain of sojourn times in the G/G/1 queue. For details on a SF approach to the above formula, we refer to [127].

Remark 9.5. It is worth noting that, in the simplest case of unconstrained optimization $\min J(\theta) \stackrel{\text{def}}{=} \mathbb{E}[X^\pi(\theta)]$, we may replace the unbiased gradient estimator $\hat{D}(\theta)$ in (9.19) by a simpler one:

$$\tilde{D}(\theta) \stackrel{\text{def}}{=} \sum_{n=0}^{\eta_\theta^\pm(X^\pi(\theta))} f(X_n^+(\theta, X^\pi(\theta)) - f(X_n^-(\theta, X^\pi(\theta)). \qquad (9.21)$$

Because $\nu_\theta, c_\theta(X^\pi(\theta))$, and $\mathbb{E}[\nu_\theta]$ are all strictly positive this implies that the corresponding vector field is a descent direction and it is also coercive for the unconstrained optimization problem. See Exercise 9.5 for a way to implement this simplified estimator when optimizing the service parameter of a $GI/G/1$ queue via stochastic optimization.

⋆9.3.1 IPA for the Stationary Problem for Discrete State Space Models

IPA is based on sample-path derivatives, which presupposes real-valued and continuous sample-path variables (because integer-valued random variables are piece wise constant as a mapping of θ and therefore the sample-path derivative is either zero or not defined). We already encountered this obstacle for IPA in Example 8.5 when we discussed the Poisson counting process $N(T)$, which has sample-path derivative zero with probability one. This makes it even more surprising that for discrete Markov chains, gradients can be evaluated on a sample-path basis without directly differentiating sample paths. In the following we illustrate this approach with the queue-length process of the M/M/1 queue, with inter arrival times with distribution $\text{Exp}(\lambda)$ and service times with distribution $\text{Exp}(\mu)$, i.e., the mean inter arrival time is $1/\lambda$ and the mean service time is $1/\mu$. For stability we assume that $\lambda < \mu$. Let X_n denote the number of customers at the queue (with the one in service included) after the occurrence of the n-th event, where an event is either a service completion or an arrival of a customer. We want to perform a perturbation analysis of the stationary distribution of X_n. Specifically, $\{X_n\}$ is a Markov chain with state space $\mathbb{N}$ and transition probability kernel

$$P = \begin{pmatrix} 0 & 1 & 0 & \cdots & \cdots \\ \frac{\mu}{\lambda+\mu} & 0 & \frac{\lambda}{\lambda+\mu} & \cdots & \cdots \\ 0 & \frac{\mu}{\lambda+\mu} & 0 & \frac{\lambda}{\lambda+\mu} & \cdots \\ \vdots & 0 & \ddots & 0 & \ddots \end{pmatrix}. \qquad (9.22)$$

Suppose that we perturb P by some matrix Q such that for small θ

$$P(\theta) = P + \theta Q,$$

where Q is such that $P(\theta)$ is again a Markov chain (which implies that Q has row sum zero). Observe that the above model implies

$$\frac{d}{d\theta} P(\theta) = Q,$$

where the derivative is taken element-wise. Let

$$d(i, j) = \mathbb{E}\left[\sum_{n=1}^{\infty} (X_n^i - X_n^j)\right],$$

where X_n^s denotes the Markov chain with initial value s. The factor $d(i, j)$ is called the *perturbation realization factor*, and the matrix $D = (d(i, j))_{i,j \in \mathbb{N}}$ is called the *realization matrix*. We have already encountered this formulations when analyzing the difference processes. When we want to express that D is evaluated for X_n with transition matrix $P(\theta)$ we write $D(\theta)$ for D. For early references on perturbation analysis via realization factors we refer to [56, 58, 190]. Let

$$\tau(\theta; i, j) = \inf\{n : X_n^i = X_n^j\}$$

denote the number of transitions until the version started in i and the version started in j hit the same state for the first time. By the strong Markov property it holds

$$d(i, j) = \mathbb{E}\left[\sum_{n=1}^{\tau(\theta;i,j)} (X_n^i - X_n^j)\right]. \qquad (9.23)$$

Let $\pi(\theta)$ denote the stationary distribution of $P(\theta)$. Choose θ_0 be such that $D(\theta_0)$ exists and that there is a neighborhood around θ_0 such that $P(\theta)$ and $\pi(\theta)$ are well-defined for all θ out of this neighborhood. It then follows that

$$\left.\frac{d}{d\theta}\right|_{\theta=\theta_0} \pi(\theta) = \pi(\theta_0) Q D(\theta_0).$$

For a derivation of this formula we refer to Section 13.3. The above derivative expression via the perturbation realization matrix is related to the corresponding MVD estimator, see [146].

In applications, the realization matrix D is estimated from simulation and the performance derivative is evaluated by the expression in (9.23). There exist a variety of approaches for evaluating D from one sample path via cut-and-paste, which makes the perturbation realization method to a sample path method. Specifically, the close relation of the realization factor representation and the derivative expression provides a link to Markov decision processes, which has resulted in series of papers [59, 61, 190, 330, 331]. For a general overview, we refer to [60].

9.4 PRACTICAL CONSIDERATIONS

Gradient estimation for an infinite horizon problem and a stationary problem are closely related (see Section 7.3.4), which often allows for choosing between different types of gradient estimators for a given problem setting. For example, in Exercise 9.4 the reader is asked to implement Algorithm 9.2 in a stochastic approximation to optimize the service time parameter in a stationary $GI/G/1$ queue. This exercise aims to integrate the gradient estimation for the stationary problem into the algorithms described in Chapter 4. Alternatively, one can elaborate on the fact that the expected value of a stationary sojourn time is obtainable as the limit of average expected sojourn times and thus use the infinite horizon gradient estimator, as presented in Section 9.1. Therefore, it is important to compare the optimization procedures; see Exercise 9.3.

9.5 EXERCISES

Exercise 9.1. Let $X_n(\theta)$ denote the number of customer at the queue (with the one in service included) after the occurrence of the n-th event, where an event is either a service completion or an arrival of a customer. Then, $\{X_n(\theta)\}$ is Markov chain with state space $\mathbb{N}$, transition probability

$$P(\theta) = \begin{pmatrix} 0 & 1 & 0 & \cdots & \cdots \\ \frac{\theta}{\lambda+\theta} & 0 & \frac{\lambda}{\lambda+\theta} & \cdots & \cdots \\ 0 & \frac{\theta}{\lambda+\theta} & 0 & \frac{\lambda}{\lambda+\theta} & \cdots \\ \vdots & 0 & \ddots & 0 & \ddots \end{pmatrix},$$

and stationary distribution π_θ for $\lambda < \theta$. Can the realization factor method be applied to estimate $\pi'(\theta)$? Note that $P(\theta)$ is element-wise differentiable and it holds that

$$P'(\theta) = \begin{pmatrix} 0 & 0 & 0 & \cdots & \cdots \\ \frac{\lambda}{(\lambda+\theta)^2} & 0 & -\frac{\lambda}{(\lambda+\theta)^2} & \cdots & \cdots \\ 0 & \frac{\lambda}{(\lambda+\mu)^2} & 0 & -\frac{\lambda}{(\lambda+\theta)^2} & \cdots \\ \vdots & 0 & \ddots & 0 & \ddots \end{pmatrix}.$$

How can the above derivative be included in this framework?

Exercise 9.2. Consider the scenario of Example 9.5, where $\{T_i\}$ are the inter arrival times of the (marked) Poisson process $N_\theta(\cdot)$ with rate $\theta > 0$, $A_i = \sum_{k=1}^{i} T_i$ are the corresponding arrival epochs, and W_i are the corresponding marks associated with each arrival. Let S be a random variable independent of this process.

(a) Show that the score function for the finite horizon problem satisfies
$$\frac{d}{d\theta}\mathbb{E}[h(T_1,\ldots,T_N)] = \mathbb{E}[h(T_1,\ldots,T_N)\mathsf{SF}_N(T_1,\ldots,T_N)], \quad \mathsf{SF}_n \stackrel{\text{def}}{=} \sum_{k=1}^{n}\left(\frac{1}{\theta} - T_k\right).$$

(b) Apply Theorem 9.2 to derive the SF estimator of the cost $C(\theta)$ and constraint $g(\theta)$ for this example:
$$C(\theta) = \mathbb{E}\left[\sum_{i=1}^{N_\theta(S)} W_i\right]; \quad g(\theta) = \mathbb{E}\left[\sum_{i=1}^{N_\theta(S)} \mathbf{1}_{\{W_i + S - A_i(\theta) > \delta\}}\right] - \alpha,$$

and explain why simply replacing N by $N_\theta(S)$ in the finite horizon formula will result in a biased estimator.

(c) Show that $\tau(\theta) = N_\theta(S) + 1$ is a random stopping time adapted to the natural filtration of the process filtration $\mathfrak{F}_n = \sigma(S; T_1, \ldots, T_n)$.

(d) Use simulation to estimate $dC(\theta)/d\theta$ (confidence intervals are explained in Appendix D). For comparison, do this first using $N_\theta(S)$ in the SF finite horizon formula, and then using $\tau(\theta)$, and discuss your results.

(e) Program the SA procedure to optimize the cost adapting the two-time scale penalty algorithm. Explain your choice of hyperparameters, as well as the distribution of S and the definition that you use for $P(\theta)$. Use first $N_\theta(S)$ in the expression for the SF estimator and compare your results with those using $\tau(\theta)$ instead.

Exercise 9.3. Consider the model of Example 9.3. This is a single server $GI/G/1$ queue where the inter arrival times $\{A_n\}$ are iid with unit mean and the service times $\{S_n(\theta)\}$ are iid with distribution F_θ. For this model we assume that θ is a scaling parameter, $\mathbb{E}[S_n(\theta)] = \theta$ and $0 < \theta < 1$. The consecutive sojourn times $X_n(\theta)$ satisfy the Lindley recursion
$$X_{n+1}(\theta) = S_{n+1}(\theta) + (X_n(\theta) - A_{n+1})_+$$

and have a stationary distribution π_θ, which is also the limiting distribution, whenever $\theta < 1$.

The cost associated with the server's speed is $C(\theta) = 1/\theta$ and the problem seeks to minimize
$$J(\theta) \stackrel{\text{def}}{=} C(\theta) + \kappa\mathbb{E}[X^\pi(\theta)],$$

over $(0, 1)$, where $X^\pi(\theta)$ has the stationary distribution π_θ. The constant κ is known as a *shadow price* in economic theory and it transforms waiting times into monetary values. Consider the target vector field
$$G(\theta) = -J'(\theta) = \frac{1}{\theta^2} - \kappa\frac{d}{d\theta}\mathbb{E}[X^\pi(\theta)].$$

(a) Argue that the problem is well-posed and that G is coercive for this problem on $(0, 1)$.
(b) Refer to Example 9.3 for notation and let
$$g(\xi_n, \theta) = \frac{1}{\theta^2} - \frac{\kappa}{\theta}\sum_{i \in \mathcal{L}_n} d_n(\theta; i).$$

Show that this function satisfies assumption (a3) of Theorem 5.4.
(c) Build a feedback estimator Y_n using means of consecutive batches of quantities $g(\xi_n, \theta_n)$. Using Example 5.7 as a basis for your calculations, show that all the assumptions of Theorem 5.4 are satisfied.

(d) Example 8.19 provides the finite horizon MVD, with pseudocode given in Algorithm 8.4. You may use this as a basis for the batch estimation. Using Algorithm 5.1 as a basis for the general stochastic approximation for the infinite horizon model, program the procedure using batch sizes $K = 10, 50, 100$. You may use $\kappa = 1$ and service times with Gamma distribution. When reporting your results, include CPU times and confidence intervals.

Exercise 9.4. Consider the same problem as in Exercise 9.3. Now use (9.19) in order to estimate the derivative of the sojourn times, as illustrated in Algorithm 9.2. That is, use the feedback

$$Y_n = \frac{1}{\theta^2} + \kappa \hat{D}(\theta_n), \qquad (9.24)$$

where $\hat{D}(\theta)$ is defined in (9.19).

(a) Show that the assumptions of Theorem 4.4 are satisfied.
(b) Program the corresponding SA using again $\kappa = 1$ and Gamma distributed service times. When reporting your results, include CPU times and confidence intervals.

Exercise 9.5. Consider again the same problem as in Exercise 9.3. This time, instead of (9.19), we aim to implement (9.21), which is simpler and requires much less running time. However, replacing $\hat{D}(\theta_n)$ by $\tilde{D}(\theta_n)$ in (9.24) will not produce a coercive field for the problem, because the (random) non negative quantity $\frac{\nu_\theta c_\theta(X^\pi(\theta))}{\mathbb{E}[\nu_\theta]}$ is not multiplying the term $1/\theta^2$.

(a) Now define the random process:

$$\tilde{X}_n(\theta) \stackrel{\text{def}}{=} \frac{1}{\theta} + \kappa X_n(\theta).$$

Show that $\tilde{X}_n(\theta)$ is also an ergodic Markov chain with stationary distribution $\tilde{\pi}_\theta$ and that its regenerative cycles coincide with those of $X_n(\theta)$. Provide the recursion satisfied by this process.

(b) Under this representation, it follows that

$$J(\theta) = \mathbb{E}[\tilde{X}^{\tilde{\pi}}], \tilde{X}^\pi \sim \tilde{\pi}_\theta.$$

Explain how do you calculate the plus and minus processes for $\{\tilde{X}_n\}$ and provide the corresponding expression for $\tilde{D}(\theta_n)$.

(c) Program the corresponding SA using again $\kappa = 1$ and Gamma distributed service times. When reporting your results, include CPU times and confidence intervals.

Part III

Selected Topics in Stochastic Approximation

Chapter Ten

Applications of Stochastic Approximation to Inventory Problems

Inventory theory addresses the problem a firm is faced with in deciding how much to order in each time period to meet demand for its products. How well the company manages this challenge has a major impact on its profitability. Inventory optimization seeks to match supply to stochastic customer demand. In this chapter we will discuss the use and application of the theory developed in Parts I and II to an inventory problem with one type of good and stochastic demand. Several variations of the problem will be treated. First, we address finding the replenishing policy that minimizes the overall cost. Next, we show how learning and root-finding techniques can be applied for calibrating the model.

10.1 OPTIMIZATION USING MVD GRADIENT ESTIMATION

In this section we study an inventory problem containing the salient characteristics of *threshold control* problems. The model and methodology in this section is inspired from the original research in [311], where the estimators were first developed using the SPA approach. In many real-life situations, it is practical to apply threshold controls that signal or trigger an alarm, calling for actions. With increasing use of sensors and the Internet of Things, it is important that such "alarms" be triggered at the correct values, called "thresholds." In inventory, the so-called (s, S) inventory is a popular threshold control policy, where the stock is replenished to maximal capacity S whenever the inventory level falls below a certain threshold s. We will use here $\theta = s$ to denote this threshold, in order to be consistent with our notation. We consider a periodic review inventory system, where at the end of each business day the inventory is inspected.

Let $X_n(\theta)$ denote the (continuous) level of inventory at the start of day n. The demand per day is denoted η_n and it is independent of the level $X_n(\theta)$. We assume that consecutive demands are iid random variables with a continuous density $f_\eta(\cdot)$ independent of θ. The (θ, S)-policy requires that a replenishment order be placed at the end of the day if the inventory level falls under the threshold value θ. In the simplest model there is no delay in delivery of an order, so at the start of the following day the level is back to S:

$$X_{n+1}(\theta) = \begin{cases} X_n(\theta) - \eta_n & \text{if } \eta_n \leq X_n(\theta) - \theta, \\ S & \text{otherwise.} \end{cases} \quad (10.1)$$

The inventory cost during stage n is composed of (i) a holding cost of h dollars (or euros) per unit of inventory level at the end of the stage n, (ii) a fixed ordering cost K when an order is placed, and (iii) a penalty cost of p of dollars (or euros) per unit of unsatisfied demand.

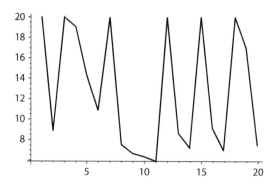

Figure 10.1. A typical trajectory of the cost $c(X_k(\theta), \eta_k)$, $0 \le k \le 20$. Here $S = 20$, $\theta = 10$, $h = 2$, $p = 8$, $K = 10$, and the demands are exponential random variables with mean 10.

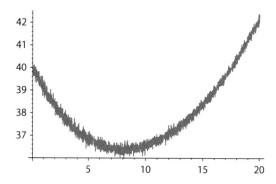

Figure 10.2. Estimated cost $J(\theta)$ with 100 replications of the simulation.

The latter cost reflects situations where the vendor must outsource stock if the demand is higher than the inventory. It is customary to assume that $p \gg h$. In mathematical notation,

$$c(X_n(\theta), \eta_n) = \begin{cases} h(X_n(\theta) - \eta_n) & \text{if } \eta_n \le X_n(\theta) - \theta, \\ h(X_n(\theta) - \eta_n) + K & \text{if } X_n(\theta) - \theta < \eta_n \le X_n(\theta), \\ p(\eta_n - X_n(\theta)) + K & \text{if } X_n(\theta) < \eta_n. \end{cases} \quad (10.2)$$

10.1.1 Finite Horizon

The goal is to find the optimal value of the threshold parameter θ^* that minimizes the total cost:

$$J_N(\theta) = \mathbb{E}\left[\frac{1}{N} \sum_{n=1}^{N} c(X_n(\theta), \eta_n)\right]. \quad (10.3)$$

Figure 10.2 shows the result of the mean cost using 100 replications of the simulation at various values of θ, using $N = 1{,}000$, $S = 10$, $\theta = 5$, $h = 2$, $p = 8$, $K = 10$ and $\eta_n \sim \text{Exp}(0.1)$. This plot took Mathematica 6,625 seconds of CPU time, it uses independent replications at each value of θ. Although it could be modified to decrease considerably the computational time, we will see that directly doing a gradient search results in a much more efficient algorithm.

For any fixed value of θ the process $\{X_n(\theta)\}$ is a Markov chain with transition kernel $P_\theta(x, dy) = \mathbb{P}(X_{n+1}(\theta) \in dy \mid X_n(\theta) = x)$, where

$$\mathbb{P}(X_{n+1}(\theta) = S \mid X_n(\theta) = x) = \mathbb{P}(\eta_n \geq x - \theta)$$

$$\mathbb{P}(X_{n+1}(\theta) \in y + dy \mid X_n(\theta) = x) = \mathbb{P}(\eta_n \in [x-y-dy, x-y)) \approx f_\eta(x-y)\, dy,$$

for $y + dy < S$.

For Markov chains the underlying measure is a product measure. For any subinterval $A \subset \mathbb{R}^+$

$$\mathbb{P}(X_{n+1}(\theta) \in A \mid X_0(\theta)) = \int \mathbb{P}(X_{n+1}(\theta) \in A \mid X_1(\theta) = x_1) P_\theta(x_0, dx_1)$$

$$= \int\int \mathbb{P}(X_{n+1}(\theta) \in A \mid X_2 = x_2) P_\theta(x_1, dx_2)\, P_\theta(x_0, dx_1)$$

$$= \int \cdots \int P_\theta(A, dx_n) \prod_{i=1}^n P_\theta(x_{i-1}, dx_i).$$

Each of the terms in the product is the measure defined by the transition kernel $P_\theta(x; dx)$. Thus, for any continuous and bounded function ϕ we may use the results of Theorem 8.6 whenever the kernel is $\mathcal{D}$-differentiable. Under this assumption (that we will check later) for any bounded and continuous function ϕ

$$\frac{d}{d\theta}\mathbb{E}[\phi(X(\theta)) \mid X_0(\theta) = x_0] = \sum_{k=1}^N \int \cdots \int \phi(x_1, \ldots, x_n)$$

$$\prod_{n=1}^{k-1} P_\theta(x_{n-1}, dx_n)\, P'_\theta(x_{k-1}, dx_k) \prod_{n=k+1}^N P_\theta(x_{n-1}, dx_n),$$

where $X(\theta) = (X_1(\theta), \ldots, X_N(\theta))$. We now find the MVD estimator for the derivative of the kernel $P'_\theta(x_{k-1}, dx_k)$ for the particular example of the inventory process. Notice that in this model we can use the demand process $\{\eta_n\}$ as the underlying process with natural filtration $\{\mathfrak{F}_n\}$, because $\{X_n(\theta)\}$ is adapted to its filtration. Here, the instantaneous cost function (10.2) depends on both $X_n(\theta)$ and η_n, and we remark that $(X_{n+1}(\theta), \eta_n)$ is a two-dimensional Markov chain. It satisfies the assumptions of Theorem 8.7 whenever the kernel is $\mathcal{D}$-differentiable. Following Definition 8.8, this kernel is indeed $\mathcal{D}$-differentiable, where $\mathcal{D}$ is the set of bounded and piecewise continuous functions on $[\theta, S] \times \mathbb{R}^+$. Following the method introduced in Example 8.20, for $c \in \mathcal{D}$ we have

$$\frac{d}{d\theta}\mathbb{E}[c(X_{k+1}(\theta), \eta_{k+1}) \mid X_k(\theta) = x_k]$$

$$= \frac{d}{d\theta}\int_0^{x_k-\theta} c(x_k - y, y) f_\eta(y)\, dy + \frac{d}{d\theta}\int_{x_k-\theta}^\infty c(S, y) f_\eta(y)\, dy$$

$$= -c(\theta, x_k - \theta) f_\eta(x_k - \theta) + c(S) f_\eta(x_k - \theta)$$

$$= f_\eta(x_k - \theta)\, (c(S, x_k - \theta) - c(\theta, x_k - \theta)).$$

The interpretation is as follows: the "plus" process considers $\eta_k^+ = \lim_{\delta \downarrow 0} = (x_k - \theta) + \delta$ and incurs an ordering cost, setting $x_{k+1}^+ = S$ (a degenerate probability), while the "minus" process considers $\eta_k^- = \lim_{\delta \downarrow 0} = (x_k - \theta) - \delta$ and does not have an ordering cost, setting $x_{k+1}^- = \theta$. The only requirement for the above formula to hold is that the density f_η be bounded.

Using Theorem 8.6 an estimator of the derivative is given by

$$\sum_{k=1}^{N} f_\eta(X_k(\theta) - \theta) \, (L_N(X^+(\theta;k), \xi) - L_N(X^-(\theta;k), \xi), \tag{10.4}$$

where $\xi = (\eta_1, \ldots, \eta_N)$ and $L_N(X, \xi) = (1/N) \sum_{n=1}^{N} c(X_n, \eta_n)$. The k-th plus and minus processes $X^\pm(\theta;k)$ are defined by

$$X_n^+(\theta;k) = \begin{cases} X_n(\theta) & \text{if } n < k, \\ S & \text{if } n = k, \end{cases} \quad \text{and} \quad X_n^-(\theta;k) = \begin{cases} X_n(\theta) & \text{if } n < k, \\ \theta & \text{if } n = k. \end{cases}$$

For $n > k$ we let $\mathbb{P}(X_n^\pm(\theta;k) \in dx \mid X_n^\pm(\theta;k)) = P_\theta(X_n^\pm(\theta;k), dx)$. That is, the future inventory in the phantom processes evolves as in (10.1).

The randomized MVD estimator, which is not efficient, is shown in pseudo code below, in order to help understand the implementation of the "splitting" trajectories of the algorithm. In Algorithm 10.1 we choose an index σ uniformly among all possible N days. Splitting occurs only from day σ onward. From day σ onward, the two processes $\{X_n^\pm, n > \sigma\}$ evolve using the same daily demands, but their initial value is different. Thus, we obtain two trajectories of parallel inventories.

Algorithm 10.1 Randomized MVD for the finite horizon inventory

---

Initialize $X_1 = x$, $X_\sigma^+ = S$, $X_\sigma^- = \theta$
RANDOM-MVD(θ, N)
Generate $\sigma \sim U\{1, \ldots, N\}$
for $(n = 1, \ldots, \sigma - 1)$ **do**
    Generate demand $\eta_n \sim f_\eta$
    $\Delta = X_n - \eta_n$
    $X_{n+1} = \Delta \mathbf{1}_{\{\Delta > \theta\}} + S \mathbf{1}_{\{\Delta \leq \theta\}}$
factor $= f_\eta(X_\sigma - \theta)$
MVD $= c(X_\sigma^+, \eta_\sigma) - c(X_\sigma^-, \eta_\sigma)$
for $(n = \sigma, \ldots, N - 1)$ **do**
    Generate demand $\eta_n \sim f_\eta$
    $\Delta^\pm = X_n^\pm - \eta_n$
    $X_{n+1}^\pm = \begin{cases} \Delta^\pm & \text{if } \Delta^\pm > \theta \\ S & \text{if } \Delta^\pm \leq \theta \end{cases}$
    MVD $\leftarrow$ MVD $+ c(X_{n+1}^+, \eta_{n+1}) - c(X_{n+1}^-, \eta_{n+1})$
return MVD $\times$ factor $/N$

The pseudocode for the full MVD algorithm is shown in Algorithm 10.2. In this code, every step n sees the "birth" of the plus and minus processes that are initialized at S and θ, respectively. At iteration n we update all of the previously generated phantom (plus and minus) processes. Thus there is a **for** loop nested into the outer loop.

Algorithm 10.2 is not optimal: many operations can be calculated in a more efficient manner, and appropriate data structures should be used, together with parallelization of code. However, this pseudocode provides the main ideas of the operations needed. Importantly, we have chosen a version that can be implemented when streaming data (e.g., reading demands from real time sources).

Algorithm 10.2 Full (parallel) MVD for the finite horizon inventory

Initialize $X_1 = x$, MVD $= 0$.
M$_{VD}$(θ, N)
for $(n = 1, \ldots, N-1)$ **do**
  Generate demand $\eta_n \sim f_\eta$
  $\Delta = X_n - \eta_n$
  $X_{n+1} = \Delta \mathbf{1}_{\{\Delta > \theta\}} + S \mathbf{1}_{\{\Delta \leq \theta\}}$
  Birth of n-th phantoms:
  $X_n^+(n) = S$, $X_n^-(n) = \theta$
  $f(n) = f_\eta(X_n - \theta)$
  $d(n) = c(X_n^+(n), \eta_n) - c(X_n^-(k), \eta_n)$
  Update all phantom trajectories with new demand:
  for $(k = 1 \ldots, n)$ **do**
    $\Delta^\pm(k) = X_n^\pm(k) - \eta_n$
    $X_{n+1}^\pm(k) = \begin{cases} \Delta^\pm(k) & \text{if } \Delta^\pm(k) > \theta \\ S & \text{if } \Delta^\pm(k) \leq \theta \end{cases}$
    $d(k) \leftarrow d(k) + c(X_{n+1}^+(k), \eta_{n+1}) - c(X_{n+1}^-(k), \eta_{n+1})$
for $(k = 1, \ldots, N)$ **do**
  MVD $\leftarrow$ MVD $+ f(k) d(k)$
return MVD$/N$

Stochastic Approximation. We now turn to designing a SA algorithm for finding the optimal threshold θ. Given is

$$\min_{\theta \in [0,S]} J_N(\theta). \tag{10.5}$$

We note that the constraint $\theta \in [0, S]$ is a hard (physical) constraint because the process does not make sense for other values of θ. However, the derivative of $J_N(\theta)$ is defined at the boundary points. Inspecting Figure 10.2 it can be seen that for the numerical setting at hand that $J_N(\theta)$ is convex on the admissible parameter range. Moreover, the location of the minimum is between 5 and 10, and thus not located on a boundary point. From this it can be argued the problem is well-posed and that $-J_N'(\theta)$ is coercive for (10.5) on $[0, S]$, where we use that the ODE driven by $-J'(\theta)$ has the location of the minimum of $J_N(\theta)$ as globally stable point provided started in $[0, S]$.

Remark 10.1. A mathematical argument can be constructed without reference to Figure 10.2. Indeed, we have shown that the transition kernel $P(s; du)$ is $\mathcal{D}$-differentiable, where $\mathcal{D}$ is the set of bounded measurable mappings. The product rule of weak differentiation then implies that $J_N(\theta)$ is differentiable. Hence, $J_N(\theta)$ is continuous and existence of a solution of the optimization problem follows from the Weierstrass theorem. Moreover, inspecting the weak derivative of $J_N(\theta)$ it is easily seen that $dJ_N(\theta)/d\theta$ is continuous on $[0, S]$, and we conclude that $J_N(\theta) \in C^1$. In the notation of Chapter 1, the feasible set $\Theta = [0, S]$ yields the constraint functions $g_1(\theta) = \theta - S \leq 0$ and $g_2(\theta) = -\theta \leq 0$. Possible KKT points are either the boundary points $0, S$ or stationary points of $J_N(\theta)$ in $[0, S]$. Hence, all KKT points are in the compact set $[0, S]$. We conclude that the problem is well-posed.

We can use here for the feedback Y_n the negative of the MVD estimator from either Algorithm 10.1 or Algorithm 10.2 obtained from a simulation of N consecutive days at parameter θ_n and then update θ_n via

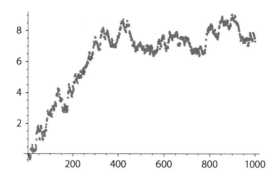

Figure 10.3. A trajectory of the stochastic approximation to find the optimal θ^* using $\epsilon = 0.05$.

$$\theta_{n+1} = \Pi_{[0,S]}(\theta_n - \epsilon_n Y_n), \theta_0 \in (0, S),$$

where we use projection on $[0, S]$ to guarantee that θ_n is feasible.

We use independent replications of the finite horizon simulation from one update to another. Specifically, we use a sequence $\{\xi_n\} = \{(\eta_{n,1}, \ldots, \eta_{n,N})\}$ of iid demand vectors. The vector ξ_n is then used to simulate the n-th finite horizon trajectory. From independence it follows that $\mathbb{E}[Y_n \mid \mathfrak{F}_{n-1}] = -J'_N(\theta_n)$ is unbiased for $\theta \in [0, S]$. Because the state space is bounded it follows that $\text{Var}(Y_n) \leq \bar{V}$ is uniformly bounded, implying uniform integrability of $\{Y_n\}$. Weak continuity of the kernels follows from the MVD analysis, so that using a stepsize sequence that satisfies $\sum_i \epsilon_i = +\infty$, $\sum_i \epsilon_i^2 < \infty$ means that Theorem 4.2 ensures that $\theta_n \to \theta^*$ a.s. When using constant stepsize, Theorem 4.4 ensures that, as $\epsilon \to 0$, the interpolated processes converge to the solution of

$$\frac{dx(t)}{dt} = \Pi_{[0,S]}(-J'_N(\theta)),$$

which, in turn, has a unique stable point at θ^*. A trajectory of the stochastic approximation using the code in Algorithm 10.1 is shown in Figure 10.3. This procedure took 21 seconds in Mathematica.

Inspecting the trajectory in Figure 10.3, it seems reasonable to assume that for $n \geq 400$ the interpolation process $x_\epsilon(t)$ of $\{\theta_n\}$ is close to its limit, which is a stationary Ornstein-Uhlenbeck process (see Theorem 6.1). Taking a mean value over the θ_n therefore gives a good approximation of θ^*; for details we refer to Chapter 6. To achieve better convergence a Polyak-Rupert type averaging [249] may be applied for $n \geq 400$, i.e., the output of the algorithm is given by

$$\bar{\theta}_m = \frac{1}{m-n+1} \sum_{i=n}^{m} \theta_i.$$

It worth noting that for Polyak-Rupert type averaging with $\epsilon_n = n^{-\gamma}$ it is required that $\gamma < 1$, i.e., the standard choice $\gamma = 1$ does not apply.

10.1.2 Infinite Horizon

The model follows the dynamics in (10.1), but we are now interested in minimizing the long-term cost rate, namely the limit of (10.3), as N grows:

$$J(\theta) = \mathbb{E}\left[\lim_{N \to \infty} \frac{1}{N} \sum_{n=1}^{N} c(X_n(\theta), \eta_n)\right]. \tag{10.6}$$

APPLICATIONS OF STOCHASTIC APPROXIMATION

In order to address this problem we aim to apply Theorem 9.1, which we do next. After that we will use the MVD for finite horizon to drive a stochastic approximation and apply Theorem 5.4 in order to establish convergence to the optimal solution.

From (10.1) it follows that the inventory level $\{X_n(\theta)\}$ itself is a Markov chain (driven by the stochastic demands). This process is uniformly bounded: $X_n(\theta) \in [\theta, S] \subset [0, S]$ w.p.1, for all n. Harris recurrence follows from a regenerative argument, as mentioned in [103]. The regeneration point for general demand distributions is $X_n(\theta) = S$, because once the stock is full to capacity, the process is independent of the past. The only requirement for this to happen infinitely often is that $\mathbb{P}(\xi_n > 0) > 0$, which is a very mild assumption. Conditions (i), (ii) and (iv) of Theorem 9.1 now follow from geometric ergodicity of this Markov chain. Condition (iii) will allow the finite horizon estimator to be implemented for the infinite horizon problem, and it will also help to bound the expected running time (and memory requirement) of the MVD algorithm. In the following we will turn to establishing condition (iii) of Theorem 9.1.

The estimator in (10.4) can be rewritten as

$$\hat{G}_N^{(MVD)} = \frac{1}{N} \sum_{i=1}^{N} f_\eta(X_i(\theta) - \theta) \sum_{n=1}^{N} d_n(\theta; i),$$

where the i-th phantom difference process is defined with

$$d_n(\theta; i) = c(X_n^+(\theta; i)) - c(X_n^-(\theta; i)).$$

By construction $d_n(\theta; i) = 0$ for all $n < i$. Define the lifetime of the i-th phantom difference process by $\tau_i(\theta) = \min(n > i: X_n^+(\theta; i) = X_n^-(\theta; i) = S)$, so that $d_n(\theta; i) = 0$ for $n \geq i + \tau_i$. With this notation we can write

$$\hat{G}_N^{(MVD)} = \frac{1}{N} \sum_{i=i}^{N} f_\eta X_i(\theta) - \theta) \sum_{n=i}^{(i+\tau_i(\theta)) \wedge N} d_n(\theta; i).$$

We now use the same approach as in Chapter 9 to express the derivative process as a non-anticipative function of the process. This expression will also allow for a more efficient code that we will use for the SA. At time n, define the list of "living" phantom differences by

$$\mathcal{L}_n(\theta) = \{j \leq n: j + \tau_j(\theta) > n\},$$

which is well defined from the history of the inventory process up to time n. Introduce now the augmented underlying process

$$\xi_n'(\theta) = (X_n(\theta), \mathcal{L}_n(\theta), (X^\pm(\theta; i), i \in \mathcal{L}_n(\theta))).$$

Then the alternative expression for the MVD estimator for finite horizon is

$$\hat{G}_N^{(MVD)} = \frac{1}{N} \sum_{n=i}^{N} f_\eta(X_n(\theta) - \theta) \sum_{i \in \mathcal{L}_n} d_n(\theta; i) \stackrel{\text{def}}{=} \frac{1}{N} \sum_{n=i}^{N} g(\xi_n'(\theta), \theta). \quad (10.7)$$

Proposition 10.1. *Assume that the demand density is bounded and that $\mathbb{P}(\eta_n > S) = p_0 > 0$. Then $\mathbb{E}[|g(\xi_n'(\theta), \theta)|] \leq K < \infty$ for all $\theta \in \Theta = [0, S]$, and assumption (iii) of Theorem 9.1 holds for g.*

Proof. The claim follows using a stochastic domination argument. Under the condition on the demand distribution the probability that both the plus and the minus processes fall below

θ is larger or equal to p_0. This implies that all of the living phantoms die at once when this condition is met. That is, the regeneration time for the process $\{\xi'_n(\theta)\}$, which is $\min(n \geq 1: \mathcal{L}_n = \varnothing)$, is stochastically bounded by a geometric random variable with parameter p_0. Thus it is finite a.s. On the other hand, we have assumed that f_ξ is also bounded, which yields the claim. With this result, Theorem 9.1 holds so $\lim_{N\to\infty} \mathbb{E}[\hat{G}_N^{(\text{MVD})}] = J'(\theta)$ and we may use $g(X_n(\theta), \xi_n)$ as a basis for the derivative estimator. □

It is worth mentioning that the expected running time of this algorithm is linear in N, due to the bound on $\tau_i(\theta)$, because all quantities $d_n(\theta; i)$ are bounded a.s.

Algorithm 10.3 Full (parallel) MVD for the finite horizon inventory using list

---

Initialize $X_1 = x$, MVD $= 0$, $\mathcal{L} = \varnothing$.
MVD-LIST(θ, N)
for $(n = 1, \ldots, N - 1)$ **do**
    Generate demand $\eta_n \sim f_\eta$
    $\Delta = X_n - \eta_n$
    $X_{n+1} = \Delta \mathbf{1}_{\{\Delta > \theta\}} + S \mathbf{1}_{\{\Delta \leq \theta\}}$
    Birth of n-th phantoms:
    $X_n^+(n) = S$, $X_n^-(n) = \theta$
    $f(n) = f_\eta(X_n - \theta)$
    $d(n) = c(X_n^+(n), \eta_n) - c(X_n^-(k), \eta_n)$
    append (n) to list $\mathcal{L}$
    Update living phantoms with new demand:
    for $(i \in \mathcal{L})$ **do**
        $\Delta^\pm(i) = X_n^\pm(i) - \eta_n$
        $X_{n+1}^\pm(i) = \begin{cases} \Delta^\pm(i) & \text{if } \Delta^\pm(i) > \theta \\ S & \text{if } \Delta^\pm(i) \leq \theta \end{cases}$
        $d(i) \leftarrow d(i) + c(X_{n+1}^+(i), \eta_{n+1}) - c(X_{n+1}^-(i), \eta_{n+1})$
        if $(X_{n+1}^+(i) == X_{n+1}^-(i))$ **then remove** $(n+1)$ from list $\mathcal{L}$
for $(k = 1, \ldots, N)$ **do**
    MVD $\leftarrow$ MVD $+ f(k) d(k)$
return MVD$/N$

Stochastic Approximation. Alternatively, consecutive batches can be used to estimate the feedback:

$$Y_n = -\frac{1}{K} \sum_{k=nK+1}^{(n+1)K} g(\xi'_k(\theta_n), \theta_n) \tag{10.8}$$

where g is defined as in (10.7), that is,

$$g(\xi'_k(\theta), \theta) = f_\eta(X_k - \theta_n) \sum_{i \in \mathcal{L}_k} d_k(\theta_n; i), \quad k = nK + 1, \ldots, (n+1)K.$$

It is very important to realize that with this formulation, the $(n + 1)$-st batch starts with the living phantoms that were born before, that is, we do not reset the estimation. This is also the reason why we have initialized the state and the list $\mathcal{L}$ *outside* the function MVD-LIST(θ, K) in Algorithm 10.3. The SA uses (10.8) to update

$$\theta_{n+1} = \Pi_{[0,S]}(\theta_n - \epsilon Y_n) \quad \theta \in [0, S], \tag{10.9}$$

so that θ_n stays feasible. We define the target vector field as the negative gradient: $G(\theta) = -J'(\theta)$, which is coercive for $\min_{\theta \in [0,S]} J(\theta)$ on $[0, S]$, where we note that $G(\theta)$ is also on the boundary points of $[0, S]$ well defined. In the following we show that Theorem 5.4 can be applied to (10.9).

The weak continuity of the Markov kernels in (a1) follows from our analysis of the MVD estimator. Assumption (a2') follows from the fact that $\{X_n(\theta)\}$ are ergodic Markov chains on a bounded state space, so the set of invariant measures is tight. Assumption (a3) follows from Theorem 9.1, as we just verified; see Proposition 10.1. Because the space is bounded, all the estimators have finite support, thus uniformly bounded variance, and this establishes both (a4') and (a5). We end this section with the pseudo-code for the SA for infinite horizon. It calls the function MVD-LIST defined in Algorithm 10.3.

Algorithm 10.4 Stochastic approximation for the infinite horizon problem
---

Read θ_1, ϵ, x, K
Initialize $X_1 = x$, MVD = 0, $\mathcal{L} = \emptyset$
for $(n = 1, \ldots, N)$ **do**
    $Y_n = -$ MVD-LIST(θ_n, K)
    $\theta_{n+1} = \theta_n + \epsilon Y_n$

There is no set rule for choosing the batch size K. Exercise 5.3 shows that the limit ODE is independent of the batch size; however, the limit variance of the interpolated process will decrease with increasing K. Clearly smaller values of K accelerate convergence. In particular, when using the procedure to optimize in real time (perhaps using streaming data) it is highly desirable to spend little time in the "exploration" phase of the algorithm. The batch size could also be chosen to vary with n, for example. Most of the time one must perform pilot experimentation in order to decide the hyper parameters.

10.2 MODEL FITTING (AN IPA APPLICATION)

In the previous section, the inventory model was used a description of an actual inventory system and used for finding the optimal order threshold in an (s, S) policy. In other words, in Section 10.1 the model represented the true model. In this section we take a different perspective. Here, we assume that we have information on the actual costs over the first N periods given to us as data vector $(c_1, \ldots, c_N)$ and we want to fit an inventory model to this data. For this we postulate an inventory and fit the parameters of the model so that the fitted postulated model yields the best available model instance given the available data). Specifically, we postulate that the model introduced in Section 10.1 provides a realistic representation of the inventory system, but following the famous dictum on models—"all models are wrong"—we want to fit the model as best as possible to the actual cost data and thereby compensate for model mismatches. In order to do this, we let the mean of the demand distribution, called θ, be a free parameter and use SA for fitting θ to the model. It is reasonable to assume that we know the true mean demand, say ϑ, from data. Hence, the difference $\theta - \vartheta$ can serve as an indicator for the mismatch that our postulated model has with reality. To summarize, by fitting θ we adjust our incorrect, but analytically treatable model, to real data. In the following we assume that θ is a scale parameter of the continuous distribution of the demands. We express this dependency in the notation by writing $\eta(\theta)$ for the demand. The underlying process is given by $\xi_n(\theta) = (\eta_{n,1}(\theta), \ldots, \eta_{n,N}(\theta))$, that is, a vector of N iid samples of an exponential random variable with mean θ. For this variation

of the inventory model in Section 10.1, the costs depend on θ through the demand. More specifically, for a sample
$$\xi(\theta) = (\eta_1(\theta), \ldots, \eta_N(\theta))$$
of demands, the inventory level evolves as follows:
$$X_{n+1} = \begin{cases} X_n - \eta_n(\theta) & \text{if } \eta_n(\theta) \le X_n - s, \\ S & \text{otherwise,} \end{cases} \tag{10.10}$$
compare with (10.1). Consequenly, the cost evolves as follows:
$$c(X_n, \xi(\theta)) = \begin{cases} h(X_n - \eta_n(\theta)) & \text{if } \eta_n(\theta) \le X_n - s, \\ h(X_n - \eta_n(\theta)) + K & \text{if } X_n - s < \eta_n(\theta) \le X_n, \\ p(\eta_n(\theta) - X_n) + K & \text{if } X_n < \eta_n(\theta), \end{cases} \tag{10.11}$$
compare with (10.2). For fitting the inventory problem to the observed costs, we let
$$J(\theta) = \mathbb{E}\left[\frac{1}{2}\sum_{k=1}^N \left(c(X_n, \xi(\theta)) - c_k\right)^2\right]$$
and solve
$$\min_{\theta \in (0, \infty)} J(\theta). \tag{10.12}$$

From the theory developed in Chapter 7, it is clear that $d\eta_n(\theta)/d\theta = \eta_n(\theta)/\theta$. Since $\mathbb{P}(X_k - \eta_k(\theta) = 0)$ is zero, the IPA derivative for the costs is
$$\frac{d}{d\theta} c(X_n, \xi(\theta)) = \begin{cases} -h\,\eta_n(\theta)/\theta & \text{if } \eta_n(\theta) \le X_k - s, \\ -h\,\eta_n(\theta)/\theta & \text{if } X_n - s \le \eta_n(\theta) < X_n, \\ p\,\eta_n(\theta)/\theta & \text{if } X_n < \eta_n(\theta). \end{cases}$$

From the above we see that the IPA derivate of the n state is bounded by $(h+p)\eta_n(\theta)/\theta$, which shows that $c(X_n, \xi(\cdot))$ is a.s. Lipschitz on a neighborhood of θ; see Lemma 8.2. This allows us to apply Theorem 8.2, which establishes that
$$\frac{d}{d\theta}J(\theta) = \frac{d}{d\theta}\mathbb{E}\left[\frac{1}{2}\sum_{k=1}^N \left(c(X_k, \xi(\theta)) - c_k\right)^2\right] = \mathbb{E}\left[\sum_{k=1}^N \left(\frac{d}{d\theta}c(X_n, \xi(\theta))\right)\left(c(X_k, \xi(\theta)) - c_k\right)\right].$$

The problem belongs to the class of root-finding problems (in this case the data vector $(c_1, \ldots, c_N)$ is the target). The well-posedness of the optimization problem follows from standard arguments.

Stochastic Approximation. We apply a fixed gain size SA to solving $\min_{\theta \in (0,\infty)} J(\theta)$. For this we let
$$\theta_{n+1}^\epsilon = \Pi_{[\delta, 1/\delta]}(\theta_n^\epsilon - \epsilon Y_n), \quad \theta_0 \in [\delta, 1/\delta],$$
for δ sufficiently small to keep θ_{n+1} feasible, where
$$Y_n = \sum_{k=1}^N \left(\frac{d}{d\theta}c(X_n, \xi(\theta))\right)\left(c(X_k, \xi(\theta)) - c_k\right).$$

As detailed above, Y_n is an unbiased estimator for $dJ(\theta)/d\theta$ on $(0, \infty)$ and $G(\theta) = -dJ(\theta)/d\theta$ is coercive for (10.12) on $[\delta, 1/\delta]$ for δ sufficiently small. To see this, observe that (10.12) is a fitting problem the solution of which neither is achieved for arbitrarily small or arbitrarily large values of θ.

Provided the variance of Y_n is uniformly bounded, it follows from Theorem 4.4 together with Theorem 6.1 (where we assume that δ is chosen so that the projection is not active in the limit) that (i) θ_n^ϵ is approximately normal distributed for n sufficiently large and (ii) the mean value of θ_n^ϵ is a proxy of the location of the minimum of $J(\theta)$. It remains to provide a sufficient condition for the variance of Y_n to be bounded. For this, note that

$$c(X_k, \xi(\theta)) \leq (h+p)S + K$$

and thus

$$|c(X_k, \xi(\theta)) - c_k| \leq (h+p)S + K + \max_{1 \leq k \leq N} c_k.$$

Moreover,

$$\left|\frac{d}{d\theta} c(X_n, \xi(\theta))\right| \leq (h+p)\frac{\eta_n(\theta)}{\theta} = (h+p)\eta_n(1),$$

where the last equality stems from the fact that θ is a scale parameter of $\eta_n(\theta)$; see Proposition 8.1. This gives

$$\left|\sum_{k=1}^{N} \left(\frac{d}{d\theta} c(X_n, \xi(\theta))\right) \left(c(X_k, \xi(\theta)) - c_k\right)\right| \leq L \sum_{k=1}^{N} \eta_n(1),$$

for an appropriate value of L. Due to the iid assumption on $\eta_n(\theta)$, it follows that the variance of Y_n is uniformly bounded provided that $\eta_n(1)$ has finite variance.

10.3 VARIATIONS OF THE MODEL

In this section we study variants of the inventory problem.

10.3.1 Finding the Optimal Replenishment Period (An SPA Application)

10.3.1.1 The Basic Model

Consider a single-item inventory system with observation period length θ, that is, every θ time-units the inventory is inspected, and it is immediately replenished up to level S. The decision variable is thus the fixed time θ between (potential) replenishment. The time between consecutive demands (called here the "orders") is a sequence of iid variables γ_k, and we denote their density by f_γ and the mean by m_γ. As in the previous sections, the demand of the k-th customer is η_k, and $\{\eta_k\}$ is an iid sequence with density f_η. To the initial inventory $X_1 = S$, the first customer arrives at time γ_1, which causes holding cost of $hX_1\gamma_1$. The demand, placed at time γ_1, is of size η_1, resulting in the new inventory level X_1. The next demand is placed γ_2 time units later and is of size η_2. The unit holding cost is c and the unit penalty cost is p. Note that we allow X_1 to become negative, which models a backlog of demand. If just before time θ, the inventory level is x, then $S-x$ items are delivered, generating a fixed cost K. The sequence of demand times is given by

$$\tau_k = \sum_{i=1}^{k} \gamma_i$$

for $k \geq 1$, and we denote by $N(\theta)$ the index of the first order placed just after (or at) θ, that is,

$$\tau_{N(\theta)-1} < \theta \leq \tau_{N(\theta)}.$$

For $k \leq N(\theta) - 1$, the process of the inventory level follows, for $k \geq 1$, the simple recursion

$$X_{k+1} = X_k - \eta_k, \quad X_0 = S, \qquad (10.13)$$

and the cost associated with the k-th order are as follows:

- if $X_k - \eta_k > 0$, then

$$C_k = hX_k\gamma_k,$$

because the level X_k was held for γ_k time units and the k-th order could be served from the present inventory,

- if $X_k > 0$ and $0 > X_k - \eta_k$, then $\eta_k - X_k$ could not be served from the inventory, causing penalty cost

$$C_k = p(\eta_k - X_k),$$

where we assume that orders can be split, i.e., the penalty is only paid for the part of the order of size η_k that was not satisfied,

- $X_k < 0$, then

$$C_k = p\eta_k.$$

The accumulated cost is given by

$$\mathbb{E}\left[\sum_{k=0}^{N(\theta)-1} C_k\right].$$

10.3.1.2 Perturbation Analysis

Perturbing θ by $\Delta > 0$ yields

$$\mathbb{E}[C(\Delta, \theta)] := \mathbb{E}\left[\sum_{k=1}^{N(\theta+\Delta)-1} C_k - \sum_{k=1}^{N(\theta)-1} C_k\right] = \mathbb{E}\left[\sum_{k=N(\theta)}^{N(\theta+\Delta)-1} C_k\right],$$

where $C(\Delta, \theta)$ denotes the change in cost. The effect of perturbation Δ is given by the cost incurred by the demand arriving between θ and $\theta + \Delta$. The event that at least one demand arrives between θ and $\theta\Delta$ is given by $\tau_{N(\theta)} \leq \theta + \Delta$. To simply the notation, we denote by γ the residual time to the next demand at θ, that is,

$$\gamma = \tau_{N(\theta)} - \theta,$$

so that

$$\tau_{N(\theta)} \leq \theta + \Delta \text{ if and only if } \gamma \leq \Delta.$$

Note that

$$\mathbb{P}_{\theta, \Delta}(\gamma \leq x) = \mathbb{P}(\gamma_{N(\theta)} + \tau_{N(\theta)-1} \leq x | \gamma_{N(\theta)} + \tau_{N(\theta)-1} \geq \theta).$$

Hence, the evaluation of $C(\Delta, \theta)$ depends on the event $\gamma \leq \Delta$, and we condition $\mathbb{E}[C(\Delta, \theta)]$ on either $\gamma \leq \Delta$ or $\gamma > \Delta$. This type of perturbation analysis belongs to family smoothed perturbation analysis (SPA), and we refer to Section 8.1.3 for a discussion on SPA. With our new notation, we can write

$$\mathbb{E}[C(\Delta, \theta)] = \mathbb{E}\left[\sum_{k=N(\theta)+1}^{N(\theta+\Delta)} C_k \middle| \gamma \leq \Delta\right] \mathbb{P}(\gamma \leq \Delta) + \mathbb{E}\left[\sum_{k=N(\theta)+1}^{N(\theta+\Delta)} C_k \middle| \gamma > \Delta\right] \mathbb{P}(\gamma > \Delta).$$

APPLICATIONS OF STOCHASTIC APPROXIMATION   289

The event $\{\gamma > \Delta\}$ refers to a non critical event: no new demand arrives in $[\theta, \theta + \Delta]$ and the sample cost is affected when increasing θ to $\theta + \Delta$ only through the cost of holding inventory $X_{N(\theta)-1}$ for Δ time units longer. We obtain for the incremental cost

$$\mathbb{E}[C(\Delta, \theta)\mathbf{1}\{X_{N(\theta)-1} > 0, \gamma > \Delta\}] = \mathbb{E}\left[(hX_{N(\theta)-1}\Delta)\mathbf{1}\{X_{N(\theta)-1} > 0\}\Big|\gamma > \Delta\right]\mathbb{P}(\gamma > \Delta).$$

Taking the limit yields

$$\lim_{\Delta \to 0} \frac{1}{\Delta}\mathbb{E}\left[(hX_{N(\theta)-1}\Delta)\mathbf{1}\{X_{N(\theta)-1} > 0\}\Big|\gamma > \Delta, X_{N(\theta)-1}\right]\mathbb{P}(\gamma > \Delta)$$
$$= hX_{N(\theta)-1}\mathbf{1}\{X_{N(\theta)-1} > 0\}.$$

The event $\{\gamma \leq \Delta\}$ refers to the critical event: at least one additional demand is admitted, which causes a discontinuity in the sample cost. More precisely, for Δ sufficiently small, the probability of only one additional demand placed at time $\theta + \gamma \leq \theta + \Delta$ is of order Δ, whereas the probability of more than one additional demand placed in $\theta + \gamma \leq \theta + \Delta$ is of order Δ^2 and this event can thus be neglected.

We proceed now to calculate the effect of the perturbation on the cost, given the critical event. For $X_{N(\theta)-1} = x$, the expected incremental cost caused by Δ sufficiently small can be decomposed into the following separate effects.

1. First, provided that $X_{N(\theta)-1} = x > 0$, $hx\gamma$ is introduced. Moreover, the additional demand of size η placed at time $\theta + \gamma$ leads to the following costs.

    (a) Provided that $X_{N(\theta)-1} - \eta = x - \eta > 0$, this additional demand does not deplete the stock, and we have new holding cost $h(x - \eta)(\Delta - \gamma)$. The cost for this event is given by

    $$hx\gamma + h(x - \eta)(\Delta - \gamma) = h(x - \eta)\Delta + h\eta\gamma.$$

    With this, the change in the expected cost can be expressed in formula as

    $$\mathbb{E}[C(\Delta, \theta)\mathbf{1}\{x - \eta > 0, x > 0\}]$$
    $$= \mathbb{E}\left[(h(x - \eta)\Delta + h\eta\gamma)\mathbf{1}\{x - \eta > 0, x > 0\}\Big|\gamma \leq \Delta\right]\mathbb{P}(\gamma \leq \Delta).$$

    For the following we assume that:
    (i) The limit

    $$\lim_{\Delta \to 0} \frac{1}{\Delta}\mathbb{P}(\gamma \leq \Delta) = r$$

    exists.
    Letting Δ tend to zero, we arrive at

    $$\lim_{\Delta \to 0} \mathbb{E}\left[(h(x - \eta)\Delta + h\eta\gamma)\mathbf{1}\{x - \eta > 0, x > 0\}\Big|\gamma \leq \Delta\right]\frac{\mathbb{P}(\gamma \leq \Delta)}{\Delta}$$
    $$= \lim_{\Delta \to 0} \frac{1}{\Delta}\mathbb{E}\left[h(x - \eta)\Delta\mathbf{1}\{x - \eta > 0, x > 0\}\Big|\gamma \leq \Delta\right]\mathbb{P}(\gamma \leq \Delta)$$
    $$+ \lim_{\Delta \to 0} \mathbb{E}\left[h\eta\gamma\mathbf{1}\{x - \eta > 0, x > 0\}\Big|\gamma \leq \Delta\right]\frac{\mathbb{P}(\gamma \leq \Delta)}{\Delta}$$
    $$= \lim_{\Delta \to 0} \frac{1}{\Delta}\mathbb{E}\left[h(x - \eta)\Delta\mathbf{1}\{x - \eta > 0, x > 0\}\Big|\gamma \leq \Delta\right]\mathbb{P}(\gamma \leq \Delta)$$

$$= \lim_{\Delta \to 0} \frac{1}{\Delta} \mathbb{E}\left[h(x-\eta)\Delta \mathbf{1}\{x-\eta > 0, x > 0\}\right]$$

$$= \mathbb{E}\left[h(x-\eta)\mathbf{1}\{x-\eta > 0, x > 0\}\right].$$

(b) Alternatively, provided that $x > 0$ and the new order of size η depletes the stock, we have an additional holding cost of $hx\gamma$ and a penalty cost $p(\eta - x)$. In formula,

$$\mathbb{E}[C(\Delta, \theta)\mathbf{1}\{x-\eta < 0, x > 0\}]$$
$$= \mathbb{E}\left[\left(hx\gamma + p(\eta - x)\right)\mathbf{1}\{x-\eta < 0, x > 0\}\Big|\gamma \leq \Delta\right]\mathbb{P}(\gamma \leq \Delta).$$

Provided that (i) holds, we obtain for the limit

$$\lim_{\Delta \to 0} \mathbb{E}\left[\left(hx\gamma + p(\eta - x)\right)\mathbf{1}\{x-\eta < 0, x > 0\}\Big|\gamma \leq \Delta\right]\frac{1}{\Delta}\mathbb{P}(\gamma \leq \Delta)$$
$$= p\, r\, \mathbb{E}\left[(\eta - x)\mathbf{1}\{x-\eta < 0, x > 0\}\right].$$

2. Second, in case $x \leq 0$, the additional order will generate an additional penalty. This leads to

$$\lim_{\Delta \to 0} \mathbb{E}\left[p\eta \mathbf{1}\{x \leq 0\}\Big|\gamma \leq \Delta\right]\frac{1}{\Delta}\mathbb{P}(\gamma \leq \Delta) = r\, p\, m_\eta \mathbf{1}\{x \leq 0\}.$$

This leads to the following SPA estimator for the cost sensitivity:

$$\lim_{\Delta \to 0} \frac{1}{\Delta} \mathbb{E}[C(\Delta, \theta)|\mathcal{F}_{N(\theta)-1}] = hX_{N(\theta)-1}\mathbf{1}\{X_{N(\theta)-1} > 0\} \quad (10.14)$$
$$+ \mathbf{1}\{X_{N(\theta)-1} > 0\}\mathbb{E}\left[h(\eta - X_{N(\theta)-1})\mathbf{1}\{\eta - X_{N(\theta)-1} \geq 0\}\right]$$
$$+ \mathbf{1}\{X_{N(\theta)-1} > 0\}r\, p\mathbb{E}\left[(\eta - X_{N(\theta)-1})\mathbf{1}\{\eta - X_{N(\theta)-1} \geq 0\}\right] + \mathbf{1}\{X_{N(\theta)-1}\}r\, p\, m_\eta$$
$$= \mathbf{1}\{X_{N(\theta)-1} > 0\}\left(hx + (h+r\, p)\mathbb{E}\left[(\eta - X_{N(\theta)-1})\mathbf{1}\{\eta - X_{N(\theta)-1} \geq 0\}\right]\right)$$
$$+ \mathbf{1}\{X_{N(\theta)-1} \leq 0\}r p m_\eta.$$

Condition (i) is illustrated with the following example.

Example 10.1. For γ_k iid exponential distributed with mean m_γ, it follows from L'Hôpital's rule (see Section A.9 in Appendix A) that

$$\lim_{\Delta \to 0} \frac{1}{\Delta}\mathbb{P}(\gamma \leq \Delta) = \lim_{\Delta \to 0} \frac{1 - e^{-\Delta/m_\gamma}}{\Delta} = \frac{1}{m_\gamma}, \quad (10.15)$$

where we make use of the memoryless property of the exponential distribution to argue that the residual time γ is exponentially distributed with mean m_γ. Hence, in this case, condition (i) is satisfied with

$$r = \frac{1}{m_\gamma}.$$

The inventory system is run over a fixed time period T. Hence, in $[0, T]$, we have (approximately) T/θ fixed period inspections and we consider

$$\min_{\theta \in (0, \infty)} J(\theta), \quad \text{with} \quad J(\theta) = \frac{T}{\theta}\mathbb{E}\left[\sum_{k=1}^{N(\theta)-1} C_k\right]. \quad (10.16)$$

APPLICATIONS OF STOCHASTIC APPROXIMATION

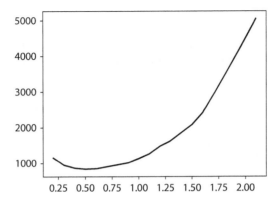

Figure 10.4. The cost function $J(\theta)$ for the inventory with inspection interval θ.

We now argue that the problem is well-posed. On the one hand, as θ tends to zero, we replenish immediately after each demand, which is desirable as penalty costs for being out of stock are avoided; unfortunately, this leads to unbounded increase in the number of inspections T/θ. On the other hand, letting θ tend to ∞, the number inspections tends to zero with the negative side effect that the penalty cost will tend to infinity. This shows that there is some optimal choice for θ balancing both effects in the best possible way. Since $J(\theta)$ is differentiable, with

$$\frac{d}{d\theta}J(\theta) = \frac{T}{\theta}\left(\frac{d}{d\theta}\mathbb{E}\left[\sum_{k=1}^{N(\theta)-1} C_k\right]\right) - \frac{T}{\theta^2}\mathbb{E}\left[\sum_{k=1}^{N(\theta)-1} C_k\right], \quad (10.17)$$

there is a global minimum of $J(\theta)$ that is attained at some finite value that is a stationary point of $dJ(\theta)/\theta$. For an illustration, see Figure 10.4, where we let $S = 10, h = 2, p = 5$, $K = 7$, and $T = 20$. Moreover, we let the time between demands and the demand size be exponentially distributed with mean valus $m_\gamma = 1 = m_\eta$.

Stochastic Approximation. We choose as target vector field the negative gradient, i.e., $G(\theta) = -dJ(\theta)/d\theta$. As can be seen from Figure 10.4, the negative gradient is coercive for (10.16) on, say, $\Theta = [0.375, 0.55]$. The underlying process is $\xi = (\eta_k, \gamma_k : 1 \leq k \leq N(\theta))$ and the problem is of a random horizon type. We apply the algorithm

$$\theta_{n+1} = \Pi_\Theta(\theta - \epsilon_n Y_n), \quad \theta_0 \in \Theta,$$

where the feedback is obtained from (10.17) elaborating on (10.14) for an unbiased SPA estimator for the accumulated cost derivative. The cost function, and thereby the SPA derivative, has relatively large variance. This poses a challenge to our SA algorithm as derivative estimates, though unbiased, can yield updates with such a high variance that the algorithm is likely to be sent in the wrong direction. For our application of SA, we compensate for the larger variance by applying a decreasing gain size algorithm, where $\epsilon_n = 1/(n+1)$, and the initial value $\theta_0 = 0.45$ is close to the location of the minimum, as indicated in Figure 10.4. Convergence of the SA toward the location of the minimum of $J(\theta)$ follows from Theorem 4.2 where we use the fact that Y_n is unbiased for the gradient and has bounded variance for $\theta \in \Theta$.

For the algorithm we updated θ with each observed period of the inventory system. We run the algorithm for $n = 10^4$ iterations yielding $\theta^* = 0.4218$ as solution. Intensive brute-force simulation using 10^6 iid replications and exploiting CRN for variance reduction showed that the location of the minimum is approximately 0.425.

10.3.2 Identifying the Optimal Scenario for Chosen Policy (An SF Application)

The model is the same as in the previous section but now we perform a perturbation analysis of the cost with respect to θ being the mean time between demands. Notation is as introduced in Section 10.3.1.1.

For this we assume that the inventory follows the same fixed time periodic review policy as detailed in Section 10.3.1, and we denote the time between inspections by T, so T is fixed in this case (recall that we let $\theta = T$ in Section 10.3.1). We apply the SF approach for random horizon problems, as detailed in Section 9.2. For this we note that the accumulated cost over n orders can be written as a measurable mapping $L_n(\gamma_1(\theta), \eta_1, \ldots, \gamma_n(\theta), \eta_n)$. Moreover, since θ is the mean of the time between demands, we have that $N(\theta) \leq N(\theta_0)$ for $\theta \geq \theta_0$. Hence, by Theorem 9.2 we have

$$\frac{d}{d\theta} \mathbb{E}\left[\sum_{k=1}^{N(\theta)-1} C_k\right] = \frac{1}{\theta^2} \mathbb{E}\left[\sum_{k=1}^{N(\theta)-1} C_k \sum_{k=1}^{N(\theta)-1} (\gamma_k(\theta) - \theta)\right], \tag{10.18}$$

where we refer to Example 8.8 for the score function of γ_k with respect to θ. As usual, we let

$$J(\theta) = \mathbb{E}\left[\sum_{k=1}^{N(\theta)-1} C_k\right]$$

and consider

$$\min_{\theta \in (0,\infty)} J(\theta). \tag{10.19}$$

In words, we want to identify the mean time between demand placements for which the chosen review period T performs best. This is an admittingly rather academic setting, which can, however, be motivated by assuming that the company may want to influence customers behavior via advertisements and marketing actions while keeping the inspection periods fixed, as changing T incurs high cost due to the internal organization of the inventory.

We argue that the problem is well-posed as follows. For θ approaching zero, the number of orders tends to infinity, creating increasing backlog costs. For θ tending to ∞, no order will be placed and the cost approaches the holding cost of the initial inventory, which is ST. It is obvious that for reasonable values of θ orders occur before time T, which decreases the inventory without causing penalty cost, resulting in overall costs lower than ST. This shows that there exists a finite θ that solves the minimization problem, and since $J(\theta)$ is differentiable throughout $(0, \infty)$ any (local) minimum is a stationary point. Optimization problem (10.19) is thus well-posed and $G(\theta) = -dJ(\theta)/d\theta$ is coercive for (10.19) on $[\delta, 1/\delta]$, for $\delta > 0$.

The underlying process is $\xi(\theta) = (\eta_k, \gamma_k(\theta) : 1 \leq k \leq N(\theta))$ and for the n-th sample of ξ, denoted by $\xi_n(\theta_n) = (\eta_k(n), \gamma_k(\theta_n; n) : 1 \leq k \leq N(\theta_n))$, we obtain as feedback

$$Y_n = -\frac{1}{\theta_n} \sum_{k=1}^{N(\theta_n)-1} C_k(\xi_n(\theta_n)) \sum_{k=1}^{N(\theta_n)-1} (\gamma_k(\theta_n; n) - \theta_n),$$

where $C_k(\xi_n(\theta_n))$ denotes the k-th cost for the input provided by $\xi_n(\theta_n)$. By Theorem 9.2, Y_n is unbiased for the negative gradient field, i.e., $G(\theta_n) = \mathbb{E}[Y_n | \mathfrak{F}_{n-1}]$, for $\theta_n \in (0, \infty)$, see also (10.18). Boundedness of the variance of $\phi^\epsilon(\xi(\theta), \theta)$ for $\theta \in [\delta, 1/\delta]$ allows for

APPLICATIONS OF STOCHASTIC APPROXIMATION 293

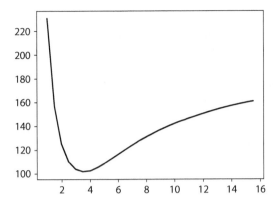

Figure 10.5. The cost function $J(\theta)$ for the inventory with inspection interval T and mean time between orders θ.

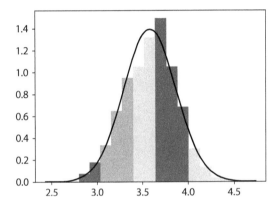

Figure 10.6. Histogram of $\theta_{100}^{0.01}$ with fitted normal density.

applying Theorem 4.4, which yields that the interpolation process of the SA

$$\theta_{n+1} = \Pi_{[\delta,1/\delta]}(\theta_n + \epsilon Y_n), \quad \theta_0 \in [\delta, 1/\delta], \tag{10.20}$$

converges weakly to the solution of the ODE driven by the negative derivative of the cost function $J(\theta)$. In turn, the only asymptotically stable point of this ODE is the solution to (10.19).

For a numerical example, we let $p = 5.0$, $h = 0.5$, $S = 8$, $T = 50.0$, and $m_\gamma = 1$. Figure 10.5 shows the cost as function of θ, where θ denotes the mean value of the time between orders.

Figure 10.5 corroborates that the negative gradient is coercive for minimizing $J(\theta)$ on $\theta \in (0, \infty)$.

For SA, we use $\theta_0 = 2.0$ as initial value, and execute 10^3 iid SA runs over $n = 100$ iterations for fixed $\epsilon = 0.01$, where for each iteration we averaged 10^2 derivative samples. The histogram of the samples of θ_n is shown in Figure 10.6. Gray shading of a bar indicates that the normal approximation is in line with the value of the bar, dark gray shading indicates that a bar is higher than expected from the normal density, and light gray shading indicates that the normal density has larger values than the particular histogram bar.

It is worth noting that Figure 10.6 shows that for $n = 100$ large and $\epsilon = 0.01$ the SA stays inside $[\delta, 1/\delta]$, for, e.g., $\delta = 0.01$, which shows that the projection in (10.20) is not required for n large. We may thus apply Theorem 6.1 to conjecture that θ_n^ϵ is approximately normally distributed for n sufficiently large, which can be statically tested by means of the Jarque-Bera test, see [175]. Inspecting Figure 10.6 (and possibly performing statistical tests) we may accept our collection of 10^3 samples of $\theta_{100}^{0.01}$ as normal distributed and can construct a confidence interval for the location of the minimum. Specifically, the location of the minimum is found to be 3.586 with a standard deviation of 0.286 resulting in an approximated 95% confidence interval for the true location of the minimum of (3.5682, 3.6037).

Chapter Eleven

Pseudo-Gradient Methods

A classical pseudo-gradient method is the Kiefer-Wolfowitz algorithm, which is based on a two-sided finite difference (FD) scheme, see Example 4.1 for the d-dimensional setting. Note that a straightforward application of FD in d dimensions would require 2^d evaluations of the function $J(\theta)$, because for each dimension we need to evaluate the perturbations along each coordinate. This can be impractical in modern applications where the dimension of the control variable θ is very large. Simultaneous perturbation stochastic approximation (SPSA) is a randomized pseudo-gradient method that is an extension of the FD principle to high dimensions; see [292] for an early reference, and the monograph [295] for more details. SPSA is introduced in Section 11.1.1. Other than FD-based pseudo-gradient methods, methods that read the performance function at randomized locations and than rescale the output, have been introduced in the literature. These methods are collectively called Gaussian smoothed functional approximation (GSFA) and will be discussed in Section 11.2. Eventually, we address the application of SPSA and GSFA to constrained optimization problems in Section 11.1.3.

11.1 SIMULTANEOUS PERTURBATION STOCHASTIC APPROXIMATION

11.1.1 First-Order SPSA

The main ingredients for SPSA are the sequence of perturbation vectors $\Delta_n \in \mathbb{R}^d$, and perturbation sizes $\eta_n \in (0, \infty)$. In this section we consider the problem of minimizing some sufficiently smooth cost function $J(\theta)$ over $\mathbb{R}^d$ (constrained optimization will be briefly discussed in Section 11.3). For our analysis, we assume that the successive random perturbation vectors satisfy the following condition:

(C1) $\{\Delta_n\}$ *is a sequence of iid random vectors, with* $\mathbb{E}[\Delta_n(k)] < \infty$ *as well as* $\mathbb{E}[1/\Delta_n(k)] < \infty$, *symmetric around zero, and the components of Δ_n are mutually independent.*

A standard choice for the perturbations is to use a binary random variable for each coordinate:

$$\mathbb{P}(\Delta_n(i) = -1) = \frac{1}{2} = \mathbb{P}(\Delta_n(i) = 1), \quad i = 1, \ldots, d \qquad (11.1)$$

for all n. SPSA uses the following pseudo-gradient:

$$(J_n)_i := (Y_n(\theta_n))_i = \frac{J(\theta_n + \eta_n \Delta_n) - J(\theta_n - \eta_n \Delta_n)}{2\eta_n \Delta_n(i)}, \quad i = 1, \ldots, d \qquad (11.2)$$

in SA-type optimization algorithms for solving $\min J(\theta)$. Because the numerator in (11.2) is independent of i, then evaluating the vector J_n requires only two samples of the feedback function J, which is surprising as the gradient of size d for arbitrary d can be estimated via two samples of the cost function.

In the following we will show that the expected value of $((J_n)_i : 1 \leq i \leq d)$ is a proxy for the gradient of $J(\theta)$. For this, we denote the third-order partial derivatives of $J(\theta)$ by

$$D\mathcal{J}(\theta) = \frac{\partial^{|\iota|}}{\partial \theta_1^{i_1} \partial \theta_2^{i_2} \cdots \partial \theta_d^{i_d}} J(\theta),$$

$\iota = (i_1, \ldots, i_d) \in \{0, 1, 2, 3\}^d$, and $|\iota| = i_1 + i_2 + \cdots + i_d = 3$. Following [293], we impose the condition

(C2) *The third-order derivatives $D\mathcal{J}(\theta)$ are continuous and uniformly bounded in θ.*

We denote by $R(x, \Delta_n)$ the third element of the Taylor series of $J(\theta + \eta_n \Delta_n)$ evaluated at $\theta = x$, which is given by

$$R(x, \Delta_n) = \sum_{\iota, |\iota|=3} D\mathcal{J}(\theta)(\Delta_n(1))^{i_1}(\Delta_n(2))^{i_2} \cdots (\Delta_n(d))^{i_d}.$$

Then working with a Taylor series expansion, like in the one-dimensional case, we get

$$\mathbb{E}[(J_n)_i] = \frac{1}{2\eta_n} \mathbb{E}\left[\frac{1}{\Delta_n(i)} \left(\nabla J(\theta_n)\eta_n \Delta_n + \frac{\eta_n^2}{2}\Delta_n^\top \nabla^2 J(\theta_n)\Delta_n + \eta_n^3 R(x, \Delta_n)\right.\right. \quad (11.3)$$

$$\left.\left. -\nabla J(\theta_n)(-\eta_n \Delta_n) - \frac{\eta_n^2}{2}\Delta_n^\top \nabla^2 J(\theta_n)\Delta_n - \eta_n^3 R(y, -\Delta_n)\right)\right]$$

$$= (\nabla J(\theta_n))_i + \sum_{k=1, k \neq i}^{d} \nabla^2 J(\theta_n) \mathbb{E}\left[\frac{\Delta_n(k)}{\Delta_n(i)}\right] + \eta_n^2 \mathbb{E}\left[\frac{1}{\Delta_n(i)}(R(x, \Delta_n) - R(y, -\Delta_n))\right],$$

for appropriate x and y. By independence, boundedness, and symmetry around zero, it follows that

$$\mathbb{E}\left[\frac{\Delta_n(k)}{\Delta_n(i)}\right] = 0,$$

for $k \neq i$, which shows that J_n is a biased estimator for the gradient. To summarize, we arrive at

$$\mathbb{E}[J_n] = \nabla J(\theta_n) + \beta_n,$$

with

$$\beta_n = \eta_n^2 \mathbb{E}\left[\frac{1}{\Delta_n(i)}(R(x, \Delta_n) - R(y, -\Delta_n))\right].$$

By condition (C2) we have that $\beta_n = O(\eta_n^2)$. The following lemma is a direct application of Theorem 4.1 (establishing a similar result for fixed gain size via Theorem 4.3 is left to the reader as an exercise).

Lemma 11.1. *Let $\theta_{n+1} = \theta_n - \epsilon_n J_n$ with J_n as in (11.2). Assume that*

- *assumption (C1) and (C2) hold,*
- *$\eta_n \to 0$, $\epsilon_n \to 0$ are such that*

$$(i) \quad \sum_n \epsilon_n = \infty, \quad (ii) \quad \sum_n \epsilon_n \eta_n^2 < \infty, \quad (iii) \quad \sum_n \frac{\epsilon_n^2}{\eta_n^2} < \infty,$$

and that

- *the ODE $G(\theta) = -\nabla J(\theta)$ is coercive for the well-posed problem $\min_\theta J(\theta)$.*

If $J(\theta)$ has unique stationary point θ^*, then θ_n converges a.s. to the location of the minimum of $J(\theta)$.

If $J(\theta)$ is, in addition, Lipschitz continuous, then (iii) can be replaced by $\sum_n \epsilon_n^2 < \infty$.

Proof. Note that $\theta_{n+1} = \theta_n - \epsilon_n J_n$ satisfies the martingale difference noise model, and we will apply Theorem 4.1 for proving the claim. We only have to argue that conditions (a2) and (a3) are satisfied. Condition (a2) on the bias is implied by $\sum_n \epsilon_n \eta_n^2 < \infty$ as $\beta_n = O(\eta_n^2)$. Following the line of argument put forward in Example 4.1 it follows that the variance of the SPSA estimator is of order $1/\eta^2$. The variance condition (a3) in Theorem 4.1 is thus satisfied if

$$\sum_n \frac{\epsilon_n^2}{\eta_n^2} < \infty,$$

By coercivity of the vector field, the trajectories track the location of the optimal value (see Lemma 2.29), which proves the first part of the claim.

We turn to the second part of the lemma. By Lipschitz continuity of J with Lipschitz constant L, J_n can be bounded by

$$|(J_n)_i| \leq \frac{|J(\theta_n + \eta_n \Delta_n) - J(\theta_n)|}{2\eta_n |\Delta_n(i)|} + \frac{|J(\theta_n) - J(\theta_n - \eta_n \Delta_n)|}{2\eta_n |\Delta_n(i)|},$$

$$\leq \frac{\|\eta_n \Delta_n\| L}{2\eta_n |\Delta_n(i)|} + \frac{\|-\eta_n \Delta_n\| L}{2\eta_n |\Delta_n(i)|}$$

$$= \frac{\|\Delta_n\|}{|\Delta_n(i)|} L,$$

for $1 \leq i \leq d$, which gives by (C1) that the variance of J_n is bounded. The variance condition (a3) in Theorem 4.1 is thus satisfied through

$$\sum_n \epsilon_n^2 < \infty. \qquad \square$$

We obtain for SPSA that the conditions on ϵ_n and η_n are satisfied if we let, for example, $\epsilon_n = 1/(n+1)$ and $\eta_n = 1/n^\alpha$, for $\alpha < 1/2$. SPSA can also be applied in fixed gain size SA, and we refer to [109, 110] for details.

We now discuss the application of SPSA to a stochastic objective function, that is, the problem is to find the minimum of a cost function $J(\theta) = \mathbb{E}[h(\xi, \theta)]$. In the following, we let $\{\xi_n\}$ denote a sequence of iid realizations of the underlying process variables. Let

$$(Z_n)_i := (Z_n(\theta_n, \xi_n))_i = \frac{h(\xi_n, \theta_n + \eta_n \Delta_n) - h(\xi_n, \theta_n - \eta_n \Delta_n)}{2\eta_n \Delta_n(i)}, \qquad (11.4)$$

for $1 \leq i \leq d$. Note that, like for the deterministic setting, Z_n requires only two samples of the feedback function h. Taking expected values in (11.4) yields

$$\mathbb{E}[(Z_n(\theta_n,\xi_n))_i|\Delta_n] = \frac{\mathbb{E}[h(\xi_n,\theta_n+\eta_n\Delta_n)|\Delta_n] - \mathbb{E}[h(\xi_n,\theta_n-\eta_n\Delta_n)|\Delta_n]}{2\eta_n\Delta_n(i)}$$
$$= \frac{J(\theta_n+\eta_n\Delta_n) - J(\theta_n-\eta_n\Delta_n)}{2\eta_n\Delta_n(i)}$$
$$= (J_n)_i,$$

for $1 \leq i \leq d$. The only difference with the deterministic setting is that we have to replace $J(\theta)$ by an estimate of $h(\xi,\theta)$, which creates additional variance in Z_n. Lemma 11.1 applies to this version of SPSA as well. For an early reference on the convergence result, we refer to [293]. As for the classical finite difference estimator put forward in Example 4.1, we obtain for SPSA that the conditions on ϵ_n and η_n are satisfied if we let, for example, $\epsilon_n = 1/(n+1)$ and $\eta_n = 1/n^\alpha$, for $\alpha < 1/2$.

Remark 11.1. The variance of SPSA stems from the randomness of the perturbation vectors Δ_n, see (C1). In [37] it is shown that, for d even, sampling of Δ_n can be replaced by a carefully chosen sequence of d deterministic perturbation vectors that are repeated in the optimization. These deterministic instances of Δ_n are then given as rows of a so-called Hadarmard matrix

$$H_2 = \begin{pmatrix} 1 & 1 \\ 1 & -1 \end{pmatrix} \quad \text{and} \quad H_{2^k} = \begin{pmatrix} H_{2^{k-1}} & H_{2^{k-1}} \\ H_{2^{k-1}} & -H_{2^{k-1}} \end{pmatrix}, \quad \text{for } k > 1.$$

11.1.2 Newton-Raphson-Type SPSA

The Newton-Raphson method is an adaptive stepsize method (see Section 1.2), where the stepsize ϵ_n is given by the inverse of the Hessian of $J(\theta)$ at θ_n. In the one-dimensional case, i.e., $\theta \in \mathbb{R}$, the second-order derivative of $J(\theta)$ can be approximated via a central finite difference by

$$\frac{d^2}{d\theta^2} J(\theta) \approx \frac{1}{2\Delta} \left(\frac{d}{d\theta} J(\theta+\Delta) - \frac{d}{d\theta} J(\theta-\Delta) \right)$$

and inserting the central finite difference approximation for the derivatives

$$\approx \frac{1}{2\Delta} \left(\frac{J(\theta+\Delta+\Delta) - J(\theta+\Delta-\Delta)}{2\Delta} - \frac{J(\theta-\Delta+\Delta) - J(\theta-\Delta-\Delta)}{2\Delta} \right) \quad (11.5)$$
$$= \frac{1}{4\Delta^2} \Big(J(\theta+2\Delta) + J(\theta-2\Delta) - 2J(\theta) \Big).$$

A straightforward finite-difference approximation of the Newton-Raphson algorithm therefore becomes

$$\theta_{n+1} = \theta_n - \frac{2\Delta_n}{(J(\theta_n+2\Delta_n) + J(\theta_n-2\Delta_n) - 2J(\theta_n))} \Big(J(\theta_n+\Delta_n) - J(\theta_n-\Delta_n) \Big), \quad (11.6)$$

for $\{\Delta_n\}$ a sequence of differences that tends to zero as n tends to ∞. When running the algorithm, we have to ensure that the approximation of the second-order derivative does not become too close to zero in order to prevent numerical instabilities.

We now turn to the SPSA algorithm in the multidimensional case. Let $\{\Delta_n\}$ and $\{\tilde{\Delta}_n\}$ be two mutually independent sequences satisfying (A1), choose $\eta_n > 0$ such that $\lim_n \eta_n = 0$, following (11.5) we introduce the matrix-valued mapping $H(\theta)$, with elements

$$(H_n(\theta))_{ij} =$$
$$\frac{\left(J(\theta+\eta_n(\Delta_n+\tilde{\Delta}_n))-J(\theta+\eta_n(\Delta_n-\tilde{\Delta}_n))-\left(J(\theta+\eta_n(-\Delta_n+\tilde{\Delta}_n))-J(\theta+\eta_n(-\Delta_n-\tilde{\Delta}_n))\right)\right)}{4\eta_n^2 \Delta_n(i)\tilde{\Delta}_n(j)},$$

as SPSA approximation for the Hessian. The SPSA version of the Newton-Raphson algorithm is given by the following adaptive stepsize algorithm:

$$\theta_{n+1} = \theta_n - \epsilon_n \hat{H}_n^{-1} J_n, \quad \hat{H}_n = f(\bar{H}_n)$$
$$\bar{H}_n = \frac{n}{n+1}\bar{H}_{n-1} + \frac{1}{n+1}H_n(\theta_n),$$

for $n \geq 0$, where $f(\cdot)$ is a mapping designed to cope with possible nonpositive definiteness of $\bar{H}_n$. Note that an update of the algorithm in (11.6) requires a total of six measurements independent of the dimension of θ. In running the algorithm, the approximation of the inverse of the Hessian potentially causes a large variance of the updates and averaging independent measures of H_n is often advisable.

A practical form for $f(\cdot)$ is to let

$$f(\bar{H}_n) = \left(\bar{H}_n \bar{H}_n\right)^{\frac{1}{2}} + \delta_n I_n,$$

with I_n the unit matrix of appropriate size, $\delta_n \geq 0$ is some small number, and the square root is the (unique) positive square root of a semidefinite square matrix. Alternatively, one may let

$$f(\bar{H}_n) = \bar{H}_n \bar{H}_n + \delta_n I_n,$$

for δ_n large enough to ensure postive definiteness of $f(\bar{H}_n)$. If the dimension of θ is large, then for the sake of numerical efficiency, it is advisable to let $f(\bar{H}_n)$ be the matrix that is zero except for the diagonal elements of $\bar{H}_n + \delta_n I_n$. Under conditions put forward in Lemma 11.1 and additional smoothness conditions, it can be shown that θ_n converges a.s. to a local minimum of $J(\theta)$; see [294, 295]. In case that $J(\theta) = \mathbb{E}[h(\xi, \theta)]$ a similar convergence result holds; see [294, 295].

11.1.3 Further Readings on SPSA

Through the years, SPSA has put itself on the map as an efficient general purpose pseudo-gradient method; see [37] for a recent monograph on SPSA and [35] for an overview of SPSA for constrained optimization. For recent results on the choice of the gain size we refer to [50, 52]. Extensions to constrained problems involving the projection operator can be found in [35, 97, 276, 324]. Additionally, [333] provides an extension by considering updates given as weighted sum of two SPSA pseudo-gradients. For an application of SPSA to classification, we refer to [135]. For recent work on SPSA for reinforcement learning, we refer to [254]. Furthermore, SPSA has been successfully applied to neural networks, where it is was shown that SPSA outperforms the industry-standard back-propagation algorithm; see [237, 291]. An extension of SPSA using updates mixed with past updates is provided in [187]. Finally, applications of SPA to discrete optimization are reported in [107, 108], where the SPSA algorithm is introduced for the optimal buffer allocation problem; notably, [251] solves a staffing problem with the help of SPSA. For general results of SPSA for discrete parameter optimization, we refer to [36]. Extension of SPSA by additional noise injection are presented in [213] and conditions under which the ensuing SPSA provably converges toward the global minimum are provided therein.

11.2 GAUSSIAN SMOOTHED FUNCTIONAL APPROXIMATION

The derivation of the GSFA presented in this section follows [76]. For ease of presentation, let θ be one dimensional. For some smooth cost function $J(\theta)$ consider the Taylor series

$$J(\theta + \eta\Delta) = J(\theta) + J'(\theta)\eta\Delta + \frac{\eta^2}{2}\Delta^2 J^{(2)}(\theta) + \frac{\eta^3}{6}\Delta^3 J^{(3)}(\theta) + \ldots$$

where $J^{(i)}$ is the i-th order derivative of the function J. Multiplying the equation by $\eta\Delta$ gives

$$J(\theta + \eta\Delta)\eta\Delta = J(\theta)\eta\Delta + J'(\theta)\eta^2\Delta^2 + \frac{\eta^3}{2}\Delta^3 J^{(2)}(\theta) + \frac{\eta^4}{6}\Delta^4 J^{(3)}(\theta) + \ldots$$

Now assume $\mathbb{E}[\Delta] = \mathbb{E}[\Delta^3] = 0$, $\mathbb{E}[\Delta^2] = 1$, and $\mathbb{E}[\Delta^k] = \rho_k \in \mathbb{R}$ for $k > 3$, then

$$\mathbb{E}\left[J(\theta + \Delta)\eta\Delta\right] = J'(\theta)\eta^2 + \rho_4 \frac{\eta^4}{6} J^{(3)}(\theta) + \ldots$$

Dividing by η^2, we have

$$\mathbb{E}\left[\frac{1}{\eta} J(\theta + \eta\Delta)\Delta\right] = J'(\theta) + \rho_4 \frac{\eta^2}{6} J^{(3)}(\theta) + \ldots \qquad (11.7)$$

In case that $J^{(3)}(\theta)$ is bounded in θ, it follows that the bias of the above righthand side as proxy for the derivative is of order $O(\eta^2)$. It is worth noting the resemblance of the above expression for the derivative to Stein's equation (or Stein's lemma) in statistics, where it is shown that $\mathbb{E}[J'(\Delta)] = \mathbb{E}[\Delta J(\Delta)]$, for Δ a standard normal random variable; see, e.g., [296].

In case $d = 1$, a suitable choice of Δ is to let it follow a standard normal distribution, and in the higher dimensional case, we may let d be vector of mutually independent standard normal variables. We arrive at the following algorithm (establishing a similar result for fixed stepsize is left to the reader as exercise).

Lemma 11.2. *Let $\{\Delta_n\}$ be an iid sequence of multivariate normal random variables, $J \in C^3$, and consider the algorithm*

$$\theta_{n+1} = \theta_n - \epsilon_n \frac{\Delta}{\eta_n} \left(J(\theta_n + \eta_n \Delta) - J(\theta_n)\right),$$

where

- *$J(\theta)$ is Lipschtiz and (C2) holds,*
- *$\eta_n \to 0$, $\epsilon_n \to 0$ such that $\sum_n \epsilon_n = \infty$, $\sum_n \epsilon_n^2 < \infty$, and $\sum_n \epsilon_n \eta_n^2 < \infty$,*
- *and the ODE $G(\theta) = -\nabla J(\theta)$ is coercive for the well-posed problem $\min_\theta J(\theta)$.*

If $J(\theta)$ has a unique stationary point θ^*, θ_n converges a.s. to θ^*, which is the global minimum of $J(\theta)$.

Proof. As Δ_n has mean zero it holds that

$$\mathbb{E}\left[\frac{1}{\eta} J(\theta + \eta\Delta)\Delta\right] = \mathbb{E}\left[\frac{\Delta}{\eta}\left(J(\theta + \eta\Delta) - J(\theta)\right)\right]. \qquad (11.8)$$

We can now refer to (11.7) for showing that the bias of $\Delta_n (J(\theta_n + \eta_n \Delta_n) - J(\theta_n))/\eta_n$ is of order η_n^2. Since the ODE, given by the negative gradient, is coercive for the minimization problem, the trajectories of the ODE trace the location of the minimum of $J(\theta)$; see Lemma 2.29. The lemma now follows directly from Theorem 4.1 as the bias is of order $O(\eta_n^2)$ and the variance is bounded. To see the latter, argue as follows. By Lipschitz continuity,

$$\left| \frac{\Delta}{\eta} (J(\theta + \eta \Delta) - J(\theta)) \right| \leq \frac{||\Delta||}{\eta} |J(\theta + \eta \Delta) - J(\theta)| \leq ||\Delta||^2 L,$$

where L denotes the global Lipschitz constant for $J(\theta)$. This shows that the variance of $\frac{1}{\eta} J(\theta + \eta \Delta) \Delta / \eta$ is bounded. □

Note that if we let $\eta_n = \epsilon_n = 1/(n+1)$ in Lemma 11.2, which we may, the algorithm becomes

$$\theta_{n+1} = \theta_n - \Delta_n (J(\theta + \eta_n \Delta_n) - J(\theta_n)),$$

for $n \geq 0$. The gradient proxy

$$\frac{\Delta_n}{\eta_n} (J(\theta_n + \eta_n \Delta_n) - J(\theta_n))$$

for the algorithm in Lemma 11.2 belongs to the class of GSFA algorithms. GSFA was introduced in [180], and has been further developed in [76, 79, 297]. The term "Gaussian smoothing" refers to the fact that in the aforementioned papers the gradient proxy was not introduced in the straightforward manner but via a smoothing approach, where the basic idea is that of replacing $J(\theta)$ by a smoothed version $\mathbb{E}[J(\theta + \eta \Delta)]$; see [228]. Other than SPSA, GSFA is not built from an FD estimator but on the basic equation (11.7). Using GFSA in SA therefore belongs to the family of so-called derivative-free algorithms; see [80].

11.3 FEASIBLE PERTURBED PARAMETER VALUES FOR SPSA AND GSFA

We conclude with a comment on the philosophy of SPSA and GSFA. While the randomization approach, which is the basis of SPSA and GSFA, sounds almost to good to be true, such a randomization comes for constrained problems at the price of being able to guarantee that the randomized parameter values $\theta \pm \eta \Delta$ are feasible, which becomes an issue when optimizing over a constrained set Θ. Indeed, the problem that a sample $\theta \pm \eta \Delta$ falls outside of Θ is inherent with any random direction algorithm. In case the solution of the minimization problem is a boundary solution, i.e., lies on the boundary of Θ, non feasible solutions are sampled with high probability even for small perturbations $\pm \eta \Delta$. While soft constraints can be dealt with through applying SPSA or GSFA in combination with, e.g., the penalty method (see Section 1.4), hard constraints pose a very challenging problem for pseudo-gradient methods such as SPSA and GSFA; see Section 1.5 for a discussion of soft and hard constrains. In the presence of hard inequality constraints, it is important to note that one cannot simply project $\pm \eta \Delta$ onto Θ because the resulting SPSA nominator $J(\Pi_\Theta(\theta + \eta \Delta)) - J(\Pi_\Theta(\theta - \eta \Delta))$ violates the condition in the proof of SPSA since

$$\mathbb{E}[\Pi_\Theta(\theta + \eta \Delta)) - \Pi_\Theta(\theta - \eta \Delta)] \neq 0,$$

in general. In the presence of inequality constraints the above problem can be dealt with by replacing the inequality, say $\theta \leq 0$ by $\theta + c \leq 0$. Choosing $\eta \leq c$, we ensure that for any

θ such that $\theta \leq -c$, the perturbed values $\theta \pm \eta\Delta$ are feasible, and SPSA can be applied. After updating θ_n in a SA the resulting value is projected back to $(\infty, -c]$. Letting c tend to zero as a mapping of n, SPSA will solve the constrained optimization problem. This approach goes back to [276] and for a recent extension we refer to [287]. Linear equality constraints can be dealt with by means of a base transformation; see [100]. We conclude this section by showing how a hard constraint can be dealt with for GSFA by means of a change of variable.

Example 11.1. Consider the mining example introduced in Example 7.1 and further analyzed in Example 7.3. A straightforward application of GSFA to the objective

$$\mathbb{E}\left[\frac{X(\theta)}{\sqrt{X(\theta)+1}}\right]$$

would give

$$\frac{\Delta}{\eta}\left(\frac{X(\theta+\eta\Delta)}{\sqrt{X(\theta+\eta\Delta)+1}} - \frac{X(\theta)}{\sqrt{X(\theta)+1}}\right)$$

as a GSFA estimator. Our assumption that $X(\theta)$ is exponentially distributed with mean 50θ, implies that $\theta > 0$ is a hard constraint for this problem. As $X(\theta)$ has no probabilistic interpretation for $\theta \leq 0$, the GSFA expression is not well defined with positive probability (i.e., on the event $\theta + \eta\Delta \leq 0$). Fortunately, we may apply a change of variable and consider the problem

$$\mathbb{E}\left[\frac{X(\theta^2)}{\sqrt{X(\theta^2)+1}}\right]$$

with ensuing GFSA

$$\frac{\Delta}{\eta}\left(\frac{X((\theta+\eta\Delta)^2)}{\sqrt{X((\theta+\eta\Delta)^2)+1}} - \frac{X(\theta^2)}{\sqrt{X(\theta^2)+1}}\right).$$

Since $(\theta + \eta\Delta)^2 > 0$ almost surely, the GSFA estimator is now well defined. For the SA, we choose $\rho > 0$ small and consider the projected SA

$$\theta_{n+1} = \Pi_{[\rho,\infty)}\left(\theta_n - \epsilon_n \frac{\Delta_n}{\eta_n}\left(\frac{X((\theta_n+\eta_n\Delta_n)^2)}{\sqrt{X((\theta_n+\eta_n\Delta_n)^2)+1}} - \frac{X(\theta_n^2)}{\sqrt{X(\theta_n^2)+1}}\right)\right),$$

with Δ_n, ρ_n and ϵ_n as in Lemma 11.2; and θ_n is then the approximation for the location of the minimum of the original problem.

It is worth noting that the change of variable has introduced a new stationary point. Indeed, by the chain rule

$$\frac{d}{d\theta}\mathbb{E}\left[\frac{X(\theta^2)}{\sqrt{X(\theta^2)+1}}\right] = 2\theta \left.\frac{d}{d\vartheta}\right|_{\vartheta=\theta^2} \mathbb{E}\left[\frac{X(\vartheta)}{\sqrt{X(\vartheta)+1}}\right],$$

which gives $\theta = 0$ as new stationary point. Projection on $[\rho, \infty)$ excludes this stationary point, which can be done without any harm as inspecting Figure 7.1 shows that the solution of the minimization problem is not altered by excluding parameter range $[0, \rho)$. ✻✻✻

Chapter Twelve

IPA for Discrete Event Systems

This chapter summarizes the known results when applying IPA to an important class of processes known as *discrete event systems*.

12.1 DISCRETE EVENT SYSTEMS

A discrete event system (DES) is a system that changes its state at discrete points in time due to the occurrence of events (for a summary of results on gradient estimation for DES systems, see [309]). An example of a DES is a service system with J service stations, where a state $s = (s_1, \ldots, s_J)$ of the DES is the vector reporting the number of jobs s_j present at station j, for $1 \le j \le J$, and the events refer to either an external arrival or a service completion at one of the J stations. DESs cover transportation networks, production systems, multi-agent systems, and communication networks. Generally speaking, queuing systems can be interpreted as DES. In the following we will provide a precise definition of a DES. Subsequently, we show that under very mild conditions, IPA estimators can be built in a generic form for standard performance measures of DES.

12.1.1 The Basic Setup

A DES is described by means of:

- The *physical state space*, denoted by S.
- A finite *set of possible events*, denoted by E. An event $e \in E$ can be *active* or *inactive*. The set of all active events in state s is called the *event list* and is denoted by $L(s)$. We assume that $L(s) \ne \emptyset$ for all $s \in S$. With each event $e \in E$ we associate a lifetime distribution $F(e)$, and we denote the n-th lifetime of event e by $X(e, n)$, that is, $\mathbb{P}(X(e, n) \le x) = F(e; x)$, for all $n \ge 1$.
- The (physical) *state transition mapping* ϕ. Due to the occurrence of an event, the system state changes. Given that the transition out of state s is triggered by the occurrence of event e, the new state is $s' = \phi(e, s)$. The state transition mapping may be a random mapping and we let

$$p(e, s; s') = \mathbb{P}(s' = \phi(e, s)),$$

for all $s, s' \in S$ and $e \in L(s)$. The set of events that stay active in s', denoted by $O(e, s; s')$, is given by

$$O(e, s; s') = L(s') \cap (L(s) \setminus \{e\}),$$

where the letter "O" refers to "old," and the set of new events in state s', denoted by $\mathcal{N}(e, s; s')$, is given by

$$\mathcal{N}(e, s; s') = L(s') \setminus (L(s) \setminus \{e\}),$$

where the letter "$\mathcal{N}$" refers to "new." It is assumed that any event $e \in L(s)$ that does not trigger the transition out of s, is still active in $\phi(e, s)$; in formula $e \in L(\phi(e', s))$ for all $e' \in L(s)$ and $e \neq e'$. This is called the *non interruption condition*.

A realization of a DES is constructed as follows. The initial state is $S(0) = s_0$. For all events $e \in L(S(0))$ we sample lifetimes $X(e, 1)$, and we store their values as *clock times* $C(e, 0) = X(e, 1)$. For $e \notin L(s(0))$, we let $C(e, 0) = \infty$. The idea is that that clock times show the residual lifetimes of the events. This gives the augmented state $Z(0) = (S(0); C(e, 0) : e \in E)$. In the following we describe the state-transition from $Z(n) = (S(n); C(e, n) : e \in E)$ to $Z(n+1)$.

The holding time $T(n)$ of the physical state $S(n)$ is given by the minimal residual/remaining lifetime of the active events

$$T(n) = \min\{C(e, n) : e \in L(S(n))\}.$$

The event (there may be several of them) with the smallest remaining clock time is the next to occur. Suppose that $e', e'' \in L(S(n))$ with $C(e', n) = C(e'', n) = T(n)$, then e' and e'' occur simultaneously, and the event $e(n)$ that triggers the transition out of $S(n)$ is selected according to a *selection rule h*. Thus, the event that triggers the state transition is uniquely defined by

$$e(n+1) = h(\arg_e \min\{C(e, n) : e \in L(S(n))\}),$$

and we record the event epoch by

$$\tau(n+1) = \sum_{k=0}^{n} T(k),$$

where we let $\tau(0) = 0$. Note that $e(n+1)$ is a.s. uniquely defined without using a selection rule if all lifetime distributions are continuous. However, in the presence of lifetime distributions with discrete components, simultaneous events may occur with positive probability and the sample path dynamics are only completely defined with the help of the selection rule h. We denote by

$$N(e, n) = \sum_{k=1}^{n} \mathbf{1}(e(n) = e)$$

the number of occurrences of e among the first n occurring events. Moreover, we denote by $r(n)$ the n-th *event pair* $(e(n), N(e(n), n))$, i.e., if $r(4) = (e, 3)$, then the third occurrence of event e constitutes the fourth overall event. The new state visited is

$$S(n+1) = \phi(e(n+1), S(n)).$$

The clock times of all events that stay active in $S(n+1)$ are decreased by the holding time

$$C(e, n+1) = C(e, n) - T(n), \quad e \in O(e(n+1), S(n); S(n+1)),$$

and for the new event(s) e in $\mathcal{N}(e(n+1), S(n); S(n+1))$ we sample clock times $X(e, N(e, n)+1)$ from the corresponding lifetime distribution $F(e)$. For $e \notin L(S(n+1))$, we set $C(e, n+1) = \infty$.

We denote by $T(e, k)$ the time of the k-th occurrence of event e, i.e., $T(e, k) = \tau(n)$ if $e(n) = e$ and $k = N(e, n)$.

The above procedure generates the sequence $(S(k), (C(e, k) : e \in E))$, $k \geq 0$, of physical states and clock times. From this sequence we construct the continuous time process

$(S_t : C_t \geq 0)$ by

$$S_t = S(n) \quad \text{if and only if} \quad \tau(n) \leq t < \tau(n+1),$$

and

$$C_t = (C_t(e) : e \in E),$$

with

$$C_t(e) = C(e,n) - (t - \tau_n) \quad \text{if and only if} \quad \tau(n) \leq t < \tau(n+1),$$

where we let $\infty - x = \infty$ for all $x \in \mathbb{R}$. The process $(S_t : 0 \geq t)$ is the physical state process, which becomes, if augmented with the clock time process $(C_t : 0 \geq t)$, a Markov process. The component $(S_t : 0 \geq t)$ is called a *generalized semi-Markov process* (GSMP) and has been intensively studied in the literature as a GSMP provides a mathematical model for a discrete event Monte Carlo simulation; see, for example, [125, 278, 279, 325].

We denote the number of state transitions until time t by $N(t)$. The accumulated performance L over the state process is then given by

$$L_T = \int_0^T L(S_t)\, dt = \sum_{k=0}^{N(T)-1} T(k)\, L(S(k)) + L(S(N(T)))(T - \tau(N(T))).$$

The accumulated performance over the first m events is obtained from

$$L_m = \int_0^{\tau(m)} L(S_t)\, dt = \sum_{k=0}^{m-1} T(k)\, L(S(k)),$$

provided that $\tau(m) < \infty$; and the accumulated performance until the k-th occurrence of e is

$$L_{e,k} = \int_0^{T(e,k)} L(S_t)\, dt = \sum_{k=0}^{N(e,k)-1} T(k)\, L(S(k)),$$

provided that $T(e,k) < \infty$.

Example 12.1. Consider a single server queue so that $(S_t : t \geq 0)$ denotes the queue length process, i.e., $S_t \in \mathbb{N}$ gives the number of jobs at the queue at time t, for $t \geq 0$. Letting $L(s) = s$, for $s \geq 0$, the average queue length over the first T time units is given by

$$\frac{1}{T} \int_0^T S_t\, dt.$$

Let β be the "service completion" event, and let $T(e,n)$ denote the time of the n-th occurrence of event e. If $S(T(\beta,n)) = 0$, i.e., the n-th departing job empties the queue, then

$$\frac{1}{T(\beta,n)} \int_0^{T(\beta,n)} S_t\, dt$$

is the sojourn time averaged over the first n jobs. The throughput over the first n departures, is obtained from $n/T(\beta,n))$.

In a queueing network consisting of J service stations, letting L denote the projection of $s = (s_1, \ldots, s_J)$ on the j coordinate, i.e., $L(s) = s_j$, we obtain the average queue length, average sojourn time, and the throughput at station j in a similar way. ✲✲✲

12.1.2 Extension of the GSMP Framework

In the GSMP framework introduced above, lifetimes of events run down at unit speed. An extension of model is to allow for varying speeds of clocks. Models with speeds allow to freeze a clock by setting the speed at which the clock runs down to zero. This extends the standard GSMP to systems with pre-emptive and resume service discipline. Moreover, with speeds state-dependent service rates can be modeled, which occur in systems with processor sharing. For details on IPA for GSMPs with speeds, we refer to [114].

GSMPs were originally introduced through supplementing the physical state with the ages of events; see [188, 214] for early references. In the following, we present this alternative GSMP formalism. For physical state s, let $(a_e : e \in E)$ denote the age variables of all events where we let $a_0 = -\infty$ if e is not active in s. The *integrated hazard function* of event e, denoted by $\Lambda_e(u)$, is given by

$$\Lambda_e(u) = \begin{cases} -\ln(1 - F(e,u)) & \text{if } u \geq 0, \\ 0 & \text{otherwise,} \end{cases}$$

where the value $\Lambda_e(u) = \infty$ is not excluded. The cdf of the lifetime of e can be recovered through

$$F(e;u) = 1 - e^{-\Lambda_e(u)}, u \geq 0.$$

We assume that $F(e)$ has pdf $f(e)$. The *hazard rate function* of e is defined by

$$\lambda(e,u) = \frac{f(e,u)}{1 - F(e,u)} = \frac{\partial}{\partial u} \Lambda_e(u), \quad u \geq 0,$$

and describes the rate with which the lifetime of event will end at time u provided that the event is u time units old. The pdf of the lifetime of e can be recovered from the hazard-rate function through

$$f(e,u) = \lambda_e(u) e^{-\int_0^u \lambda_e(v)\,dv}, u \geq 0.$$

The state description of the GSMP with ages is given by the physical state s, the holding time of the state, denoted by T, and the age vector $(a_e : e \in E)$. Including the holding time of a state has the advantage that the event that triggers the transition out of s is equal to e^* with probability

$$\frac{\lambda_{e^*}(T)}{\sum_{e \in E} \lambda_e(T)}.$$

Occurrence of e^* triggers the transition to the new physical state s^*. Next, the ages are updates where, for example, the age of all events in $O(s, e^*; s^*)$ are updated through $a_e^* := a_e + T$. The holding time of the new state s^* has then density

$$\left(\sum_{e \in E} \lambda_e(u + a_e^*)\right) \exp\left(-\int_0^u \sum_{e \in E} \lambda_e(w + a_e^*) dw\right).$$

In case of exponential lifetimes, the above formulae simplify considerably. Generally speaking, the age GSMP construction is less useful for the actual construction of the system process $(S_t : t \geq 0)$ but offers interesting properties for sensitivity analysis as the triggering event as well as the holding time density become differentiable. We refer to [245] for details on gradient estimation using the age representation, and to [114] for the exponential case. As a historical note, we mention that this GSMP framework was introduced by [189].

12.2 THE COMMUTING CONDITION

Next we introduce a fundamental concept in the analysis of GSMPs.

Definition 12.1. A GSMP is said to satisfy the commuting condition (CC) if

$$\forall s, \forall e, e' \in L(S): \phi(e, \phi(e', s)) = \phi(e', \phi(e, s)).$$

The CC expresses the structural property that the state reached by the execution of active events is independent of the order of their execution.

Example 12.2. Consider a single server queue with infinite buffer. The physical state space $S = \mathbb{N}$ gives the number of jobs at the queue. As events we have α as the event of an arrival of a new job, and β as the event of a service completion; i.e., $E = \{\alpha, \beta\}$ and $L(0) = \{\alpha\}$, $L(s) = \{\alpha, \beta\}$, for $s = 1, 2, \ldots$. Moreover, we have for $s \in \{1, 2, \ldots\}$,

$$\phi(\alpha, s) = s+1 \text{ and } \phi(\beta, s) = s-1,$$

and $\phi(\alpha, 0) = 1$. For $\alpha, \beta \in L(s)$, i.e., $s \geq 1$, we have $\phi(\alpha, s) = s+1$ and $\phi(\beta, s+1) = s$, so that $\phi(\beta, \phi(\alpha, s)) = s$. Switching the sequence of α and β again, yields $\phi(\alpha, \phi(\beta, s)) = s$. This shows that the GSMP associated with the single server queue with infinite buffer satisfies the CC. Now suppose that the single server has B buffer places so that in total $B+1$ jobs can be at the queue. Furthermore, assume that a job arriving to a full queue is lost. Let $s = B+1$, i.e., consider a full queue. Then, due to the loss of the arriving job, $\phi(\alpha, B+1) = B+1$, whereas $\phi(\beta, B+1) = B$. This gives

$$\phi(\beta, \phi(\alpha, B+1)) = B \neq B+1 = \phi(\alpha, \phi(\beta, B+1)).$$

Hence, the single server queue with finite buffer and loss, does not satisfy the CC. ✳✳✳

As the next lemma shows, the commuting condition is sufficient for the GSMP to be independent of the selection rule.

Lemma 12.1. *If the commuting condition holds, then the GSMP $(S_t : 0 \leq t)$ is independent of the selection rule.*

Proof. Let $e, e' \in S(n)$ and, assume for the sake of simplicity, that

$$\{e, e'\} = \arg_e \min\{C(e, n) : e \in L(S(n))\}.$$

Provided that the CC holds, it follows that $S(n+2) = \phi(e, \phi(e', n)) = \phi(e', \phi(e, n))$. Since $C(e, n) = C(e', n)$, it follows that $T(n+1) = 0$. Hence, the order in which e and e' occur does not effect the continuous time process S_t as the holding time of the intermediate state $S(n+1)$ is zero. The proof follows by induction. □

Example 12.3. Consider a tandem queueing system with J single servers with infinite buffer capacity. Customers arrive to the system from the outside at station 1. After service completion at station j, jobs advance to station $j+1$, for $1 \leq j \leq J-1$, after service completion at server J jobs leave the system. The state $s = (s_1, \ldots s_J) \in S = \mathbb{N}^J$ represents that s_j jobs are present at station j, for $1 \leq j \leq J$. The event list is $E = \{\alpha, \beta_1, \ldots, b_J\}$, where α is the arrival event and β_j is the event of a service completion at station j, for $1 \leq j \leq J$.

The state transition mapping is given by

$$\phi(\alpha, (s_1, \ldots, s_J)) = (s_1 + 1, \ldots, s_J),$$

$$\phi(\beta_j, (s_1, \ldots, s_J)) = (s_1, \ldots, s_j - 1, s_{j+1} + 1, \ldots, s_J), \quad 1 \leq j \leq J - 1,$$

and

$$\phi(\beta_J, (s_1, \ldots, s_J)) = (s_1, \ldots, s_{J-1}, \ldots, s_J - 1).$$

The network satisfies the commuting condition. Indeed, let $\alpha, \beta_j \in L(s)$, then

$$\phi(\beta_j, \phi(\alpha, (s_1, \ldots, s_J))) = \phi(\beta_j, (s_1 + 1, \ldots, s_J))$$
$$= (s_1 + 1, \ldots, s_j - 1, s_{j+1} + 1 \ldots, s_J)$$
$$= \phi(\alpha, (s_1, \ldots, s_j - 1, s_j, \cdots, s_J))$$
$$= \phi(\alpha, \phi(\beta_j, (s_1, \ldots, s_J))).$$

For other event combinations the results follow the same way. ✻✻✻

Example 12.4. We revisit the tandem line from Example 12.3 with finite buffer places. More precisely, station 1 has infinite buffer capacity, and, for $2 \leq j \leq J$, station j has B_j buffer places, with $0 \leq B_j < \infty$, that is, at most $B_j + 1$ jobs can be present at station $2 \leq j \leq J$. We assume blocking after service (BAS) discipline, that is, a job that has finished service at station j only leaves the station if a place is available at station $j+1$. We extend the state space to $(s_1, s_2, b_2, s_3, b_3, \ldots, s_J, b_J)$, where $b_j = 1$ indicates that server j is blocked and $b_i = 0$ otherwise. Following the same line of computation is put forward in Example 12.3, it is easily seen that the system satisfies the commuting condition. ✻✻✻

Example 12.5. Consider a network of J single server infinite capacity queues with Markovian routing. Like for the other examples, we let $s = (s_1, \ldots s_J) \in S = \mathbb{N}^J$ represent that s_j jobs are present at station j, for $1 \leq j \leq J$. The event set is $E = \{\alpha, \beta_1, \ldots, \beta_J\}$, where α is the arrival event and β_j is the event of a service completion at station j, for $1 \leq j \leq J$. Routing is governed by a Markov matrix P, where $P(i, j)$ is the probability that a job leaving station i proceeds to station j. In this case ϕ is a random mapping and

$$\mathbb{P}\big(\phi(\beta_i, (s_1, \ldots s_J)) = (s_1, \ldots, s_i - 1, \ldots, s_j + 1, \ldots, s_J)\big) = P(i, j).$$

Following the line of argument put forward in Example 12.3 it is easily seen that the CC holds. ✻✻✻

We conclude our discussion of the commuting condition by presenting structural conditions implying the commuting condition for a general class of networks, called a general queueing network (GQN). A GQN is a network of servers that may be single, multi-, or infinite servers, service time distributions may be arbitrary, and queues may have finite capacity. Customers/jobs move through the network according to Markovian routing, and may belong to one of several classes. Service is organized in FCFS discipline and blocking is organized with the BAS discipline; see Example 12.4. Then, a GQN satisfies the commuting condition if

- all stations visited by more than one class of customers are fed by a single source;
- all stations with a finite number of service places are visited by only one class of customers;

- all stations with finite capacity are fed by a single source;
- in case the network is open, all stations fed directly by the external source have infinite capacity.

See [144] for details and proofs. For more on the analysis of GSMPs via events, we refer to [115, 124].

12.3 UNBIASEDNESS OF IPA

We turn to gradient estimation for DESs. The theory developed here relies on the IPA techniques introduced in Chapters 7 and 8 of Part II. As will be detailed in the following, we assume that one (or several) lifetime distribution(s) depend on a some system parameter $\theta \in \Theta \subset \mathbb{R}^d$, where Θ denotes the set of admissible values. IPA gradient estimators for performance measures introduced in Section 12.1.1 will be derived, and sufficient conditions for their unbiasedness will be presented.

The jump epoch $\tau(n)$ is determined by the occurrence of the n-th event, which is termination of the k-th lifetime of event $e = e(n)$, denoted by $X(e, k)$. This lifetime $X(e, k)$ was generated upon the occurrence of some event $e' = e(n')$, for some $n' < n$. The lifetime $X(e', n')$ was again generated upon the occurrences of some event e''. Working backward this way, $\tau(n)$ can be written as sum of lifetimes of appropriate event pairs. This connection via the event pairs is recorded by the triggering indicators introduced next.

Definition 12.2. The triggering indicators $\eta(\cdot, \cdot; \cdot, \cdot)$ are equal to zero except for the following cases:

- For every $e \in E$ and $k \geq 1$, we let $\eta(e, k; e, k) = 1$.
- If the k-th clock time of event e is set at the j-th occurrence of event e', we let $\eta(e, k; e', j) = 1$.
- If $\eta(e_1, k_1; e_2, j_2) = 1$ and $\eta(e_2, k_2; e_3, j_3) = 1$, then $\eta(e_1, k_1; e_3, j_3) = 1$.

The triggering indicators allow for the representation of jump times as a sum of appropriate life times; see the following lemma.

Lemma 12.2. *For all n it holds*

$$\tau(n) = \sum_{k=1}^{n} X(r(k))\eta(r(n); r(k)),$$

and

$$T(e, k) = \sum_{e', j} X(e', j)\eta(e, k; e', j).$$

Proof. The proof follows by backward induction. □

For ease of reference we introduce the following assumptions on the dependency of the life times on θ:

(a1) For all $e \in E$, $n \geq 1$, $X(e, n)$ are path wise differentiable on Θ.
(a2) For all $e \in E$, the lifetime distributions are continuous for all $\theta \in \Theta$ and have zero mass in 0.

For establishing an IPA theory for DES, we will apply Theorem 8.2 to the specific performance measures introduced in Section 12.1.1. Note that condition (i) in

Theorem 8.2 is equivalent to condition (a1). The following theorem shows that (a1) implies also a.s. differentiability for the DES performance measures. Put differently, under (a1), conditions (i) and (ii) in Theorem 8.2 are satisfied for the performance measures in Section 12.1.1.

Theorem 12.1. *Provided that (a1) and (a2) hold, we have*

$$\frac{d}{d\theta}\tau(n) = \sum_{k=1}^{n}\left(\frac{d}{d\theta}X(r(k))\right)\eta(r(n);r(k)),$$

and

$$\frac{d}{d\theta}T(e,k) = \sum_{e',j}\left(\frac{d}{d\theta}X(e',j)\right)\eta(e,k;e',j).$$

Moreover,

$$\frac{d}{d\theta}L_T = \sum_{k=1}^{N(T)}\frac{d}{d\theta}\tau(k)\bigl(L(S(k-1)) - L(S(k))\bigr),$$

$$\frac{d}{d\theta}L_m = \sum_{k=0}^{m-1}\left(\frac{d}{d\theta}\tau(k+1) - \frac{d}{d\theta}\tau(k)\right)L(S(k)),$$

provided that $\tau(n) < \infty$; *and*

$$\frac{d}{d\theta}L_{e,k} = \sum_{k=0}^{N(e,k)-1}\left(\frac{d}{d\theta}\tau(k+1) - \frac{d}{d\theta}\tau(k)\right)L(S(k)),$$

provided that $T(e,k) < \infty$.

Proof. Fix θ_0. By (a2), the triggering events are a.s. uniquely defined and for Δ sufficiently small the event sequence is invariant on $(\theta_0 - \Delta, \theta_0 + \Delta)$, and therefore the triggering indicators are constant. By (a1), we may differentiate the lifetimes with respect to θ at θ_0, and the first two statements follow directly from Lemma 12.2. As for L_T, using the fact that the sequence of physical state is a.s. constant on $(\theta_0 - \Delta, \theta_0 + \Delta)$ we have

$$\frac{d}{d\theta}L_T = \sum_{k=0}^{N(T)}L(S(k))\frac{d}{d\theta}T(k) - L(S(N(T)))\frac{d}{d\theta}\tau(N(T)).$$

Using that

$$\frac{d}{d\theta}\tau(k+1) - \frac{d}{d\theta}\tau(k) = \frac{d}{d\theta}T(k),$$

rearranging terms yields the third statement of the theorem. The remaining statements follow from similar arguments. □

As the proof of Theorem 12.1 shows, pathwise differentiability follows in a straightforward way from that of the lifetimes. As it is known from the IPA theory in Chapter 8, the key step in establishing unbiasedness of IPA sample-path derivative is to establish a.s. Lipschitz continuity of the performamce measure. For the performance measures of DES as introduced in Section 12.1.1, continuity of the jump epochs is implied by the commuting condition together with (a1). Once, we have continuity of $\tau(n)$, a.s. Lipschitz continuity requires the following condition:

(a3) For all $e \in E$, $n \geq 1$, $X(e,n)$ are pathwise differentiable and continuous on Θ with

$$\mathcal{L}(e,n) = \sup_{\theta \in \Theta} \left| \frac{d}{d\theta} X(e,n) \right| < \infty \quad a.s.,$$

satisfying $\mathbb{E}[\mathcal{L}(e,n)] < \infty$.

Put differently, condition (iii) in Theorems 8.1 and 8.2 is satisfied for the performance measures in Theorem 12.1.

The key technical means for analysis is that the event sequence $e(n, \theta)$ as a mapping of θ is piecewise constant; where we call a mapping $f : \Theta \to \Xi$, for Ξ some finite set, piecewise constant if the set of discontinuity points (i.e., jump locations) is a non-dense subset of Θ. For ease of reference, we introduce the following definition.

Definition 12.3. A GSMP is said to satisfy the event condition (EC) if for all n, $e(n, \cdot) : \Theta \mapsto E$ is piecewise constant.

We now turn to the prove of unbiasedness of the IPA derivative. In order to apply Theorem 8.2 for establishing unbiasedness, it remains to establish condition (iii) on the a.s. Lipschitz continuity of the DES performance measures.

Theorem 12.2. *Assume that (a1), (a2), (a3), the CC, and the EC hold, and assume that L is bounded.*

(i) If $\mathbb{E}[\sup_{\theta \in \Theta} N^2(T)] < \infty$, then

$$\frac{d}{d\theta} \mathbb{E}[L_T] = \mathbb{E}\left[\frac{d}{d\theta} L_T \right].$$

(ii) If $\mathbb{E}[\sup_{\theta \in \Theta} \tau(n)] < \infty$ for all n, then

$$\frac{d}{d\theta} \mathbb{E}[L_m] = \mathbb{E}\left[\frac{d}{d\theta} L_m \right].$$

(iii) If $\mathbb{E}[\sup_{\theta \in \Theta} T^2(e,k)] < \infty$ and $\mathbb{E}[\sup_{\theta \in \Theta} N^2(T(e,k))] < \infty$, then

$$\frac{d}{d\theta} \mathbb{E}[L_{e,k}] = \mathbb{E}\left[\frac{d}{d\theta} L_{e,k} \right].$$

Proof. We only prove the first part of the theorem. For a detailed proof of the other parts we refer to [114]. The proof has several parts. First, we show that (a2) together with the EC imply that $\tau(n)$ are continuous on Θ. Fix θ and denote by n^* the smallest index such that the event sequence changes at θ; i.e.,

$$\lim_{\delta \downarrow 0} e(n^*, \theta + \delta) \neq \lim_{\delta \downarrow 0} e(n^*, \theta - \delta).$$

Note that the limits are well-defined by the EC. Let the two events that change the order of their occurrence be the k-th occurrence of e and the k'-th occurrence of e'. By Lemma 12.2 and (a2) together with the fact the event sequence does not change up to the n^*-th state transition, we see that the time that e occurs for the k-th time, denoted by $\tau(e,k)$, and the time that e' occurs for the k'-th time, denoted by $\tau(e', k')$ are both continuous. This gives that

$$\tau(n^*, \theta) = \min(\tau(e,k), \tau(e', k')) \quad \text{and} \quad \tau(n^*+1, \theta) = \max(\tau(e,k), \tau(e', k')) \quad (12.1)$$

are both continuous. Proceeding this way, we see that $\tau(n)$ is continuous at θ for all n. As θ was arbitrary, this shows continuity of $\tau(n)$ throughout Θ.

As the next step, we show that continuity of L_T follows from continuity of $\tau(n)$ together with (CC). Note that $T(n)$ is continuous for $n < n^*$, with n^* the smallest index such that the event sequence changes at θ, as before. By continuity of $\tau(n)$, the holding time

$$T(n^*, \theta) = \tau(n^* + 1, \theta) - \tau(n^*, \theta)$$
$$= \max(\tau(e, k), \tau(e', k')) - \min(\tau(e, k), \tau(e', k'))$$

is continuous. By (CC),

$$\lim_{\delta \downarrow 0} S(n^* + 1, \theta + \delta) = \lim_{\delta \downarrow 0} S(n^* + 1, \theta - \delta),$$

$$\lim_{\delta \downarrow 0} L(S(n^*, \theta + \delta)) T(n^*, \theta + \delta) = 0 = \lim_{\delta \downarrow 0} L(S(n^*, \theta - \delta)) T(n^*, \theta - \delta),$$

and

$$\lim_{\delta \downarrow 0} L(S(n^* + 1, \theta + \delta)) T(n^*, \theta + \delta) = \lim_{\delta \downarrow 0} L(S(n^* + 1, \theta - \delta)) T(n^*, \theta - \delta).$$

We now carry out an induction over the indices of event order changes at θ. As θ was arbitrary, this shows continuity of $T(n)$ and of $L(S(n))T(n)$ throughout Θ. This shows that L_T is continuous as well.

For the sake of completeness we address continuity of $L(S(N(T)))(T - \tau(N(T)))$. By the EC, $N(T)$ is piecewise constant as a mapping of θ, and we write $N_\theta(T)$ to express the dependency of $N(T)$ on θ. On any interval $(\theta_a, \theta_b) \subset \Theta$ where $N_\theta(T)$, $L(S(N_\theta(T)))$ are constant, $\tau(N_\theta(T), \theta)$ is continuous, which implies that $L(S(N_\theta(T)))(T - \tau(N_\theta(T), \theta))$ is continuous on (θ_a, θ_b). Now let $\theta_0 \in \Theta$ be such that $N_\theta(T)$ jumps at θ for given realization ω, i.e.,

$$\lim_{\theta \uparrow \theta_0} N_\theta(T) \neq \lim_{\theta \downarrow \theta_0} N_\theta(T).$$

In words, under θ_0 there is an event that occurs at time T for ω. Observe that this implies that

$$\lim_{\theta \uparrow \theta_0} \tau(N_\theta(T), \theta) = T = \lim_{\theta \downarrow \theta_0} \tau(N_\theta(T), \theta).$$

Continuity now follows from

$$\lim_{\theta \uparrow \theta_0} L(S(m))(\tau(N_\theta(T), \theta) - T) = 0 = \lim_{\theta \downarrow \theta_0} L(S(m))(\tau(N_\theta(T), \theta) - T).$$

This completes the proof of continuity of L_T.

We now turn to the third part of the proof and establish a Lipschitz modulus. Let

$$N^*(T) = \sup_{\theta \in \Theta} N(T),$$

then, by part three of Theorem 12.1,

$$\sup_{\theta \in \Theta} \left| \frac{d}{d\theta} L_T \right| \leq 2\mathbf{L} \sum_{k=1}^{N^*(T)} \sup_{\theta \in \Theta} \left| \frac{d}{d\theta} \tau(k) \right|,$$

where $\mathbf{L}$ denotes the bound on L. Next, we note that

$$\sup_{\theta \in \Theta} \left| \frac{d}{d\theta} \tau(k) \right| \leq \sum_{m=1}^{N^*(T)} \sup_{\theta \in \Theta} \left| \frac{d}{d\theta} X(r(m)) \right|$$

$$\leq \sum_{m=1}^{N^*(T)} \sum_{e \in E} \mathcal{L}(e,m), \tag{12.2}$$

where $\mathcal{L}(e,m)$ is the Lipschitz modulus of $X(e,m)$, which are mutually independent as the instance of $X(e,m)$ are. Hence,

$$\sup_{\theta \in \Theta} \left| \frac{d}{d\theta} L_T \right| \leq 2\mathbf{L} \sum_{k=1}^{N^*(T)} \sum_{m=1}^{N^*(T)} \sum_{e \in E} \mathcal{L}(e,m).$$

Using Wald's equality we obtain by independence of the event lifetimes,

$$\mathbb{E}\left[\sup_{\theta \in \Theta} |L_T|\right] \leq 2\mathbf{L}\mathbb{E}\left[(N^*(T))^2\right] \mathbb{E}\left[\sum_{e \in E} \mathcal{L}(e,1)\right].$$

Hence,

$$2\mathbf{L} \sum_{k=1}^{N^*(T)} \sum_{m=1}^{N^*(T)} \sum_{e \in E} \mathcal{L}(e,m) \tag{12.3}$$

is a Lipschitz modulus for L_T. To summarize, L_T is a.s. differentiable, continuous throughout Θ with Lipschitz modulus as in (12.3). The proof then follows from Theorem 8.2. □

Note that for $L \equiv 1$, Theorem 12.2 provides sufficient conditions for

$$\frac{d}{d\theta}\mathbb{E}\left[\tau(n)\right] = \mathbb{E}\left[\frac{d}{d\theta}\tau(n)\right].$$

Let $\Theta = [a,b]$ and assume that the assumptions put forward in Theorem 12.2 hold.

- *Fix* realization ω and $n \in \mathbb{N}$. Then, there exits by the EC a finite sequence of points $C(\omega) := (\theta_k(\omega) : 1 \leq k \leq K(\omega))$ such that $\tau(i)$, $1 \leq i \leq n$, are continuous throughout Θ and differentiable on $\Theta \setminus C(\omega)$. By Lemma 12.2, this, together with condition (a3), yields Lipschitz continuity of $\tau(n)$ on Θ with integrable modulus.
- For *fixed* θ, the set
$$\{\omega \in \Omega : \tau(n) \text{ fails to be differentiable at } \theta\}$$
is a null set, i.e., $\tau(n)$ is a.s. differentiable at any θ.

Why does differentiability of $\tau(n)$ fail at $\theta_k(\omega)$, for $1 \leq k \leq K(\omega)$? Inspecting the proof of Theorem 12.2 we see that this stems from the fact that for ω at θ two events occur simultaneously. Specifically, revisit (12.1) and use the notation introduced in the proof of Theorem 12.2, assuming that

$$\frac{d}{d\theta}\tau(n,\theta) = \begin{cases} \frac{d}{d\theta}\tau(e,k) & \text{if } \tau(e,k) < \tau(e',k') \text{ at } \theta, \\ \frac{d}{d\theta}\tau(e',k') & \text{if } \tau(e',k') < \tau(e,k) \text{ at } \theta, \end{cases} \tag{12.4}$$

and

$$\frac{d}{d\theta}\tau(n+1,\theta) = \begin{cases} \frac{d}{d\theta}\tau(e,k) & \text{if } \tau(e,k) > \tau(e',k') \text{ at } \theta, \\ \frac{d}{d\theta}\tau(e',k') & \text{if } \tau(e',k') > \tau(e,k) \text{ at } \theta. \end{cases} \tag{12.5}$$

As can be seen by (12.4) and (12.5), choosing an appropriate order in which (e,k) and (e',k') occur, one-sided differentiability of $\tau(n)$ and $\tau(n+1)$ can be achieved even in case that $\tau(e,k) = \tau(e',k')$ at θ. This motivates the definition of the following selection rules $h^{\pm}$:

$$e^+(n) := h^+(S(n), C(n)) = \arg_e \min \left\{ \frac{d^+}{d\theta^+} T(e, N(e, n)) : e \in L(S(n)) \right\},$$

and

$$e^-(n) := h^-(S(n), C(n)) = \arg_e \max \left\{ \frac{d^-}{d\theta^-} T(e, N(e, n)) : e \in L(S(n)) \right\}.$$

Indeed, using selection rules h^+ (resp. h^-) implies a.s. right (resp. left)-sided differentiability of the jump epochs *throughout* Θ.

In case of continuous lifetimes, we argue that at a specific θ a non-differentiability occurs with probability zero, and therefore the IPA derivative are a.s. well defined at θ. If some of the lifetime distribution are discrete, this reasoning does no longer hold, as now the probability of two events occurring simultaneously may be larger than zero. As argued in [143], using h^+ for the constructing of the GSMP, IPA yields unbiased estimators for the right-sided derivatives, the same holds under h^- for left-sided derivatives. This leads to the following IPA result.

Corollary 12.1. *Provided that (a1), (a3), the CC, and the EC hold, the statements put forward in Theorem 12.1 and in Theorem 12.2 hold for right-sided derivatives if selection rule h^+ is applied, and the statements hold for left-sided derivatives if selection rule h^- is applied.*

12.4 SUFFICIENT CONDITIONS FOR THE EVENT CONDITION

Consider a closed queuing network with two single servers, i.e., $J = \{1, 2\}$. Both servers have no additional buffer space, i.e., there is at most one customer at each server. Two indistinguishable customers cycle in the system, i.e., we always see one customer at server 1 and the other at server 2, respectively. The jobs change servers simultaneously. After such a transition the system restarts independently from the past. Let γ denote the event "service completion at server 1" and β the event "service completion at server 2." Choose $\Theta := (-1, 1)$ and set

$$f(m, \theta) := \begin{cases} \theta^m \sin(1/\theta) + 1 & \text{if } \theta \neq 0, \\ 1 & \text{else,} \end{cases} \quad (12.6)$$

for $m > 1$. Suppose that the service time distribution at server 1, i.e., the lifetime distribution of event γ, is given by

$$F(\gamma, x) := \frac{1}{2} G(x) + \frac{1}{2} \delta_1(x),$$

where $\delta_y(x)$ denotes the Dirac measure in y, and $G(x)$ is a continuous cdf. Moreover, the service time distribution at station 2, i.e., for event β, is given by

$$F_\theta(\beta, x) := \delta_{f(m, \theta)}(x).$$

Suppose that the first lifetime of γ equals 1, i.e., $X(\gamma, 1) = 1$, which happens with probability $1/2$. The lifetimes of β depend on θ through $X_\theta(\beta, n) = f(m, \theta)$, for $n \in \mathbb{N}^+$. Note that $X_\theta(\beta, n)$ is, for $m > 3k$, k times differentiable with respect to θ on Θ. The event $e(1, \theta)$, which triggers the first state transition, is given by

$$e(1,\theta) = \begin{cases} \beta & \text{if } f(m,\theta) < 1, \\ h(\{\beta,\gamma\}) & \text{if } f(m,\theta) = 1, \\ \gamma & \text{if } f(m,\theta) > 1. \end{cases}$$

By definition, $f(m,\theta)$ oscillates around 1 in the neighborhood of $\theta = 0$. Therefore, the one-sided limits of $e(1,\theta)$ with respect to θ do not exist in 0. Thus, $e(1,\theta)$ fails to be a piecewise constant function in the neighborhood of 0 with probability $1/2$. Since $e(2,\theta)$ is completely determined by $e(1,\theta)$ via

$$e(2,\theta) = \begin{cases} \beta & \text{if } e(1,\theta) = \gamma, \\ \gamma & \text{if } e(1,\theta) = \beta, \end{cases}$$

$e(2,\theta)$ fails to be a piecewise constant function with respect to θ, as well. After the occurrence of γ and β, the system restarts independently from the past and we obtain for all $n \in \mathbb{N}^+$

$$\mathbb{P}\Big(e(n,0) \text{ fails to be a piecewise constant function}\Big) \geq \frac{1}{2}.$$

In the presence of time distributions with discrete components, piecewise constancy of the event sequence, i.e., the event condition is not a mere consequence of continuity or differentiability of the lifetimes, i.e., (a1) to (a3). The key insight is that we have to avoid the possibility that time epochs of occurring events oscillate around one another. More specifically, it can be shown that the EC holds if θ is a scale or location parameter. Moreover, it is easily seen that all distributions discussed in Chapter 7 (e.g., see Examples 7.4–7.6), and Chapter 8 (e.g., see Example 8.1), do not show oscillating behavior and the EC holds provided that these type of distributions are present in the DES. For more details, we refer to [143].

12.5 CONCLUDING REMARKS

In the light of Part II, IPA for DES as presented in this chapter can be seen as a worked out application of IPA. In particular, the result put forward in part (ii) of Theorem 12.2 can be seen as an instance of Theorem 8.2 in Section 8.1.2 expoiting the specific form of the cost function and the event structure of a DES. Parts (i) and (iii) of Theorem 12.2 are instances of IPA for the random horizon problem; see Section 7.3.3 and Chapter 9.

Chapter Thirteen

A Markov Operator Approach

In this chapter we provide an operator approach to differentiation of Markov chains. For basic definitions and concepts on Markov chains we refer to Appendix C. For ease of presentation, we assume that the Markov chains live on the same measurable space $(S, \mathcal{S})$, and assume in the following that S is a subset of $\mathbb{R}^m$, for some $m \geq 1$. Before we can turn to developing the theory, we will have to prepare some tools required for our analysis. Specifically, we will extend the concept of Lipschitz continuity and weak differentiability of measures to Markov kernels, which requires us to introduce the concept of a transition kernel. Moreover, we will extend the v-norm to transition kernels. To begin with, we introduce the concept of a transition kernel on $(S, \mathcal{S})$, which is a generalization of the concept a Markov kernel on $(S, \mathcal{S})$.

Definition 13.1. *R* is called a *transition kernel* if $R(x, \cdot)$ is a finite (possibly signed) measure for all $x \in S$, and $R(\cdot, B)$ is a measurable mapping for all measurable sets $B \in \mathcal{S}$. If, in addition, R is a probability measure for all x, then R is called a Markov kernel.

If R is a Markov kernel, then the sequence $\{X_n\}$ on $(S, \mathcal{S})$ satisfying

$$\mathbb{P}(X_{n+1} \in B | X_n = x) = R(x, B), \quad n \geq 0,$$

for $X_0 = x_0 \in S$ and all $x \in S, B \in \mathcal{S}$, is called the Markov chain associated to R.

Next we extend the definition of the v-norm in Definition 8.6 to transition kernels, where the v-norm of a transition kernel is defined through the operator norm associated to v.

Definition 13.2. The v-norm of a transition kernel R (and thereby of a Markov kernel) on $(S, \mathcal{S})$ is defined by:

$$||R||_v = \sup_{x \in S} \frac{1}{v(x)} \int_S v(y) |R(x, dy)|, \quad \text{for } v(x) \geq 1 \text{ for all } x \in S.$$

In the following we drop denoting the measure space if this causes no confusion.

Note that with this definition we have the following bounding technique for $R(x;)$ integrals:

$$|(Rh)(x)| = \left| \int h(y) R(x, dy) \right| = \int |h(y)| \, |R(x, dy)|$$

$$\leq \int \left| \frac{h(y)}{v(y)} \right| v(y) |R(x, dy)|$$

$$\leq \int \left(\sup_x \frac{|h(y)|}{v(y)} \right) v(y) |R(x, dy)|$$

$$\leq \|h\|_v \int v(y)|R(x,dy)|$$

$$= v(x)\|h\|_v \frac{1}{v(x)} \int v(y)|R(x,dy)|$$

$$\leq v(x)\|h\|_v \|R\|_v, \tag{13.1}$$

for all x. The above implies a very useful property of the v-norm, which we call *norm preservation*, namely, that for $\|h\|_v < \infty$ and $\|R\|_v < \infty$ it holds that the integrated value of $h(y)$ with respect to $R(x,dy)$, given by $\int h(y)R(x,dy)$, as a mapping of x has again finite v-norm.

We now introduce weak differentiability and Lipschitz-continuity of Markov kernels. For general sufficient conditions for the existence of a $\mathcal{D}_\alpha$-derivative of Markov kernel we refer to [150].

Definition 13.3. Let $v(x) \geq 1$ for all x. We say that a Markov kernel P_θ is v-Lipschitz continuous with modulus M_P if

$$\int v(y)|P_{\theta+\Delta}(x,dy) - P_\theta(x,dy)| \leq |\Delta| M_P \, v(x),$$

for all x, for some finite M_P.

Let $\mathcal{D}$ be a set of measurable mappings from $(S,\mathcal{S})$ to $\mathbb{R}$ equipped with the Borel field. We say that a Markov kernel P_θ is $\mathcal{D}$-differentiable with respect to θ if a transition kernel P'_θ exists such that for all x and $h \in \mathcal{D}$ it holds that

$$\frac{d}{d\theta} \int h(y) P_\theta(x,dy) = \int h(y) P'_\theta(x,dy).$$

Furthermore, we call the triple $(c_\theta(\cdot), P_\theta^+, P_\theta^-)$ a $\mathcal{D}$-derivative of P_θ if for all x and all $h \in \mathcal{D}$

$$\int h(y) P_\theta(x,dy) = c_\theta(x) \left(\int h(y) P_\theta^+(x,dy) - \int h(y) P_\theta^-(x,dy) \right),$$

where $c_\theta(\cdot)$ is a measurable mapping and $P_\theta^\pm$ are Markov kernels.

Recall that we have already worked with the v-norm in Chapter 8. In particular, we have worked with

$$v_\alpha(x) := v_{\alpha,k}(x) = \alpha^{|x_1|+|x_2|+\cdots|x_k|}, \quad \alpha \geq 1, \text{ and } x \in \mathbb{R}^k.$$

A property that will become very handy for the analysis in this chapter is that

$$v_\alpha(x,y) = v_{\alpha,k+m}(x,y) = v_{\alpha,k}(x) v_{\alpha,m}(y) = v_\alpha(x) v_\alpha(y), \quad x \in \mathbb{R}^k, y \in \mathbb{R}^m. \tag{13.2}$$

We denote by $\mathcal{D}_\alpha$ the set of all measurable mappings $g : \mathbb{R}^k \mapsto \mathbb{R}$ with finite v_α-norm, i.e., $g \in \mathcal{D}_\alpha$ if and only if

$$|g(x)| \leq \|g\|_{v_\alpha} v_\alpha(x),$$

for all $x \geq 0$. To simplify notation, we use $\|g\|_\alpha$ to denote the v_α-norm. Moreover, we suppress denoting the dimension k as it is understood that the dimension is taken corresponding to the argument, i.e., for $g \in \mathcal{D}_\alpha$ with $g : \mathbb{R}^k \mapsto \mathbb{R}$, the v_α norm of g is computed via $v_{\alpha,k}$. Typically, $\mathcal{D}_\alpha = \{g \in \mathcal{D} : \|g\|_{v_\alpha} < \infty\}$ is the subset of mappings g with finite v_α-norm for g out of some *base set* $\mathcal{D}$. In applications, we often take the base set to be the set of measurable mappings, and then $\mathcal{D}_\alpha$ is the set of all measurable mappings g with finite v_α-norm. It might, however, also be the case that we have to work with a more restrictive base set, such as, the set of all continuous mappings, then $\mathcal{D}_\alpha$ is the set of all continuous mappings g with finite v_α-norm. In the following we will, unless stated otherwise, use for $\mathcal{D}_\alpha$ the set of measurable mappings as base set.

The inequality in (13.1) shows that for P with v_α-norm and $h \in \mathcal{D}_\alpha$ it follows that $\int h(y)P(\cdot, dy)$ is again in $\mathcal{D}_\alpha$. For later reference we summarize this result in the following lemma.

Lemma 13.1. *If $\|R\|_\alpha < \infty$ and $h \in \mathcal{D}_\alpha$, then $\int h(y)R(\cdot, dy) \in \mathcal{D}_\alpha$, i.e., R is $\mathcal{D}_\alpha$-preserving.*

We will illustrate the Markovian setup and v_α-norm concept with the sojourn time example. Specifically, we provide sufficient conditions for the transition kernel of the sojourn times to be v_α-norm preserving.

Example 13.1. We revisit our basic sojourn times example with the notation as introduced in Example 7.16. Recall that the n-th interarrival time is denoted by A_n and the n-th service time by $S_n(\theta)$. We assume that the service system starts initially empty and that the typical independence assumptions apply. Moreover, we denote the density of A_n by f_A and the density of $S_n(\theta)$ by $f_S(\theta)$, and we assume that $f_S(\theta, s)$ is differentiable as a mapping in θ for all $s \geq 0$. Eventually, we assume that the service and interarrival times have finite moment generation functions. Consecutive sojourn times $X_n(\theta)$ then follow the recursive relation (7.30):

$$X_{n+1}(\theta) = \max(0, X_n(\theta) - A_{n+1}(\theta)) + S_{n+1}(\theta), \quad n \geq 1, \tag{13.3}$$

and $X_1(\theta) = S_1(\theta)$. We denote the transition kernel of the sojourn time chain by P_θ, i.e.,

$$P_\theta(x, \mathcal{A}) = \mathbb{P}(X_{n+1}(\theta) \in \mathcal{A} | X_n(\theta) = x)$$

for $x \geq 0$ and $\mathcal{A} \subset [0, \infty)$ a (Borel) measurable set, or, more specifically

$$P_\theta(x, \mathcal{A}) = \int_0^\infty \left(\int_0^\infty 1_{\{\max(x-a,0)+s \in \mathcal{A}\}} f_S(\theta, s) ds \right) f_A(a) da. \tag{13.4}$$

It is easily seen that

$$\begin{aligned} |g(\max(x-a, 0) + s)| &\leq \|g\|_\alpha v_\alpha(\max(x-a, 0) + s) \\ &= \|g\|_\alpha \alpha^{\max(x-a,0)+s} \\ &\leq \|g\|_\alpha \alpha^{x+a+s} \\ &= \|g\|_\alpha v_\alpha(x) v_\alpha(a) v_\alpha(s), \quad x, a, s \geq 0. \end{aligned} \tag{13.5}$$

Elaborating the power law $\alpha^{x+y} = \alpha^x \alpha^y$, we obtain

$$\begin{aligned} \int g(y) P_\theta(x, dy) &= \int_0^\infty \left(\int_0^\infty g(\max(x-a, 0) + s) f_S(\theta, s) ds \right) f_A(a) da \\ &\stackrel{(13.5)}{\leq} \int_0^\infty \left(\int_0^\infty \|g\|_\alpha \alpha^{x+a+s} f_S(\theta, s) ds \right) f_A(a) da \\ &\leq \alpha^x \|g\|_\alpha \int_0^\infty \left(\int_0^\infty \alpha^{a+s} f_S(\theta, s) ds \right) f_A(a) da \\ &= \alpha^x \|g\|_\alpha \mathbb{E}\left[\alpha^{A_1}\right] \mathbb{E}\left[\alpha^{S_1(\theta)}\right] < \infty, \end{aligned} \tag{13.6}$$

where the last inequality follows from the fact that interarrival and service times have a finite moment generating function. Hence, $\|P_\theta\|_\alpha \leq \|g\|_\alpha \mathbb{E}\left[\alpha^{A_1}\right] \mathbb{E}\left[\alpha^{S_1(\theta)}\right]$ and P_θ is therefore v_α-norm preserving, i.e.,

$$\int g(y) P_\theta(\cdot, dy) \in \mathcal{D}_\alpha.$$

Moreover, we obtain from (13.6)

$$\|P_\theta\|_\alpha \leq \sup_{x \geq 0} \frac{1}{\alpha^x} \int \alpha^y P_\theta(x, dy)$$
$$\leq \mathbb{E}\left[\alpha^{A_1}\right] \mathbb{E}\left[\alpha^{S_1(\theta)}\right] < \infty.$$

The above computations illustrates the usefulness of the choice of the power form $v(x) = \alpha^x$ for the v-norm. Finally, provided that

$$\sup_{\hat{\theta} \in \Theta_0} \mathbb{E}\left[\alpha^{S_1(\theta)}\right] < \infty,$$

for Θ_0 a neighborhood of θ, it furthermore follows that

$$\sup_{\hat{\theta} \in \Theta_0} \|P_{\hat\theta}\|_\alpha < \infty. \qquad \text{✷✷✷}$$

In the following example we illustrate the relation between condition $\mathbb{E}[\alpha^X] < \infty$, for some random variable X, and the existence of the moment generation function of X.

Example 13.2. Suppose that X has a finite moment generating function (mgf) on $(-t, t)$, i.e., assume that $\mathbb{E}[e^{uX}]$ is finite for $|u| < t$ where the case $t = \infty$ is not excluded (and note that this implies that $\mathbb{E}[X^n e^{uX}] < \infty$, for all $n \in \mathbb{N}$, provided that $|u| < t$). This shows that $\mathbb{E}[\alpha^X] < \infty$ for $\ln(\alpha) < t$. If, for example, X is exponentially distributed with mean m, the mgf is given by $1/(1-(u/m))$ and it follows that $\mathbb{E}[\alpha^X] < \infty$ as well as $\mathbb{E}[X\alpha^X] < \infty$ for $\alpha < e^m$. In case the mgf exists for all u (i.e., $t = \infty$), like for the uniform distribution on $[a, b]$, $\mathbb{E}[\alpha^X] < \infty$ as well as $\mathbb{E}[X\alpha^X] < \infty$ for any $\alpha \geq 0$. ✷✷✷

Before we establish the product rule of differentiation for Markov kernels, we will illustrate the Markovian setup together with Lipschtiz continuity and weak differentiability with the sojourn time example. In particular we illustrate in the following example how to compute this derivative using the $\mathcal{D}_\alpha$-derivative of the service time distribution.

Example 13.3. We use the notation as introduced in Example 13.1. Represent $df_S(\theta)/d\theta$ by the triple $(c_\theta, f_S^+(\theta), f_S^-(\theta))$; see (7.22) for the case of the exponential density and (7.21) for the general principle. As detailed in Example 7.11,

$$\frac{\partial}{\partial \theta} f_S(\theta, x) = \frac{1}{\theta}\left(f_S^\gamma(\theta, x) - f_S(\theta, x)\right),$$

where $f^\gamma(\theta, x) = \frac{x}{\theta^2} e^{-\frac{x}{\theta}}$ denotes the pdf of Gamma$(2, 1/\theta)$ distribution.

For given x, the transition kernel $P_\theta(x, dy)$ in (13.4) depends on the product measure of the exponential distribution with mean θ, denoted here by μ_θ, and the distribution of the inter arrival times ν. Note that μ_θ is $\mathcal{D}_\alpha$-differentiable and ν is $\mathcal{D}_\alpha$-differentiable as

ν is independent of θ (take as ν' as the null measure and $(1, \nu, \nu)$ as weak derivative). For $g \in \mathcal{D}_\alpha$ we let
$$h_g(x, a, s) = g(\max(x - a, 0) + s),$$
for $x, a, s \in [0, \infty)$. Since $g \in \mathcal{D}_\alpha$, it follows from (13.5) that
$$|h_g(x, a, s)| \leq \|g\|_\alpha v_\alpha(x) v_\alpha(a) v_\alpha(s), \quad x, a, s \geq 0,$$
which implies that $h_g(x, \cdot, \cdot)$ has finite v_α-norm. Then, elaborating on the product technique for the score-function method (see Example 8.12) and applying Theorem 8.5, we compute
$$\frac{d}{d\theta} \int g(y) P_\theta(x, dy) = \frac{d}{d\theta} \int \int h_g(x, a, s) \mu_\theta(ds)\, \nu(da)$$
$$= \frac{d}{d\theta} \int_0^\infty \left(\int_0^\infty h_g(x, a, s) f_S(\theta, s)\, ds \right) f_A(a)\, da$$
$$= \frac{1}{\theta} \left(\int_0^\infty \left(\int_0^\infty h_g(x, a, s) f_S^\gamma(\theta, s)\, ds \right) f_A(a)\, da \right.$$
$$\left. - \int_0^\infty \left(\int_0^\infty h_g(x, a, s) f_S(\theta, s)\, ds \right) f_A(a)\, da \right),$$

and we obtain as $\mathcal{D}_\alpha$-derivative of P_θ the triple $(1/\theta, P_\theta^+, P_\theta^-)$, with $P_\theta^- = P_\theta$ and
$$P_\theta^+(x, A) = \int_0^\infty \left(\int_0^\infty 1_{\{\max(x-a,0)+s \in A\}} f_S^\gamma(\theta, s)\, ds \right) f_A(a)\, da,$$
for $x \geq 0$ and $A \in \mathcal{S}$.

We now turn to establishing Lipschitz continuity of P_θ. Following (8.20) in Example 8.8, we have for $\theta \in [\theta_0, \theta_1]$
$$\left| \frac{\partial}{\partial \theta} f_S(\theta, x) \right| \leq \frac{\theta_1}{\theta_0^2} \left(\frac{x}{\theta_0} + 1 \right) f_S(\theta_0, x).$$

Hence,
$$\left| \frac{d}{d\theta} \int \alpha^y P_\theta(x, dy) \right| = \frac{\theta_1}{\theta_0^2} \int_0^\infty \left(\int_0^\infty \left(\frac{s}{\theta_0} + 1 \right) \alpha^{\max(x-a,0)+s} f_S(\theta_0, s)\, ds \right) f_A(a)\, da.$$

Note that
$$\left(\frac{s}{\theta_0} + 1 \right) \alpha^{\max(x-a,0)+s} \leq \alpha^s \left(\frac{s}{\theta_0} + 1 \right) \alpha^x \alpha^a,$$
for $x, s, a \geq 0$. Following the line of argument put forward in Example 13.1, where we now use that for exponentially distributed service times $\mathbb{E}[S(\theta)\alpha^{S(\theta)}] < \infty$, it follows that
$$\sup_{\theta \in [\theta_0, \theta_1]} \left| \frac{d}{d\theta} \int \alpha^y P_\theta(x, dy) \right| \leq M_P \alpha^x,$$
where
$$M_P = \frac{\theta_1}{\theta_0^2} \int_0^\infty \alpha^a \left(\int_0^\infty \left(\frac{s}{\theta_0} + 1 \right) \alpha^s f_S(\theta_0, s)\, ds \right) f_A(a)\, da.$$

Hence, by the mean-value theorem it holds that for $\theta, \theta + \Delta \in [\theta_0, \theta_1]$ that

$$\int \alpha^y |P_{\theta+\Delta}(x, dy) - P_\theta(x, dy)| \leq |\Delta| M_P \alpha^x. \tag{13.7}$$

In words, P_θ is v_α-Lipschitz continuous with modulus M_P. ✱✱✱

13.1 THE FINITE HORIZON PROBLEM

In the following we will establish the differentiation rule for n-fold product of Markov kernels yielding gradient estimators for the finite horizon problem. Let P_θ and Q_θ be transition kernels on the same state space. Then we use $P_\theta Q_\theta$ as shorthand notation for the product of the two kernels, that is,

$$\int Q_\theta(y, B) P_\theta(x, dy) = (Q_\theta P_\theta)(x, B),$$

for all $x \in S$ and measurable sets $B \in \mathcal{S}$. For sake of simplicity, we present the result for products of identical Markov kernels. For this, we denote $P_\theta^n = P_\theta P_\theta^{n-1}$ for $n \geq 2$. Moreover, we denote by P_θ^0 the identity operator. It is worth noting that P_θ^n allows for two interpretations.

- First, as defined in the above way, P_θ^n is the transition probability from $X_\theta(0)$ to $X_\theta(n)$ and thus a conditional probability measure of the form

$$\mathbb{E}[h(X_\theta(n))|X_\theta(0) = x] = \int h(y) P_\theta^n(x, dy),$$

for any measurable mapping h and initial value x such that the above expression is finite, and thus modeling the *n-step transition dynamic* of the Markov chain associated with Markov kernel P_θ, i.e., P_θ^n is transition kernel on $(S, \mathcal{S})$.
- Second, P_θ^n can be interpreted as a *probability measure on the sample path* $(X_\theta(1), \ldots, X_\theta(n))$ for given initial value $X_\theta(0) =$, i.e.,

$$\mathbb{E}[h(x, X_\theta(1), \ldots, X_\theta(n))|X_\theta(0) = x] = \int_{\mathbb{R}^n} h(y) P_\theta^n(x, dy)$$
$$= \int \left(\int \cdots \left(\int h(x, x_1, \ldots, x_n) P_\theta(x_{n-1}, dx_n) \right) P_\theta(x_{n-2}, dx_{n-1}) \cdots \right)$$
$$P_\theta(x, dx_1),$$

for any measurable mapping such that the above expressions are finite. For given initial state x, P_θ^n is thus a probability measure on $(S^n, \mathcal{S}^{\otimes n})$, where $(S^n, \mathcal{S}^{\otimes n})$ denotes the n-fold product of the measurable space $(S, \mathcal{S})$.

We now provide a proof of the product rule of measure-valued differentiation for products of Markov kernels for the n-step transition interpretation. The result we present is not the most general as we restrict ourselves to v_α for the v-norm, $S \subset \mathbb{R}^k$, for some $k \geq 1$, and products of identical Markov kernels.

Theorem 13.1. *Let P_θ be a Markov kernel on $(S, \mathcal{S})$ defined on some open neighborhood Θ of θ, and let $\mathcal{D}_\alpha$ be a set of real-valued measurable mappings on S with finite v_α-norm. Suppose that for $\alpha > 1$*

(i) $\|P_\theta\|_\alpha$ *is finite,*

(ii) P_θ is v_α-Lipschitz on Θ with modulus in M_P, and
(iii) P_θ is $\mathcal{D}_\alpha$-differentiable.

Then, P^n is $\mathcal{D}_\alpha$ differentiable a transition kernel on $(S, \mathcal{S})$ at θ with $\mathcal{D}_\alpha$-derivative

$$(P_\theta^n)' = \sum_{j=0}^{n-1} P_\theta^{n-j-1} P_\theta' P_\theta^j.$$

If, in addition, P_θ has $\mathcal{D}_\alpha$-derivative $(c_\theta(\cdot), P_\theta^+, P_\theta^-)$, then, it holds that for any $h \in \mathcal{D}_\alpha$ that

$$\int h(y) P_\theta^n(x, dy) = \sum_{j=0}^{n-1} \int \left(\int h(z) \left(P_\theta^{n-j-1} P_\theta^+ \right)(y, dz) \right.$$
$$\left. - h(z) \left(P_\theta^{n-j-1} P_\theta^- \right)(y, dz) \right) c_\theta(y) P_\theta^j(x, dy),$$

or, equivalently,

$$\frac{d}{d\theta} \int h(y) P_\theta^n(x, dy) = 2n \mathbb{E}\left[\int \left(\int h(z) \left(P_\theta^{n-\eta} P_\theta^{[\sigma]} \right)(y, dz) \right) c_\theta(y) P_\theta^{\eta-1}(x, dy) \right],$$

where σ is uniformly distributed on $\{1, \ldots, 2n\}$ independent of everything else, $\eta = \sigma$ mod n,

$$P_\theta^{[\sigma]} = \begin{cases} P_\theta^+(y, dz) & \text{if } \sigma \leq n, \\ -P_\theta^-(y, dz) & \text{if } \sigma > n. \end{cases}$$

Proof. We prove the statement for $n=2$ for the n-step transition interpretation. By condition (i) we have that

$$\int |h(y)| P_\theta(x, dy) \leq v_\alpha(x) ||P_\theta||_\alpha \in \mathcal{D}_\alpha,$$

see Lemma 13.1.

We split the overall difference quotient in three parts:

$$\lim_{\Delta \to 0} \frac{1}{\Delta} \left(\int \left(\int h(z) P_{\theta+\Delta}(y, dz) \right) Q_{\theta+\Delta}(x, dy) \right.$$
$$\left. - \int \left(\int h(z)(P_\theta(y, dz)) Q_\theta \right)(x, dy) \right) \quad (13.8)$$
$$= \lim_{\Delta \to 0} \frac{1}{\Delta} \int \left(\int h(z)(P_{\theta+\Delta}(y, dz) - P_\theta(y, dz)) \right) Q_\theta(x, dy)$$
$$+ \lim_{\Delta \to 0} \frac{1}{\Delta} \int \left(\int h(z) P_\theta(y, dz) \right) (Q_{\theta+\Delta}(x, dy) - Q_\theta(x, dy))$$
$$+ \lim_{\Delta \to 0} \frac{1}{\Delta} \int \left(\int h(z)(P_{\theta+\Delta}(y, dz) - P_\theta(y, dz)) \right) (Q_{\theta+\Delta}(x, dy) - Q_\theta(x, dy)).$$

We turn to the first limit on the above righthand side. By v_α-Lipschitz continuity of P_θ, see (ii), we have

$$\int |h(z)| |P_{\theta+\Delta}(y, dz) - P_\theta(y, dz)| \leq |\Delta| M_P ||h||_\alpha v_\alpha(y), \quad (13.9)$$

and since $\int v_\alpha(y) Q_\theta(x, dy) < \infty$, we may interchange the limit with the Q_θ integral. From $\mathcal{D}_\alpha$-differentiability of P_θ, condition (iii), we arrive at

$$\lim_{\Delta \to 0} \frac{1}{\Delta} \int \left(\int h(z)(P_{\theta+\Delta}(y, dz) - P_\theta(y, dz)) \right) Q_\theta(x, dy)$$
$$= \int \left(\int h(z) P'_\theta(y, dz) \right) Q_\theta(x, dy).$$

For the second limit, we argue as follows:

$$\int h(z) P_\theta(y, dz) =: \hat{h}(y)$$

such that by (ii) $\hat{h} \in \mathcal{D}_\alpha$. Then, we obtain from $\mathcal{D}$-differentiability of Q_θ that

$$\lim_{\Delta \to 0} \frac{1}{\Delta} \int \left(\int h(z) P_\theta(y, dz) \right) (Q_{\theta+\Delta}(x, dy) - Q_\theta(x, dy))$$
$$= \int \left(\int h(z) P_\theta(y, dz) \right) Q'_\theta(x, dy).$$

We now show that the third limit is equal to zero, where again we fix x. By (13.9) for P_θ and Q_θ, we may apply the dominated convergence theorem

$$\lim_{\Delta \to 0} \frac{1}{\Delta} \int \left(\int |h(z)| |P_{\theta+\Delta}(y, dz) - P_\theta(y, dz)| \right) |Q_{\theta+\Delta}(x, dy) - Q_\theta(x, dy)|$$
$$\leq \lim_{\Delta \to 0} \int \left(\int |h(z)| \frac{1}{|\Delta|} |P_{\theta+\Delta}(y, dz) - P_\theta(y, dz)| \right) |Q_{\theta+\Delta}(x, dy) - Q_\theta(x, dy)|$$
$$\leq \lim_{\Delta \to 0} \int M_P \alpha^y |Q_{\theta+\Delta}(x, dy) - Q_\theta(x, dy)|$$
$$\leq \lim_{\Delta \to 0} |\Delta| M_P M_Q v_\alpha(x)$$
$$= 0.$$

Hence, for all h such that $\|h\|_\alpha < \infty$,

$$\int h(z)(P_\theta Q_\theta)'(x, dz) = \int h(z)(P'_\theta Q_\theta)(x, dz) + \int h(y)(P_\theta Q'_\theta)(x, dz),$$

which concludes the proof of the first part of the theorem. The other parts of the theorem follow by standard arguments and the proof is therefore omitted. □

When expanding the proof to the sample-path interpretation, we exploit the following monotonicity property of the v_α-norm:

$$\mathcal{D}_{\hat{\alpha}} \subset \mathcal{D}_\alpha, \quad \hat{\alpha} \leq \alpha.$$

This is a simple implication of the fact that $\mathcal{D}_{\hat{\alpha}}$ contains less elements than $\mathcal{D}_\alpha$. In the same way we see that if P_θ is v_α-Lipschitz, it is also $v_{\hat{\alpha}}$-Lipschitz for any $\alpha > \hat{\alpha}$, and that if P_θ is $\mathcal{D}_\alpha$-differentiable it is also $\mathcal{D}_{\hat{\alpha}}$-differentiable for any $\alpha > \hat{\alpha}$. We now present the product rule for the path measure interpretation.

Lemma 13.2. *Let P_θ be a Markov kernel on $(S, \mathcal{S})$ defined on some open neighborhood Θ of θ, and let $\mathcal{D}_\alpha$ be a set of real-valued measurable mappings on S with finite v_α-norm.*

Suppose that for $\alpha > 1$

(i) $\|P_\theta\|_\alpha$ *is finite,*
(ii) P_ϑ *is* v_α-*Lipschitz on* Θ *with modulus in* M_P, *and*
(iii) P_ϑ *is* $\mathcal{D}_\alpha$-*differentiable at* $\vartheta = \theta$.

Then, P^n *is, for given initial state, a* $\mathcal{D}_{\sqrt[n]{\alpha}}$ *differentiable probability measure on* $(S^n, S^{\otimes n})$ *at* θ *with* $\mathcal{D}_{\sqrt[n]{\alpha}}$-*derivative*

$$(P^n_\theta)' = \sum_{j=0}^{n-1} P^{n-j-1}_\theta P'_\theta P^j_\theta.$$

If, in addition, P_ϑ *has* $\mathcal{D}_\alpha$-*derivative* $(c_\theta(\cdot), P^+_\theta, P^-_\theta)$ *at* θ, *then, it holds that for any* $h \in \mathcal{D}_{\sqrt[n]{\alpha}}$ *that*

$$\int_{\mathbb{R}^n} h(x,y) P^n_\theta(x,dy) = \sum_{j=0}^{n-1} \int_{\mathbb{R}^j} \left(\int_{\mathbb{R}} \int_{\mathbb{R}^{n-j-1}} h(x,y,z,r) P^{n-j-1}_\theta(z,dr) P^+_\theta(y,dz) \right.$$

$$\left. - h(x,y,z,r) P^{n-j-1}_\theta(z,dr) P^-_\theta(y,dz) \right) c_\theta(y) P^j_\theta(x,dy),$$

or, equivalently,

$$\frac{d}{d\theta} \int h(y) P^n_\theta(x,dy) = 2n\mathbb{E} \left[\int_{\mathbb{R}^{\eta-1}} \left(\int_{\mathbb{R}^{n-\eta}} h(x,x_1,\ldots,x_n) \left(P^{n-\eta}_\theta P^{[\sigma]}_\theta \right)(y,dz) \right) \right.$$
$$\left. c_\theta(y) P^{\eta-1}_\theta(x,dy) \right],$$

where σ *is uniformly distributed on* $\{1,\ldots,2n\}$ *independent of everything else,* $\eta = \sigma$ *mod* n,

$$P^{[\sigma]}_\theta = \begin{cases} P^+_\theta(y,dz) & \text{if } \sigma \leq n, \\ -P^-_\theta(y,dz) & \text{if } \sigma > n. \end{cases}$$

Proof. We prove the statement for $n = 2$ for the probability on a sample-path interpretation. The proof follows the same line of argument as the one for the n-step transition interpretation, and, in the following, we only discuss the parts of the proof that differ from that of Theorem 13.1.

We do this for the first part of the decomposition (13.8) in the proof of Theorem 13.1 for the path-dependent version, i.e.,

$$\lim_{\Delta \to 0} \frac{1}{\Delta} \int \left(\int h(x,y,z)(P_{\theta+\Delta}(y,dz) - P_\theta(y,dz)) \right) Q_\theta(x,dy).$$

By $v_{\sqrt{\alpha}}$-Lipschitz continuity of P_θ, which follows from (ii) together with

$$v_{\sqrt{\alpha}}(y) v_{\sqrt{\alpha}}(y) = v_\alpha(y),$$

we have

$$\int |h(x,y,z)| |P_{\theta+\Delta}(y,dz) - P_\theta(y,dz)|$$
$$\leq \int \|h\|_{\sqrt{\alpha}} v_{\sqrt{\alpha}}(x) v_{\sqrt{\alpha}}(y) v_{\sqrt{\alpha}}(z) |P_{\theta+\Delta}(y,dz) - P_\theta(y,dz)|$$

$$= \|h\|_{\sqrt{\alpha}} v_{\sqrt{\alpha}}(x) \, v_{\sqrt{\alpha}}(y) \int v_{\sqrt{\alpha}}(z) |P_{\theta+\Delta}(y,dz) - P_\theta(y,dz)|$$
$$\leq |\Delta| \, \|h\|_{\sqrt{\alpha}} v_{\sqrt{\alpha}}(x) \, v_{\sqrt{\alpha}}(y) \, v_{\sqrt{\alpha}}(y) \, M_P$$
$$\leq |\Delta| \, \|h\|_{\sqrt{\alpha}} v_{\sqrt{\alpha}}(x) \, v_\alpha(y) \, M_P.$$

Since $\int v_\alpha(y) Q_\theta(x,dy) < \infty$, we may interchange the limit with the Q_θ integral. From $h \in \mathcal{D}_{\sqrt{\alpha}}$ with $\mathcal{D}_{\sqrt{\alpha}}$-differentiability of P_θ, condition (iii), we arrive at

$$\lim_{\Delta \to 0} \frac{1}{\Delta} \int \left(\int h(x,y,z)(P_{\theta+\Delta}(y,dz) - P_\theta(y,dz)) \right) Q_\theta(x,dy)$$
$$= \int \left(\int h(x,y,z) P'_\theta(y,dz) \right) Q_\theta(x,dy).$$

The arguments for the other parts follow in a similar way. □

In the following example we illustrate the application of Lemma 13.2 to the sojourn time example.

Example 13.4. We continue with the analysis of the sojourn time analysis by checking that the conditions in Lemma 13.2 are satisfied. Following Example 13.2, $\alpha < e^{\theta_1}$ implies that $\mathbb{E}[(1 + S_1(\theta))\alpha^{S_1(\theta)}] < \infty$ for $\theta \in [\theta_0, \theta_1]$. Moreover $\mathbb{E}[(1 + A_1)\alpha^{A_1}] < \infty$ as A_1 is uniformly distributed. Condition (i) follows from Example 13.1. Conditions (ii) and (iii) have been established in Example 13.3. Hence, the finite horizon MVD estimator can be applied to any $h \in \mathcal{D}_{\sqrt[n]{\alpha}}$.

We now work out the specific case for $S(\theta)$ uniformly distributed on $[1, \theta]$ for $\theta \geq 2$. The uniform distribution with θ-dependent support cannot be dealt with by the score function, and we now show how a gradient estimator can be obtained from Lemma 13.2. We establish $\mathcal{D}_\alpha$ differentiability of the sojourn time kernel, where we take for $\mathcal{D}_\alpha$ the set of all continuous mappings with finite v_α-norm. By computation,

$$\frac{d}{d\theta} \int g(y) P_\theta(x,dy) = \frac{d}{d\theta} \int_0^\infty \frac{1}{\theta-1} \left(\int_1^\theta g(\max(x-a,0)+s) ds \right) f_A(a) da$$
$$= \frac{1}{\theta-1} \int_0^\infty g(\max(x-a,0)+\theta) f_A(a) da$$
$$- \frac{1}{(\theta-1)^2} \int_0^\infty \left(\int_1^\theta g(\max(x-a,0)+s) ds \right) f_A(a) da.$$

In words, the P^+ kernel has the service time fixed to θ and the P^- kernel is equal to P, with $c_\theta = 1/(\theta-1)$, compare with Example 8.20.

For a specific instance of the estimator, consider the sequence $X_i(\theta)$, $1 \leq i \leq n$, of sojourn times; see (13.3). Suppose we are interested in the deviation of the mean sojourn time from the maximum, i.e., consider

$$h(X_1(\theta), \ldots, X_n(\theta)) = \max(X_1(\theta), \ldots, X_n(\theta)) - \frac{1}{n} \sum_{i=1}^n X_i(\theta).$$

In the literature, a performance function that, like $h(X_1(\theta), \ldots, X_n(\theta))$, measures the deviation of the sample-extreme from the sample mean is called *jitter*. Let

$$h_i(X_1(\theta), \ldots, X_n(\theta); \theta) = h(X_1(\theta), \ldots, X_{i-1}(\theta), \theta, X_{i+1}(\theta), \ldots, X_n(\theta)),$$

for $1 \leq i \leq n$. As h has finite $v_{n\sqrt{\alpha}}$-norm for any n and $\alpha \geq 1$, we may apply Lemma 13.2 and obtain, for example,

$$\frac{d}{d\theta}\mathbb{E}[h(X_1(\theta), \ldots, X_n(\theta))]$$
$$= \frac{1}{(\theta-1)}\mathbb{E}\left[\sum_{i=1}^{n} h_i(X_1(\theta), \ldots, X_n(\theta); \theta) - nh(X_1(\theta), \ldots, X_n(\theta))\right].$$

※※※

13.2 THE RANDOM HORIZON PROBLEM

The random horizon problem can be analyzed in a structural manner provided that the horizon τ is a first entrance time of a Markov chain. To see this, let $\{X_n\}$ denote a Markov chain. Select a set $A \in \mathcal{S}$, and denote by $\tau(A, s)$ the first entrance time of the chain in A started in s:

$$\tau(A, s) = \inf\left\{n \geq 1 : X_n \in A \text{ and } X_i \notin A, \ 1 \leq i \leq n-1\right\}.$$

Hitting times such as the first entrance time introduced above can be analyzed in a very neat way by means of the taboo kernel to be introduced presently. Define $_AP$ as

$$_AP(s, B) = P(s, B \cap A^c),$$

for all $s \in S$, $A \in \mathcal{S}$, and A^c denoting the complement of A. Then

$$\mathbb{P}(X_{n+1} \in B \text{ and } X_{n+1} \notin A | X_n = x) = {_AP}(x, B)$$

is a deficient Markov transition kernel avoiding entering the set A, hence the name taboo kernel. Denote the n-fold convolution of $_AP$ by $(_AP)^n$, i.e., for $n = 2$

$$(_AP)^2(s, B) = \int_S {_AP}(r, B) {_AP}(s, dr).$$

The main tool of analysis is rewriting expressions containing hitting times as an infinite expression where the hitting does not appear anymore as random variable. The following lemma shows the relation between the taboo kernel representation and the framework for regenerative processes.

Lemma 13.3. *Consider a Harris recurrent Markov chain with atom A and regeneration points $\{\tau_k\}$ (see Section C.1 in Appendix C for definitions). Provided that $||_AP||_v < 1$ it holds that*

$$\mathbb{E}\left[\sum_{n=0}^{\tau_1-1} f(X_n) \bigg| X_0 \in A\right] = \left(\sum_{n=0}^{\infty}(_AP)^n f\right)(X_0) < \infty,$$

$$\sum_{n=0}^{\infty}(_AP)^n = (I - {_AP})^{-1},$$

and

$$\pi f = \frac{\left(\sum_{n=0}^{\infty}(_AP)^n f\right)(X_0)}{\left(\sum_{n=0}^{\infty}(_AP)^n \mathbf{1}\right)(X_0)},$$

for $X_0 \in A$ and all f such that $||f||_v < \infty$. Moreover,

$$||\pi||_v \leq \frac{1}{1 - ||_A P||_v}.$$

Proof. For $s \in S$ and $B \cap A = \emptyset$,

$$\mathbb{E}\left[\mathbf{1}_{X_2(\theta) \in B} \mathbf{1}_{X_1 \notin A} | X_0 = s\right] = \int_{r \in S \setminus A} P(s, dr) P(r, B)$$

$$= \int_{r \in S \setminus A} {}_A P(s, dr) {}_A P(r, B).$$

By induction we obtain for any cost function $f : S \mapsto \mathbb{R}$, with $||f||_v < \infty$, that

$$\mathbb{E}\left[\sum_{n=0}^{\tau(A,s)-1} f(X_n) \Big| X_0 = s\right] = \mathbb{E}\left[\sum_{n=0}^{\infty} f(X_n) \mathbf{1}_{\{X_i \notin A, 1 \leq i \leq n\}} \Big| X_0 = s\right]$$

$$= \left(\sum_{n=0}^{\infty} ({}_A P)^n f\right)(s),$$

whenever the expression on the left hand side exists. Provided that $||_A P|| < 1$, the infinite sum yields the inverse of $I - {}_A P$ and we obtain

$$\mathbb{E}\left[\sum_{n=0}^{\tau(A,s)-1} f(X_n) \Big| X_0 = s\right] = \left((I - {}_A P)^{-1} f\right)(s).$$

The next statement follows directly from renewal theory (see Theorem C.2 in appendix C). For the final part of the proof, observe that $\tau(A, s) \geq 1$ and thus $\mathbb{E}[\tau(A, s)] \geq 1$. Hence,

$$||\pi_v|| \leq \sum_{n=0}^{\infty} ||({}_A P)^n||_v \leq \sum_{n=0}^{\infty} (||_A P||_v)^n = \frac{1}{1 - ||_A P||_v}.$$

□

The following example explains the main technique for a Markov chain with discrete state space.

Example 13.5. Consider the embedded queue length process $\{X_n\}$ of the M/M/1 queue with transition matrix P as introduced in Section 9.3.1. Let $\tau(i)$ denote the first time the queue becomes empty when started in state i. Letting $A = \{0\}$, the embedded queue length process is classical regenerative with regenerative set A, and we have $\tau(i) = \tau(\{0\}, i)$. Let

$$_0 P(i, j) = \begin{cases} P(i, j) & \text{if } j > 0, \\ 0 & \text{if } j = 0. \end{cases}$$

For illustration purposes, consider the embedded queue length process $\{X_n\}$ of the M/M/1 queue with four places, i.e., a buffer of size three and one service place. Customers that arrive at a full queue are lost. The Markov kernel and the taboo kernel then read

$$P = \begin{pmatrix} 0 & 1 & 0 & 0 \\ \frac{\mu}{\lambda+\mu} & 0 & \frac{\lambda}{\lambda+\mu} & 0 \\ 0 & \frac{\mu}{\lambda+\mu} & 0 & \frac{\lambda}{\lambda+\mu} \\ 0 & 0 & 1 & 0 \end{pmatrix}, \quad _0 P = \begin{pmatrix} 0 & 1 & 0 & 0 \\ 0 & 0 & \frac{\lambda}{\lambda+\mu} & 0 \\ 0 & \frac{\mu}{\lambda+\mu} & 0 & \frac{\lambda}{\lambda+\mu} \\ 0 & 0 & 1 & 0 \end{pmatrix},$$

see (9.22), where $_0P$ is obtained from P be setting $P(i,0)$ to zero for all i. Then,

$$(I-{_0P})^{-1} = \begin{pmatrix} 1 & -1 & 0 & 0 \\ 0 & 1 & -\frac{\lambda}{\lambda+\mu} & 0 \\ 0 & -\frac{\mu}{\lambda+\mu} & 1 & -\frac{\lambda}{\lambda+\mu} \\ 0 & 0 & -1 & 1 \end{pmatrix}^{-1},$$

and letting $\lambda/(\lambda+\mu) = \mu/(\lambda+\mu) = 0.5$, we can numerically solve this as,

$$(I-{_0P})^{-1} = \begin{pmatrix} 1 & 2 & 2 & 1 \\ 0 & 2 & 2 & 1 \\ 0 & 2 & 4 & 2 \\ 0 & 2 & 4 & 3 \end{pmatrix}.$$

Letting $f(i) = i$, then $(I-{_0P})^{-1}f$ is the accumulated queue length over a cycle from the queue starting with i customers until the queue is emptied:

$$h = (I-{_0P})^{-1}f = \begin{pmatrix} 1 & 2 & 2 & 1 \\ 0 & 2 & 2 & 1 \\ 0 & 2 & 4 & 2 \\ 0 & 2 & 4 & 3 \end{pmatrix} \begin{pmatrix} 0 \\ 1 \\ 2 \\ 3 \end{pmatrix} = \begin{pmatrix} 9 \\ 9 \\ 16 \\ 19 \end{pmatrix}.$$

Specifically, for $i = 0$, the accumulated queue length over a cycle can be obtained from

$$\mathbb{E}\left[\sum_{n=0}^{\tau(0)-1} f(X_n) \bigg| X_0 = 0\right] = h(0) = 9.$$

※※※

The following examples showcases an application of the taboo kernel for a problem with a general state space.

Example 13.6. An insurance company receives premiums from clients at some constant rate $r > 0$ while claims $\{Y_i(\theta) : i \geq 1\}$ arrive according to a Poisson process with rate $\lambda > 0$. Let $\{X_i : i \geq 1\}$ denote the inter arrival times of the Poisson process and let N_τ denote the number of claims recorded up to some fixed time horizon $\tau > 0$. Assume further that the values of claims are iid following a Pareto type I distribution π_θ, i.e.,

$$\pi_\theta(dx) = \frac{\beta\theta^\beta}{x^{\beta+1}} 1_{x \in (\theta,\infty)} dx,$$

and that the values of the claims are independent of the Poisson process. Note that θ models the support of the claim distribution. We assume that the mean claim size is strictly smaller than the mean revenue:

$$\mathbb{E}[rX_n] > \mathbb{E}[Y_n(\theta)], \qquad (13.10)$$

for all n and $\theta \leq \theta_0$.

Let $V_0(\theta) \geq 0$ denote the initial credit of the insurance company. The credit (resp. debt) of the company right after the n-th claim, denoted by $V_\theta(n)$, follows the recurrence relation

$$\forall n \geq 0: V_{n+1}(\theta) = V_n(\theta) + rX_{n+1} - Y_{n+1}(\theta).$$

Ruin occurs before time τ if at least one $n \leq N_\tau$ exists such that $V_n(\theta) < 0$. We are interested in estimating the derivative w.r.t. θ of the expected time until ruin, that is, for

$$\eta_\theta = \inf\{n \geq 0 : V_n(\theta) \leq 0\}$$

we want to estimate

$$\frac{d}{d\theta}\mathbb{E}[\eta_\theta].$$

Taboo set is $A = (-\infty, 0)$ and

$$\mathbb{E}[\eta_\theta] = \sum_{n=1}^{\infty} {}_A P_\theta^n(x, dy),$$

where

$$P_\theta(x, B) = \mathbb{E}[\mathbf{1}_{V_{n+1}(\theta) \in B} | V_n(\theta) = x],$$

for $x \in \mathbb{R}$ and any measurable set $B \subset \mathbb{R}$. It is easily seen that

$$\int_{-\infty}^{\infty} {}_A P_\theta(x, dy) = \int_0^{\infty} P_\theta(x, dy) = \mathbb{P}(x + rX_{n+1} \geq Y_{n+1}(\theta))$$

and therefore, letting $\alpha = 1$ for $v(x) = \alpha^x$,

$$||{}_A P_\theta||_1 = \sup_{x \geq 0} \mathbb{P}(x + rX_{n+1} \geq Y_{n+1}(\theta)) \leq \mathbb{P}(rX_{n+1} \geq Y_{n+1}(\theta)) < 1.$$

Moreover, by (13.10)

$$\sup_{\theta \leq \theta_0} ||{}_A P_\theta||_1 < 1.$$

Recall that π_θ has measure-valued derivative $(\beta/\theta, \pi_\theta, \delta_\theta)$, with δ_θ denoting the Dirac measure in θ; see Table 8.1. Thus,

$$\frac{d}{d\theta}\int_{-\infty}^{\infty} {}_A P(x, dy) = \frac{\beta}{\theta}\left(\mathbb{P}(x + rX_{n+1} \geq y | Y_{n+1} = y)\pi_\theta(dy) - \mathbb{P}(x + rX_{n+1} \geq \theta)\right)$$

and we see that ${}_A P$ is $\mathcal{D}_\alpha$ differentiable for $\alpha = 1$.

※※※

We now present the main differentiation results for taboo kernels.

Theorem 13.2. *Let P_ϑ be a Markov kernel on $(S, \mathcal{S})$ defined on some open neighborhood Θ of θ, and let $\mathcal{D}_\alpha$ be a set of real-valued measurable mappings on S with finite v_α-norm. Assume that*

(i) *$\sup_{\vartheta \in \Theta} ||P_\vartheta||_\alpha < 1$,*
(ii) *P_ϑ is v_α-Lipschitz continuous on Θ with modulus M, and*
(iii) *P_ϑ is $\mathcal{D}_\alpha$-differentiable at $\vartheta = \theta$.*

Then for any $f \in \mathcal{D}_\alpha$ it holds that for the derivative at θ that

$$\left(\sum_{n=0}^{\infty} P_\theta^n f\right)' = \sum_{n=0}^{\infty}\sum_{m=0}^{\infty} P_\theta^n P_\theta' P_\theta^m f,$$

or, equivalently,

$$\left(\sum_{n=0}^{\infty} P_\theta^n f\right)' = (I - P_\theta)^{-1} P_\theta' (I - P_\theta)^{-1} f.$$

Proof. By algebra,

$$\sum_{n=0}^{m} P_{\theta+\Delta}^n - P_\theta^n = \sum_{n=0}^{m} \sum_{k=0}^{n-1} P_{\theta+\Delta}^k (P_{\theta+\Delta} - P_\theta) P_\theta^{n-k-1}.$$

By (i), $\rho < 1$ exists such that

$$\sup_{\vartheta \in \Theta} \|P_\vartheta\|_\alpha \leq \rho.$$

This together with (ii) implies for $\theta, \theta + \Delta \in \Theta$ that

$$\|P_{\theta+\Delta}^k (P_{\theta+\Delta} - P_\theta) P_\theta^{n-k-1}\|_\alpha \leq |\Delta| \, M \, \rho^{n-1},$$

where M is v_α-norm Lipschitz modulus of P_θ. By (i), (ii), together with (iii), it follows that

$$\lim_{\Delta \to \infty} \frac{1}{\Delta} P_{\theta+\Delta}^k (P_{\theta+\Delta} - P_\theta) P_\theta^{n-k-1} f = P_\theta^k P_\theta' P_\theta^{n-k-1} f,$$

where we use Lemma 13.1, see the proof of Theorem 13.1 for details.

Noting that

$$\sum_{n=0}^{m} \sum_{k=0}^{n-1} \rho^{n-1} = \sum_{n=0}^{m} n \rho^{n-1}$$

implies that

$$\sum_{n=0}^{\infty} \sum_{k=0}^{\infty} \rho^{n+m} < \infty,$$

and the proof of the theorem follows from the dominated convergence theorem. □

Remark 13.1. The interpretation of the derivative expression in Theorem 13.2 in terms of simulation estimators requires close inspection of $_A P'$ as the derivative of the taboo kernel may, for example, contain only positive values. To see this, consider

$$P = \begin{pmatrix} 1 - \theta & \theta \\ 1 - \frac{\theta}{2} & \frac{\theta}{2} \end{pmatrix},$$

for $\theta \in (1/2, 1)$. Letting $A = \{0\}$, gives

$$_0 P = \begin{pmatrix} 0 & \theta \\ 0 & \frac{\theta}{2} \end{pmatrix}$$

and

$$_0 P' = \begin{pmatrix} 0 & 1 \\ 0 & \frac{1}{2} \end{pmatrix}.$$

Hence, letting $c_P(0) = 1$ and $c_P(1) = 1/2$, we simulate $_0 P'$ by deterministically letting the chain jump to state 1, and rescale by c_P.

We illustrate an application of Theorem 13.2 with the queue length and the insurance example. For more examples and variations of the estimator we refer to [154].

Example 13.7. We revisit Example 13.5 and study the embedded queue length process $\{X_n(\theta)\}$ of the M/M/1 queue with four places, i.e., a buffer of size three and one service place. Letting $\theta = \lambda$, we have

$$\frac{d}{d\lambda} {}_0 P = \begin{pmatrix} 0 & 0 & 0 & 0 \\ 0 & 0 & -\frac{\mu}{(\lambda+\mu)^2} & 0 \\ 0 & \frac{\mu}{(\lambda+\mu)^2} & 0 & -\frac{\mu}{(\lambda+\mu)^2} \\ 0 & 0 & 0 & 0 \end{pmatrix},$$

where λ is the arrival rate and μ is the service rate. As the Markov chain has a finite state space the conditions in Theorem 13.2 are easily checked. Let $\lambda = 1, \mu = 2$. Form the theorem it then follows that

$$\frac{\partial}{\partial \lambda} \mathbb{E}\left[\sum_{n=0}^{\tau_\theta(0)-1} f(X_n(\theta)) \,\Big|\, X_0(\theta) = 0 \right] = \hat{f}(0) = 6,$$

where

$$\hat{f} = (I - {}_0P)^{-1} {}_0 P' (I - {}_0P)^{-1} f$$

$$= \begin{pmatrix} 1 & 2 & 2 & 1 \\ 0 & 2 & 2 & 1 \\ 0 & 2 & 4 & 2 \\ 0 & 2 & 4 & 3 \end{pmatrix} \begin{pmatrix} 0 & 0 & 0 & 0 \\ 0 & 0 & -2/9 & 0 \\ 0 & 2/9 & 0 & -2/9 \\ 0 & 0 & 0 & 0 \end{pmatrix} \begin{pmatrix} 1 & 2 & 2 & 1 \\ 0 & 2 & 2 & 1 \\ 0 & 2 & 4 & 2 \\ 0 & 2 & 4 & 3 \end{pmatrix} \begin{pmatrix} 0 \\ 1 \\ 2 \\ 3 \end{pmatrix}$$

$$= \frac{1}{9} \begin{pmatrix} 54 \\ 54 \\ 160 \\ 160 \end{pmatrix}.$$

✳✳✳

Example 13.8. Recall the definitions put forward in Example 13.6. Let $V(i, x, j)$ denote the credit value after $i + j$ claims where the value of claim i is set to x, i.e., $Y_i(\theta) = x$. Denote by

$$\eta_\theta(i, y) = \inf\{i + j \in \mathbb{N} : V(i, y, j) \leq 0 \text{ and } i \leq \eta_\theta\}.$$

In words, $\eta_\theta(i, y)$ is time until ruin if the i-th claim is set to y. In case the set on the above right-hand side is empty (e.g., ruin occurred before i), we let $\eta_\theta(i, y) = 0$.

Taking $h = 1$ as cost function, we obtain from Theorem 13.2,

$$\frac{d}{d\theta} \mathbb{E}[\eta_\theta] = \frac{\beta}{\theta} \mathbb{E}\left[\sum_{i=1}^{\eta_\theta - 1} (\eta_\theta(i, Y(\theta)) - \eta_\theta) \right],$$

where $Y(\theta)$ is distributed according to the claim-size distribution independent of everything else.

✳✳✳

An Extended Example: The Sojourn Times in the G/G/1 Queue

In this section, we revisit the sojourn time example in Example 13.1 notations are as introduced therein. We take as taboo set $A = \{0\}$, and, following (13.4), the taboo kernel of the sojourn times becomes

$$\forall x \geq 0 : ({}_0 P_\theta h)(x) := \mathbb{E}_\theta \left[h(x + S(\theta) - A) \cdot \mathbb{I}_{\{x+S(\theta) > A\}} \right],$$

for any h measurable so that the expected value on the above right-hand side is finite. The atom $\{0\}$ carries positive mass for $P_\theta(x, \cdot)$ for any $x \geq 0$. As the following lemma shows, removing the mass on $\{0\}$ is enough to decrease the v_α-norm of P_θ below 1 for appropriate α. Note that the $G/G/1$ queue is stable if $\mathbb{E}[(S_n(\theta)] < \mathbb{E}[A_n]$, and we denote the stability set by

$$\Theta := \{ \theta > 0 : \mathbb{E}[S_n(\theta)] < \mathbb{E}[A_n] \},$$

for any n. As the next lemma shows we have $\|P_\theta\|_\alpha < 1$, for some $\alpha > 1$. More specifically, given $\theta_1, \theta_2 \in \Theta_s$, such that $\theta_1 < \theta_2$, there exist sufficiently small $\alpha > 1$ such that $\|P_\theta\|_v < 1$, uniformly in $\theta \in [\theta_1, \theta_2]$. The precise statement is as follows.

Lemma 13.4. Let $\theta_1, \theta_2 \in \Theta$ with $\theta_1 < \theta_2$. Assume that A_1 and $S_1(\theta)$, for $\theta_1 \leq \theta \leq \theta_2$, have finite moment generating functions on $(-t, t)$. Then

$$\|{}_0 P_\theta\|_\alpha = \mathbb{E}[\alpha^{S_1(\theta) - A_1}]$$

and there exists $\bar{\alpha} > 1$, such that $\ln(\bar{\alpha}) < t$, so that for $\alpha \in (1, \bar{\alpha})$ it holds that

$$\sup_{\theta \in [\theta_1, \theta_2]} \|{}_0 P_\theta\|_{v_\alpha} < 1. \tag{13.11}$$

Proof. Following (7.30), we obtain for the taboo kernel of the sojourn times

$$\|{}_0 P_\theta\|_\alpha = \sup_x \frac{1}{\alpha^x} \mathbb{E}\left[\mathbf{1}_{\{x - A_1 > 0\}} \alpha^{x - A_1 + S_1(\theta)} \right] = \sup_x \mathbb{E}\left[\mathbf{1}_{\{x - A_1 > 0\}} \alpha^{S_1(\theta) - A_1} \right].$$

Observing that $\sup_x \mathbf{1}_{\{x - A_1 > 0\}} = 1$, we arrive at

$$\|{}_0 P_\theta\|_\alpha = \mathbb{E}\left[\alpha^{S_1(\theta) - A_1} \right].$$

We now turn to the proof of the second part. Let $F : [1, \infty) \times [\theta_1, \theta_2] \to \mathbb{R} \cup \{\infty\}$ be defined as

$$\forall \alpha, \theta : F(\alpha, \theta) = \|{}_0 P_\theta\|_{v_\alpha} = \mathbb{E}\left[\alpha^{S_1(\theta) - A_1} \right] = \mathbb{E}\left[e^{\ln(\alpha)(S_1(\theta) - A_1)} \right]. \tag{13.12}$$

Note that, by hypothesis, we may choose α^*, such that $1 < \alpha^*$ and $\ln(\alpha^*) < t$, and then have that $F(\alpha, \theta) < \infty$ for all $\alpha \in [1, \alpha^*]$ and $\theta \in [\theta_1, \theta_2]$. Moreover, for $\alpha \in [1, \alpha^*)$ and $\theta \in [\theta_1, \theta_2]$ we have

$$\forall n \geq 0 : \sup_{y \in \mathbb{R}} |y|^n e^{(\ln(\alpha) - \ln(\alpha^*))y} = \left[\frac{n}{(\ln(\alpha^*) - \ln(\alpha))e} \right]^n.$$

Therefore, it follows that

$$\forall n \geq 0, y \in \mathbb{R} : |y|^n e^{\ln(\alpha) y} \leq \left[\frac{n}{(\ln(\alpha^*) - \ln(\alpha))e} \right]^n e^{\ln(\alpha^*) y} \tag{13.13}$$

and by letting $y = S_1(\theta) - A_1$ in (13.13) and taking expected values, we arrive at

$$\mathbb{E}_\theta \left[|S_1(\theta) - A_1|^n e^{\ln(\alpha)(S_1(\theta) - A_1)} \right] \leq \left[\frac{n}{(\ln(\alpha^*) - \ln(\alpha))e} \right]^n \mathbb{E}_\theta \left[(\alpha^*)^{S_1(\theta) - A_1} \right] < \infty. \tag{13.14}$$

For $\theta \in [\theta_1, \theta_2]$ we have $F(1, \theta) = 1$, $\lim_{\alpha \uparrow \infty} F(\alpha, \theta) = \infty$ and

$$\lim_{\alpha \downarrow 1} \frac{d}{d\alpha} F(\alpha, \theta) = \lim_{\alpha \downarrow 1} \mathbb{E}\left[\frac{1}{\alpha}(S_1(\theta) - A_1)e^{\ln(\alpha)(S_1(\theta) - A_1)}\right] = \mathbb{E}[S_1(\theta) - A_1] < 0.$$

Moreover, the second derivative with respect to α satisfies

$$\forall \alpha \in (1, \alpha^*), \theta \in [\theta_1, \theta_2]: \frac{d^2}{d\alpha^2} F(\alpha, \theta)$$
$$= \mathbb{E}_\theta\left[\frac{1}{\alpha^2}(S_1(\theta) - A_1)^2 + (A_1 - S_1(\theta))e^{\ln(\alpha)(S_1(\theta) - A_1)}\right] > 0,$$

where finiteness of the second derivative follows from (13.14) and we use that $\mathbb{E}[A_1 - S_1(\theta)] > 0$ for stable queues. Hence, we conclude that F is strictly convex in α and consequently for each $\theta \in [\theta_1, \theta_2]$ there exist an unique $\alpha > 1$ satisfying $F(\alpha, \theta) = 1$. If we denote this value by α_θ it holds that

$$\forall \alpha \in [1, \alpha_\theta): F(\alpha, \theta) < 1.$$

Continuity of F in both α and θ implies continuity of the implicit function $\theta \mapsto \alpha_\theta$; see, e.g., [192]. Therefore, we have $\inf\{\alpha_\theta : \theta \in [\theta_1, \theta_2]\} > 1$. Letting

$$\bar{\alpha} = \min\{\alpha^*, \inf\{\alpha_\theta : \theta \in [\theta_1, \theta_2]\}\}$$

concludes the proof. $\square$

Following the line of argument put forward in Example 13.1, it follows that $_0P_\theta$ is $\mathcal{D}_\alpha$-differentiable and ν_α-Lipschitz. We denote by $\tau(\theta)$ the number of transitions the sojourn time sequence, started in zero, takes until returning to zero. Hence, by Lemma 13.4, it follows from Theorem 13.2 that the derivative of the accumulated costs over a cycle is given by

$$\frac{d}{d\theta}\mathbb{E}\left[\sum_{n=0}^{\tau(\theta)-1} h(X_n(\theta)) \bigg| X_0(\theta) = 0\right]$$
$$= \frac{1}{\theta}\mathbb{E}\left[\sum_{n=0}^{\tau(\theta)-1}\sum_{m=1}^{\tau^+(\theta,n)} \left(h(X^+(\theta; n, m) - h(X_{n+m}(\theta))\right) \bigg| X_0(\theta) = 0\right],$$

where $X^+(\theta; n, m)$ is $(n+m)$-th sojourn time where for the transition from $X_n(\theta)$ to the next sojourn time the service time is replaced by the sum of two independent copies of the service time, and $\tau^+(\theta, n)$ is the first time that the perturbed sojourn time processes couples with the nominal one. Constructing $X_n(\theta)$ and $X^+(\theta; n, m)$ on a common probability space so that both sequences are constructed with same sequence of interarrival times and service times, $\tau^+(\theta, n)$ is given by

$$\tau^+(\theta, n) = \{k \geq 1 : X_{n+k}(\theta) = X^+(\theta; n, k)\},$$

where we let $\tau^+(\theta, n) = \infty$ if the set on the above right-hand side is empty. Note that under the conditions put forward in Theorem 13.2 $\tau^+(\theta, n)$ has finite expected value and is thus a.s. finite.

13.3 THE STATIONARY PROBLEM

Consider a Markov process $\{X_n(\theta)\}$ with transition probability P_θ and unique stationary distribution π_θ. Existence assumed, we denote by Π_θ the ergodic projector of a Markov kernel, i.e., for any distribution μ it holds that $\mu \Pi_\theta = \pi_\theta$. Note that for P_θ on a countable state space with unique stationary distribution, Π_θ is a matrix with rows identical to π_θ. We denote by

$$D_\theta = \Pi_\theta + \sum_{n=0}^{\infty}(P_\theta^n - \Pi_\theta) = (I - P_\theta + \Pi_\theta)^{-1}$$

the *deviation matrix* of P_θ. Before we present the differentiation formula, we provide a sufficient condition for the existence of the deviation matrix.

Lemma 13.5. *Consider a Harris recurrent Markov chain with atom A (see Section C.1 in Appendix C for definitions). Provided that $||_A P||_v < 1$ it holds that $||D||_v < \infty$.*

Proof. As shown in [1, 178] it holds that

$$D = (I - \Pi)\sum_{n=0}^{\infty}(_A P)^n (I - \Pi).$$

For the proof we apply the v-norm on both sides of the above equation. By Lemma 13.3, $||\Pi||_v$ is finite and thus $||I - \Pi||_v < \infty$ as well. Moreover, it follows from the assumption $||_A P||_v$ that $||\sum_{n=0}^{\infty}(_A P)^n||_v < \infty$. Hence, $||D||_v < \infty$. □

In the following we present the general result on weak differentiation of ergodic Markov kernels, for some extensions we refer to [150]. For sufficient conditions for the normed ergodicity, we refer to Section C.2 in Appendix C. To simplify the notation we write f'_θ for the derivative of f_ϑ with respect to ϑ at $\vartheta = \theta$.

Theorem 13.3. *Let P_ϑ be a Markov kernel on $(S, \mathcal{S})$ defined on some open neighborhood Θ of θ such that P_ϑ has unique stationary distribution π_ϑ and that D_ϑ exist for $\vartheta \in \Theta$. Let $\mathcal{D}_\alpha$ be a set of real-valued measurable mappings on S with finite v_α-norm. If*

(i) *P_ϑ is $\mathcal{D}_\alpha$-differentiable at θ and v_α-Lipschitz continuous on Θ,*
(ii) *$\sup_{\vartheta \in \Theta} ||\pi_\vartheta||_\alpha$ is finite, and*
(iii) *$||D_\theta||_\alpha$ is finite,*

then π_ϑ is $\mathcal{D}$-differentiable at θ with $\mathcal{D}$-derivative

$$\pi'_\theta = \pi_\theta \sum_{n=0}^{\infty} P'_\theta P_\theta^n,$$

or, equivalently,

$$\pi'_\theta = \pi_\theta P'_\theta \sum_{n=0}^{\infty}(P_\theta^n - \Pi_\theta).$$

Proof. By simple algebra,

$$I - \Pi_\vartheta = (I - P_\vartheta)D_\vartheta,$$

which is the Poisson equation in matrix format; see Section C.3 in Appendix C for a proof. For Δ sufficiently small it holds that $\Pi_{\vartheta+\Delta}\Pi_\vartheta = \Pi_\vartheta$, and multiplying the above equation by

$\Pi_{\vartheta+\Delta}$ we obtain
$$\Pi_{\vartheta+\Delta} - \Pi_\vartheta = \Pi_{\vartheta+\Delta}(I - P_\vartheta)D_\vartheta.$$

Using that $\Pi_{\vartheta+\Delta}P_{\vartheta+\Delta} = \Pi_{\vartheta+\Delta}$, gives
$$\Pi_{\vartheta+\Delta} - \Pi_\vartheta = \Pi_{\vartheta+\Delta}(P_{\vartheta+\Delta} - P_\vartheta)D_\vartheta.$$

Switching to vector notation, we arrive at
$$\pi_{\vartheta+\Delta} - \pi_\vartheta = \pi_{\vartheta+\Delta}(P_{\vartheta+\Delta} - P_\vartheta)D_\vartheta. \tag{13.15}$$

We now turn to the proof of the theorem. By (i), the Markov kernel P_ϑ is v_α-Lipschitz continuous on a neighborhood of θ, i.e.,
$$||P_{\theta+\Delta}(x,\cdot) - P_\theta(x,\cdot)||_\alpha \leq |\Delta|Mv_\alpha(x) \tag{13.16}$$

for some finite constant M. Applying v_α-norms to (13.15) for $\theta = \vartheta$ yields
$$\begin{aligned}||\pi_{\theta+\Delta} - \pi_\theta||_\alpha &\leq |\Delta|\,||\pi_{\theta+\Delta}||_\alpha\,M||D_\theta||_\alpha \\ &\leq |\Delta|\,M\,||D_\theta||_\alpha \sup_{\vartheta \in \Theta} ||\pi_\vartheta||_\alpha,\end{aligned} \tag{13.17}$$

where we make use of conditions (ii) and (iii).

We now turn to the proof of differentiability. By (13.15),
$$\frac{1}{\Delta}(\pi_{\theta+\Delta} - \pi_\theta) = \pi_\theta \frac{1}{\Delta}(P_{\theta+\Delta} - P_\theta)D_\theta + (\pi_{\theta+\Delta} - \pi_\theta)\frac{1}{\Delta}(P_{\theta+\Delta} - P_\theta)D_\theta. \tag{13.18}$$

By $\mathcal{D}_\alpha$-differentiability of P_θ it follows from (i) for $f \in \mathcal{D}_\alpha$ that
$$\lim_{\Delta \to 0} \frac{1}{\Delta}(P_{\theta+\Delta} - P_\theta)D_\theta f = P'_\theta D_\theta f, \tag{13.19}$$

where we use (13.1) to show that (iii) implies $D_\theta f \in \mathcal{D}_\alpha$ for $f \in \mathcal{D}_\alpha$. Moreover, by v_α-Lipschitz continuity of P_θ together with (iii), we obtain
$$\left|\frac{1}{\Delta}\int (P_{\theta+\Delta}(x,dy) - P_\theta(x,dy))\int D_\theta(y,dz)f(z)\right| = \left|\frac{1}{\Delta}((P_{\theta+\Delta} - P_\theta)D_\theta f)(x)\right|$$
$$\leq v_\alpha(x)\,M||f||_\alpha ||D_\theta||_\alpha,$$

where we make use of (13.1). Dominated convergence yields
$$\lim_{\Delta \to 0} \pi_\theta \frac{1}{\Delta}(P_{\theta+\Delta} - P_\theta)D_\theta = \pi_\theta P'_\theta D_\theta.$$

Using the bound in (13.17) together with v_α-Lipschitz continuity of P_θ, we have
$$\left\|(\pi_{\theta+\Delta} - \pi_\theta)\frac{1}{\Delta}(P_{\theta+\Delta} - P_\theta)D_\theta\right\|_\alpha \leq ||(\pi_{\theta+\Delta} - \pi_\theta)||\,M\,||D_\theta||_\alpha.$$

Hence,
$$\lim_{\Delta \to 0}(\pi_{\theta+\Delta} - \pi_\theta)\frac{1}{\Delta}(P_{\theta+\Delta} - P_\theta)D_\theta = 0,$$

which concludes the proof. □

Conditions (ii) and (iii) put forward in Theorem 13.3 can be established by means of stability theory of Markov chains; see Section C.2 in Appendix C.

Remark 13.2. If we replace condition (iii) in Theorem 13.3 by the stronger condition,

$$(iii') \qquad \sup_{\vartheta \in \Theta} \|D_\vartheta\|_\alpha < \infty,$$

then it follows from (13.17) in the proof of Theorem 13.3 that π_ϑ is v_α-Lipschitz continuous on Θ.

Example 13.9. Revisit the G/G/1 sojourn time example. Take $\Theta = (0, \theta_0)$ where $\mathbb{E}[A_1] > \mathbb{E}[S(\theta)]$ for all $\theta \in \Theta$. Moreover, assume that A_1 and $S(\theta)$, for all $\theta \in \Theta$, have finite moment generating functions. Then, the sojourn time process is classical recurrent with regeneration set $\{0\}$ (i.e., the sojourn time process is Harris recurrent with atom $\{0\}$), and the stationary distribution of the sojourn time exists. We turn to establishing the conditions put forward in Theorem 13.3. Sufficient conditions for condition (i) have already been provided in Example 13.1 and Example 13.3. As shown in Lemma 13.4, $\sup_{\vartheta \in [\theta_1, \theta_2]} \|_0 P_\vartheta\| < 1$. This, together with Lemma 13.3, establishes condition (ii), where we choose $\Theta \subset (\theta_1, \theta_2)$. Finally, as $\|_0 P\|_v < 1$, condition (iii) is a direct consequence of Lemma 13.5. ✱✱✱

For a discussion of the various gradient estimators that ensue from the differentiation result present in Theorem 13.3, we refer to Section 9.3.

13.4 THE INFINITE HORIZON PROBLEM

In this section we will combine the finite-horizon result put forward in Theorem 13.1 and the differentiation formula for the stationary distribution presented in Theorem 13.3 for achieving an infinite horizon result. We will first present the general result and then explain its interpretation in terms of gradient estimation.

Theorem 13.4. *Let P_ϑ be a Markov kernel on $(S, \mathcal{S})$ defined on some open neighborhood Θ of θ such that P_ϑ has unique stationary distribution π_ϑ and that D_ϑ exists for $\vartheta \in \Theta$. Let $\mathcal{D}_\alpha$ be a set of real-valued measurable mappings on S with finite v_α-norm. If*

(i) $\|P_\vartheta\|_\alpha$ *is finite, for* $\vartheta \in \Theta$,
(ii) P_ϑ *is $\mathcal{D}_\alpha$-differentiable at θ and v_α-Lipschitz continuous on Θ,*
(iii) $\sup_{\vartheta \in \Theta} \|\pi_\vartheta\|_\alpha$ *is finite, and*
(iv) $\|D_\theta\|_\alpha$ *is finite,*

then for any $h \in \mathcal{D}_\alpha$ it holds for the derivate at θ that

$$\lim_{N \to \infty} \left(\frac{1}{N} \sum_{n=1}^{N} P_\theta^n h \right)' = \pi'_\theta h.$$

Proof. By the finite-horizon result in Theorem 13.1 we have for any $\vartheta \in \Theta$, we have

$$\left(\frac{1}{N} P_\vartheta^n \right)' = \frac{1}{N} \sum_{k=0}^{N-1} P_\vartheta^{N-k+1} P'_\vartheta P_\vartheta^k, \qquad (13.20)$$

where P'_ϑ denotes that $\mathcal{D}_\alpha$-derivative of P_ϑ. Letting Π_ϑ denote the ergodic projector of P_ϑ (so that, for any distribution μ on $(S, \mathcal{S})$ is holds that $\mu \Pi_\vartheta = \pi_\vartheta$), we have

$$\lim_{N \to \infty} \frac{1}{N} \sum_{n=1}^{N} P_\vartheta^n = \Pi_\vartheta, \tag{13.21}$$

for $\vartheta \in \Theta$. Moreover, for $k \geq 0$,

$$P'_\vartheta P_\vartheta^k = P'_\vartheta (P_\vartheta^k - \Pi_\vartheta),$$

which gives

$$\sum_{k=0}^{\infty} P'_\vartheta P_\vartheta^k = \sum_{k=0}^{\infty} P'_\vartheta (P_\vartheta^k - \Pi_\vartheta) = P'_\vartheta D_\vartheta. \tag{13.22}$$

Taking the limit in (13.20) gives

$$\lim_{N \to \infty} \left(\frac{1}{N} P_\vartheta^n \right)' = \lim_{N \to \infty} \frac{1}{N} \sum_{n=1}^{N} \sum_{k=0}^{n-1} P_\vartheta^{N-k+1} P'_\vartheta P_\vartheta^k$$

and inserting (13.22) and (13.21) we arrive at

$$\lim_{N \to \infty} \left(\frac{1}{N} P_\vartheta^n \right)' = \Pi_\vartheta P'_\vartheta D_\vartheta,$$

and evoking Theorem 13.3, we have

$$\lim_{N \to \infty} \left(\frac{1}{N} P_\vartheta^n \right)' = \Pi'_\vartheta,$$

which proves the claim. $\square$

Letting $\{\xi_n(\theta)\}$ denote a Markov chain with transition probabilities given by P_θ, and assuming that the conditions put forward in Theorem 13.4 are satisfied, then Theorem 13.4 establishes for any $h \in \mathcal{D}_\alpha$

$$\lim_{N \to \infty} \frac{1}{N} \sum_{n=1}^{N} \frac{d}{d\theta} \mathbb{E}[h(\xi_n(\theta))] = \lim_{N \to \infty} \frac{1}{N} \frac{d}{d\theta} \mathbb{E}[L_N(\xi_1(\theta), \ldots, \xi_N(\theta))] = \frac{d}{d\theta} \mathbb{E}[h(\xi(\theta))],$$

where $\xi(\theta)$ is distributed according to π_θ, and

$$L_N(\xi_1(\theta), \ldots, \xi_N(\theta)) = \sum_{n=1}^{N} h(\xi_n(\theta)).$$

Hence, the limit of the finite-horizon gradients is the gradient of the stationary performance. While Theorem 13.4 is based on weak differentiation, and thus at a first glance it seems only to be applicable to distributional parameters, it is worth noting that the result can be applied to IPA (and SF) as well. To see this, consider a parameter θ that can be interpreted as sample-path parameter as well as a distributional parameter (e.g., the mean of an exponential; see Section 7.4 for more detail). Provided that an IPA estimator, denoted by ϕ_N^{IPA}, as well as an MVD estimator, denoted by ϕ_N^{MVD}, are unbiased, it holds

$$\frac{d}{d\theta}\mathbb{E}[L_N(\xi_1(\theta),\ldots,\xi_N(\theta))] = \mathbb{E}\left[\phi_N^{\text{IPA}}(L_N(\xi_1(\theta),\ldots,\xi_N(\theta)))\right]$$
$$= \mathbb{E}\left[\phi_N^{\text{MVD}}(L_N(\xi_1(\theta),\ldots,\xi_N(\theta)))\right]$$

and evoking Theorem 13.4 we arrive at

$$\lim_{N\to\infty}\frac{1}{N}\mathbb{E}\left[\phi_N^{\text{IPA}}(L_N(\xi_1(\theta),\ldots,\xi_N(\theta)))\right] = \frac{d}{d\theta}\mathbb{E}[h(\xi(\theta))].$$

The above line of argument provides an alternative approach to the asymptotic unbiasedness result put forward in Theorem 9.1. We conclude by illustrating the application of Theorem 13.4 to the steady-state waiting times in the G/G/1 queue.

Example 13.10. We revisit the waiting times of the G/G/1. The conditions put forward in Theorem 13.4 can be established as discussed in Example 13.9. This shows that the finite horizon gradient estimators, ϕ_N^{IPA} or ϕ_N^{MVD} put forward in Part II, are asymptomatically unbiased. When applied in a stochastic approximation algorithm $\theta_{n+1} = \theta_n + \epsilon_n Y_n$ we can take, for example, $Y_n = \phi(\hat{\xi}_n(\theta_n), \theta_n) = \phi_N^{\text{IPA}}(\hat{\xi}_n(\theta_n))$ for any $N \geq 1$, where $\hat{\xi}_n(\theta_n)$ is the underlying process containing the information over N successive waiting times and their derivative process; see Example 8.6. While $N = 1$ is feasible by theory, choosing $N > 1$ appropriately may have benefits as variance reduction technique. ✱✱✱

Chapter Fourteen

Stochastic Approximation in Statistics

Generally speaking, stochastic approximation is used in statistics for building algorithms for dynamic model fitting; see, e.g., Example 3.4 or [137]. In this chapter we focus on the application of the score function from Part II in maximum likelihood estimation, and the use of SA in the iterative generalized method of moments.

14.1 THE SCORE FUNCTION IN STATISTICS

In statistics the score function is used in maximum likelihood estimation (MLE). MLE is the most popular statistical technique for estimating unknown parameters from data; see, e.g., [96, 219, 320] for background on MLE. It is worth noting that the use of the SF in statistics is different from its use in optimization. As we explain in the following, in MLE, the logarithm of a density is the objective function to be maximized and pathwise differentiation of the logarithmic density objective yields the score function,[1] whereas in the score function method for gradient estimation, as discussed in Part II of this monograph, the SF is used to estimate the gradient for minimizing some expected cost function. This chapter is on applications of the SF in statistics rather than the score function method.

Given observations $x_1, \ldots x_t$ of an (possibly multi variate) outcome X with pdf $f_\theta(x)$, MLE finds the parameter value $\theta \in \Theta \subset \mathbb{R}^d$ for which the observed data become most probable in the model represented by $f_\theta(x)$. Under the assumption that the data is obtained from iid observations of X, the overall likelihood of observing $x_1, \ldots x_t$ is given by the product $f_\theta(x_1) \cdots f_\theta(x_t)$. For numerical purposes the product is translated into a summation by applying the logarithm, and MLE solves the optimization problem

$$\arg\max_\theta \ell_t(\theta, x_1, \ldots, x_t) \qquad (14.1)$$

for

$$\ell_t(\theta, x_1, \ldots, x_t) = \sum_{i=1}^{t} \ln(f_\theta(x_i)),$$

which translates to the root-finding problem

$$\nabla \ell_t(\theta^*, x_1, \ldots, x_t) = 0. \qquad (14.2)$$

---

[1]From the point of view taken in this monograph, the SF function in the MLE setting is actually the IPA derivative of the log-likelihood.

Remark 14.1. Connecting to the notation developed for the score function method in the previous chapters, we may write for one dimensional case ($d = 1$)

$$\frac{\partial}{\partial \theta} \ell_t(\theta, x_1, \ldots, x_t) = \sum_{i=1}^{t} \text{SF}(\theta, x_i),$$

where $\text{SF}(\theta, x_i) = \partial \ln(f_\theta(x_i))/\partial \theta$, for $1 \leq i \leq t$.

Suppose that θ^* solves equation (14.2). If the Hessian of the objective function $\ell_t(\theta, x_1, \ldots, x_t)$ in (14.1) is negative semi-definite at θ^* for fixed data vector $x_1, \ldots x_t$, then θ^* maximizes the likelihood of observing this particular data vector; see Theorem 1.1 in Chapter 1. Fortunately, most common probability densities and in particular the exponential family are logarithmic concave, which implies that the Hessian is negative semi-definite at θ^* and thus guarantees that the solution of the above root-finding problem indeed solves the MLE problem, see; e.g., [179].

Because the solution θ^* depends on the observed data, i.e., $\theta^* = \theta^*(x_1, \ldots, x_t)$, a natural variation of (14.1) is to consider

$$\arg\max_\theta \mathbb{E}\left[\ell_t(\theta; X_1, \ldots, X_t) \mid \theta\right], \qquad (14.3)$$

where X_i, $1 \leq i \leq t$, are iid samples of X, and the conditioning on θ expresses the fact that X may depend on θ (we follow here the standard notation in MLE).

The Hessian matrix of the expected scores has in this case elements

$$I_{ij} = \mathbb{E}\left[-\frac{\partial^2}{\partial \theta_i \partial \theta_j} \ell_t(\theta, X_1, \ldots, X_t) \,\Big|\, \theta\right]$$

and is called the one-step *Fisher information matrix*. Under appropriate smoothness conditions, taking the limit as $t \to \infty$ in both problems eventually leads eventually to the same solution, so that solving θ^* from a large data set via (14.1) is also the best choice for solving the problem in the expected value version; see, e.g., [282]. Next to classical statistical applications, MLE has also been applied to stochastic process and queuing models; see, e.g., [19, 20, 98, 88, 246, 262, 319].

Solving the problem in (14.1) with an iterative procedure leads to the standard decreasing stepsize ascent algorithm

$$\theta_{n+1} = \theta_n + \epsilon_n \nabla \ell_t(\theta_n; x_1, \ldots, x_t). \qquad (14.4)$$

Since the data vector $x_1, \ldots, x_t$ is fixed, the Newton-Raphson method (see Chapter 1 for details and adjust for maximization) is also used in MLE leading to

$$\theta_{n+1} = \theta_n + (H_{\theta_n})^{-1} \nabla \ell_t(\theta_n; x_1, \ldots, x_t), \qquad (14.5)$$

where H_θ is the negative Hessian of $\ell_t(\theta_n; x_1, \ldots, x_t)$. To avoid the numerically costly evaluation of mixed derivatives in the Hessian, the Hessian can be replaced by the outer product of the gradient given by

$$\sum_{i=1}^{d} (\nabla \ell_t(\theta_n; x_1, \ldots, x_t))^\top \nabla \ell_t(\theta_n; x_1, \ldots, x_t),$$

see [28, 232]. Also, finite difference approximations of the Hessian discussed in Section 11.1.2 may be used here.

STOCHASTIC APPROXIMATION IN STATISTICS

For the stochastic version of the MLE problem in (14.3), we refer to SA as developed in Part I, where we adjust the theory to maximization. More specifically, we can write (14.4) as

$$\theta_{n+1} = \theta + \epsilon_n Y_n(\xi_n),$$

with ξ_n is an iid copy of $(X_1, \ldots, X_t)$ under θ_n, and

$$Y_n(\xi_n) = \nabla \ell_t(\theta_n; (\xi_n)_1, \ldots, (\xi_n)_t).$$

Taking as target vector field

$$G(\theta) = \nabla \mathbb{E}\left[\ell_t(\theta; (\xi_n)_1, \ldots, (\xi_n)_t) \mid \theta\right],$$

we see that Y_n provides an unbiased estimator for $G(\theta)$ under quite general conditions; see the IPA theory developed in Part II for details. Hence, if for each update a new data sample is used, then the MLE problem fits into the framework of Chapter 4. This setting covers applications to data streaming.

Moreover, the convergence results presented in Chapter 6 show that the Fisher information matrix or its sample-path counter part provides valuable information on the asymptotic correlation between the components of θ^*. Recent work by Peng et al. [238] connects MLE to simulation analytics in cases where ∇f_θ and f_θ cannot be obtained in a closed analytical form.

Example 14.1. A data vector $(x_1, \ldots, x_t)$ of iid observations is available for fitting an exponential distribution via MLE. Let $f_\theta(x) = \theta e^{-\theta x}$, $x \geq 0$, denote the density of the exponential distribution with rate θ (and mean $1/\theta$). Then,

$$\ln f_\theta(x) = \ln(\theta) - \theta x$$

and

$$\ell_t(\theta; x_1, \ldots, x_t) = t \ln(\theta) - \theta \sum_{i=1}^t x_i.$$

The MLE problem then reads in the standard setting

$$\min_{\theta > 0} J(\theta), \quad \text{for} \quad J(\theta) = -\ell_t(\theta; x_1, \ldots, x_t),$$

and we let

$$\min_{\delta \leq \theta \leq (1/\delta)} J(\theta), \quad \text{for} \quad J(\theta) = -\ell_t(\theta; x_1, \ldots, x_t),$$

for some small δ. From $d^2 J(\theta)/d\theta^2 = -t/\theta^2 < 0$, it follows that $J(\theta)$ is convex on $[\delta, 1/\delta]$ and the problem is thus well-posed. Moreover, for δ sufficiently small, we argue that the constraints are not active. The unique stationary point of $J(\theta)$ is solved from $J'(\theta) = 0$ as

$$\theta^* = \left(\frac{1}{t}\sum_{i=1}^t x_i\right)^{-1},$$

which is the only KKT point and solves the original MLE problem. The problem can be iteratively solved by the steepest decent algorithm

$$\theta_{n+1} = \theta_n - \epsilon_n \left(\sum_{i=1}^t x_i - \frac{t}{\theta_n}\right),$$

for appropriate choice of ϵ_n. The Newton-Raphson version of the algorithm is obtained from letting $\epsilon_n = \theta_n^2/t$.

Alternatively, consider the MLE in (14.3), which reads in our standard setting as

$$\min_{\delta \leq \theta \leq (1/\delta)} \tilde{J}(\theta), \quad \text{for} \quad \tilde{J}(\theta) = -\mathbb{E}\left[\ell_t(\theta; X_1(\theta), \ldots, X_t) \mid \theta\right].$$

We argue for (14.1) that the problem is well-posed, and that the negative gradient is coercive for minimizing $\tilde{J}(\theta)$. Let $\xi_n = (X_1(n), \ldots, X_t(n))$ be a sample of the data vector, then

$$\theta_{n+1} = \theta_n - \epsilon_n Y_n(\xi_n),$$

where

$$Y_n(\xi_n) = \left(\sum_{i=1}^{t} X_i(n) - \frac{t}{\theta_n}\right).$$

By construction Y_n is an unbiased estimator for $\tilde{J}'(\theta_n)$. Since the variance of ξ_n is independent of θ, the variance of Y_n is uniformly bounded. Let $\epsilon_n = \epsilon - 1/(n+1)$, for $n \geq 0$. By Theorem 4.1 it follows that θ_n converges a.s. to the solution of (14.3). ✳✳✳

14.2 GENERALIZED METHOD OF MOMENTS

In statistics, SA is applied for simultaneously fitting the first k moments of a distribution to observed data. This approach is called *method of moments* (MoM) and *generalized method of moments* (GMoM). In the following, we explain the basic setup of GMoM. For more details we refer to [138], and for applications we refer to [139, 170]. Consider observations $(X_1, \ldots, X_N)$, where each observation X_t is a (possibly multivariate) outcome of a certain statistical model X, and let $h(\cdot, \theta) \in \mathbb{R}^k$ denote a mapping that "reads information from X," for $\theta \in \Theta \subset \mathbb{R}^d$. The goal of GMoM is to find θ^* such that

$$m(\theta^*) := \mathbb{E}[h(X_t, \theta^*)] = \mathbf{0},$$

where $\mathbf{0}$ is a zero-vector of length k. The sample path counterpart of the above root-finding problem is solving θ^* out of

$$\hat{m}(\theta^*) = \frac{1}{N} \sum_{i=1}^{N} h(X_i, \theta^*) = \mathbf{0}.$$

Note that, for $\Theta \subset \mathbb{R}^d$ and $d > k$, θ^* is not directly identifiable via the above root-finding problem, as there are fewer equations than variables to fit. Casting the problem into an optimization problem overcomes this drawback. For this, introduce the weighted norm

$$||\hat{m}(\theta)||_W = (\hat{m}(\theta))^\top W \hat{m}(\theta),$$

for some positive definite $k \times k$ square matrix W. The rationale behind introducing W is that it reflects that components of $m(\theta)$ are typically of different scale. The GMoM procedure requires to solve

$$\theta^* = \arg\min_{\theta} \left(\frac{1}{N} \sum_{i=1}^{N} h(X_i, \theta)\right)^\top W \left(\frac{1}{N} \sum_{i=1}^{N} h(X_i, \theta)\right),$$

which is the sample-path counterpart of

$$\theta^* = \arg\min_\theta \mathbb{E}[\|h(X,\theta)\|_W] = \arg\min_\theta \mathbb{E}[h(X,\theta)^\top W h(X,\theta)].$$

It can be shown that the optimal choice for W is

$$W^{-1} = \mathbb{E}[h(X,\theta^*)h(X,\theta^*)^\top], \tag{14.6}$$

see [140]. As the optimal choice of W involves the unknown value θ^*, many approximate versions of W avoiding explicit knowledge of θ^* exist, and we refer to the standard literature for details; see, e.g., [230]. Choosing for W the identity matrix reduces GMoM to the classical MoM.

With the notation used in our monograph, we let $J(\theta) = \mathbb{E}[\|h(X,\theta)\|_W]$, and GMoM solves the problem

$$\min_\theta J(\theta),$$

which leads to the following steepest descent SA algorithm, called *iterated GMoM*,

$$\theta_{n+1} = \theta_n - \epsilon_n Y_n,$$

where

$$Y_n(\xi_n) = 2\left(\nabla \frac{1}{N}\sum_{i=1}^N h((\xi_n)_i,\theta_n)\right)^\top W \left(\frac{1}{N}\sum_{i=1}^N h((\xi_n)_i,\theta_n)\right),$$

with ξ_n an iid sample of X of length N under θ_n. While, due to the dependencies, Y_n is a biased estimator for $\nabla J(\theta) = G(\theta) = \mathbb{E}[(\nabla h(X,\theta))W h(X,\theta)]$, the bias tends to zero with increasing N. A simple heuristic for dealing with the fact that W is not known in the above SA is as follows. Initially W is set to the identity matrix and SA is run for a number of k iterations. Assuming that θ_k is already "close" to the true and unknown θ^*, one estimates W be simulation with θ^* replaced by θ_k. Then, SA is continued where W is determined by the auxiliary simulation experiment is used.

Example 14.2. Consider the student-t distribution with θ degrees of freedom. This distribution has mean 0, variance $\theta/(\theta-2)$, for $\theta > 2$, and fourth moment $3\theta^2/(\theta-2)(\theta-4)$, for $\theta > 4$. Suppose we have data $X = (X_1, \ldots, X_N)$ available, and we want to apply GMoM for fitting a student-t distribution with θ degrees of freedom to the data via the second and fourth moments of the distribution. Let

$$h(X,\theta) = h(X_1, \ldots, X_N, \theta) = \begin{pmatrix} \frac{1}{N}\sum_{i=1}^N X_i^2 - \frac{\theta}{\theta-2} \\ \frac{1}{N}\sum_{i=1}^N X_i^4 - \frac{3\theta^3}{(\theta-2)(\theta-4)} \end{pmatrix}.$$

Then, for ξ_n an iid sample of data vector X, we have

$$Y_n(\xi_n) = 2\left(\frac{2}{(\theta_n-2)^2}, -\frac{3\theta_n^2(\theta_n^2 - 12\theta_n + 24)}{(\theta_n-2)^2(\theta_n-4)^2}\right)^\top$$

$$\times W \begin{pmatrix} \frac{1}{N}\sum_{i=1}^N ((\xi_n)_i)^2 - \frac{\theta_n}{\theta_n-2} \\ \frac{1}{N}\sum_{i=1}^N ((\xi_n)_i)^4 - \frac{3\theta_n^3}{(\theta_n-2)(\theta_n-4)} \end{pmatrix},$$

and the ensuing SA becomes $\theta_{n+1} = \theta_n - \epsilon_n Y_n$, where we assume that W has been determined via the heuristic described above. ✻✻✻

Chapter Fifteen

Stochastic Gradient Techniques in AI and Machine Learning

In this chapter, we discuss applications of the SA theory in AI and machine learning to gradient-based methods and reinforcement learning.

15.1 GRADIENT-BASED APPROACHES

Gradient descent is one of the main techniques in AI and machine learning for supervised and unsupervised learning. Using our terminology, generally speaking, machine learning is mainly focused on the finite horizon problem; see Section 7.3. More formally, one solves

$$\min_{\theta} J(\theta), \qquad (15.1)$$

with

$$J(\theta) = \begin{cases} \mathbb{E}[l(\xi(\omega), \theta)], & \text{the probabilistic setting} \\ \sum_{k=1}^{N} l_k(\theta), & \text{the deterministic setting,} \end{cases}$$

where for the probabilistic setting l is some cost function and $\xi(\omega)$ a random input, and for the deterministic setting l_k, $1 \leq k \leq N$ is a collection of functions.

In data fitting applications, one typically has a (large) collection of pairs $(z_k = (x_k, y_k) : 1 \leq k \leq N)$, for N large, where x_k is a data input vector and y_k is the corresponding true outcome for x_k. The error incurred by predicting y_k through a θ-parameterized model is denoted by $l_k(\theta) = l(\theta, z_k)$, and the problem of optimally fitting a postulated model to the data points is given in (15.1). The fact that N is typically large (and $N \gg d$) makes evaluating the gradient of $J(\theta)$ numerically challenging, and puts the focus on dealing with the fact that the objective function is a sum of a large number of individual, smooth functions. We note that for N large and iid data,

$$\frac{1}{N} \sum_{k=1}^{N} l_k(\theta) = \frac{1}{N} \sum_{k=1}^{N} l(\theta, z_k) \approx \mathbb{E}[l(\xi(\omega), \theta)], \qquad (15.2)$$

which relates the deterministic and probabilistic settings. This is similar to the SAA approach in Section 3.5. Facilitating (15.2), the gradient is often replaced by an estimate calculated from a randomly selected batch; see, e.g., [134].

Standard settings are the least square fit for a linear regression for data $x_k \in \mathbb{R}^m$ and $y_k \in \mathbb{R}$, where the loss function is of the form $l(\theta, z_k) = (\theta_0 + \theta^\top x_k - y_k)^2$ so that $(\theta_0, \theta) \in \mathbb{R}^{(m+1)}$. In logistic regression, the loss is given by $l(\theta, z_k) = (z_k - 1/(1 + e^{\theta_0 + \theta x_k}))^2$. Finally, for $l(\theta, z_k) = -\nabla_\theta \log f_\theta(z_k)$ we obtain the recursive maximum likelihood for a density f_θ over $\mathbb{R}^n$; see also Chapter 14. The goal is to fit regression line h parametrized by θ to data points (x, z), where x is the independent and z is the dependent variable. With this notation, the deterministic setting views the available data set as random but given and the optimization is carried on this fixed data set. One either assumes that the data set is exhaustive, or argues along (15.2) by making sure that the data set is large enough so that the average approaches the mean. The former leads to the fitting model discussed in Example 3.3 and the latter to the setting in Example 3.4. In the probabilistic setting, like in Example 3.4, the learning problem is cast into a minimization of an expected value, and gradient estimation in combination with SA, as developed in Parts I and II, can be applied. For a detailed discussion of these variations for an inventory problem we refer to Chapter 10.

The setting in machine learning refers to fitting man-made exploratory functions to data, and one has thus control on the complexity of the fitting function. This allows for developing variants of the basic algorithm tailored to different instances of fitting functions; see, for example, [223], and problem sizes (in terms of the size of the parameter vector). For recent overviews on variants of the basic SA algorithm, we refer to [41, 42, 136, 221].

Developing efficient SA algorithms for specific settings that are meaningful in applications is a field of active research. A prominent example is the Adam algorithm (see [171, 183]) developed for deep neural network training. Next to gradient methods and their variations, pseudo-gradient methods are frequently used in machine learning contexts, and we refer to our discussion of pseudo-gradient methods in Chapter 11. In addition, we mention that there is considerable work on gradient-free methods; see, e.g., [229] and [62, 172].

In the following we provide a brief overview on the main directions in this field of research related to SA.

15.1.1 Incremental Gradient Methods

When the cost function is a sum of many functions we can use randomization. Consider the deterministic setting, and let σ be uniformly distributed on $\{1, \ldots, N\}$, then

$$J(\theta) = \sum_{k=1}^{N} \nabla l_k(\theta) = N \mathbb{E}\left[\nabla l_\sigma(\theta)\right].$$

A naive approach would be to use $N\nabla l(\theta, z_\sigma)$ as unbiased feedback. Moreover, dropping N, the expected value of the resulting feedback is a descent direction and the theory of Part I applies to $\nabla l(\theta, z_\sigma)$ as well. While this is the only setting that can be mathematically analyzed, in practice, randomly choosing the component out of large data set renders the optimization slow and one uses $\nabla l(\theta, z_i)$ where z_i runs in a deterministic way through the data set. Moreover, "batching" is applied as the feedback becomes numerically more stable if

$$\sum_{i \in I_n} \nabla l(\theta, z_i)$$

is used, where I_n is the n-th batch of data.

While this feedback may serve (in the randomized case) as a descent direction for $\nabla J(\theta)$ it leads to low convergence rates as all mappings other than l_σ do not contribute to the update. An improvement is to use for the update a locally updated version of the old gradient. More specifically, note that

$$\nabla J(\theta) = \nabla J(\theta') + \nabla J(\theta) - \nabla J(\theta') = \nabla J(\theta') + \mathbb{E}\left[(N\nabla l_\sigma(\theta') - N\nabla l_\sigma(\theta))\right].$$

The local update is combined with the incremental improvement in the following way (where we again drop N). First, the number m of incremental steps is chosen. Then, θ_0 is chosen and the full gradient $\nabla J(\theta_0)$ is computed. For $n = mk$, $k \geq 0$, to $n = mk + m - 1$ we perform m incremental updates:

$$\theta_{n+i} = \theta_n - \epsilon \left(\nabla J(\theta_n) + \nabla l_{\sigma_i}(\theta_{n+i}) - \nabla l_{\sigma_i}(\theta_n) \right),$$

$i = 1, \ldots, m-1$. Then, at $n = m(k+1)$, $\nabla J(\theta_n)$ is evaluated and the full gradient is updated, after which we perform a again $m-1$ incremental updates. Algorithms of this kind are subsumed under the name (semi-) stochastic gradient methods; see [210, 211, 212, 226] for early references, and for recent active research see [173, 280, 281, 332].

Other than via random updating, the algorithm can also be applied in a deterministic version. For this, the gradients of component functions l_i are used for the feedback in cyclic order. This variation of the algorithm is know as a mini-batch algorithm in training of neural networks. Indeed, here one typically uses only a subset of the m data samples for updating the gradient.

A variation of the incremental gradient algorithms are the so-called *momentum*-based methods. These types of algorithms use the past update as the driving force in the updates. An example of a momentum-type algorithm is the following:

$$\theta_{n+1} = \theta_n - \epsilon \nabla J(\theta_n) + \beta(\theta_n - \theta_{n-1}),$$

and in its randomized form (with N dropped)

$$\theta_{n+1} = \theta_n - \epsilon \nabla l_\sigma(\theta_n) + \beta(\theta_n - \theta_{n-1}),$$

for some constant $\beta \in [0, 1)$, σ uniformly on $\{1, \ldots, N\}$, and $\theta_{-1} = \theta_0$; see [30, 208, 302]. Like above, in applications, random selection is replaced by letting the algorithm pass through (deterministic) batches of data.

The above is an instance of Nesterov's accelerated gradient method. See [227] for an early reference, and we refer to [271] for an overview on variations of the original algorithm; for an account of Nesterov's accelerated gradient method in the stochastic setting we refer to [17].

15.1.2 Large Dimensional Case: Coordinate Descent Methods

Evaluating the gradient becomes numerically challenging when the size of the parameter vector is large. Again, like in Section 15.1.1, randomization can be be applied. Indeed, let σ be uniformly distributed on $\{1, \ldots, d\}$ and denote by e_i the i-th unit vector in $\mathbb{R}^d$, then

$$\nabla J^\top(\theta) = \sum_{i=1}^d e_i \frac{\partial}{\partial \theta_i} J(\theta) = d \, \mathbb{E} \left[e_\sigma \frac{\partial}{\partial \theta_\sigma} J(\theta) \right].$$

The gradient-descent algorithm becomes

$$\theta_{n+1} = \theta_n - \epsilon_n e_{\sigma_n} (Y_n)_{\sigma_n}, \quad \text{with} \quad (Y_n)_i = e_i \frac{\partial}{\partial \theta_i} J(\theta_n),$$

where $\{\sigma_n\}$ is an iid sequence of uniformly $\{1, \ldots, d\}$ distributed random variables. It is standard to drop the factor d, which we may as the resulting update is a descent direction; indeed it holds that $\mathbb{E}[Y_n | \mathcal{F}_{n-1}] = (1/d) \nabla J(\theta_n)$. Unfortunately, the above modifications renders the algorithm slow as only one component is updated per iteration. Of course, we can modify the algorithm by randomly selecting subsets of components to update but this is counter-effective to keeping the numerical burden low. In deterministic optimization such methods are known as block-coordinate descent methods and are well studied (see [21, 207]), and extensions to the stochastic case are provided in [186, 318, 338].

An alternative approach is to use for the update a locally updated version of the old gradient like in the incremental gradient methods. More specifically, note that

$$\nabla J(\theta) = \nabla J(\theta') + \nabla J(\theta) - \nabla J(\theta') = \nabla J(\theta') + d\,\mathbb{E}\left[e_\sigma\left(\frac{\partial}{\partial \theta_\sigma}J(\theta) - \frac{\partial}{\partial \theta_\sigma}J(\theta')\right)\right].$$

Hence, in the SA algorithm we update every component at each iteration but only one component is updated according to the "correct" partial derivatives, the other components are updated according to a "faulty" partial derivative. Under appropriate smoothness conditions on the gradient it is conceivable that the old/faulty partial derivatives are close to the true one. The ensuing SA becomes

$$\theta_{n+1} = \theta_n - \epsilon_n \hat{Y}_n,$$

where

$$(\hat{Y}_n)_i = \begin{cases} (\hat{Y}_{n-1})_i & \sigma_n \neq i \\ \frac{\partial}{\partial \theta_i}J(\theta_n) & \sigma_n = i. \end{cases}$$

In words, the feedback requires sampling the partial derivative of J with respect to component σ_n at parameter values θ_n and θ_{n-1}. We conclude by mentioning that the extension to block-coordinate descent methods, where a subset of elements of the gradient is used per update, is quite natural for the above method.

15.2 Q-LEARNING AND REINFORCEMENT LEARNING

Consider a control problem of the following form. A dynamical system evolves in discrete time passing through a number of possible *states* in a finite space S. At each (time) *stage n*, if the system is at state $i \in S$ then there is a finite number of *actions* that can be chosen from finite set $\mathcal{A}(i)$. Call $\xi_n \in S$ the state at stage n, and $u_n \in \mathcal{A}(i)$ the action taken and consider the stochastic process $\{(\xi_n, u_{n-1})\}$ with natural filtration $\mathfrak{F}_n$. It is assumed that the process satisfies the Markovian property:

$$\mathbb{P}(\xi_{n+1} = j \mid \mathfrak{F}_n) = \mathbb{P}(\xi_{n+1} = j \mid \xi_n, u_n),$$

and we denote the transition probabilities by

$$p_{ij}(a) = \mathbb{P}(\xi_{n+1} = j \mid \xi_n = i, u_n = a) = p_{ij}(a), \quad \text{for } a \in \mathcal{A}(i). \tag{15.3}$$

At every stage n, when the system is at state $\xi_n = i$, if action $u_n \in \mathcal{A}(i)$ is taken, then there is an *instantaneous reward* $r(i,a)$. The total reward is a function of the individual rewards at every step. In some formulations the total number of steps N is constant, in others it is a random horizon. The problem can also be formulated for infinite horizon models either using a stationary model, or using a discounted reward. For our presentation here we consider the case where N is an a.s. finite stopping time of the form $N = \min(n : \xi_n \in B)$ for some subset $B \subset S$, with no discounting. One of the most popular applications of MDP and Q-learning is training robots to find the shortest path in an environment which they must discover by traveling through it. Shortest path problems define a stage every time a step is taken to move, and so the trajectory ends when the destination is reached, which happens after a random number of steps.

Let the total reward be defined by

$$R(\xi_0) = \mathbb{E}\left[\sum_{n=0}^{N-1} r(\xi_n, u_n) + \bar{r}_N(\xi_N)\right],$$

where $\bar{r}_N(\cdot)$ is a terminal reward. The problem is to chose the *optimal strategy* of consecutive actions $\{u_n\}$ to maximize the reward J. We note here that this is a *control problem*, rather than an optimization problem. The optimization here is done not over a variable $\theta \in \mathbb{R}^d$, but over the possible actions. To reflect the fact that when making decisions the "agent" (controller or decision maker) does not know the future, it is imposed that the optimization be made over the *admissible* control policies. In mathematical terms, this means that u_n must be an $\mathfrak{F}_n$-measurable function (possibly random).

Solving such problems can be very challenging. The monograph [252] presents a comprehensive study of various methodologies for solving MDPs. In particular, when the system has the Markovian dynamics (15.3) it can be shown that optimal policies can be restricted to Markovian (deterministic) policies of the form $u_n = \phi(\xi_n)$. This is a "table"-type solution, where the optimal action depends only on the current state of the system, and not on the past trajectories or the stage n. Optimal gaming strategies for blackjack or backgammon can be stated in this way, but the "state" needs to consider the cards already played, so it can be very large. Even in this case where the optimal policy is restricted to be a function $\phi(\xi)$ the optimization must take place over a set of functions. In what follows we explain how the now popular Q-learning method transforms the control problem into an optimization problem for which SA can be used.

15.2.1 Formulation for Shortest Path Problems

Following Bellman's original approach [23], we start "at the end" (much like solving a labyrinth problem becomes much easier when starting at the exit). Define:

$$J_N(\xi_N) = r_N(\xi_N)$$
$$J_n(\xi_n) = \max_{u_n \in \mathcal{A}(\xi_n)} \left(r(\xi_n, u_n) + \mathbb{E}[J_{n+1}(\xi_{n+1})]\right), \quad n < N$$

where $N = \min(n: \xi_n \in B)$ for some subset $B \subset S$. Then the optimal reward is $R^*(\xi_0) = J_0(\xi_0)$. This equation expresses the idea that at the "present" stage n the optimal revenue is the instantaneous revenue plus the so-called "revenue-to-go" (expected future revenue).

Bellman's equation follows from the optimality principle and states that

$$R^*(\xi) = \max_{u \in \mathcal{A}(\xi)} \left(r(\xi, u) + \mathbb{E}[R^*(\xi_{n+1}) \mid (\xi_n, u_n) = (\xi, u)]\right) \quad (15.4)$$

This optimality principle is at the basis of the well-known dynamic programming algorithms.

Define the Q-factors as

$$Q^*(i, a) = r(i, a) + \sum_{j \in S} p_{ij}(a) R^*(j), \quad (15.5)$$

which denotes the maximal revenue at state i when action a is taken. Bellman's equation (15.4) implies that

$$R^*(i) = \max_{u \in \mathcal{A}(i)} Q^*(i, u)$$

and using (15.5), this in turn yields

$$Q^*(i,a) = r(i,a) + \sum_{j \in S} p_{ij}(a) \left(\max_{u \in \mathcal{A}(i)} Q^*(j,u) \right). \tag{15.6}$$

Define the operator T on matrices:

$$(TQ)(i,a) = r(i,a) + \sum_{j \in S} p_{ij}(a) \left(\max_{u \in \mathcal{A}(i)} Q(j,u) \right)$$

then it can be shown [203] that this is a contraction operator with a unique fixed point Q^*, so it follows from Example 4.5 that $TQ - Q$ is a coercive field for this unique fixed point. Specifically, the ODE

$$\frac{dx(t)}{dt} = Tx(t) - x(t) \tag{15.7}$$

has a unique equilibrium point at Q^* and it is asymptotically stable. This leads to the recursive algorithm

$$Q_{n+1}(i,a) = Q_n(i,a) + \epsilon_n \left((TQ_n)(i,a) - Q_n(i,a) \right), \tag{15.8}$$

that converges to the desired fixed point. We remark that, once the matrix Q^* is known, then the optimal policy is a deterministic Markov policy of the form $u_n^* = \phi(\xi_n)$, with

$$\phi(i) = \arg \max_{u \in \mathcal{A}(i)} Q^*(i,u).$$

In machine learning literature, the feedback $TQ_k - Q_k$ in (15.8) is known as the *temporal difference* because it is the difference in value of the Q factors from one iteration to the next and it reflects the "new" knowledge learned by the algorithm at every step. We remark that in our notation, the Q-factors are the control variable θ.

Evaluating (15.8) presents practical challenges because in many situations the transition probabilities $\{p_{ij}(a)\}$ are not known, or because the dimension of the state/action space is very large and each update requires a prohibitive amount of CPU time. To address these problems, the standard form of the algorithm randomizes the feedback. Specifically, for each (i, a), let

$$Y_n(i,a) = r(i,a) + \max_{u \in \mathcal{A}(i)} Q_n(j,u) - Q_n(i,a); \quad j \sim p_{i,\cdot}(a).$$

In this case, by construction $\mathbb{E}[Y_n(i,a)] = TQ_n(i,a) - Q_n(i,a)$, for each state/action pair (i, a). Now the problem is how to generate the state j without knowing the transition probabilities. There are two approaches depending on the problem at hand. If the transition probabilities are not known analytically but given (i, a) the next state can be generated via simulation, then we can estimate $Y_n(i,a)$. This approach is best understood in the case where all state/action pairs (i, a) are used to create the feedback matrix for the n-th update, that is the matrix Y_n, because in this case $\mathbb{E}[Y_n \mid Q_n] = TQ_n - Q_n$. Because the state/action pairs are finite and the rewards $r(i, a)$ are finite, it follows that the feedback has bounded variance on compact sets (in Q) and all assumptions of Theorem 4.2 (for decreasing learning rate) or Theorem 4.4 (for constant learning rate) can be verified directly.

15.2.2 Randomization

It is not necessary to evaluate the feedback Y_n for all the state/action pairs at each update of the algorithm. Rather, it is common to assume that for the n-th feedback one chooses a random state $\xi_{n-1} = i$ and depending on the implementation of the algorithm, either a random action $u_{n-1} \in \mathcal{A}(i)$ is used, or all possible actions are used in parallel to generate (or observe) ξ_{n-1} and create the feedback vector $Y_n(i, \cdot)$. Thus, one analyzes the algorithm by components, each converging to the corresponding component of the ODE (15.7). To illustrate the method of analysis, assume that the algorithm chooses state/action pair (i, a) with probability $\pi(i, a)$ (independent of n). Imagine that there are *control agents* assigned to each state/action pair (i, a), each in charge of updating the corresponding Q-factor. Seen this way, this is an example of what is known as *distributed* computation. When the algorithm chooses an agent, the corresponding feedback is calculated and the corresponding component of the Q-matrix is updated. Notice that here ξ_n is the state *visited by the algorithm*. Using this approach we obtain:

$$Y_{n+1}(i,a) = \begin{cases} r(\xi_n, u_n) + \max_{u \in \mathcal{A}(\xi_n)} Q_n(J, u) - Q_n(\xi_n, u_n) & \text{if } i = \xi_n, a = u_n \\ 0 & \text{otherwise,} \end{cases}$$

where $J \sim p_{\xi_n,\cdot}(u_n)$ is generated according to the transition probabilities in (15.3). So the control agents take turns to update the Q-matrix component by component. Using this scheme, it follows that

$$\mathbb{E}[Y_{n+1}(i,a) \mid Q_n] = \pi(i,a)\,(TQ_n(i,a) - Q_n(i,a)),$$

so the corresponding limit ODE is

$$\frac{dx_t(i,a)}{dt} = \pi(i,a)\,(Tx_t(i,a) - x_t(i,a)). \tag{15.9}$$

The difference between this ODE and (15.7) can be interpreted in terms of a time scale change: components with higher values of $\pi(i,a)$ will update more frequently. This methodology was introduced in [305] for decentralized and asynchronous control, and developed further in [310]. The crucial observation is that the drift vector field in (15.9) is also a contraction mapping with same fixed point Q^* as that of (15.7) if all the scaling factors are strictly positive, that is, $\pi(i,a) > 0$, which means that a proper time scale change relates the two dynamics. Therefore, for the algorithm to converge globally to the optimal Q-factors it is commonly assumed that all state/action pairs are visited by the algorithm infinitely often; see [33].

15.2.3 Reinforcement Learning

The second approach is more popular in computer science and robotics and it is based on real-time control (online) streaming data from the actual system. In this approach one assumes that the controller actually takes action u_n at stage n and observes the next state ξ_{n+1}, which is used to calculate Y_{n+1}:

$$Y_{n+1}(i,a) = \begin{cases} r(\xi_n, u_n) + \max_{u \in \mathcal{A}(\xi_n)} Q_n(\xi_{n+1}, u) - Q_n(\xi_n, u_n) & \text{if } i = \xi_n, a = u_n \\ 0 & \text{otherwise,} \end{cases}$$

where, by assumption, $\mathbb{P}(\xi_{n+1} = j \mid \xi_n, u_n) = p_{\xi_n, j}(u_n)$. This means that only the component of the Q-factor that is "visited" by the system is updated. In other words, this is a concurrent scheme for estimation and control (the states "visited" by the algorithm are the same as the states visited by the real process). This is known in machine learning as *reinforcement learning*, which is a special case of unsupervised learning. In the example of robots learning an environment, it means that robots can "crash" against walls or obstacles during the process of learning. Once the terminal time N is reached the process starts again. This usually means resetting the robots in their labyrinth to keep learning about the best steps to make, or continuing to stream data from various processes (for example clients). Because learning is a continual activity, it is common to consider constant learning rates ϵ. The appropriate results for decreasing ϵ_n can be obtained using similar arguments.

It is common to assume that the original process is what is known as a *unichain* (see Section C.3 in Appendix C for a definition), where each possible state is reachable from any other state (this is a property related to ergodicity of a Markov chain for the Markov decision process in question). Consecutive observations of the process $\{(\xi_n, u_n)\}$ are used to build the feedback. However, in order to implement the scheme one must first specify how the consecutive actions u_n are taken during the algorithm. As discussed before, use of parallel simulations using all possible actions for u_n in parallel is sometimes possible [203] and it may increase efficiency, updating the vector $Q_n(i, \cdot)$ at every iteration of the SA. But this is not always practical when implementing the controls in the real world.

Suppose that given Q, the action u_n is only a (possibly random) function of the state ξ_n (known as a random Markov policy), so that $\mathbb{P}(u_n = a \mid \xi_n = i) = P_Q(i, a)$ is independent of n. If this is the case, then choosing the "greedy" policy

$$u_n = \max_{u \in \mathcal{A}(\xi_n)} Q(\xi_n, u). \tag{15.10}$$

An argument similar to the time scale argument before establishes that

$$\mathbb{E}[Y_{n+1}(i, a) \mid Q_n, \xi_n] = \mathbf{1}_{\{\xi_n = i\}} P_{Q_n}(i, a) \left(T Q_n(i, a) - Q_n(i, a) \right).$$

We now build the SA following the model of Chapter 5. Call $\theta = Q$ the Q-factors and consider the process when θ is fixed. Because the control policy is Markovian, it follows that the state process $\{\xi_n(\theta)\}$ is a regular Markov chain with well-defined transition probabilities:

$$\mathbb{P}(\xi_{n+1}(\theta) = j \mid \xi_n(\theta) = i) = \sum_{a \in \mathcal{A}(i)} P_\theta(i, a) \, p_{ij}(a).$$

Under the uni-chain assumption and the construction of the process that re-starts whenever reaching N, there is a stationary measure $\pi_\theta(i)$. The fact that this measure is well defined and tight for compact sets Θ follows from the construction of the randomized policies and the fact that the state and action pairs are finite. All assumptions of Theorem 5.4 are then satisfied, identifying the target vector field as the stationary expectation of the feedback, that is:

$$G(\theta)(i, a) = \pi_\theta(i) \, P_\theta(i, a) \, (T\theta(i, a) - \theta(i, a)).$$

The unichain assumption ensures that, in the long run, $\pi(i) > 0$ for all $i \in S$. There is, however, no assurance that all the policies will be visited with positive probability unless we explicitly impose this condition. That is, the vector field $G(\theta)$ may have equilibrium points for values θ such that for some $i \in S$, $P_\theta(i, a) = 0$ for a non-null subset of actions. In words, if an action is not explored early and its Q-value is zero, then using, for example, (15.10) means that this action will never be chosen in the sequel as soon as some other action has a positive Q value for the same state. What we need is for the vector field to have unique equilibrium points θ^* that are the same as the fixed points Q^* of the contraction mapping T.

This paradigm of reinforcement learning is known in the literature as the *exploration versus exploitation* scheme. An example of it [13, 299] is known as the "ϵ-greedy" scheme.[1] At iteration n, when at state ξ_n and with the current estimate of the Q-factors being Q_n, one uses

$$u_n = \begin{cases} \arg\max_{u \in \mathcal{A}(\xi_n)} Q_n(\xi_n, u), & \text{with probability } 1 - \rho, \\ \text{Uniform on } \mathcal{A}(\xi_n), & \text{with probability } \rho. \end{cases}$$

Under this policy assignment, given $Q_n = Q$ we have that

$$P_Q(i, a) = \frac{\rho}{\|\mathcal{A}(i)\|} + (1 - \rho)\mathbf{1}_{\{a = \arg\max_{u \in \mathcal{A}(i)} Q(i,u)\}} > 0.$$

Therefore, for this scheme the only equilibrium point of the vector field $G(\theta)$ is indeed the fixed point of the mapping T, which is what we were seeking. This in turn implies that the limit ODE converges to the optimal Q-factors.

As is the case for all reinforcement learning schemes that use exploration versus exploitation steps (such as multi-armed bandit or MAB problems), one defines a "regret" function as the difference between the expected actual reward and $R^*(\xi_0)$. Exploration steps usually deviate from optimal policies and these happen with probability ρ. Yet, as explained above, not exploring may lead to suboptimality (the robot has no chance to learn a good action if it never takes it). During the exploitation steps the actions may also be suboptimal because the agents are learning the optimal Q-factors, all of which contribute to a difference in actual expected revenue and the optimal one. It has been shown (see [13]) that the fastest convergence of the regret (to zero) is a cooling scheme where $\rho_n = O(1/n)$.

---

[1] The name follows because the Greek letter ϵ is commonly used to denote the exploration probability, but the present monograph consistently uses ϵ to denote the learning rate, so we will use here "ρ" to denote the exploration probability.

Part IV

Appendixes

Appendix A

Analysis and Linear Algebra

For a mapping $f: D_f \subset \mathbb{R}^d \mapsto \mathbb{R}^n$, such that f is defined on D_f and otherwise not, we call D_f the domain of f, and the set $R_f = \{f(x) : x \in D_f\}$ the range of f. Moreover, the set $G_f = \{(x, y) : x \in D_f, f(x) \leq y\}$ is called the upper graph of f.

A.1 CONVEXITY

A set $G \subset \mathbb{R}^n$ is called convex if for $x, y \in G$ it holds that $\alpha x + (1 - \alpha)y \in G$ for $\alpha \in [0, 1]$.

Definition A.1. A mapping f from $\mathbb{R}^d$ to $\mathbb{R}^n$ with domain D_f is called *convex* if G_f is a convex set. In addition a mapping is called *concave* if $-f$ is convex. Alternatively, f is called convex if for any $x, y \in D_f$ it holds that

$$f(\alpha x + (1 - \alpha)y) \leq \alpha f(x) + (1 - \alpha) f(y),$$

for all $\alpha \in [0, 1]$. In case the above inequality holds for "$<$," then f is called strictly convex. Moreover, f is called *concave* if the above holds with "$\geq$" in place of "$\leq$," and if the inequality holds for "$>$," then f is called strictly concave.

A.2 MULTIDIMENSIONAL DERIVATIVES

Definition A.2 (First Order (Total) Derivative). Let $f: D_f \subset \mathbb{R}^d \mapsto \mathbb{R}$ be defined on an open neighborhood of a point c. We call f differentiable at c if there exists $f'(c) \in \mathbb{R}^d$ such that

$$f(c + h) = f(c) + f'(c) \cdot h + r(h),$$

with $r(h) = o(h)$. In this case, $f'(x)$ is called the (total) derivative of f at c, which is also denoted by $Df(c)$.

If f is differentiable at a point, then f is also continuous in that point. Indeed, differentiability implies $f(c + h) - f(c) = f'(c) \cdot h + r(h)$, which yields continuity.

Definition A.3 (Directional Derivative). Let $f : D_f \subset \mathbb{R}^d \mapsto \mathbb{R}$ be defined on an open neighborhood of a point c. Let $a \in \mathbb{R}^d$. We call f differentiable at c in direction a if

$$\lim_{t \to 0} \frac{f(c + ta) - f(c)}{t}$$

exists. In this case the derivative of f at c in direction a is denoted by $D_a f(c)$.

Definition A.4 (Partial Derivatives). Let $f : D_f \subset \mathbb{R}^d \mapsto \mathbb{R}$ be defined on an open neighborhood of $c \in D_f$. For $1 \leq k \leq d$, the fixture of f to all but the k-th component of c is defined by
$$g_k(x_k) = f(c_1, \ldots, c_{k-1}, x_k, c_{k+1}, \ldots, c_d).$$

If g_k is differentiable with respect to x_k, then f is said to be partially differentiable at c with respect to x_k. The derivative $g'_k(c_k)$ is called the partial derivative of f with respect to x_k at c and is denoted by $\frac{\partial f(c)}{\partial x_k}$. We call f partially differentiable at c if $\frac{\partial f(c)}{\partial x_k}$, $1 \leq k \leq d$, exist, and let
$$\mathrm{grad}_f(c) = \left(\frac{\partial f(c)}{\partial x_1}, \ldots, \frac{\partial f(c)}{\partial x_d} \right).$$

The row vector $\mathrm{grad}_f(c)$ is called the gradient.[1] We call f partially differentiable if f is partially differentiable in every point of D_f.

Theorem A.1. *Let $f : D_f \subset \mathbb{R}^d \mapsto \mathbb{R}$ and let c be an interior point of D_f. If f is partially differentiable in c and if the partial derivatives are continuous at c, then f is differentiable at c and $Df(c)$ is given by the gradient of f at c; in formula*
$$Df(x) = \mathrm{grad}_f(x), \quad x = c.$$

In addition, the directional derivative at x in direction $a \in \mathbb{R}^d$, with $|a| = 1$, is given by the inner product of the derivative of f with the direction of the derivative:
$$D_a f(x) = \mathrm{grad}_f(x) a.$$

Definition A.5 (Second Order Derivative). Let $f : D_f \subset \mathbb{R}^d \mapsto \mathbb{R}$ be defined on an open neighborhood of x and twice continuously partially differentiable at x. Then the $d \times d$ matrix $Hf(x)$, given by

$$Hf(x) = \begin{pmatrix} \frac{\partial}{\partial x_1}\left(\frac{\partial}{\partial x_1} f(x)\right) & \frac{\partial}{\partial x_1}\left(\frac{\partial}{\partial x_2} f(x)\right) & \cdots & \frac{\partial}{\partial x_1}\left(\frac{\partial}{\partial x_d} f(x)\right) \\ \frac{\partial}{\partial x_2}\left(\frac{\partial}{\partial x_1}(x)\right) & \frac{\partial}{\partial x_2}\left(\frac{\partial}{\partial x_2} f(x)\right) & \cdots & \frac{\partial}{\partial x_2}\left(\frac{\partial}{\partial x_d} f(x)\right) \\ \vdots & \vdots & & \vdots \\ \frac{\partial}{\partial x_d}\left(\frac{\partial}{\partial x_1} f(x)\right) & \frac{\partial}{\partial x_d}\left(\frac{\partial}{\partial x_2} f(x)\right) & \cdots & \frac{\partial}{\partial x_d}\left(\frac{\partial}{\partial x_d} f(x)\right) \end{pmatrix}$$

or

$$(Hf(x))_{ij} = \frac{\partial}{\partial x_i}\left(\frac{\partial}{\partial x_j} f(x)\right)$$

is called the Hessian of f at x.

Note that, using the gradient $f'(x) = \mathrm{grad}_f(x)$, the Hessian can be written as

$$Hf(x) = \begin{pmatrix} \frac{\partial}{\partial x_1} f'(x) \\ \frac{\partial}{\partial x_2} f'(x) \\ \vdots \\ \frac{\partial}{\partial x_n} f'(x) \end{pmatrix}. \tag{A.1}$$

---

[1] If $f : D_f \mapsto \mathbb{R}^m$, then the matrix J met $J_{ij} = \partial f_i / \partial x_j$ is called the Jacobian of f.

Lemma A.1. *Let $f : D_f \subset \mathbb{R}^d \mapsto \mathbb{R}$ be defined on an open neighborhood of x so that f is twice continuously partially differentiable at x, then*

$$\frac{\partial^2 f}{\partial x_i \partial x_j}(x) = \frac{\partial}{\partial x_i}\left(\frac{\partial f}{\partial x_j}(x)\right),$$

for $1 \leq i, j \leq d$, and $i \neq j$. Put differently, $Hf(x)$ is symmetric.

A.3 GEOMETRIC INTERPRETATION OF THE GRADIENT

The gradient of a mapping $f(x)$ from $\mathbb{R}^d$ onto $\mathbb{R}$ evaluated at a point is a vector in the parameter space $\mathbb{R}^d$ pointing in the direction if the steepest ascent of $f(x)$. It is worth noting that the graph of f lies in $\mathbb{R}^{d+1}$ while the gradient lies in $\mathbb{R}^d$. We now discuss how the relation between the gradient and the tangent hyperplane extends the well-known fact from one-dimensional analysis that the derivative at a point is the slope of the tangent at the function in that point. For ease of presentation we discuss the case $d = 3$.

For $a, b \in \mathbb{R}^3$ the cross-product of a and b is defined by

$$a \times b = (a_2 b_3 - a_3 b_2, a_3 b_1 - a_1 b_3, a_1 b_2 - a_2 b_1).$$

The cross-product vector $a \times b$ is perpendicular to the plane given by a and b. Moreover, $|a \times b| = |a| |b| \sin(\eta)$, where η is the angle between a and b, which implies that $|a \times b| = 0$ if a and b are orthogonal.

Example A.1. *Consider vectors $a = (1, 2, 3)^\top$ and $b = (4, 5, 6)^\top$. Then,*

$$(a \times b)^\top = (1, 2, 3) \times (4, 5, 6) = (2 \times 6 - 3 \times 5, 3 \times 4 - 1 \times 6, 1 \times 5 - 2 \times 4) = (-3, 6, -3).$$
※※※

Let P_1, P_2, and P_3 be three points on hyperplane Q. Then, Q is the set of all points satisfying

$$P_1 + s(P_2 - P_1) + t(P_3 - P_1), \tag{A.2}$$

for $s, t \in \mathbb{R}$. Alternatively, a hyperplane can be characterized by a vector perpendicular to the plane, called a *normal vector*, and a point on the plane. More specifically, we can write Q in (A.2) alternatively as the set of points x such that

$$Nx = NP_1 \quad \Leftrightarrow \quad N(x - P_1) = 0, \tag{A.3}$$

where

$$N = (P_2 - P_1) \times (P_3 - P_1).$$

Let $N = (N_x, N_y, N_z)$ and $P_1 = (P_{1x}, P_{1y}, P_{1z})$, then $(x, y, z)^\top$ in Q if

$$N_x x + N_y y + N_z z = N_x P_{1x} + N_y P_{1y} + N_z P_{1z}.$$

We call (A.2) the *parameter* representation and (A.3) the *normal* representation with N denoting the normal vector.

Example A.2. *Let $P_1 = (0, 0, 1)^\top$, $P_2 = (2, 0, 0)^\top$ and $P_3 = (0, 3, 0)^\top$. Then $P_2 - P_1 = (2, 0, -1)^\top$ and $P_3 - P_1 = (0, 3, -1)^\top$, and*

$$N = (P_2 - P_1) \times (P_3 - P_1) = (3, 2, 6)^\top, \quad |N| = 7.$$

Hence, the hyperplane given by $(P_3 - P_1)$ and $(P_3 - P_1)$ going through P_1 (this is the unique hyperplane defined by P_1, P_2 en P_3) is the set points $(x, y, z)^\top \in \mathbb{R}^3$ satisfying

$$3x + 2y + 6z = 6.$$

✼✼✼

Consider the function $F(x, y) : \mathbb{R}^2 \to \mathbb{R}$ and let

$$FL_{(x_0, y_0)}(x, y) = F(x_0, y_0) + \nabla F(x_0, y_0)(x - x_0, y - y_0)^\top, \tag{A.4}$$

for $(x_0, y_0) \in \mathbb{R}^2$. The *tangent hyperplane* is then the collection of all points $(x, y, FL(x, y))^\top$ for $(x, y)^\top \in \mathbb{R}^2$.

Lemma A.2. *The normal vector N to the tangent hyperplane in $(x_0, y_0)^\top$ is given by*

$$N = (F_x(x_0, y_0), F_y(x_0, y_0), -1) = (\nabla F(x_0, y_0), -1).$$

Proof. Given x and y, the equation for the z-coordinate of a point of FL is

$$z = F(x_0, y_0) + \nabla F(x_0, y_0)(x - x_0, y - y_0)^\top,$$

or, equivalently,

$$z = F(x_0, y_0) + F_x(x_0, y_0)(x - x_0) + F_y(x_0, y_0)(y - y_0)$$
$$\Leftrightarrow z - F_x(x_0, y_0)x - F_y(x_0, y_0)y = F(x_0, y_0) - F_x(x_0, y_0)x_0 - F_y(x_0, y_0)y_0$$
$$\Leftrightarrow F_x(x_0, y_0)x + F_y(x_0, y_0)y - z = -F(x_0, y_0) + F_x(x_0, y_0)x_0 + F_y(x_0, y_0)y_0$$
$$\Leftrightarrow F_x(x_0, y_0)x + F_y(x_0, y_0)y - z = F_x(x_0, y_0)x_0 + F_y(x_0, y_0)y_0 - F(x_0, y_0),$$

which can be rewritten as

$$N(x, y, z)^\top = N(x_0, y_0, F(x_0, y_0))^\top.$$

The point $(x_0, y_0, F(x_0, y_0))^\top$ lies on the *tangent hyperplane* and the above is the normal representation of the tangent hyperplane. □

Lemma A.2 shows that the gradient is the projection of the normal to the tangent hyperplane on the parameter space.

Example A.3. (Geometric Interpretation of the Gradient) Let $F : \mathbb{R}^2 \to \mathbb{R}$ with $F(x, y) = xy$. Then the gradient of F is given by

$$\nabla F(x, y) = (y, x).$$

The tangent hyperplane at the point (x_0, y_0) is then the set of all points $(x, y, z)^\top \in \mathbb{R}^3$ such that

$$(y_0, x_0, -1)^\top (x - x_0, y - y_0, z - x_0 y_0)^\top = 0,$$

see (A.3) or, equivalently,

$$z = xy_0 + yx_0 - x_0 y_0.$$

Hence, the tangent plane in point $(\frac{1}{2}, 2)$ is given by

$$z = 2x + \frac{y}{2} - 1,$$

see (A.4).

✼✼✼

APPENDIX A

Let F be defined on $\mathbb{R}^2$ and assume that the level curve $F(x,y) = c$ is differentiable on $\mathbb{R}^2$. The incline of the level curve at point (x, y) is given by

$$F(x, y) = c \Rightarrow y' = -\frac{\frac{\partial}{\partial x} F(x, y)}{\frac{\partial}{\partial y} F(x, y)}.$$

Hence, the incline of the tangent in point (x_0, y_0) is

$$-\frac{\frac{\partial}{\partial x} F(x_0, y_0)}{\frac{\partial}{\partial y} F(x_0, y_0)}$$

and the equation for the tangent becomes

$$y - y_0 = -\frac{\frac{\partial}{\partial x} F(x_0, y_0)}{\frac{\partial}{\partial y} F(x_0, y_0)} (x - x_0)$$

or, equivalently,

$$y = -\frac{\frac{\partial}{\partial x} F(x_0, y_0)}{\frac{\partial}{\partial y} F(x_0, y_0)} (x - x_0) + y_0$$

(a point (x, y) lies thus on the tangent if (x, y) satisfies the above equation). This can be rewritten as

$$\frac{\partial}{\partial x} F(x_0, y_0)(x - x_0) + \frac{\partial}{\partial y} F(x_0, y_0)(y - y_0) = 0$$

or, equivalently,

$$\left(\frac{\partial}{\partial x} F(x_0, y_0), \frac{\partial}{\partial y} F(x_0, y_0) \right) (x - x_0, y - y_0)^\top = 0 \Leftrightarrow \nabla F(x_0, y_0)(x - x_0, y - y_0)^\top = 0.$$

The gradient is thus perpendicular to the tangent at the level curve.

Example A.4. We revist Example A.3. The tangent to the level curve $xy = 1$ in point $(\frac{1}{2}, 2)$ is given by

$$\nabla F(\frac{1}{2}, 2) \left(x - \frac{1}{2}, y - 2 \right)^\top = \left(2, \frac{1}{2} \right) \left(x - \frac{1}{2}, y - 2 \right)^\top$$

or

$$2x - 1 + \frac{1}{2} y - 1 \Leftrightarrow y = -4x + 4.$$

❋❋❋

A.4 WEIERSTRASS THEOREM

The Weierstrass theorem establishes the existence of a maximum/minimum for a continuous mapping f over a compact set.

Theorem A.2 (Weisterass). *Let $X \subset \mathbb{R}^d$ be compact and $f : X \mapsto \mathbb{R}$ continuous on X. Then, there exist x^* and x_* in X so that*

$$f(x^*) = \sup\{f(x) : x \in X\} \quad \text{and} \quad f(x_*) = \inf\{f(x) : x \in X\}.$$

If f is differentiable on X, then x^* and x_* are stationary points.

Theorem A.2 can be used to argue for the existence of a global minimum as follows. Suppose that f is continuous on $\mathbb{R}$ such that $\lim_{||x||\to\infty} f(x) = \infty$. Hence, we may choose M such that $f(x) > f(0)$ for all x such that $||x|| > M$. It thus suffices to search for the global minimum on the set $\{x : ||x|| \leq M\}$. Since this set is compact, the existence of the global minimum follows from Theorem A.2.

A.5 POSITIVE AND NEGATIVE DEFINITE MATRICES

A matrix $A \in \mathbb{R}^{n \times n}$ is called symmetric if and only if $A_{ij} = A_{ji}$ for $1 \leq i, j \leq n$.

Definition A.6. A symmetric matrix $A \in \mathbb{R}^{n \times n}$ is called

- *positive definite:* if for each $x \in \mathbb{R}^n$, $x \neq 0$: $x^\top A x > 0$
- *positive semi-definite:* if for each $x \in \mathbb{R}^n$, $x \neq 0$: $x^\top A x \geq 0$
- *negative definite:* if for $x \in \mathbb{R}^n$, $x \neq 0$: $x^\top A x < 0$
- *negative semi-definite:* if for each $x \in \mathbb{R}^n$, $x \neq 0$: $x^\top A x \leq 0$
- *indefinite:* if there exist $x_1 \in \mathbb{R}^n$ such that $x_1^\top A x_1 > 0$ and $x_2 \in \mathbb{R}^n$ such that $x_2^\top A x_2 < 0$.

Note that A is positive (semi-) definite if and only if $-A$ is negative (semi-) definite. A numerical way of checking the definiteness is by means of computing the eigenvalues.

Theorem A.3. *A matrix $A \in \mathbb{R}^{n \times n}$ is positive definite if and only if all eigenvalues of A are strictly larger than zero; if all eigenvalues of $A \in \mathbb{R}^{n \times n}$ are larger than zero and a least one eigenvalue is equal to zero, then A is positively semi-definite. In a similar way, A is negative definite if and only if all eigenvalues are strictly smaller than zero. If all eigenvalues of $A \in \mathbb{R}^{n \times n}$ are smaller than zero and a least one eigenvalue is equal to zero, then A is negatively semi-definite. If some eigenvalues of $A \in \mathbb{R}^{n \times n}$ are smaller than zero and some are greater than zero, then A is indefinite.*

A.6 NORMED SPACES AND EQUICONTINUITY

Let S be a non-empty set with zero element 0_S. A *norm* is a mapping $||\cdot|| : S \mapsto [0, \infty)$ having the properties (i) $0 < ||x|| < \infty$ for $x \neq 0_S$ and $||0_S|| = 0$, (ii) $||\alpha x|| = |\alpha|\,||x||$ for $\alpha \in \mathbb{R}$, and (iii) $||x + y|| \leq ||x|| + ||y||$ (triangle inequality), for any $x, y \in S$.

A *metric* is a mapping $d : S \times S \to [0, \infty)$ having the properties (i) $d(x, y) = d(y, x)$, (ii) $d(x, y) = 0 \Leftrightarrow x = y$, and (iii) $d(x, z) \leq d(x, y) + d(y, z)$ (triangle inequality), for any $x, y, z \in S$. If (i) and (iii) hold but $d(x, y) = 0$ is possible when $x \neq y$, we call d a *pseudo metric*. A *metric space* (S, d) is a set S paired with metric d.

A subset A of S is called *open* in (S, d) if for each $s \in A$ a $\delta > 0$ exists such that $\{x \in S : d(x, s) < \delta\} \subset A$. The complement of an open set is called a *closed* set.

Definition A.7. The collection of all open subsets of S is denoted by $\mathcal{T}(d)$, and $(S, \mathcal{T}(d))$ is called a *topological space* if $\mathcal{T}(d)$ is a topology, that is, $\emptyset, S \in \mathcal{T}(d)$, and unions and intersection over arbitrary collections of elements of $\mathcal{T}(d)$ are elements of $\mathcal{T}(d)$ as well.

For vector spaces S, a norm on S induces a metric and vice versa. Indeed, for $s, u \in S$, $s - u$ is an element of S as well, which allows us to define the metric $d(s, u) = ||s - u||$

induced by a norm $||\cdot||$ in S, and for $s \in S$, we have $||s|| = d(s, 0)$ as the norm induced by the metric d op S.

Definition A.8. A sequence $\{f_n\}$ in a normed vector space $\mathcal{S} = (S, ||\cdot||)$ is called a *Cauchy sequence* if for every $\epsilon > 0$ there exists $n_0 \in \mathbb{N}$ such that $||f_n - f_{n+m}|| < \epsilon$, for $n, n+m \geq n_0$.

Definition A.9. A normed metric space $\mathcal{S} = (S, ||\cdot||)$ is called

- *complete* if every Cauchy sequence converges to a point in S,
- *sequentially compact* if every infinite sequence in S has a convergent subsequence, and all accumulation points are in S.

A complete normed space $\mathcal{S} = (S, ||\cdot||)$ is called *Baire space*. If S is in particular a vector space, then $\mathcal{S}$ is also called *Banach space*.

Definition A.10. Let $(S, \mathcal{T}(d))$ be a topological space. If there is a countable collection of open subsets such that any open subset of S can be written as union of these sets, then $\mathcal{T}(d)$ is said to have a *countable basis*. A topological space $(S, \mathcal{T})$ is called a *Polish space* if (i) its topology is defined by a complete metric (that is, there exists a metric d such that $\mathcal{T} = \mathcal{T}(d)$), and (ii) $\mathcal{T}$ has a countable basis.

Theorem A.4 (Heine-Borel). *Let $S = \mathbb{R}^d$ be equipped with the Euclidean norm. Then a non-empty set $K \subset S$ is compact if and only if it is closed and bounded.*

Theorem A.5. *If $K \subset S$ is a closed subset of a complete metric space S, then K is sequentially compact if and only if K is totally bounded (all elements $x \in K$ satisfy $||x|| < \bar{b} < \infty$, where $||\cdot||$ is the norm induced by the metric of S).*

Let $C(n; a, b)$ denote set of continuous mappings from $(a, b) \subset \mathbb{R}$ to $\mathbb{R}^n$, and let $||\cdot||_\infty$ denote the supremum norm.

Proposition A.1. *For $n \in \mathbb{N}$, the normed space $(C(n; a, b), ||\cdot||_\infty)$ is complete, where $a = -\infty$ or $b = \infty$ is possible.*

Definition A.11. The set $\mathcal{F} = \{f_n\} \subset C(n; a, b)$ is called *equicontinuous at $s \in (a, b)$* if for each $\varepsilon > 0$ there exists $\delta_{s,\varepsilon} > 0$ such that

$$\forall f \in \mathcal{F}: |s - t| < \delta_{s,\varepsilon} \text{ implies that } |f(t) - f(s)| < \varepsilon.$$

The set $\mathcal{F} = \{f_n\} \subset C(n; a, b)$ is called *equicontinuous* if it is equicontinuous at each point in (a, b). Moreover, it is called *uniformly equicontinuous* if $\delta_{s,\varepsilon}$ can be chosen independently of s.

A.7 LIPSCHITZ AND UNIFORM CONTINUITY

Without loss of generality, we may think of $|\cdot|$ as the Euclidean norm.

Definition A.12. Let $X \subseteq \mathbb{R}^d$ be a connected set. A mapping $f : X \mapsto \mathbb{R}$ is called *Lipschitz continuous* if $K \in \mathbb{R}$ exists such that for any $x, x + \Delta \in X$ is holds that

$$|f(x) - f(x+\Delta)| \leq K |\Delta|.$$

The constant K is called *Lipschitz constant*.

Theorem A.6 (Mean Value Theorem). *For $X = [a, b] \subset \mathbb{R}$, let $f : X \mapsto \mathbb{R}$ be continuous on $[a, b]$ and differentiable on $]a, b[$. Then $\xi \in]a, b[$ exists such that*

$$\frac{f(b) - f(a)}{b - a} = f'(\xi),$$

where f' denotes the derivative of f.

The mean value theorem generalizes to mappings $f : D_f \subset \mathbb{R}^d \mapsto \mathbb{R}$ as follows. Let D_f be connected and assume that f is differentiable on D_f. Then, for $x, y \in D_f$, there exists a point $\xi = (1 - c)x + cy$, for $c \in [0, 1]$, such that

$$\frac{|f(x) - f(y)|}{|x - y|} \leq |\nabla f(\xi)|.$$

For differentiable mappings, a sufficient condition for Lipschitz continuity can be found by the mean value theorem. The precise statement is given in the following.

Result: For $X = [a, b] \subset \mathbb{R}$, let $f : X \mapsto \mathbb{R}$ be continuous on $[a, b]$ and differentiable on $]a, b[$. If

$$\sup_{x \in]a,b[} |f'(x)| \stackrel{\text{def}}{=} K < \infty,$$

then f is Lipschitz continuous on (a, b) with Lipschitz constant K. Moreover, a differentiable mapping f, with $f : \mathbb{R} \mapsto \mathbb{R}$, is Lipschitz continuous with constant K if and only if

$$\sup_{x} |f'(x)| \stackrel{\text{def}}{=} K < \infty.$$

Let $\mathcal{F}$ denote a set of mappings from $X = [a, b] \subset \mathbb{R}$ to $\mathbb{R}$. $\mathcal{F}$ is called *uniformly bounded* if $M \in [0, \infty)$ exists such that, for any $f \in \mathcal{F}$,

$$|f(x)| \leq M, \quad x \in X.$$

A mapping $f : X = [a, b] \subset \mathbb{R}$ is called *uniformly continuous* if for any $\eta > 0$ a $\delta > 0$ exists such that, for any $x_1, x_2 \in X = [a, b] \subset \mathbb{R}$ with $|x_1 - x_2| < \delta$, it holds that

$$|f(x_1) - f(x_2)| < \eta.$$

The set $\mathcal{F}$ of mappings is called *uniformly continuous* if for any $\eta > 0$ a $\delta > 0$ exists such that, for any $x_1, x_2 \in X = [a, b] \subset \mathbb{R}$ with $|x_1 - x_2| < \delta$ and any $f \in \mathcal{F}$, it holds that

$$|f(x_1) - f(x_2)| < \eta.$$

Theorem A.7 (Ascoli-Arzelà). *Let $\mathcal{F}$ be a (at least) countable set of mappings from $X = [a, b] \subset \mathbb{R}$ to $\mathbb{R}$. If $\mathcal{F}$ is uniformly bounded and uniformly continuous, then one can choose a uniformly convergent sequence out of $\mathcal{F}$.*

A.8 TAYLOR SERIES EXPANSIONS

For ease of presentation we summarize here the essential facts on Taylor series expansions in the one-dimensional case. Fix $a, b \in \mathbb{R}$ and $a < b$, let $f : [a, b] \mapsto \mathbb{R}$ be an $(n + 1)$ times differentiable mapping on (a, b) with all (higher order) derivatives continuous on $[a, b]$.

APPENDIX A

Then, it holds for $x_0, x_0 + \Delta \in (a, b)$ that

$$f(x_0 + \Delta) = \sum_{m=0}^{n} \frac{\Delta^m}{m!} \left.\frac{d^m}{dx^m}\right|_{x=x_0} f(x) + R_{n+1}(x_0) \qquad (A.5)$$

and $\nu \in (0, 1)$ exists such that

$$R_{n+1}(x_0) = \frac{\Delta^{n+1}}{(n+1)!} \left.\frac{d^{n+1}}{dx^{n+1}}\right|_{x=x_0+\nu\Delta} f(x).$$

The above remainder term is called *Lagrange* remainder. An alternative way of expressing R_{n+1} is the *Cauchy* remainder:

$$R_{n+1}(x_0) = \frac{\Delta^{n+1}}{n!} (1-\nu')^n \left.\frac{d^{n+1}}{dx^{n+1}}\right|_{x=x_0+\nu'\Delta} f(x),$$

where ν' is again a number in $(0, 1)$. In addition to that, R_{n+1} can be expressed using integration as follows:

$$R_{n+1}(x_0) = \frac{1}{n!} \int_{x_0}^{x_0+\Delta} (x_0 + \Delta - t)^n \left.\frac{d^{n+1}}{dx^{n+1}}\right|_{x=t} f(x)\, dt. \qquad (A.6)$$

The expression on the right-hand side of (A.5) is called a *Taylor polynomial for f of degree n at x_0*. If

$$\sum_{m=0}^{\infty} \frac{\Delta^m}{m!} \left.\frac{d^m}{dx^m}\right|_{x=x_0} f(x) \qquad (A.7)$$

exists, for given x_0 and Δ, then this series is called *Taylor series* or *Taylor series expansion* for f at x_0 evaluated at Δ. In the particular case $x_0 = 0$, (A.7) is also called *MacLaurin series*. The radius of convergence of a Taylor series, denoted by $r(x_0)$, is the largest Δ such that the sum in (A.7) exists and is finite. Because Taylor series are power series, they converge absolutely if they converge at all. Hence, if the radius of convergence of the Taylor series expansion for f at x_0 is $r(x_0) > 0$, then the series converges for any $x_0 + \Delta$, with $|\Delta| \leq r(x_0)$. The radius of convergence of the Taylor series for f at x_0 is given by the formula of Cauchy-Hadamard:

$$\frac{1}{r(x_0)} = \limsup \left(\frac{1}{n!} \left| \left.\frac{d^n}{dx^n}\right|_{x=x_0} f(x) \right| \right)^{\frac{1}{n}}, \qquad (A.8)$$

where $r(x_0) = 0$ if the lim sup equals ∞ and $r(x_0) = \infty$ if the lim sup equals 0.

A.9 L'HÔPITAL'S RULE

Let f, g be real-valued functions that are differentiable on some open internal $\Theta \subset \mathbb{R}$ except for some point $\theta_0 \in \Theta$. L'Hôpital's rule states that if

(i)
$$\lim_{\theta \to \theta_0} f(\theta) = \lim_{\theta \to \theta_0} g(\theta) = H,$$

for H either 0 or $\pm\infty$,

(ii)
$$g(\theta) \neq 0, \quad \text{for all } \theta \in \Theta \setminus \{\theta_0\},$$

(iii)
$$\lim_{\theta \to \theta_0} \frac{\frac{d}{d\theta}f(\theta)}{\frac{d}{d\theta}g(\theta)} \in \mathbb{R},$$

then
$$\lim_{\theta \to \theta_0} \frac{f(\theta)}{g(\theta)} = \lim_{\theta \to \theta_0} \frac{\frac{d}{d\theta}f(\theta)}{\frac{d}{d\theta}g(\theta)} \in \mathbb{R}.$$

A.10 CESÀRO LIMITS

A real-valued sequence $\{x_n\}$ is called *Cesàro summable* if

$$\lim_{n \to \infty} \frac{1}{n} \sum_{m=1}^{n} x_m$$

exists. If

$$\lim_{n \to \infty} x_n = x$$

exists, then

$$\lim_{n \to \infty} \frac{1}{n} \sum_{m=1}^{n} x_m = x.$$

In words, any convergent sequence is Cesàro summable. The converse is, however, not true. To see this, consider the sequence $x_n = (-1)^n$, $n \in \mathbb{N}$.

Appendix B

Probability Theory

B.1 INFORMATION STRUCTURE

Definition B.1. Let $S \neq \emptyset$ be a set. A *σ-field* $\mathcal{S}$ on S is a collection of subsets of S with the following properties:

(a) $S \in \mathcal{S}$,
(b) if $A \in \mathcal{S}$, then $A^c \in \mathcal{S}$, where $A^c = \{s \in S : s \notin A\}$, and
(c) if $A_i \in \mathcal{S}$, for $i \in \mathbb{N}$, then $\bigcup_{i \in \mathbb{N}} A_i \in \mathcal{S}$.

Elements of a σ-field are often referred to as "events" in probability theory.

Definition B.2. Let $\mathcal{A}$ denote a collection of subsets of S. The σ-field *generated* by $\mathcal{A}$ is the smallest σ-field that contains $\mathcal{A}$.

Definition B.3. Let $(S, \mathcal{T})$ be a topological space (see Definition A.7). The *Borel field* of S, denoted by $\mathcal{B}$, is the σ-field generated by the collection of open sets $\mathcal{T}$, in formula: $\mathcal{B} = \sigma(\mathcal{T})$.

Definition B.4. Given a set $S \neq \emptyset$ and a σ-field $\mathcal{S}$ on S, the pair $(S, \mathcal{S})$ is called a *measurable space*.

Definition B.5. Let $(S, \mathcal{S})$ and $(R, \mathcal{R})$ be two measurable spaces. A mapping $g : S \mapsto R$ is said to be *measurable* (w.r.t. $\mathcal{S}$) if for any $A \in \mathcal{R}$ it holds true that $\{s \in S : g(s) \in A\} \in \mathcal{S}$. A measurable mapping is also called a *random variable*.

Definition B.6. The *σ-field generated* by a random variable X on a measurable space $(\Omega, \mathfrak{F})$, denoted $\sigma(X)$ is the smallest σ-field with respect to which X is measurable. Similarly, if $X_1, \ldots, X_n$ are random variables defined on a common space $(\Omega, \mathfrak{F})$, then $\sigma(X_1, \ldots, X_n)$ is the smallest σ-field w.r.t which every $X_i, i = 1, \ldots, n$, is measurable.

Result: Suppose that X is a random variable on $(\Omega, \mathfrak{F})$. If $\sigma(X) \subset \mathcal{G} \subset \mathfrak{F}$, then X is also $\mathcal{G}$-measurable.

In this monograph we study time dynamics by referring to appropriate information structures defined through the "arrow of time." The history of the process is represented by having more refined σ-fields as time increases. Knowledge increases only in one direction, that of increasing time. That is, all of the events that are described at any given time are also described in future times.

Definition B.7. Given a set $T \subset \mathbb{R}$ (or $T \subset \mathbb{N}$) and a measurable space $(\Omega, \mathfrak{F})$ a *filtration* $\mathbb{F}$ is a sequence of increasing σ-fields $\mathbb{F} = \{\mathfrak{F}_t ; t \in T\}$ satisfying $\mathfrak{F}_s \subset \mathfrak{F}_t \subset \mathfrak{F}$, for every $s < t; s, t \in T\}$.

Definition B.8. Given a measurable space $(\Omega, \mathfrak{F})$ and a filtration $\mathbb{F} = \{\mathfrak{F}_t; t \in T\}$ on it, a *stochastic process* $\{X_t; t \in T\}$ is a collection of random variables on the common measurable space, where for each $t \in T$, X_t is $\mathfrak{F}_t$-measurable. The particular case where $\mathfrak{F}_t = \sigma(X_t)$ is called the *natural filtration* of the process.

Definition B.9. A *random stopping time* τ adapted to the filtration $\{\mathfrak{F}_t; t \in T\}$ on $(\Omega, \mathfrak{F})$ is a random variable on $(\Omega, \mathfrak{F})$ that satisfies

$$\{\omega : \tau(\omega) \leq t\} \in \mathfrak{F}_t; \quad \forall t \in T.$$

The σ-field $\mathfrak{F}_\tau$ contains all events A such that $A \cup \{\tau(\omega) \leq t\} \in \mathfrak{F}_t$.

B.2 (PROBABILITY) MEASURES

Definition B.10. A *measure* μ on a measurable space $(S, \mathcal{S})$ is a mapping $\mu : \mathcal{S} \mapsto \mathbb{R} \cup \{-\infty, \infty\}$ such that for any sequence $\{A_n\}$ of mutually disjoint elements of $\mathcal{S}$ it holds that

$$\mu\left(\bigcup_{n=1}^{\infty} A_n\right) = \sum_{n=1}^{\infty} \mu(A_n).$$

The measure m on $(\mathbb{R}, \mathcal{B})$, where $\mathcal{B}$ denotes the Borel field on $\mathbb{R}$, assigning $m((a, b]) = b - a$ to an interval $(a, b]$ is called a *Lebesgue measure*. It generalizes the notion of length in geometry and is the case closest to everyday intuition.

Definition B.11. Let $(S, \mathcal{S})$ be a measurable space and μ a measure on $(S, \mathcal{S})$, then the collection $(S, \mathcal{S}, \mu)$ is called *measure space*.

Definition B.12. A measure μ is called *signed* if $\mu(A) < 0$ for some $A \in \mathcal{S}$ and otherwise it is called *non-negative*. Furthermore, a measure μ is called *finite* if $\mu(A) \in \mathbb{R}$ for any $A \in \mathcal{S}$. A non-negative measure μ is called σ-*finite* if there exist countably many sets A_i in $\mathcal{S}$ such that $\mu(A_i) < \infty$ and $\bigcup_i A_i = S$.

We denote the set of signed measures on $(S, \mathcal{S})$ by $\mathcal{M}$. Let $\mu \in \mathcal{M}$ be non-negative. Then for any measurable mapping $g : S \mapsto \mathbb{R}$ the μ-*integral* of g, denoted by

$$<g, \mu> = \int_S g(s) \mu(ds),$$

is defined although it may take values in $\{-\infty, \infty\}$. In particular, for any $A \in \mathcal{S}$,

$$<1_A, \mu> = \mu(A),$$

where $1_A : S \mapsto \mathbb{R}$ is defined by $1_A(s) = 1$ for $s \in A$ and $1_A(s) = 0$ otherwise.

For any signed measure $\mu \in \mathcal{M}$ a measurable set S_μ^+ exists such that, for any $A \in \mathcal{S}$, it holds that $\mu(A \cap S_\mu^+) \geq 0$, whereas $\mu(A \cap (S \setminus S_\mu^+)) \leq 0$; see, for example, [158] for a proof. The positive part of μ is defined by

$$[\mu]^+(A) = \mu(A \cap S_\mu^+), \quad A \in \mathcal{S},$$

and the negative part by

$$[\mu]^-(A) = -\mu(A \cap (S \setminus S_\mu^+)), \quad A \in \mathcal{S}.$$

APPENDIX B

The pair $([\mu]^+, [\mu]^-)$ is called *Hahn-Jordan decomposition*. The absolute measure $|\mu|$ is defined by $|\mu| = [\mu]^+ + [\mu]^-$. Integration with respect to a signed measure is defined by

$$<g, \mu> = <g, [\mu]^+> - <g, [\mu]^->$$

and integration with respect to an absolute measure is defined by

$$<g, |\mu|> = <g, [\mu]^+> + <g, [\mu]^->, \qquad (B.1)$$

provided that the terms on the right-hand side of the above formulas are finite. The Hahn-Jordan decomposition is unique in the sense that if $\hat{G}$ is another set, such that $\mu(A \cap \hat{G}) \geq 0$ and $\mu(A \cap \hat{G}^c) \leq 0$ for any $A \in \mathcal{S}$, then $\mu(A \cap G) = \mu(A \cap S_\mu^+)$ for any $A \in \mathcal{S}$. A signed measure $\mu \in \mathcal{M}$ is finite if $[\mu]^+(S)$ and $[\mu]^-(S)$ are finite.

Definition B.13. A *probability measure* μ is a non-negative measure such that $\mu(S) = 1$ (which already implies that $\mu(\emptyset) = 0$). If μ is a probability measure on $(S, \mathcal{S})$, then the collection $(S, \mathcal{S}, \mu)$ is called *probability space*.

Theorem B.1 (Continuity Theorem for Decreasing (Increasing) Events). *Let* $(\Omega, \mathfrak{F}, \mathbb{P})$ *be a probability space. If* $\{A_n\}$ *is a sequence of decreasing (increasing) events in* $\mathfrak{F}$ *(that is, $A_{n+1} \subset A_n$ in case of decreasing, and $A_n \subset A_{n+1}$ in case of increasing), then* $\mathbb{P}(\lim_n A_n) = \lim_n \mathbb{P}(A_n)$.

Definition B.14. Let μ and ν be σ-finite measures on a measurable space $(S, \mathcal{S})$. The measure μ is said to be *absolutely continuous* with respect to ν (denoted $\mu \ll \nu$) if $\nu(A) = 0$, for $A \in \mathcal{S}$, implies $\mu(A) = 0$.

Theorem B.2 (Radon-Nikodym). *Two σ-finite measures on a measurable space* $(\Omega, \mathfrak{F})$ *satisfy $\mu \ll \nu$ if, and only if, there exists a non-negative measurable mapping* $d\mu/d\nu : S \mapsto \mathbb{R}$ *such that*

$$\mu(A) = \int_A \frac{d\mu}{d\nu}(s) \, \nu(ds), \quad \text{for every } A \in \mathfrak{F}.$$

The random variable $d\mu/d\nu$ is called ν-density of μ, or Radon-Nikodym derivative.

Remark B.1. In the particular case that $\nu = m$ is the Lebesgue measure on $\mathbb{R}$ and μ is a probability measure, the Radon-Nikodym derivative is the usual pdf.

Let $(S, \mathcal{S}, \mu)$ and $(T, \mathcal{T}, \nu)$ be two probability spaces. The product of μ and ν on $S \times T$, denoted by $\mu \times \nu$, is a measure such that

$$\forall A \in \mathcal{S}, B \in \mathcal{T}: \quad (\mu \times \nu)(A \times B) = \mu(A) \, \nu(B)$$

and *Fubini's theorem* states that

$$\int_{S \times T} f(s,t) \, (\mu \times \nu)(ds, dt) = \int_T \left(\int_S f(s,t) \, \mu(ds) \right) \nu(dt)$$
$$= \int_S \left(\int_T f(s,t) \, \nu(dt) \right) \mu(ds),$$

for any measurable mapping $f : S \times T \mapsto \mathbb{R}$.

Definition B.15. Let $(S, \mathcal{S}, \mu)$ be a measure space and $g : S \mapsto \mathbb{R}$ a measurable mapping from $(S, \mathcal{S})$ to $(\mathbb{R}, \mathcal{R})$, where $\mathcal{R}$ is a σ-field over $\mathbb{R}$. The *induced measure* of g, denoted by

μ^g, is defined as follows:

$$\mu^g(A) = \mu\big(\{s \in S : g(s) \in A)\}\big), \quad A \in \mathcal{R}.$$

Definition B.16. The cdf of a real-valued random variable X defined on a probability space $(S, \mathcal{S}, \mu)$ is the function $F : [-\infty, \infty] \mapsto [0, 1]$, where

$$F(x) = \mu^X((-\infty, x]), \quad -\infty \leq x \leq \infty.$$

We take the domain $[-\infty, \infty]$ since it is natural to assign the values 0 and 1 to $F(-\infty)$ and $F(\infty)$. A cdf has the decomposition $F(x) = F'(x) + F''(x)$, where $F'(x)$ is positive only on a set of Lebesgue measure zero, and $F''(x)$ is absolutely continuous with respect to the Lebesgue measure. The Radon-Nikodym derivative of F'' with respect to the Lebesgue measure exists and is called the probability density function. If f is the pdf of the cdf F, then it holds that $f(x) = dF(x)/dx$ except for a set of Lebesgue measure zero.

Definition B.17. Two random variables X, Y on a common probability space $(\Omega, \mathfrak{F})$ said to be *equal almost surely* (a.s., or w.p.1) if $\mathbb{P}(\omega : X(\omega) = Y(\omega)) = 1$. Two random variables X and Y not necessarily defined on a common probability space are said to be *equal in distribution* (denoted $X \stackrel{d}{=} Y$) if $\mathbb{P}(X \leq x) = \mathbb{P}(Y \leq x)$ for every $x \in \mathbb{R}$.

Theorem B.3 (Skorokhod Representation: Random Variables). *For any real-valued random variable X on a probability space $(\Omega, \mathfrak{F}, \mathbb{P})$ there exists a real-valued random variable $\tilde{X}$ on the canonical probability space $([0, 1], \mathcal{B}([0, 1]), m)$ (where m is the Lebesgue measure) such that $X \stackrel{d}{=} \tilde{X}$.*

Let μ be a finite measure on $(S, \mathcal{S})$, where S is a locally compact Hausdorff space[1] and $\mathcal{S}$ contains the Borel field on S. The measure μ is called *regular* if

$$\mu(A) = \inf\{\mu(U) : U \text{ open in } S, A \subset U\}, \quad A \in \mathcal{S},$$

and for any open set $U \subset S$ it holds

$$\mu(U) = \sup\{\mu(F) : F \text{ is compact in } S, F \subset U\}.$$

B.3 EXPECTATIONS AND CONDITIONING

Definition B.18. Let $\mathbb{P}$ be a probability measure and let X be a random variable on $(\Omega, \mathfrak{F})$. The *expectation* of X is defined as the integral:

$$\mathbb{E}[X] = \int_\Omega X(\omega)\, \mathbb{P}(d\omega) = \int_\mathbb{R} x F(dx),$$

where F is the cdf of X.

Definition B.19. Let X be a random variable on $(\Omega, \mathfrak{F}, \mathbb{P})$. The *variance* of X is defined as

$$\text{Var}(X) = \mathbb{E}[X^2] - (\mathbb{E}[X])^2.$$

---

[1] A topological space is called Hausdorff if for any two distinct points an open neighborhood for each of them can be found so that theses sets are disjoint.

Theorem B.4 (Markov Inequality). *Let X be a non-negative random variable on the probability space $(\Omega, \mathfrak{F}, \mathbb{P})$. Then for any positive real number $a \in \mathbb{R}^+$*

$$\mathbb{P}(X > a) \leq \frac{\mathbb{E}[X]}{a}.$$

A corollary to this theorem is known as Chebyshev's inequality:

$$\mathbb{P}(|X - \mathbb{E}(X)| \geq a) \leq \frac{\text{Var}(X)}{a^2}.$$

Theorem B.5 (Wald's Equality). *Let $\{X(n)\}$ be an iid sequence such that $\mathbb{E}[X(1)]$ is finite. Furthermore, let η be a non-negative integer-valued random variable with finite mean. If for all $m \geq 0$ the event $\{\eta = m\}$ is independent of $\{X(m+n) : n \geq 1\}$, then*

$$\mathbb{E}\left[\sum_{i=1}^{\eta} X(i)\right] = \mathbb{E}[\eta]\,\mathbb{E}[X(1)] < \infty.$$

Definition B.20. Let X be a random variable on the probability space $(\Omega, \mathfrak{F}, \mathbb{P})$ such that $\mathbb{E}[|X|] < \infty$, and let $\mathcal{G} \subset \mathfrak{F}$ be another σ-field on Ω. The *conditional expectation* of X given $\mathcal{G}$, is a random variable Z satisfying:

(a) Z is $\mathcal{G}$-measurable, and
(b) for all $B \in \mathcal{G}$, $\mathbb{E}[Z\mathbf{1}_{\{B\}}] = \mathbb{E}[X\mathbf{1}_{\{B\}}]$.

We often denote Z by $\mathbb{E}[X \mid \mathcal{G}]$.

Definition B.21. Given two random variables on $(\Omega, \mathfrak{F}, \mathbb{P})$, the *conditional expectation* of X given Y is defined as $\mathbb{E}[X \mid Y] = \mathbb{E}[X \mid \sigma(Y)]$.

Theorem B.6 (Summary of Properties of Conditional Expectations). *Consider a probability space $(\Omega, \mathfrak{F}, \mathbb{P})$ and sub σ-algebras $\mathcal{G}, \hat{\mathcal{G}} \subset \mathfrak{F}$.*

(a) *If X is $\mathcal{G}$-measurable, then $\mathbb{E}[X \mid \mathcal{G}] = X$ a.s.*
(b) *Law of the Iterated Expectation: If $\hat{\mathcal{G}} \subset \mathcal{G} \subset \mathfrak{F}$ then $\mathbb{E}[\mathbb{E}[X \mid \mathcal{G}] \mid \hat{\mathcal{G}}] = \mathbb{E}[\mathbb{E}[X \mid \hat{\mathcal{G}}] \mid \mathcal{G}] = \mathbb{E}[X \mid \hat{\mathcal{G}}]$.*
(c) $\mathbb{E}[X \mid \{\Omega, \emptyset\}] = \mathbb{E}[X]$.
(d) *For any real valued function h such that $\mathbb{E}[h(X)] < \infty$, $\mathbb{E}[\mathbb{E}[h(X) \mid \mathcal{G}]] = \mathbb{E}[h(X)]$.*
(e) *If Y is $\mathcal{G}$-measurable and both $\mathbb{E}[X]$ and $\mathbb{E}[Y]$ exist, then $\mathbb{E}[XY \mid \mathcal{G}] = Y\,\mathbb{E}[X \mid \mathcal{G}]$ a.s.*
(f) *Jensen's inequality: if h is a convex function and $\mathbb{E}|h(X)| < \infty$, then $h(\mathbb{E}[X]) \leq \mathbb{E}[h(X)]$.*

B.4 CONVERGENCE OF RANDOM SEQUENCES

Let $X, X_n, n \geq 0$, be real-valued random variables defined on a common probability space $(\Omega, \mathfrak{F}, \mathbb{P})$ with state space $S \subset \mathbb{R}$ and let S be equipped with the Borel field $\mathcal{B}(S)$. When we talk about "convergence" there must be an implicit norm defined on an appropriate space to study "how close" the sequence gets to the limit. For random variables, there are several ways to define the notion of being "close." Recall that two random variables may be equal in distribution yet they may be defined on entirely different measurable spaces. The following are the most common concepts of convergence for random sequences.

Definition B.22. The sequence $\{X_n\}$ converges *almost surely* (or w.p.1) to X as n tends to ∞ if for any $\delta > 0$

$$\mathbb{P}\left(\lim_{n\to\infty} \sup_{m\geq n} |X_m - X| > \delta\right) = 0.$$

Another equivalent condition is that the event $\{\lim_{n\to\infty} X_n = X\}$ has probability one. This is denoted $X_n \to X$ a.s.

Using the continuity theorem for decreasing events, if $X_n \to X$ a.s., then it holds that $\lim_{n\to\infty} \mathbb{P}\left(\sup_{m\geq n} |X_m - X| > \delta\right) = 0$.

Definition B.23. The sequence $\{X_n\}$ converges *in probability* to X as n tends to ∞ if for any $\delta > 0$

$$\lim_{n\to\infty} \mathbb{P}(|X_n - X| \geq \delta) = 0,$$

or, equivalently,

$$\lim_{n\to\infty} \mathbb{P}\left(|X_n - X| > \delta\right) = 0.$$

This is denoted $X_n \xrightarrow{P} X$.

Almost sure convergence of $\{X_n\}$ to X implies convergence in probability of $\{X_n\}$ to X. On the other hand, convergence in probability of $\{X_n\}$ to X implies a.s. convergence of a subsequence of $\{X_n\}$ to X.

Definition B.24. The sequence $\{X_n\}$ converges *in the r-th mean* to X as n tends to ∞ if

$$\mathbb{E}[|X_n - X|^r] \to 0.$$

This is denoted $X_n \xrightarrow{r} X$.

Theorem B.7. *If* $\lim_{m,n\to\infty} X_{m+n} - X_n = 0$ *a.s., or in the r-th mean, or in probability, then there exists a random variable X such that $X_n \to X$ in the same sense.*

Definition B.25. Let $C_b(\mathbb{R})$ denote the set of bounded continuous mapping from S onto $\mathbb{R}$. A sequence $\{\mu_n\}$ of measures on S is said to *converge weakly* to a distribution μ if

$$\lim_{n\to\infty} \int_S f \, d\mu_n = \int_S f \, d\mu, \quad \text{for any } f \in C_b(\mathbb{R}).$$

Definition B.26. Let F_n denote the distribution of X_n and F the distribution of X. If $\{F_n\}$ converges weakly to F as n tends to ∞, then we say that $\{X_n\}$ *converges in distribution* to X. Equivalently, $\lim_n F_n(x) = F(x)$ for every point of continuity of F. This is denoted $X_n \xRightarrow{d} X$.

Convergence in probability implies convergence in distribution but the converse is not true.

Definition B.27. The *total variation norm* of a (signed) measure μ on S is defined by

$$\|\mu\|_{tv} = \sup_{\substack{f \in C_b(\mathbb{R}) \\ |f| \leq 1}} \left| \int_S f \, d\mu \right|,$$

or, equivalently, $\|\mu\|_{tv} = [\mu]^+(S) + [\mu]^-(S)$.

APPENDIX B

Definition B.28. Let μ_n denote the distribution of X_n and μ the distribution of X. If $\{\mu_n\}$ converges in total variation to μ as n tends to ∞ (that is, if $\lim_{n\to\infty} \|\mu_n - \mu\|_{tv} = 0$) then we say that $\{X_n\}$ *converges in total variation* to X. Convergence in total variation of $\{X_n\}$ to X can be expressed equivalently by

$$\lim_{n\to\infty} \sup_{A \in S} |\mathbb{P}(X_n \in A) - \mathbb{P}(X \in A)| = 0.$$

Convergence in total variation implies convergence in distribution (or, weak convergence) but the converse is not true.

Theorem B.8 (Continuous Mapping Theorem). *Let $h: \mathbb{R} \mapsto \mathbb{R}$ be measurable with discontinuity points confined to a set D_h, where $\mu(D_h) = 0$. If μ_n converges weakly toward μ as n tends to ∞, then μ_n^h tends weakly to μ^h as n tends to ∞, or, equivalently,*

$$\lim_{n\to\infty} \int f(h(x)) \mu_n(dx) = \int f(h(x)) \mu(dx), \quad f \in C_b(\mathbb{R}).$$

Hence, if $\{X_n\}$ converges weakly and h is continuous, then $\{h(X_n)\}$ converges weakly.

A random sequence may converge a.s. and yet not in expectation, as the following example illustrates. Consider the canonical probability space $([0,1], \mathcal{B}([0,1]), m)$ and let

$$X_n(\omega) = \begin{cases} n & \omega \in [0, 1/n), \\ 0 & \text{otherwise,} \end{cases}$$

and define $X(\omega) \equiv 0$. Clearly, for every $\omega > 0$ there is an integer n such that $X_n(\omega) = 0$. Because $m(0) = 0$, then the event $\{\omega : \lim_{n\to\infty} X_n(\omega) = 0\}$ has measure 1. Thus, $X_n \to X$ a.s. However, notice that for each n,

$$\mathbb{E}[X_n] = \int_0^1 X_n(\omega) m(d\omega) = \int_1^{1/n} n\, m(d\omega) \equiv 1,$$

so that $\mathbb{E}[X_n] \not\to \mathbb{E}[X] = 0$. The problem with the example stems from the fact that on very small sets the random variables X_n can become large without bound. The following theorem states the condition under which a.s. convergence also ensures convergence in expectation.

Theorem B.9 (Dominated Convergence). *Let $(S, \mathcal{S}, \mu)$ be a probability space. Let $f_n : S \mapsto \mathbb{R}$, for $n \in \mathbb{N}$, be measurable and assume that $f, g : S \mapsto \mathbb{R}$ are measurable mappings such that, for any $n \in \mathbb{N}$, the set of points $s \in S$ with $|f_n(s)| \leq g(s)$ and $\lim_{n\to\infty} f_n(s) = f(s)$ has μ-measure one. If*

$$\int_S |g(s)| \mu(ds) < \infty,$$

then

$$\lim_{n\to\infty} \int_S f_n(s) \mu(ds) = \int_S f(s) \mu(ds).$$

Equivalently, if $X_n \to X$ a.s. and there is a random variable Z on $(\Omega, \mathfrak{F}, \mathbb{P})$ with $\mathbb{E}[Z] < \infty$ and $|X_n| \leq Z$ for every n, then $\mathbb{E}[X_n] \to \mathbb{E}[X]$.

Theorem B.10 (Monotone Convergence). *Let $\{X_n\}$ be a sequence of random variables on $(\Omega, \mathfrak{F}, \mathbb{P})$ increasing a.s. to a limit, that is, $X_{n+1} \geq X_n$ a.s. and $X = \lim_{n\to\infty} X_n$ exists a.s. Then we can interchange the limit and expectation: $\lim_{n\to\infty} \mathbb{E}[X_n] = \mathbb{E}[X]$.*

The point-wise limit of a sequence of distributions is not necessarily a distribution. To see this, consider the distribution functions:

$$F_n(x) = \begin{cases} 0 & \text{if } x < n, \\ 1 & \text{if } x \geq n, \end{cases}$$

so that $F_n(x)$ tends to zero as n tends to infinity for any $x \geq 0$. The limit function (zero) is not a probability distribution and no subsequence converges to a probability distribution. This shows that the set of distribution function is not bounded (and therefore not compact) with respect to the point-wise limit (refer to Section B.5 for the fact that the set of distributions becomes a Banach space when equipped with an appropriate norm). The problem here is that the "mass" where the probability is concentrated floats away to infinity. The following material establishes conditions under which we can characterize the compact sets of distributions.

Definition B.29. Let μ_n denote the measure on $\mathcal{B}(S)$ induced by the random variable X_n. The family of measures $\{\mu_n\}$ is said to be *tight* if

$$\lim_{K \to \infty} \sup_n \mu_n([-K, K]^c) = \lim_{K \to \infty} \sup_n \mathbb{P}(|X_n| \geq K) = 0.$$

Sometimes we say that the sequence of distributions is *mass preserving*, or that the sequence of random variables $\{X_n\}$ is *tight* (also referred to as uniformly bounded in probability). In particular, a sequence of random variables $\{X_n\}$ is tight if for each $\alpha > 0$ there are N_α, K_α such that

$$\mathbb{P}(|X_n| \geq K_\alpha) \leq \alpha \quad \text{for every } n \geq N_\alpha.$$

A sufficient condition for a sequence of measures $\{\mu_n\}$ is to be tight is that for some f, such that $f(s) \geq 0$, for all s, and $\lim_{|s| \to \infty} f(s) = +\infty$, it holds that

$$\sup_n \int_S f(s) \mu_n(ds) < C,$$

for some finite number C. For ease of reference we note the following straightforward consequence of this result.

Lemma B.1. *A sequence of probability measures on $\mathcal{B}(S)$ is tight if the variances are uniformly bounded.*

Theorem B.11. *If the sequence of probability measures $\{\mu_n\}$ is tight then it is relatively compact; that is, it contains at least one weakly convergent subsequence and all accumulation points are probability measures.*

Theorem B.12 (Skorokhod Representation: Convergence). *Consider the space $(\Omega, \mathfrak{F}, \mathbb{P})$ and let $X_n; X: \Omega \mapsto S$ be S-valued random variables, where S is a complete separable metric space with metric $\hat{d}(\cdot, \cdot)$. Suppose that $X_n \stackrel{d}{\Longrightarrow} X$. Then there is a probability space $(\tilde{\Omega}, \tilde{\mathfrak{F}}, \tilde{\mathbb{P}})$ with associated random variables $\tilde{X}_n; \tilde{X}$ such that $X_n \stackrel{d}{=} \tilde{X}_n$ for every n, $\tilde{X} \stackrel{d}{=} X$ and*

$$\lim_{n \to \infty} \hat{d}(\tilde{X}_n, \tilde{X}) = 0 \quad \tilde{\mathbb{P}}-a.s.$$

that is, $\tilde{X}_n \to \tilde{X}$ with ($\tilde{\mathbb{P}}$)-probability one.

APPENDIX B

Definition B.30. A sequence of random variables $\{X_n\}$ is *uniformly integrable* if

$$\sup_n \lim_{K \to \infty} \mathbb{E}[|X_n| \mathbf{1}_{\{|X_n| > K\}}] = 0.$$

Definition B.31. A continuous time stochastic process $\{\vartheta(t)\}$ on $\mathbb{R}^d$ is called a *cadlag* process if it is right-continuous with left limits at every point t. The space $D^d[0, \infty]$ is the space of piecewise constant cadlag processes with the Skorokhod topology (a modified sup-norm; see [38] for details).

Definition B.32. Let $X(t)$ be a continuous time stochastic process on $(\Omega, \mathfrak{F}, \mathbb{P})$ with natural filtration $\mathbb{F}$. We say that X is *locally Lipschitz continuous with probability 1* if for each $T > 0$ there is a $\mathfrak{F}_T$-measurable random variable $K(T) > 0$ with finite expectation and such that for $s > 0$

$$|X(t+s) - X(t)| \leq s K(T), \quad \text{for all } t \leq t+s \leq T.$$

Theorem B.13 (Proposition 1 in [312]). *Let $\{Y_n^\epsilon; n \geq 1, \epsilon > 0\}$ be a family of $\mathbb{R}^d$-valued random variables on $(\Omega, \mathfrak{F}, \mathbb{P})$ which is uniformly integrable and define the piecewise constant function:*

$$\vartheta^\epsilon(t) = \vartheta^\epsilon(0) + \epsilon \sum_{n=1}^{\lfloor \epsilon/t \rfloor} Y_n^\epsilon.$$

Then the sequence $\{\vartheta^\epsilon\}$ of processes is tight in $D^d[0, \infty]$ and all its accumulation points (as $\epsilon \to 0$) ϑ^* are Lipschitz continuous with probability 1.

B.5 v-NORM CONVERGENCE OF MEASURES

Let (S, d) be a separable metric space and denote the set of continuous real-valued mappings on S by $C(S)$. Let $v : S \mapsto \mathbb{R}$ be a measurable mapping such that

$$\inf_{s \in S} v(s) \geq 1.$$

The set of mappings from S to $\mathbb{R}$ can be equipped with the so-called v-*norm* introduced presently. For $g : S \mapsto \mathbb{R}$, the v-norm of g, denoted by $\|g\|_v$, is defined by

$$\|g\|_v \stackrel{\text{def}}{=} \sup_{s \in S} \frac{|g(s)|}{v(s)}.$$

If g has finite v-norm, then $|g(s)| \leq c v(s)$ for any $s \in S$ and some finite constant c. For example, the set of real, continuous v-dominated functions, defined by

$$\mathcal{D}_v(S) \stackrel{\text{def}}{=} \{g \in C(S) \mid \exists c > 0 : |g(s)| \leq cv(s), \forall s \in S\}, \tag{B.2}$$

can be characterized as the set of all continuous mappings $g : S \mapsto \mathbb{R}$ having finite v-norm. Note that $C^b(S)$ is a particular $\mathcal{D}_v(S)$-space, obtained for $v = const$. Moreover, the condition that $\inf_{s \in S} v(s) \geq 1$ implies that $C^b(S) \subset \mathcal{D}_v(S)$ for any choice of v.

The v-norm of a measure μ on $(S, \mathcal{S})$, with $\mathcal{S}$ the Borel-field, is defined through

$$\|\mu\|_v \stackrel{\text{def}}{=} \sup_{\|g\|_v \leq 1} \left| \int_S g(s) \mu(ds) \right|,$$

or, more explicitly,
$$\|\mu\|_v = \sup_{|g| \le v} \left| \int_S g(s) \mu(ds) \right|.$$

In particular, it holds that
$$\|\mu\|_v = \int v(s) |\mu|(ds), \tag{B.3}$$

see (B.1). Let $\{\mu_n\}$ be a sequence of measures on $(S, \mathcal{S})$ and let μ be a measure on $(S, \mathcal{S})$. We say that μ_n converges in v-norm toward μ if
$$\lim_{n \to \infty} \|\mu_n - \mu\|_v = 0.$$

It can be shown that the set $\mathcal{D}_v(S)$ endowed with the v-norm is a Banach space. This last remark indicates the following fact: for each measure μ with $\int v(s)\mu(ds)$ finite, the mapping $T_\mu \colon \mathcal{D}_v(S) \mapsto \mathbb{R}$ defined through
$$T_\mu(g) \stackrel{\text{def}}{=} \int g d\mu,$$

is a continuous linear functional on the Banach space $\mathcal{D}_v(S)$ and the operator norm of T_μ satisfies $\|T_\mu\| = \|\mu\|_v$. The Cauchy-Schwartz inequality thus holds for v-norms, i.e.,
$$\left| \int g(s)\mu(ds) \right| \le \|g\|_v \cdot \|\mu\|_v,$$

for all $g \in \mathcal{D}_v(S)$ and μ such that v is μ-integrable.

Let $\{\mu_n\}$ be a sequence of measures on $(S, \mathcal{S})$ and let μ be a measure on $(S, \mathcal{S})$. We say that μ_n converges weakly in $\mathcal{D}_v(S)$-sense toward μ if
$$\lim_{n \to \infty} \int_S g(s) \mu_n(ds) = \int_S g(s) \mu(ds),$$

for all $g \in \mathcal{D}_v(S)$; in symbols $\mu_n \stackrel{\mathcal{D}_v(S)}{\Longrightarrow} \mu$.

Remark B.2. Note that v-norm convergence implies $\mathcal{D}_v(S)$-convergence. This can be seen as follows. According to Cauchy-Schwartz inequality, for each $g \in \mathcal{D}_v(S)$ it holds that
$$\left| \int g(s)\mu_n(ds) - \int g(s)\mu(ds) \right| = \left| \int g(s)(\mu_n - \mu)(ds) \right| \le \|g\|_v \cdot \|\mu_n - \mu\|_v.$$

Hence, $\|\mu_n - \mu\|_v \to 0$ implies that the left-hand side in the above relation converges to 0 as $n \to \infty$.

B.6 MARTINGALE PROCESSES

Let $(\Omega, \mathfrak{F}, \mathbb{P})$ be a probability space and $\mathbb{F} = \{\mathfrak{F}_t; t \in T\}$ a filtration on it. Let $\{M_t; t \in T\}$ be a stochastic process adapted to $\mathbb{F}$ (we may, without loss of generality, consider the natural filtration of the process). When $T = \mathbb{R}^+$ it is a continuous time process, and when $T = \mathbb{Z}^+$ it is a discrete time process.

Definition B.33. If $\mathbb{E}[|M_t|] < \infty$ for all t, then we say that

- $\{M_t\}$ is a $\mathbb{F}$-*martingale* if $M_t = \mathbb{E}[M_{t+s} \mid \mathfrak{F}_t]$ a.s. for all $t, s \geq 0$.
- $\{M_t\}$ is a $\mathbb{F}$-*submartingale* if $M_t \leq \mathbb{E}[M_{t+s} \mid \mathfrak{F}_t]$ a.s. for all $s \geq 0$.
- $\{M_t\}$ is a $\mathbb{F}$-*supermartingale* if $M_t \geq \mathbb{E}[M_{t+s} \mid \mathfrak{F}_t]$ a.s. for all $s \geq 0$.

Vector-valued martingales are processes where each component is a martingale.

Definition B.34. Consider a martingale process $\{M_n\}$. The process $\{\delta M_n\}$, where $\delta M_n = M_{n+1} - M_n$ is called the *martingale difference process*.

Proposition B.1. *If* $\mathbb{E}[|M_n|^2] < \infty$ *then the martingale differences are uncorrelated:* $\mathbb{E}[\delta M_n \, \delta M_m] = 0$ *for all* $m \neq n$.

Let Y be a $\mathfrak{F}$-measurable random variable with $\mathbb{E}[|Y|] < \infty$. Then the random process $M_n = \mathbb{E}[Y \mid \mathfrak{F}_n]$ forms a martingale, because $\mathbb{E}[M_{n+1} \mid \mathfrak{F}_n] = \mathbb{E}[\mathbb{E}[Y \mid \mathfrak{F}_n]] = \mathbb{E}[Y \mid \mathfrak{F}_n] = M_n$. This is called *Doob's martingale* and it models the case where information about Y is "accumulated" as the number of observations grow.

Example: Let $\{Y_n\}$ be a sequence of random variables and construct the natural filtration with $\mathfrak{F}_n = \sigma(Y_1, \ldots, Y_n)$. Express

$$Y_n = Y_n - \mathbb{E}[Y_n \mid \mathfrak{F}_{n-1}] + \mathbb{E}[Y_n \mid \mathfrak{F}_{n-1}] = \delta M_n + \mathbb{E}[Y_n \mid \mathfrak{F}_{n-1}].$$

The "unpredictable" term in Y_n is a martingale difference with corresponding martingale process given by

$$M_n = \sum_{j=1}^{n}(Y_j - \mathbb{E}[Y_j \mid \mathfrak{F}_{j-1}]).$$

The following proposition is a corollary to Kolmogorov's inequality for submartingales.

Proposition B.2. *Let* M_n *be a square-integrable martingale with* $c = \mathbb{E}[M_0]$. *For any constant* $\Delta > 0$ *and any finite time* N

$$\mathbb{P}\left(\sup_{m \leq N} |M_m - c| \geq \Delta\right) \leq \frac{\mathrm{Var}(M_N)}{\Delta^2}.$$

Proposition B.3. *Let* $\{M_n\}$ *be a* $\mathbb{F}$-*martingale and* $q(\cdot)$ *a nondecreasing non-negative convex function (commonly, we use* $q(x) = |x|, x^2$, *or* $e^{ax}, a > 0$). *Then for all* $n < N$ *and* $\lambda > 0$

$$\mathbb{P}\left(\sup_{n \leq m \leq N} |M_m| \geq \lambda \,\Big|\, \mathfrak{F}_n \right) \leq \frac{\mathbb{E}[q(M_N) \mid \mathfrak{F}_n]}{q(\lambda)}, \tag{B.4}$$

$$\mathbb{E}\left[\sup_{n \leq m \leq N} |M_m|^2 \,\Big|\, \mathfrak{F}_n \right] \leq 4\,\mathbb{E}\left[|M_n|^2 \mid \mathfrak{F}_n\right], \tag{B.5}$$

$$\mathbb{P}\left(\sup_{n \leq m \leq N} M_m \geq \lambda \,\Big|\, \mathfrak{F}_n \right) \leq \frac{M_n}{\lambda}. \tag{B.6}$$

The first expression is a generalization of Proposition B.2. The third expression is a generalization of the Markov inequality; it is known as Kolmogorov submartingale inequality and it is also true if the process is a submartingale.

Definition B.35. Let $\{M_n\}$ be a $\mathbb{F}$-martingale and let τ be a $\mathbb{F}$-adapted stopping time. The *stopped martingale* $\{M_{\tau \wedge n}\}$ is defined by

$$M_{\tau \wedge n} = \begin{cases} M_n & \text{if } n < \tau, \\ M_\tau & \text{if } n \geq \tau. \end{cases}$$

Proposition B.4. *Let τ be a $\mathbb{F}$-adapted stopping time, then*

- *If $\{M_n\}$ is a martingale then $\{M_{\tau \wedge n}\}$ is a martingale.*
- *If $\{M_n\}$ is a submartingale then $\{M_{\tau \wedge n}\}$ is a submartingale.*
- *If $\{M_n\}$ is a supermartingale then $\{M_{\tau \wedge n}\}$ is a supermartingale.*

In general, the final value of a stopped process depends mostly on the stopping criterion, for example if a stochastic process X_t is stopped at $\tau = \min(t : X_t = 10)$ and we know that $\tau < N$ a.s. for some value of N, then $X_\tau = 10$. So, in general, it is not true that $\mathbb{E}[X_\tau] = \mathbb{E}[X_0]$ for a martingale process. The following result establishes the conditions under which this is true.

Theorem B.14 (Optional Sampling Theorem). *Let $\{M_n\}$ be a discrete-time martingale and τ a stopping time with values in $\mathbb{N} \cup \{\infty\}$, both with respect to the filtration $\mathbb{F}$. Assume that one of the following three conditions holds:*

(a) *The stopping time τ is almost surely bounded, i.e., there exists a constant $c \in \mathbb{N}$ such that $\tau \leq c$ a.s.*
(b) *The stopping time τ has finite expectation and the conditional expectations of the absolute value of the martingale increments are almost surely bounded, more precisely, $\mathbb{E}[\tau] < \infty$ and there exists a constant c such that $\mathbb{E}\big[|M_{t+1} - M_t| \big| \mathcal{F}_t\big] \leq c$ almost surely on the event $\{\tau > t\}$ for all $t \in \mathbb{N}$.*
(c) *There exists a constant c such that $|M_{\tau \wedge n}| \leq c$ a.s. for all $n \in \mathbb{N}$.*

Then M_τ is an almost surely well-defined random variable and $\mathbb{E}[M_\tau] = \mathbb{E}[M_0]$.

Similarly, if the stochastic process M is a submartingale or a supermartingale and one of the above conditions holds, then $\mathbb{E}[M_\tau] \geq \mathbb{E}[M_0]$, for a submartingale, and $\mathbb{E}[M_\tau] \leq \mathbb{E}[M_0]$, for a supermartingale.

Theorem B.15. *Let $\{M_n\}$ be a real-valued submartingale with $\sup_n \mathbb{E}[|M_n|] < \infty$. Then $\{M_n\}$ converges w.p.1 as $n \to \infty$. If $\{M_n\}$ is a real-valued supermartingale with $\mathbb{E}[\max(0, -M_n)] < \infty$ then supermartingale converges w.p.1.*

Theorem B.16. *A continuous time martingale $\{M_t\}$ whose paths are locally Lipschitz continuous w.p.1 on each bounded time interval is a constant w.p.1.*

In order to use any of the martingale convergence theorems, one must verify that the process is a martingale. The following is a useful characterization of a martingale.

Theorem B.17 (Martingale Characterization). *Consider a process $\{\vartheta(t); t \geq 0\}$ on $(\Omega, \mathfrak{F}, \mathbb{P})$ adapted to a filtration $\mathbb{F} = \{\mathfrak{F}_t\}$. Let $\{M(t)\}$ be another $\mathbb{F}$-adapted process and suppose that for each $t, r \geq 0$ and for each $p \in \mathbb{N}$, $s_i \leq t$, $i = 1, \ldots, p$ and any bounded and continuous real valued function $h_p : \mathbb{R}^p \mapsto \mathbb{R}$ the following holds:*

$$\mathbb{E}\Big[h_p(\vartheta(s_i); i = 1, \ldots, p)(M(t+r) - M(t))\Big] = 0.$$

Then $\{M(t)\}$ is a martingale.

APPENDIX B

B.7 REGENERATIVE PROCESSES

Let $\{X_n\}$ denote a stochastic process on $(\Omega, \mathfrak{F}, \mathbb{P})$ with state space S and natural filtration $\{\mathfrak{F}_n\}$. The process $\{X_n\}$ is called *classical regenerative* or *regenerative* if there exists a sequence of stopping times $\{\tau_k : k \geq 0\}$, with $\tau_0 = 0$, also called *regeneration times* such that

- $\{\tau_{k+1} - \tau_k, k \geq 0\}$ is an iid sequence;
- the processes $\{X_n : n \geq 0\}$ and $\{X_{\tau_k+n} : n \geq 0\}$ have the same distribution for all $k \geq 1$, and $\{X_n : n \geq \tau_k\}$ is independent of $\mathfrak{F}_{\tau_k}$, for all $k \geq 1$.

Thus, in a regenerative process, the regeneration epochs $\{\tau_k : k \geq 0\}$ cut the process into iid *cycles* of the form $\{X_n : \tau_k \leq n < \tau_{k+1}\}$. The length of cycles $T_k \stackrel{\text{def}}{=} \tau_k - \tau_{k-1}, k \geq 1$ are called the *renewal times* and form a renewal process.

Definition B.36. A distribution function is called *lattice* if it assigns probability one to a set of the form $\{0, \delta, 2\delta, \ldots\}$, for some $\delta > 0$, and it is called *non-lattice* otherwise.

Theorem B.18 (see [288]). *Let $\{X(n)\}$ be a regenerative process such that the distribution of $\tau_{k+1} - \tau_k$ is non-lattice. Consider a measurable mapping $f : S \mapsto \mathbb{R}$. If*

$$\mathbb{E}[\tau_1] < \infty \quad \text{and} \quad \mathbb{E}\left[\sum_{n=0}^{\tau_1-1} |f(X_n)|\right] < \infty,$$

then

$$\lim_{N \to \infty} \frac{1}{N} \sum_{n=1}^{N} f(X_n) = \lim_{N \to \infty} \frac{1}{N} \sum_{n=1}^{N} \mathbb{E}[f(X_n)] = \frac{\mathbb{E}\left[\sum_{n=0}^{\tau_1-1} f(X_n)\right]}{\mathbb{E}[\tau_1]} \quad a.s.$$

Moreover, X_n converges weakly as well as in total variaition to some random variable X_∞, where

$$\mathbb{P}(X_\infty \in B) = \frac{\mathbb{E}\left[\sum_{n=0}^{\tau_1-1} \mathbf{1}_{\{X_n \in B\}}\right]}{\mathbb{E}[\tau_1]},$$

for all measurable sets B.

Appendix C

Markov Chains

C.1 HARRIS RECURRENCE

Let $(S, \mathcal{S})$ denote a Polish state space, where $\mathcal{S}$ denotes the Borel field of S. The mapping $P: S \times \mathcal{S} \mapsto [0, 1]$ is a *Markov kernel* (on $(S, \mathcal{S})$) if

(a) $P(s; \cdot)$ is a probability measure on $(S, \mathcal{S})$, for all $s \in S$; and
(b) $P(\cdot; B)$ is $\mathcal{S}$ measurable for all $B \in \mathcal{S}$.

The product of Markov kernels is again a Markov kernel. Specifically, let P, Q be two Markov kernels on $(S, \mathcal{S})$, then the product of P and Q is defined as follows: for $s \in S$ and $B \in \mathcal{S}$ set $QP(s; B) = (Q \circ P)(s, B) = \int_S Q(z; B) P(s; dz)$. Moreover, write $P^n(s; \cdot)$ for the measure obtained by the n fold product of P in the above way.

When an initial distribution μ is given, P defines a Markov chain $\{X(n)\}$ with state space $(S, \mathcal{S})$:

$$\mathbb{P}(X(n) \in B) = \int \mu(ds) P^n(s; B),$$

where $\mathbb{P}$ denotes the underlying probability measure on $(S, \mathcal{S})$.

Let ϕ be a σ-finite measure on $(S, \mathcal{S})$. A Markov chain with transition kernel $P(x; B)$ is ϕ-*irreducible* if

$$\mathbb{P}\left(\bigcup_{n=1}^{\infty} X(n) \in B \,\middle|\, X(0) = x\right) > 0, \quad x \in S, B \in \mathcal{S},$$

whenever $\phi(B) > 0$.

A Markov chain $\{X(n)\}$ is called *uniformly ϕ-recurrent* if there exists a nontrivial measure ϕ on $(S, \mathcal{S})$ such that for each $A \in \mathcal{S}$, with $\phi(A) > 0$,

$$\lim_{k \to \infty} \sum_{m=1}^{k} {}_A P^m(x, A) = 1$$

uniformly in x, where ${}_A P^m(x, A)$ is the *taboo probability* defined by

$${}_A P^m(x, A) = \mathbb{P}(\{X(m) \in A, X(0) = x, X_i \notin A, 1 \leq i \leq m-1\}), \quad A \in \mathcal{S}.$$

A uniformly ϕ-recurrent Markov chain is also called *Harris recurrent*.

A *d-cycle* of a ϕ-irreducible chain $\{X(n)\}$ is a collection $\{S_1, \ldots, S_d\}$ of disjoint subsets of S such that $\phi(S : \cup_{i=1}^d S_i) = 0$ and $P(s; S_{i+1}) = 1$ for $s \in S_i$ and $1 \leq i \leq d$ (take $S_{i+1} = S_1$ when $i = d$). At least one d-cycle exists for a ϕ-irreducible chain and the *period* of the chain is the smallest d for which a d-cycle exists. The chain is called *aperiodic* if it is ϕ-irreducible and $d = 1$; otherwise it is called periodic. Observe that aperiodicity of a chain

APPENDIX C

already implies its ϕ-irreducibility. A uniformly ϕ-recurrent and aperiodic Markov chain is also called *Harris ergodic*.

Theorem C.1. *A uniformly ϕ-recurrent and aperiodic (resp. Harris ergodic) Markov chain converges, for any initial distribution, weakly toward a unique stationary regime π. Moreover, for any measurable mapping $f : S \mapsto \mathbb{R}$, with $\int_S f(s)\,\pi(ds)$ finite, it holds that*

$$\lim_{N \to \infty} \frac{1}{N} \sum_{n=1}^{N} f(X(n)) = \int_S f(s)\pi(ds) \quad \text{a.s.}$$

Let $\{X(n)\}$ be a Harris ergodic Markov chain. For $B \in S$, let τ_k denote the k-th hitting time of $X(n)$ on B, where we set $\tau_k = \infty$ if $X(n)$ doesn't visit B for at least k times. Hence, τ_k is a stopping time. A set $B \in S$ is called a *regeneration set* if, with probability one, $\tau_k < \infty$, for any $k \in \mathbb{N}$, and with probability one:

$$\lim_{k \to \infty} \tau_k = \infty.$$

A regeneration set B is called *atom* if the regeneration points $\{\tau_k: k \geq 0\}$ cut the Markov chain into iid cycles of the form $\{X(n): \tau_k \leq n < \tau_{k+1}\}$. Thus, whenever $X(n)$ hits B it starts independent from the past. In particular, if we consider two versions of $X(n)$, where one version is started according to an initial distribution μ and the other according to an initial distribution ν, then both versions couple when they simultaneously hit B, which occurs after a.s. finitely many transitions.

We say that there is *strong coupling convergence in finite time* (or, merely strong coupling) of a sequence $\{X_n\}$ to a stationary sequence $\{Y_n\}$ if both sequences can be constructed on a joint probability space such that

$$N^0 = \inf\{n \geq 0 \mid \forall k \geq 0: X_{n+k} = Y_{n+k}\}$$

is finite w.p.1.

Theorem C.2. *A Harris ergodic Markov chain $\{X(n)\}$ with atom converges, for any initial distribution, in strong coupling to its unique stationary regime. In addition to that, let B denote an atom of $\{X(n)\}$, then it holds that*

$$\int_S f(s)\,\pi(ds) = \frac{\mathbb{E}\left[\sum_{n=0}^{\tau_1 - 1} f(X(n)) \Big| X(0) \in B\right]}{\mathbb{E}[\tau_1 \mid X(0) \in B]},$$

for any measurable mapping $f: S \mapsto \mathbb{R}$ such that $\int_S f(s)\,\pi(ds)$ is finite, where τ_1 denotes the first hitting time of $X(n)$ on B.

C.2 NORMED ERGODICITY AND CENTRAL LIMIT THEOREM

We extend the v-norm in Section B.5 to Markov kernels on a state space S by

$$\|P\|_v = \sup_{s \in S} \sup_{\|f\|_v \leq 1} \frac{\left| \int f(z)\, P(s; dz) \right|}{|v(s)|}.$$

Definition C.1. A Markov kernel P is called v-*norm geometrically ergodic* if there exists a finite constant c and $\rho \in (0, 1)$ such that for all n

$$\|P^n - \Pi\|_v \leq c\rho^n,$$

where

$$\Pi = \lim_{n \to \infty} \frac{1}{n} \sum_{i=1}^{n} P^n$$

denotes the *ergodic projector* of P.

If P is aperiodic and irreducible with countable state space, Π is a matrix the rows of which are given by the stationary distribution of P.

For geometrically ergodic Markov chains, central limit theorems can be established. For an overview of variations of the basic results we refer to [157, 174]. We state here Theorem 9 from [174] for illustration. Let $\mathbf{X} = \{X(n)\} = \{X(s,n)\}$ be the Markov chain with initial state s and transition kernel P. The state space of $\mathbf{X}$ is denoted by $(S, \mathcal{S})$.

Theorem C.3. *Let $\mathbf{X}$ be uniformly ϕ-recurrent and aperiodic (resp. Harris ergodic) Markov chain, and denote the unique stationary regime of $\mathbf{X}$ by π. If P is v-norm geometrically ergodic and $\mathbb{E}_\pi[|f(X)|^{2+\delta}] < \infty$ for some $\delta > 0$ (with X distributed according to π), then for any initial distribution*

$$\sqrt{n}\left(\frac{1}{n}\sum_{i=1}^{n} f(X_n) - \mathbb{E}_\pi[f(X)]\right) \stackrel{d}{\Longrightarrow} \mathcal{N}(0, \sigma^f),$$

for some finite constant σ^f.

Note that a sufficient condition for $\mathbb{E}_\pi[|f(X)|^{2+\delta}] < \infty$ to hold for a v-norm geometrically ergodic Markov chains is that $\|f^{2+\delta}\|_v < \infty$. Next we provide sufficient conditions for v-norm geometric ergodicity.

(C1) There exists a function $g(s) \geq 0$, $s \in S$, such that

$$\mathbb{E}\big[g(X(s,m))\big] - g(s) \leq -\varepsilon + c\,I_V(s), \qquad (C.1)$$

for some $m \geq 1$, $\varepsilon > 0$ and $c < \infty$, where for some $d < \infty$

$$V = \{s \in S : g(s) \leq d\}.$$

(C2) There exist $n \geq 0$, $\phi(\cdot)$ a probability measure on $(S, \mathcal{S})$, and $p \in (0,1)$ such that

$$\inf_{x \in V} P\big(X(x,n) \in B\big) \geq p\,\phi(B),$$

for all $B \in \mathcal{S}$.

Assume that (C1) holds and let

$$\xi(s) \triangleq g(X(s,1)) - g(s), \quad s \in S,$$

and introduce the following condition:

APPENDIX C

(C3) The r.v. $\xi(s)$ is uniformly integrable in s, and there exist $\lambda > 0$ such that $\xi(s)\, e^{\lambda \xi(s)}$ are uniformly integrable.

Recall that uniform integrability of $\xi(s)$ in s is defined as

$$\lim_{c \to \infty} \sup_s \int_{|t|>c} P(\xi(s) \in dt) = 0,$$

and similarly the uniform integrability of $\xi(s) e^{\lambda \xi(s)}$ requires that

$$\lim_{c \to \infty} \sup_s \int_{|t|>c} P(\xi(s) e^{\lambda \xi(s)} \in dt) = 0.$$

The following result shows that conditions (C1)–(C3) imply that convergence of the Markov chain toward its stationary distribution happens at a geometrical rate for an appropriate choice of v.

Theorem C.4 ([40, 149]). *Provided that conditions (C1), (C2), and (C3) hold, then for λ small enough in the definition of v in Defintion C.1 it holds that P is geometrically v-norm ergodic, that is, there exist $c < \infty$ and $0 < \rho < 1$*

$$\|P^n - \Pi\|_v \le c\,\rho^n. \tag{C.2}$$

Let $\mathbf{X}(\theta) = \{X_\theta(n)\} = \{X_\theta(s,n)\}$, for $\theta \in \Theta$, be the Markov chain with initial state s and transition kernel P_θ. The joint state space of $\mathbf{X}(\theta)$, $\theta \in \Theta$, is denoted by S.

Lemma C.1 (Lemma 4.1 in [149]). *Let conditions (C1) and (C3) hold at θ for $\theta \in \Theta$. If the function g is the same for each θ in (C1) and the uniform convergence in (C3) can be extended to $\xi_\theta(s)$ and $\xi_\theta(s)\, e^{\lambda \xi_\theta(s)}$ for $\lambda > 0$ sufficiently small and independent of θ, then*

$$\sup_{\hat\theta \in \Theta} \|P_{\hat\theta}\|_v < \infty.$$

C.3 THE POISSON EQUATION FOR MARKOV CHAINS IN DISCRETE TIME

We present the basic results for homogenous Markov chains on a countable state space, which allows as to work in the following with matrix algebra. Using an operator approach and appropriate norm arguments (c.f. Chapter 13), the results are extendable to P living on a general state space.

A Markov chain on discrete state space S is called *irreducible* if for each state pair $i, j \in S$, there exists an n such that $P^n(i, j) > 0$, and it is called *reducible* otherwise. Moreover, P is a called *aperiodic* if for each state pair $i, j \in S$ the greatest common divisor of the set $\{n: P^n(i,j) > 0\}$ is one. Denoting by $\tau(i,j)$ the expected number of transitions it takes to reach j from i (for the first time), a set of states E is called a positive recurrent class if $\tau(i,j) < \infty$ for each state pair $i, j \in E$. An irreducible, aperiodic, and positive recurrent Markov chain is called ergodic. A Markov chain is called positive recurrent if S is a union of positive recurrent classes. Any positive recurrent P has unique ergodic projector Π, that is, $P\Pi = \Pi = \Pi P$, where

$$\lim_{n \to \infty} \frac{1}{n} \sum_{k=1}^n P^n = \Pi.$$

Note that as any finite Markov chain is positive recurrent, Π exists for any finite Markov chain. In case that P is reducible, P is called a multi chain and Π is of block diagonal structure (the diagonal blocks correspond to the positive recurrent classes and each block has identical rows representing the unique stationary distribution of P restricted to that particular ergodic class), whereas an irreducible P is called a unichain the ergodic projector of which is a matrix with all rows equal to π (which also implies that π is the unique stationary distribution as well as the unique limiting distribution of P). Moreover, a positive recurrent unichain visits all states infinetly often. For any cost vector h defined on the state space of P, simple algebra shows that

$$P \sum_{k=0}^{n-1} P^k h + h = \sum_{k=0}^{n} P^k h = P^n h + \sum_{k=0}^{n-1} P^k h. \tag{C.3}$$

We now subtract on both sides of (C.3) n times the vector Πh, and obtain

$$P \sum_{k=0}^{n-1} (P^k - \Pi) h + h = P^n h + \sum_{k=0}^{n-1} (P^k - \Pi) h.$$

Assuming that P is aperiodic and letting n tend to infinity, we arrive at

$$PDh + h = \Pi h + Dh, \tag{C.4}$$

where

$$D = \sum_{k=0}^{\infty} (P^k - \Pi)$$

is called the *deviation matrix*, and it is easily seen that

$$PD = \sum_{k=0}^{\infty} P(P^k - \Pi) = \sum_{k=0}^{\infty} (P^k - \Pi) P = DP. \tag{C.5}$$

The deviation matrix can alternatively be written as

$$D = \sum_{k=0}^{\infty} (P^k - \Pi) = (I - P + \Pi)^{-1} - \Pi, \tag{C.6}$$

provided that it exists. The matrix $(I - P + \Pi)^{-1}$ is called the *fundamental matrix* (potential) of P. Letting $A^{\#}$ denote the group inverse of the matrix $A = I - P$, see [7, 215], it holds that $D = A^{\#}$ if the deviation matrix exists. Conditions for existence of the deviation matrix and its related properties have been extensively studied in the literature (see [81, 300]) and these conditions are related to geometric ergodicity; see Definition C.1. For finite Markov chains, the deviation matrix is an instance of the generalized inverse of $I - P$; see [215] for an early reference. To summarize, we found vectors that $\phi = \Pi h$ and $v = Dh$ solve

$$\phi = P\phi$$

$$\phi + v = h + Pv,$$

which is called the *Poisson equation for Markov chains in discrete time*. In Markov decision theory v is called the *value function*.

APPENDIX C

The solution to the Poisson equation relates an observation of a cost vector h to the stationary expected value. To see this, let $\{\xi_n\}$ denote the Markov chain with transition probability matrix P. Then,

$$\sum_{k=0}^{n} \mathbb{E}[h(\xi_k)|\xi_0 = x] = \sum_{k=0}^{n} (P^k h)(x).$$

Rearranging (C.4), we obtain

$$(I - P)Dh = h - \Pi h, \tag{C.7}$$

and we now can apply (C.7) to find an expression for the difference between $h(\xi_n)$ and its long run average πh as

$$h(\xi_n) - \pi h = ((I - P)Dh)(\xi_n) = v(\xi_n) - \mathbb{E}[v(\xi_{n+1})|\xi_n],$$

with v denoting the value function as above. Hence, the above right-hand side yields an expression for the bias introduced when considering $h(\xi_n)$ as proxy for πh.

Appendix D

Confidence Intervals

This appendix provides some useful reviews of statistical estimation, mostly applied to simulation experiments. Much of the work has been developed in detail in [3].

D.1 INDEPENDENT AND IDENTICALLY DISTRIBUTED RANDOM VARIABLES

Let X be a random variable on $(\Omega, \mathfrak{F}, \mathbb{P})$ with $\mathbb{E}[X] = \mu$ and $\text{Var}(X) = \sigma^2 < \infty$. Suppose that a collection $(X_1, X_2, \ldots)$ of iid random variables is available, all with the same distribution as X. We will consider $\mathfrak{F}_n = \sigma(X_1, \ldots, X_n)$ to be the natural filtration, and the estimators $\hat{\mu}_n$ of μ to be functions of the "samples," that is, $\hat{\mu}_n = \phi(X_1, \ldots, X_n)$, where ϕ is $\mathfrak{F}_n$-measurable.

Definition D.1. As estimator $\hat{\mu}_n$ of μ is called an *unbiased* estimator if $\mathbb{E}[\hat{\mu}_n] = \mu$. Otherwise, the difference $\hat{\mu}_n - \mu$ is called the *bias* of the estimator.

Definition D.2. A sequence of random variables on a common probability space $(\Omega, \mathfrak{F}, \mathbb{P})$ with common expectation $\mathbb{E}[X_i] = \mu$ is said to satisfy

- the *weak law of large numbers* if the sample mean converges in probability, that is for all $\epsilon > 0$

$$\lim_{n \to \infty} \mathbb{P}\left(\left|\frac{1}{n}\sum_{i=1}^n X_i - \mu\right| > \epsilon\right) = 0,$$

- the *strong law of large numbers* (SLLN) if the sample average converges w.p.1, that is

$$\mathbb{P}\left(\omega : \frac{1}{n}\sum_{i=1}^n X_i(\omega) = \mu\right) = 1.$$

Theorem D.1 (Central Limit Theorem). *Let $(X_1, X_2, \ldots)$ be iid with finite mean μ and variance σ^2. Then the sample average*

$$\hat{\mu}_n \stackrel{\text{def}}{=} \frac{1}{n}\sum_{i=1}^n X_i$$

satisfies

$$\frac{\sqrt{n}(\hat{\mu}_n - \mu)}{\sigma} \stackrel{d}{\Longrightarrow} \mathcal{N}(0, 1).$$

APPENDIX D

Definition D.3. A *confidence interval* (CI) at significance level α is an interval $I_\alpha(\hat{\mu}_n)$, which is $\mathfrak{F}_n$-measurable satisfying

$$\mathbb{P}(\mu \in I_\alpha(\hat{\mu}_n)) = 1 - \alpha.$$

Example D.1. For iid samples with finite variance, the *sample average* is an unbiased estimator of μ:

$$\hat{\mu}_n = \frac{1}{n} \sum_{k=1}^{n} X_k, \qquad (D.1)$$

and $\mathrm{Var}(\hat{\mu}_n) = \sigma^2/n$. In addition, the *sample variance*

$$S_n^2 = \frac{1}{n-1} \sum_{k=1}^{n} (X_k - \hat{\mu}_n)^2 \qquad (D.2)$$

is an unbiased estimator of the variance $\sigma^2 = \mathbb{E}[S_n^2]$.

(a) Suppose that $X \sim \mathcal{N}(\mu, \sigma^2)$ has a normal distribution. Then it is possible to build an exact CI. As is customary, we will write z_q to denote the q-th quantile of the standard normal distribution. Then the symmetric CI

$$I_\alpha(\hat{\mu}_n) = \hat{\mu}_n \pm z_{1-\alpha/2} \sqrt{\frac{\sigma^2}{n}},$$

satisfies by construction that $\mathbb{P}(\mu \in I_\alpha(\hat{\mu}_n)) = 1 - \alpha$. CIs are not unique and other criteria may be suitable for specific problems. For instance when using estimation for ranking and selection, or in multi-armed bandit problems, upper confidence intervals may be preferred, seeking $\mathbb{P}(\mu \leq \hat{\mu}_n - z_\alpha \sqrt{\sigma^2/n}) = \alpha$.

(b) Now suppose that σ^2 is not known. Then we can use Slutski's theorem to ascertain that

$$\frac{\sqrt{n}\,(\hat{\mu}_n - \mu)}{\sqrt{S_n^2}} \sim t_{(n-1)},$$

where t_d is the student-t distribution with d degrees of freedom. Using the standard notation $t_{d,q}$ for the q-th quantile of this distribution, it then follows that

$$I_\alpha(\hat{\mu}_n) = \hat{\mu}_n \pm t_{(n-1), 1-\alpha/2} \sqrt{\frac{S_n^2}{n}}$$

satisfies $\mathbb{P}(\mu \in I_\alpha(\hat{\mu}_n)) = 1 - \alpha$ and it is thus a symmetric CI at level α. Figure D.1 shows the pdf of the normal distribution on top in black, together with the densities of the student-t distribution for n degrees of freedom. As n grows, it converges to the normal density. Once $n > 10$ there is little difference between the quantiles of the student-t distribution and those of the normal distribution.

(c) Finally, suppose that X has a distribution that is not necessarily normal, with mean $\mathbb{E}[X] = \mu$ and finite (unknown) variance σ^2. This is the most common situation in practice. Under the assumptions, the CLT ascertains that

$$\frac{\sqrt{n}(\hat{\mu}_n - \mu)}{\sqrt{S_n^2}} \xrightarrow{d} \mathcal{N}(0, 1), \quad \text{as } n \to \infty.$$

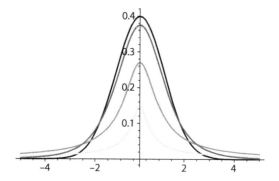

Figure D.1. Student-t density for n degrees of freedom, $n = 0.1, 0.5, 4$ and for the normal density.

In this case, we may build an *approximate* CI based on the asymptotic behavior, provided that n is "large enough." In this case, this approximate CI is

$$I_\alpha(\hat{\mu}_n) = \hat{\mu}_n \pm z_{1-\alpha/2}\sqrt{\frac{S_n^2}{n}}, \quad \text{for large } n \qquad (D.3)$$

※※※

This procedure extends to the multidimensional case as follows. Let $X \sim \mathcal{N}(\mu, V)$, where V is the variance-covariance matrix in d dimensions of θ_n^ϵ for n large and ϵ small. Let $Q(\alpha, \chi_d^2)$ be the α-quantile of the χ^2-distribution with d degrees of freedom. We know that

$$(X - \mu)^\top V^{-1}(X - \mu) \sim \chi_d^2.$$

Given an observation X, define the ellipse

$$C = \{x \in \mathbb{R}^d : (X - x)^\top V^{-1}(X - x) = Q(\alpha, \chi_d^2)\}.$$

Then, by construction, $\mathbb{P}(\mu \in \bar{C}) = 1 - \alpha$, where $\bar{C}$ contains the interior of the ellipse.

The easiest way to construct this confidence region is as follows. Call $\rho_i, \mathbf{e}_i$ the eigenvalues and eigenvectors of V, for $i = 1, \ldots, d$. Then the axes of the ellipse are in the directions $\mathbf{e}_i$, and each of the diameters are given by $\pm \mathbf{e}_i \sqrt{\epsilon Q(\alpha, \chi_d^2) \lambda_i}$.

Figure D.2 illustrates the resulting confidence interval centered on the estimate θ_n^ϵ. The width along each of the ellipse's axes is $\sqrt{\epsilon Q(\alpha, \chi_d^2) \lambda_i}$. Because this is an approximation, the actual value of θ^* is shown together with the true region, indicated by the dashed line, for which the probability is exactly $1 - \alpha$.

Definition D.4. Given an *approximate* CI (based, for example in asymptotic normality) of the form $I_\alpha(\hat{\mu}_n)$ at significance level α, the *coverage* of the interval, denoted by $p_\alpha(I_\alpha(\hat{\mu}_n))$, is the (true) probability that it contains the parameter μ, that is,

$$p_\alpha(I_\alpha(\hat{\mu}_n)) = \mathbb{P}(\mu \in I_\alpha(\hat{\mu}_n)).$$

Ideally we wish to have $p_\alpha(I) \approx 1 - \alpha$, but in many cases this requires prohibitively large sample size n. The actual coverage of an interval is rarely easy to calculate; but the fact that $1 - \alpha$ is an approximate level should be kept in mind when designing and reporting your simulation or sampling experiments.

APPENDIX D

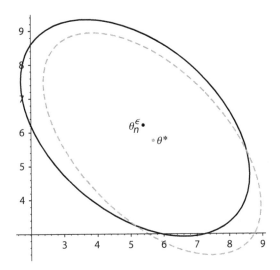

Figure D.2. Example of a confidence region centered on $\theta_n^\epsilon \in \mathbb{R}^2$. The dashed line is the true confidence region centered around θ^*.

When using simulation or sampling data, it is important to report the estimated values $\hat{\mu}_n$ as well as a CI that provides insight about the error in the approximation. Three scenarios are common in practice: (a) fixed sample size n, (b) fixed interval width (called the *error* in the approximation), and (b) fixed relative error.

Fixed Sample Size. When the sample size (or computational budget) is fixed, then common practice is to apply a CLT result and use a normal approximation to estimate $I_\alpha(\hat{\mu}_n)$ via (D.3).

Fixed Interval Width. In some applications it is required to provide estimates within a certain "precision." That is, an *error tolerance* ρ is specified in advance, and the sample size n must be chosen so that the interval is of the form $I_\alpha(\hat{\mu}_n) = \hat{\mu}_n \pm \rho$. Because the same size is not fixed at the outset, it becomes a random stopping time and estimation is performed sequentially, as we describe below (for details, see [3]).

Let $\rho > 0$ be a given tolerance error. With initial values $\hat{\mu}_1 = X_1$ and $S_1^2 = 1$, a recursive algorithm to evaluate the sample average (D.1) and sample variance (D.2) is

$$\hat{\mu}_{n+1} = \hat{\mu}_n + \frac{1}{n}(X_{n+1} - \hat{\mu}_n),$$

$$S_{n+1}^2 = \left(1 - \frac{1}{n}\right) S_n^2 + (n+1)(\hat{\mu}_{n+1} - \hat{\mu}_n)^2.$$

Define τ as the smallest sample size that achieves the required accuracy, that is,

$$\tau = \min\left(n \geq 2 : z_{1-\alpha/2} \sqrt{\frac{S_n^2}{n}} \leq \rho\right),$$

then, by construction, τ is a random stopping time adapted to the natural filtration $\{\mathfrak{F}_n\}$ of the observation process $\{X_n\}$. Even if $\mathbb{E}[S_n^2] = \sigma^2$, there is no guarantee that $\sqrt{\tau}((\hat{\mu}_\tau - \mu)/S_\tau^2$ is anywhere close to $\mathcal{N}(0, 1)$ in distribution, because τ is random. The following result provides a justification for the algorithm commonly used in practice (see [3]).

Theorem D.2. *Let $\{X_n\}$ be a sequence of iid random variables with $\mathbb{E}[X_i] = \mu$ and $\mathrm{Var}(X_1) = \sigma^2 < \infty$. Given a tolerance error $\rho > 0$, define the random stopping time*

$$\tau = \min\left(n \geq 2 : \delta(n,\alpha) \leq \sqrt{\frac{n-1}{n}\rho^2 - \frac{t_{n-1,1-\alpha/2}}{n(n-1)}}\right); \quad \delta(n,\alpha) \stackrel{\text{def}}{=} t_{n-1,1-\alpha/2}\sqrt{\frac{S_n^2}{n}}.$$

Then $\lim_{\rho \to 0} \mathbb{P}(\mu \in (\hat{\mu}_n \pm \rho)) = 1 - \alpha$.

The above result justifies the approximation of the CI (D.3) with τ samples, because when ρ is very small, then τ is likely to be "very large." As $n \to \infty$ the student-t quantile converges to that of the normal distribution: $t_{n-1,q} \to z_q$, and the bound $\sqrt{\frac{n-1}{n}\rho^2 - \frac{t_{n-1,1-\alpha/2}}{n(n-1)}} \to \rho$. Good practice would impose a stopping time of the form $\max(\tau, N)$, for N a fixed, "large" value. What is large enough often depends on the particular problem, because the variance of the estimation determines the precision of the estimator. Thus, in practice it may be simpler to do the estimation in two-stages: the first one to estimate σ^2 and the second to estimate how large τ should be.

In the two-stage method one starts with $n_0 \gg 2$ (typically ≈ 100) and use $S_{n_0}^2$ as an estimate of the variance. Then use the recursive algorithm to calculate $\hat{\mu}_n, S_n^2$ to calculate

$$\tau_1 = \min\left(n > n_0 : z_{1-\alpha/2}\sqrt{\frac{S_{n_0}^2}{n}} \leq \rho\right).$$

This provides an estimate of how many more samples would be required to achieve the desired precision, thus the remaining $n - n_0$ samples are used in the second stage for the final estimation.

Fixed Relative Error. A fixed relative error is required when the desired answer is of the form $\hat{\mu}_n \pm \rho\mu$, for $\rho \in (0,1)$. For example, if the quantity to be estimated may itself be very small, a relative error could require the error in the estimation to be no more than 1% (at confidence level α). In this case $\rho = 0.01$ and the estimation procedure from the fixed width interval must be adapted. The details are in [3].

Consider a process $\{\xi_n; n \geq 0\}$ and let τ be a (possibly random) stopping time. Suppose that we wish to estimate

$$\mu = \mathbb{E}[\phi(\xi_1, \ldots, \xi_\tau)],$$

for a well-defined functional of the process ϕ. We assume here that the process can be simulated, or that several *replications* of the process can be observed. Assuming statistical independence between replications, we can build a sequence of iid estimators. Indeed, let $X_k = \phi(\xi_1^{(k)}, \ldots, \xi_{\tau_k}^{(k)})$ be the k-th observation of the functional ϕ, then all methods described above are applicable to estimate CIs, provided that the variance is bounded: $\mathrm{Var}(\phi(\xi_1, \ldots, \xi_\tau)) < \infty$.

D.2   STATIONARY PROCESSES

In this section we focus on the methods for estimating CIs when the observations $\{\xi_n\}$ are correlated. We start with the simpler case that the process is *stationary*. Next we will address steady state estimation, where the process is Markovian and only approaches stationarity in the long-term (asymptotically).

Definition D.5. A process $\{\xi_n, n \geq 0\}$ on $(\Omega, \mathfrak{F}, \mathbb{P})$ is called

- *weakly stationary* if $\mathbb{E}[\xi_n] = \mu$ and $\text{Var}(\xi_n) = \sigma^2$ are constant, and the auto covariance function satisfies
$$\text{Cov}(\xi_n, \xi_{n+m}) = C(m), \forall n, m,$$

- *strongly stationary* if the joint distribution of the shifted process is homogeneous in time, specifically if for any $k, n \in \mathbb{N}$ and any increasing sequence $m_1, \ldots, m_k \in \mathbb{N}$
$$(\xi_{n+m_1}, \xi_{n+m_2}, \ldots, \xi_{n+m_k}) \stackrel{d}{=} (\xi_{m_1}, \xi_{m_2}, \ldots, \xi_{m_k}).$$

If a process is strongly stationary then it is also weakly stationary and we often refer to this case simply as "stationary." Let $\hat{\mu}_n$ be the sample average

$$\hat{\mu}_n = \frac{1}{n} \sum_{k=1}^{n} \xi_k.$$

Then it follows that $\mathbb{E}[\hat{\mu}_n] = \mu$ and

$$V_n \stackrel{\text{def}}{=} \text{Var}(\hat{\mu}_n) = \frac{\sigma^2}{n}(1 + \gamma_n), \quad \gamma_n = 2 \sum_{k=1}^{n-1} \left(1 - \frac{k}{n}\right) \frac{C(k)}{\sigma^2}. \tag{D.4}$$

Notice that in order for the SLLN to hold it is necessary that $\lim_{n \to \infty} \gamma_n < \infty$, and then we can ascertain that $\hat{\mu}_n \to \mu$ as $n \to \infty$ a.s. For many processes it possible to establish a CLT (or a FCLT in case that we work with interpolated processes), for example

$$\sqrt{n}(\hat{\mu}_n - \mu) \stackrel{d}{\Longrightarrow} \mathcal{N}(0, \sigma_\infty^2),$$

for some limiting variance σ_∞^2. Sometimes it is only possible to establish that

$$\frac{\sqrt{n}(\hat{\mu}_n - \mu)}{\sqrt{V_n}} \stackrel{d}{\Longrightarrow} \mathcal{N}(0, 1),$$

or some other limit distribution. However, the main problem here is that the sample variance (D.2) is biased for V_n in (D.4). There are two approaches for the estimation of the asymptotic variance that we discuss next.

Independent Replications. A first approach would be to evaluate *independent* samples of $\hat{\mu}_n$ and use the CI in (D.3). This is typically only feasible when we generate the process via simulations, in which case we perform independent *replications* of the simulation of the process, each of length b. Call $\{\xi_k^{(i)}; k = 1, \ldots, b\}$ the i-th replication of the process and define

$$X_{i,b} = \frac{1}{b} \sum_{k=1}^{b} \xi_k^{(i)}.$$

By construction, $X_i \stackrel{d}{=} \hat{\mu}_b$, so for "large" b its variance V_b should be "close" to σ_∞^2. This leads to the common method for estimation:

$$\hat{\mu}_{N,b} = \frac{1}{N} \sum_{i=1}^{N} X_{i,b} \tag{D.5}$$

$$S_{N,b}^2 = \frac{1}{N-1} \sum_{i=1}^{N} (X_{i,b} - \hat{\mu}_{N,b})^2. \tag{D.6}$$

Batch Means Method. The second approach does not require restarting the simulation, but rather using a long simulation of length bN and dividing the estimation by *batches* of size b each. Although not independent, when b is large and the covariance process decrease fast, the consecutive sample means are approximately independent. Mathematically, we consider the process $\{\xi_n; n \geq 0\}$ (not necessarily from simulation, but it could be streaming observations from sensor data). Define now

$$X_{i,b} = \frac{1}{b} \sum_{k=bi+1}^{b(i+1)} \xi_k,$$

and then use (D.5) and (D.6) to estime the corresponding CI. We will now provide rigorous results that justify the estimation by batches; for details we refer to [3].

Theorem D.3. *Let $\{\xi_n, n \geq 0\}$ be a stationary process such that $\mathbb{E}[\xi_n] = \mu$ and suppose that there is a finite constant $\sigma_\infty^2 < \infty$ such that the following CLT holds*

$$\frac{\sqrt{n}(\hat{\mu}_n - \mu)}{\sigma_\infty^2} \overset{d}{\Longrightarrow} \mathcal{N}(0, 1).$$

Let the number of batches N be constant but use increasing batch sizes $b(n) \to \infty$. Then

$$\frac{\sqrt{n}(\hat{\mu}_{N,b(n)} - \mu)}{\sqrt{S_{N,b(n)}^2}} \overset{d}{\Longrightarrow} t_{N-1}.$$

The theorem above justifies estimation by batches, using "large" batch size b. Other results can be shown (see [3] for details) that justify the "square root rule" for batches: given a total sample size n, let $N = \lfloor \sqrt{n} \rfloor$, so that $b \approx \sqrt{n}$ as well.

D.3 MARKOV CHAINS: LONG-TERM AND STATIONARY ESTIMATION

Let $\{\xi_n\}$ be a positive recurrent irreducible Markov chain on a countable state space $\xi_n \in S$ with a unique stationary measure $\pi(\cdot)$. Suppose that we wish to estimate the stationary average of a function $f(\cdot)$, that is,

$$\mu = \lim_{N \to \infty} \mathbb{E}\left[\frac{1}{N} \sum_{n=1}^{N} f(\xi_n)\right] = \int f(\xi)\, \pi(d\xi).$$

The Markov chain is called *uniformly (geometrically) ergodic* if there are constants $0 < \rho < 1$ and $\kappa < \infty$ such that for all initial states ξ_0,

$$\sup_{y \in S} |\mathbb{P}(\xi_n = y \mid \xi_0) - \pi(y)| \leq \kappa \rho^n.$$

The following result is from [174] and it provides the CLT for estimating μ via the sample averages. Compare with Theorem C.3 for an alternative version.

Theorem D.4. *Let $\{\xi_n\}$ be a uniformly geometrically ergodic Markov chain with stationary distribution π and assume that $\mathbb{E}_\pi[f^2(\xi)] < \infty$. Let*

$$\hat{\mu}_n = \frac{1}{n} \sum_{k=1}^{n} f(\xi_k)$$

APPENDIX D

be the sample average. Then for any initial distribution for ξ_0,

$$\sqrt{n}(\hat{\mu}_n - \mu) \stackrel{d}{\Longrightarrow} \mathcal{N}(0, \sigma_\infty^2), \qquad (D.7)$$

where $\sigma_\infty^2 = \mathrm{Var}_\pi(f(\xi)) + 2 \sum_{k=1}^{\infty} \mathrm{Cov}_\pi(f(\xi_0), f(\xi_k))$ is the asymptotic variance.

Replication/Deletion Method. In order to estimate the limit variance σ_∞^2 for Markov processes, it is usually recommended to "remove" the initial transient (as n grows, the distribution of ξ_n approximates the stationary distribution $\pi(\cdot)$). There are many statistical tests that can be used to decide if the process is in transient states [3]. One of the most commonly used is the Welch statistics, named after their creator Peter Welch. The idea is to make several initial replications of the process and evaluate window averages of the observations to obtain a visual test for the time when the process is approximately stationary. If using a limit theorem, those can also help to find approximate times when the process is close to stationary. Change point detection techniques may also be useful for testing this hypothesis, but as far as we know, implementing it efficiently is still an open problem.

Other experimental approaches are also convenient to have a good guess of the number N_0 of initial steps that belong to the transient phase of the process. The replication/deletion method for simulation then requires N independent replications of the process where the first N_0 observations are deleted. It is assumed that the remaining observations are stationary.

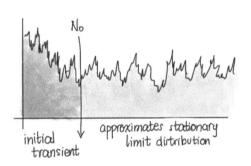

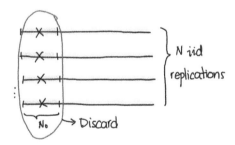

Batch Means Method. Following the same ideas as for stationary processes, a method that saves computational time is to consider consecutive batches for the estimation of the variance, having discarded only the initial transient. Once the process is approximately stationary, one can then implement the batch means method explained above.

It is important to realize that when estimating the variance, unless the samples are independent, it is not possible to use only one sample path, but several replications or batches must be used for good estimation of CIs, which in general will impose more computational effort.

Regenerative Method. A positive recurrent Markov process is a particular case of a *regenerative process* (see Section B.7), where every state $i \in S$ is a regenerative point. This section covers the estimation of confidence intervals for regenerative processes in general.

Let $\{\tau_k\}$ denote the sequence of regeneration times, and let $T_k \stackrel{\text{def}}{=} \tau_k - \tau_{k-1}$ be the consecutive cycle lengths. We will use here the result in Theorem B.18, namely that for any function $f: S \mapsto \mathbb{R}$ with finite cycle expectation,

$$\theta \stackrel{\text{def}}{=} \lim_{n \to \infty} \frac{1}{N} \sum_{n=1}^{N} f(X_n) = \frac{\mathbb{E}\left[\sum_{n=0}^{T_1} f(X_n)\right]}{\mathbb{E}[T_1]} \quad \text{a.s.}$$

The natural way to build the estimator is to use n cycles to build

$$\theta_n = \frac{\sum_{k=1}^{n} Y_k}{\sum_{k=1}^{n} T_k}; \quad Y_k \stackrel{\text{def}}{=} \sum_{j=\tau_{k-1}}^{\tau_k} f(X_j),$$

with Y_n the cost over the n-th cycle. Notice, however, that this estimator is *biased*, although strongly consistent. When estimating ratios of expectations one must recall that for any two (non-degenerate) random variables X_1, X_2, $\mathbb{E}[X_1/X_2] \neq \mathbb{E}[X_1]/\mathbb{E}[X_2]$.

Although θ_n is biased, a CLT can be obtained using the fact that $Z_n \stackrel{\text{def}}{=} Y_n - \theta T_n$ are iid random variables with $\mathbb{E}[Z_n] = \mathbb{E}[Y_n] - \theta \mathbb{E}[T_n] = 0$ and variance

$$\sigma_Z^2 \stackrel{\text{def}}{=} \text{Var}(Z_n) = \sigma_Y^2 - 2\theta \text{Cov}(Y_n, T_n) + \theta^2 \sigma_T^2,$$

Assuming that $\mathbb{E}[Z_n] < \infty$ and $\sigma_Z^2 < \infty$, then the CLT implies $\sqrt{n}\bar{Z}_n/\sigma_Z \stackrel{\text{d}}{\Longrightarrow} \mathcal{N}(0, 1)$, where $\bar{Z}_n$ is the sample average of $(Z_1, \ldots, Z_n)$. Call $\bar{Y}_n, \bar{T}_n$ the corresponding sample averages for the cycle cost and the cycle length, respectively. Then by construction, $\bar{Y}_n = \theta_n \bar{T}_n$ and simple substitution yields

$$\frac{\sqrt{n}(\theta_n - \theta) \bar{T}_n}{\sigma_Z} \stackrel{\text{d}}{\Longrightarrow} \mathcal{N}(0, 1).$$

A straightforward way to build an approximate CI at confidence level α uses

$$\theta_n \pm z_{1-\alpha/2} \frac{\widehat{\sigma_Z}}{\sqrt{n}\bar{T}_n}.$$

The reference [3] mentions that a better estimator for the variance of θ_n can be obtained by the method known as "jackknife" estimation. This yields confidence intervals with better coverage.

Bibliography

[1] Abbas, K., J. Berkhout, and B. Heidergott. A critical account of perturbation analysis of Markov chains. *Markov Processes and Related Fields* 22, no. 2 (2016): 227–266.

[2] Adan, I., and J. Resing. *Queueing Systems*. Eindhoven, Netherlands: Department of Mathematics and Computing Science, Eindhoven University of Technology, 2015.

[3] Alexopoulos, C., A. Seila, and J. Banks. Output data analysis. In *Handbook of Simulation*, edited by Jerry Banks, 225–272. New York: John Wiley & Sons, 1998.

[4] Andradóttir, S. A stochastic approximation algorithm with varying bounds. *Operations Research* 43, no. 6 (1995): 1037–1048.

[5] Andradóttir, S. A scaled stochastic approximation algorithm. *Management Science* 42, no. 4 (1996): 475–498.

[6] Andrieu, C., and M. Vihola. Markovian stochastic approximation with expanding projections. *Bernoulli* 20, no. 2 (2014): 545–585.

[7] Armijo, L. Minimization of functions having Lipschitz continuous first partial derivatives. *Pacific Journal of Mathematics* 16, no. 1 (1966): 1–3.

[8] Arrow, K., and L. Hurwicz. *Reduction of Constrained Maxima to Saddle-Point Problems*. Technical report. Santa Monica, CA: RAND, 1955.

[9] Arrow, K., and L. Hurwitz. Gradient methods for constrained maxima. *Operations Research* 5, no. 2 (1957): 258–265.

[10] Asmussen, S. Conjugate processes and the simulation of ruin problems. *Stochastic Processes and Their Applications* 20, no. 2 (1985): 213–229.

[11] Asmussen, S., and P. Glynn. *Stochastic Simulation: Algorithms and Analysis*. New York: Springer, 2007.

[12] Asmussen, S., and H. Nielsen. Ruin probabilities via local adjustment coefficients. *Journal of Applied Probability* 32, no. 3 (1995): 736–755.

[13] Auer, P., N. Cesa-Bianchi, and P. Fischer. Finite-time analysis of the multiarmed bandit problem. *Machine Learning* 47 (2002): 235–256.

[14] Baccelli, F., G. Cohen, G. Olsder, and J. Quadrat. *Synchronization and Linearity: An Algebra for Discrete Event Systems*. New York: Wiley, 1992.

[15] Baccelli, F., and D. McDonald. Mellin transforms for TCP throughput with applications to cross layer optimization. In *2006 40th Annual Conference on Information*

Sciences and Systems, 32–37. Piscataway, NJ: Institute of Electrical and Electronics Engineers, 2006.

[16] Bakhteev, O., and V. Strijov. Comprehensive analysis of gradient-based hyperparameter optimization algorithms. *Annals of Operations Research* 289 (2020): 51–65.

[17] Barakat, A., P. Bianchi, W. Hachem, and S. Schechtman. Stochastic optimization with momentum: Convergence, fluctuations, and traps avoidance. *Electronic Journal of Statistics* 15, no. 2 (2021): 3892–3947.

[18] Bartle, R. *The Elements of Integration and Lebesgue Measure*. New York: Wiley, 1995.

[19] Basawa, I. *Statistical Inferences for Stochasic Processes: Theory and Methods*. London: Academic Press, 2014.

[20] Basawa, I., U. Bhat, and J. Zhou. Parameter estimation using partial information with applications to queueing and related models. *Statistics & Probability Letters* 78, no. 12 (2008): 1375–1383.

[21] Beck, A., and L. Tetruashvili. On the convergence of block coordinate descent type methods. *SIAM Journal on Optimization* 23, no. 4 (2013): 2037–2060.

[22] Behnen, K., and G. Neuhaus. *Grundkurs Stochastik: Eine integrierte Einführung in Wahrscheinlichkeitstheorie und mathematische Statistik*. Berlin: Springer, 2013.

[23] Bellman, R. "On the Theory of Dynamic Programming." *Proceedings of the National Academy of Sciences of the United States of America* 38, no. 8 (August 1952): 716–719.

[24] Benaïm, M. Dynamics of stochastic approximation algorithms. In *Seminaire de Probabilites XXXIII*, 1–68. New York: Springer, 1999.

[25] Benaïm, M., and M. Hirsch. Learning processes, mixed equilibria and dynamical systems arising from repeated games. *Games and Economic Behavior* 29 (1999): 36–72.

[26] Benveniste, A., M. Metivier, and P. Priouret. *Adaptive Algorithms and Stochastic Approximations*. New York: Springer, 1990.

[27] Bercu, B., and P. Fraysse. A Robbins–Monro procedure for estimation in semiparametric regression models. *The Annals of Statistics* 40, no. 2 (2012): 666–693.

[28] Berndt, E., B. Hall, R. Hall, and J. Hausman. Estimation and inference in nonlinear structural models. In *Annals of Economic and Social Measurement* 3, no. 4 (1974): 653–665.

[29] Bertsekas, D. On the Goldstein-Levitin-Polyak gradient projection method. *IEEE Transactions on Automatic Control* 21, no. 2 (1976): 174–184.

[30] Bertsekas, D. Incremental least squares methods and the extended Kalman filter. *SIAM Journal on Optimization* 6, no. 3 (1996): 807–822.

[31] Bertsekas, D. *Nonlinear Programming*. Belmont, MA: Athena Scientific, 1999.

[32] Bertsekas, D., and J. Tsitsiklis. *Parallel and Distributed Computation: Numerical Methods*. Englewood Cliffs, NJ: Prentice-Hall, 1989.

[33] Bertsekas, D., and J. Tsitsiklis. *Neuro-Dynamic Programming*. Belmont, MA: Athena Scientific, 1996.

[34] Bertsekas, D., and J. Tsitsiklis. Gradient convergence in gradient methods with errors. *SIAM Journal on Optimization* 10, no. 3 (2000): 627–642.

[35] Bhatnagar, S., N. Hemachandra, and V. Mishra. Stochastic approximation algorithms for constrained optimization via simulation. *ACM Transactions on Modeling and Computer Simulation (TOMACS)* 21, no. 3 (2011): 1–22.

[36] Bhatnagar, S., V. Mishra, and N. Hemachandr. Stochastic algorithms for discrete parameter simulation optimization. *IEEE Transactions on Automation Science and Engineering* 8, no. 4 (2011): 780–793.

[37] Bhatnagar, S., H. Prasad, and L. Prashanth. *Stochastic Recursive Algorithms for Optimization: Simultaneous Perturbation Methods*. New York: Springer, 2013.

[38] Billingsley, P. *Convergence of Probability Measures*. New York: John Wiley & Sons, 2013.

[39] Borkar, S. *Stochastic Approximation: A Dynamical Systems Viewpoint*. Cambridge: Cambridge University Press, 2008.

[40] Borovkov, A., and A. Hordijk. Characterization and sufficient conditions for normed ergodicity of Markov chains. *Advances in Applied Probability* 36, no. 1 (2004): 227–242.

[41] Bottou, L., F. Curtis, and J. Nocedal. Optimization methods for large-scale machine learning. *SIAM Review* 60, no. 2 (2018). https://doi.org/10.1137/16M1080173.

[42] Bottou, L., F. Curtis, and J. Nocedal. Optimization methods for large-scale machine learning (3rd version). arXiv.org, arXiv:1606.04838, 2018.

[43] Boyle, J., and R. Dykstra. A method for finding projections onto the intersection of convex sets in Hilbert spaces. In *Advances in Order Restricted Statistical Inference: Proceedings of the Symposium on Order Restricted Statistical Inference held in Iowa City, Iowa, September 11–13, 1985*, 28–47. New York: Springer, 1986.

[44] Breiman, L. *Probability*. 1968. Philadelphia: Society for Industrial and Applied Mathematics, 1992.

[45] Brémaud, P. Maximal coupling and rare perturbation sensitivity analysis. *Queueing Systems* 11 (1993): 307–333.

[46] Brémaud, P., and W. Gong. Derivatives of likelihood rations and smoothed perturbation analysis for the routing problem. *ACM Trans. Modelling and Computer Simulation* 3, no. 2 (1993): 134–161.

[47] Brémaud, P., and F. Vázquez-Abad. On the pathwise computation of derivatives with respect to the rate of a point process: The phantom RPA method. *Queueing Systems* 10, no. 3 (1992): 249–270.

[48] Brezinski, C., and M. Zaglia. *Extrapolation Methods: Theory and Practice*. Amsterdam: Elsevier, 2013.

[49] Brinkhuis, J., and V. Tikhomirov. *Optimization: Insights and Applications*. Princeton, NJ: Princeton University Press, 2005.

[50] Broadie, M., D. Cicek, and A. Assaf. An adaptive multidimensional version of the Kiefer-Wolfowitz stochastic approximation algorithm. In *Proceedings of the 2009 Winter Simulation Conference (WSC)*, 601–612. Piscataway, NJ: Institute of Electrical and Electronics Engineers, 2009.

[51] Broadie, M., D. Cicek, and A. Zeevi. General bounds and finite-time improvement for the Kiefer-Wolfowitz stochastic approximation algorithm. *Operations Research* 59, no. 5 (2011): 1211–1224.

[52] Broadie, M., D. Cicek, and A. Zeevi. Multidimensional stochastic approximation: Adaptive algorithms and applications. *ACM Transactions on Modeling and Computer Simulation (TOMACS)* 24, no. (2014): 1–28.

[53] Brown, G. Iterative solution of games by fictitious play. In *Activity Analysis of Production and Allocation*, edited by T. C. Koopmans, 374–376. New York: Wiley, 1951.

[54] Bru, B. Problème de l'efficaciteté du tir a l'école d'artillerie de Metz. *Mathématiques, informatique et sciences humaines* 136 (1996): 29–42.

[55] Cai, J., and J. Garrido. Asymptotic forms and bounds for tails of convolutions of compound geometric distributions, with applications. In *Recent Advances in Statistical Methods*, edited by Yogendra P. Chaubey, 114–131. River Edge, NJ: World Scientific, 2002.

[56] Cao, X.-R. Realization probability in closed Jackson queueing networks and its application. *Advances in Applied Probability* 19, no. 3 (1987): 708–738.

[57] Cao, X.-R. Sensitivity estimates based on one realization of a stochastic system. *Journal of Statistical Computation and Simulation* 27, no. 3 (1987): 211–232.

[58] Cao, X.-R. *Realization Probabilities: The Dynamics of Queueing Networks*. New York: Springer, 1994.

[59] Cao, X.-R. From perturbation analysis to Markov decision processes and reinforcement learning. *Discrete Event Dynamic Systems* 13 (2003): 9–39.

[60] Cao, X.-R. *Stochastic Learning and Optimization: A Sensitivity-Based Approach*. New York: Springer, 2007.

[61] Cao, X.-R. State classification and multiclass optimization of continuous-time and continuous-state Markov processes. *IEEE Transactions on Automatic Control* 64, no. 9 (2019): 3632–3646.

[62] Cartis, C., and K. Scheinberg. Global convergence rate analysis of unconstrained optimization methods based on probabilistic models. *Mathematical Programming* 169, no. 2 (2018): 337–375.

[63] Cassandras, C., W. Gong, and J. Lee. Robustness of perturbation analysis estimators for queueing systems with unknown distributions. *Journal of Optimization Theory and Applications* 70, no. 3 (1991): 191–519.

[64] Cassandras, C., and Y. Ho. An event driven formalism for sample path perturbation analysis of discrete event dynamic systems. *IEEE Transactions on Automatic Control* 30, no. 12 (1995): 1217–1221.

[65] Cassandras, C., and V. Julka. Scheduling policies using marked/phantom slot algorithms. *Queueing Systems* 20 (1995): 207–254.

[66] Cassandras, C., and L. Lafortune. *Introduction to Discrete Event Systems*. 2nd ed. New York: Springer, 2008.

[67] Censor, Y., and A. Cegielski. Projection methods: An annotated bibliography of books and reviews. *Optimization* 64, no. 11 (2015): 2343–2358.

[68] Chan, K.-H., L. Hong, and H. Wan. Stochastic trust region gradient-free method (STRONG)—A new response-surface-based algorithm in simulation optimization. In *2007 Winter Simulation Conference*, edited by S. G. Henderson et al., 346–354. Piscataway, NJ: Institute of Electrical and Electronics Engineers, 2007.

[69] Chan, K.-H., L. Hong, and H. Wan. Stochastic trust-region response-surface method (STRONG)—A new response-surface framework for simulation optimization. *INFORMS Journal on Computing* 25, no. 2 (2013): 230–243.

[70] Chau, M., M. Fu, and H. Qu. *Multivariate Stochastic Approximation Using a Secant Tangents Averaged (STAR) Gradient Estimation*. Technical report. College Park, MD: University of Maryland, 2014.

[71] Chau, M., M. Fu, H. Qu, and I. Ryzhov. Simulation optimization: A tutorial overview and recent developments in gradient-based methods. In *Proceedings of the Winter Simulation Conference 2014*, 21–35. Piscataway, NJ: Institute of Electrical and Electronics Engineers, 2014.

[72] Chau, M., H. Qu, and M. Fu. A new hybrid stochastic approximation algorithm. *IFAC Proceedings Volumes* 47, no. 2 (2014): 241–246.

[73] Chen, H. *Stochastic Approximation and its Applications*. New York: Kluwer, 2003.

[74] Chen, H., L. Guo, and A. Gao. Convergence and robustness of the Robbins-Monro algorithm truncated at randomly varying bounds. *Stochastic Processes and their Applications* 27 (1987): 217–231.

[75] Chen, H., and Y. Zhu. Stochastic approximation procedures with randomly varying truncations. *Science in China Series A-Mathematics, Physics, Astronomy & Technological Science* 29, no. 9 (1986): 914–926.

[76] Chen, X., and C. Gao. On the convergence of smoothed functional stochastic optimization algorithms. *IFAC-PapersOnLine* 48, no. 28 (2015): 229–233.

[77] Chen, Y., and X. Ye. Projection onto a simplex. *University of Florida Digital Collection*, arXiv:1101.6081v2, 2011.

[78] Chicone, C. *Ordinary Differential Equations with Applications*, vol. 34. New York: Springer Science & Business Media, 2006.

[79] Chin, D. Comparative study of stochastic algorithms for system optimization based on gradient approximations. *IEEE Transactions on Systems, Man, and Cybernetics, Part B (Cybernetics)* 27, no. 2 (1997): 244–249.

[80] Conn, A., K. Scheinberg, and L. Vicente. *Introduction to Derivative-Free Optimization*. Philadelphia: Society for Industrial and Applied Mathematics, 2009.

[81] Coolen-Schrijner, P., and E. Van Doorn. The deviation matrix of a continuous-time Markov chain. *Probability in the Engineering and informational Sciences*, May 22, 2002, 351–366.

[82] Costa, A., and F. Vázquez-Abad. Adaptive stepsize selection for tracking in a regime-switching environment. *Automatica* 43, no. 11 (2007): 1896–1908.

[83] Dai, L., Perturbation analysis via coupling. *IEEE Transactions on Automatic Control* 45, no. 4 (2000): 614–628.

[84] Dantzig, G., and M. Thapa. *Linear Programming 1: Introduction*. New York: Springer, 1997.

[85] Dantzig, G., and M. Thapa. *Linear Programming 2: Theory and Extensions*. New York: Springer, 2003.

[86] Delyon, B., Stochastic approximation with decreasing gain: Convergence and asymptotic theory. Unpublished lecture notes, Université de Rennes, 2000.

[87] Dickey, D., and A. Fuller. Distribution of the estimators for autoregressive time series with a unit root. *Journal of the American Statistical Association* 74, no. 366 (1979): 427–431.

[88] Dieleman, N., B. Heidergott, and Y. Peng. Data-driven fitting of the M/G/1 queue. In *2019 16th International Conference on Service Systems and Service Management (ICSSSM)*, 1–5. Piscataway, NJ: Institute of Electrical and Electronics Engineers, 2019.

[89] Dieuleveut, A., A. Durmus, and F. Bach. Bridging the gap between constant step size stochastic gradient descent and Markov chains. *The Annals of Statistics* 48, no. 3 (2020): 1348–1382.

[90] Dong, H., and M. Nakayama. A tutorial on quantile estimation via Monte Carlo. In *Monte Carlo and Quasi-Monte Carlo Methods*, edited by B. Tuffin and P. L'Ecuyer, 3–30. Cham, Switzerland: Springer, 2020.

[91] Dufresne, D., and F. Vázquez-Abad. Cobweb theorems with production lags and price forecasting. *Economics: The Open-Access, Open-Assessment E-Journal* 7 (2013): 1–49.

[92] Dufresne, F., and H. Gerber. Three methods to calculate the probability of ruin. *ASTIN Bulletin: The Journal of the IAA* 19, no. 1 (1989): 71–90.

[93] Dunford, N., and J. Schwartz. *Linear Operators, Part I: General Theory*. New York: Wiley, 1966.

[94] Dupuis, P., and A. Nagurney. Dynamical systems and variational inequalities. *Annals of Operations Research* 44 (1993): 7–42.

[95] Dussault, J., D. Labrecque, P. L'Ecuyer, and R. Rubinstein. Combining the stochastic counterpart and stochastic approximation methods. *Discrete Event Dynamic Systems* 7, no. 1 (1997): 5–28.

[96] Eliason, S. *Maximum Likelihood Estimation: Logic and Practice*. Newbury Park, CA: Sage, 1993.

[97] Fabian, V. Stochastic approximation. In *Optimizing Methods in Statistics*, 439–470. Amsterdam: Elsevier, 1971.

[98] Fearnhead, P., and L. Meligkotsidou. Exact filtering for partially observed continuous time models. *Journal of the Royal Statistical Society: Series B (Statistical Methodology)* 66, no. 3 (2004): 771–789.

[99] Frank, M., and P. Wolfe. An algorithm for quadratic programming. *Naval Research Logistics Quarterly* 3 (1956): 95–110.

[100] Franssen, C., A. Zocca, and B. Heidergott. A first-order gradient approach for the connectivity analysis of weighted graphs. arXiv preprint, arXiv:2403.11744, 2024.

[101] Fu, M. Stochastic gradient estimation. In *Handbooks in Operations Research and Management Science: Simulation*, edited by S. G. Henderson and B. L. Nelson, chap. 19. Amsterdam: Elsevier, 2008.

[102] Fu, M. *Handbook of Simulation Optimization*. Vol. 216. New York: Springer, 2015.

[103] Fu, M., and J.-Q. Hu. (s, S) inventory systems with random lead times: Harris recurrence and its implications in sensitivity analysis. *Probability in the Engineering and Informational Sciences* 8, no. 3 (1994): 355–376.

[104] Fu, M., and J.-Q. Hu. *Smoothed Monte Carlo: Gradient Estimation and Optimization Applications*. Boston: Kluwer, 1997.

[105] Fudenberg, F., and D. Levine. *The Theory of Learning in Games*. Vol. 2. Cambridge, MA: MIT Press, 1998.

[106] Gerber, H. *An Introduction to Risk Theory*. Philadelphia: Huebner Foundation for Insurance Education, 1979.

[107] Gerencsér, L., S. Hill, and Z. Vago. Optimization over discrete sets via SPSA. In *Proceedings of the 31st Conference on Winter Simulation: Simulation—A Bridge to the Future*, vol. 1, 466–470. New York: Association for Computing Machinery, 1999.

[108] Gerencsér, L., S. Hill, Z. Vágó, and Z. Vincze. Discrete optimization, SPSA and Markov chain Monte Carlo methods. In *Proceedings of the 2004 American Control Conference*, vol. 4, 3814–3819. Piscataway, NJ: Institute of Electrical and Electronics Engineers, 2004.

[109] Gerencsér, L., and Z. Vágó. SPSA in noise free optimization. In *Proceedings of the 2000 American Control Conference. ACC (IEEE Cat. No. 00CH36334)*, vol. 5, 3284–3288. Piscataway, NJ: Institute of Electrical and Electronics Engineers, 2000.

[110] Gerencsér, L., and Z. Vágó. The mathematics of noise-free SPSA. In *Proceedings of the 40th IEEE Conference on Decision and Control (Cat. No. 01CH37228)*, vol. 5, 4400–4405. Piscataway, NJ: Institute of Electrical and Electronics Engineers, 2001.

[111] Ghadimi, S., and G. Lan. Optimal stochastic approximation algorithms for strongly convex stochastic composite optimization, I: A generic algorithmic framework. *SIAM Journal on Optimization* 22, no. 4 (2012): 1469–1492.

[112] Ghadimi, S., and G. Lan. Optimal stochastic approximation algorithms for strongly convex stochastic composite optimization, II: Shrinking procedures and optimal algorithms. *SIAM Journal on Optimization* 23, no. 4 (2013): 2061–2089.

[113] Ghosh, J. A new proof of the Bahadur representation of quantiles and an application. *Annals of Mathematical Statistics* 42, no. 6 (1971): 1957–1961.

[114] Glasserman, P. *Gradient Estimation via Perturbation Analysis*. Boston: Kluwer, 1991.

[115] Glasserman, P. Structural conditions for perturbation analysis of queueing systems. *Journal of the ACM (JACM)* 38, no. 4 (1991):1005–1025.

[116] Glasserman, P. Smoothing complements and randomized score functions. *Annals of Operations Research* 39 (1992): 41–67.

[117] Glasserman, P. Stationary waiting time derivatives. *Queueing Systems* 12, no. 3–4 (1992): 369–389.

[118] Glasserman, P. Regenerative derivatives of regenerative sequences. *Advances in Applied Probability* 25, no. 1 (1993): 116–139.

[119] Glasserman, P. Perturbation analysis of production networks. In *Stochastic Modeling and Analysis of Manufacturing Systems*, edited by D. Yao, 233–280. New York: Springer, 1994.

[120] Glasserman, P., and P. Glynn. Gradient estimation for regenerative processes. In *Proceedings of the 24th Winter Simulation Conference (WSC)*, edited by J. Swain, D. Goldsman, R. Crain, and J. Wilson, 280–288. New York: Association for Computing Machinery, 1992.

[121] Glasserman, P., and W. Gong. Smoothed perturbation analysis for a class of discrete-event systems. *IEEE Transactions on Automatic Control* 35, no. 11 (1990): 1218–1230.

[122] Glasserman, P., J. Hu, and G. Strickland. Strongly consistent steady-state derivative estimates. *Probability in the Engineering and Informational Sciences* 5, no. 4 (1991): 391–413.

[123] Glasserman, P., J. Hu, and S. Strickland. Strong consistency of sample path derivative estimates. In *29th IEEE Conference on Decision and Control*, 2853–2854. Piscataway, NJ: Institute of Electrical and Electronics Engineers, 1990.

[124] Glasserman, P., and D. Yao. *Monotone Structure in Discrete-Event Systems*. New York: Wiley, 1994.

[125] Glynn, P. A GSMP formalism for discrete event systems. *Proceedings of the IEEE* 77, no. 1 (1989): 14–23.

[126] Glynn, P. Note—Pathwise convexity and its relation to convergence of time-average derivatives. *Management Science* 38, no. 9 (1992): 1360–1366.

[127] Glynn, P., and P. L'Ecuyer. Likelihood ratio gradient estimation for stochastic recursions. *Advances in Applied Probability* 27, no. 4 (1995): 1019–1053.

[128] Glynn, P., and W. Whitt. The asymptotic efficiency of simulation estimators. *Operations Research* 40, no. 3 (1992): 505–520.

[129] Glynn, W. Construction of process-differentiable representations for parametric families of distributions. Technical report, University of Wisconsin, Mathematics Research Center, 1986.

[130] Goldstein, A. Convex programming in Hilbert space. *Bulletin of the American Mathematical Society* 70, no. 5 (1964): 709–710.

[131] Goldstein, A. *Constructive Real Analysis*. New York: Harper & Row, 1967.

[132] Gong, W. Smoothed perturbation analysis algorithm for a G/G/1 routing problem. In *Proceedings of the 20th Winter Simulation Conference (WSC)*, 525–531. New York: Association for Computing Machinery, 1988.

[133] Gong, W., and Y. C. Ho. Smoothed (conditional) perturbation analysis of discrete event dynamical systems. *IEEE Transactions on Automatic Control* 32, no. 10 (1987): 858–866.

[134] Goodfellow, I., Y. Bengio, and A. Courville. *Deep Learning*. Cambridge, MA: MIT Press, 2016.

[135] Granichin, O., and N. Amelina. Simultaneous perturbation stochastic approximation for tracking under unknown but bounded disturbances. *IEEE Transactions on Automatic Control* 60, no. 6 (2014): 1653–1658.

[136] Grathwohl, W., D. Choi, Y. Wu, G. Roeder, and D. Duvenaud. Backpropagation through the void: Optimizing control variates for black-box gradient estimation. ArXiv preprint, abs/1711.00123, 2017.

[137] Györfi, L., and H. Walk. On the averaged stochastic approximation for linear regression. *SIAM Journal on Control and Optimization* 34, no. 1 (1996): 31–61.

[138] Hall, A. Generalized method of moments. In *Handbook of Research Methods and Applications in Empirical Macroeconomics*. Cheltenham, UK: Edward Elgar, 2013.

[139] Hansen, B., and K. West. Generalized method of moments and macroeconomics. *Journal of Business & Economic Statistics* 20, no. 4 (2002): 460–469.

[140] Hansen, L. Large sample properties of generalized method of moments estimators. *Econometrica: Journal of the Econometric Society* 50, no. 4 (1982): 1029–1054.

[141] He, B., H. Yang, Q. Meng, and D. Han. Modified Goldstein-Levitin-Polyak projection method for asymmetric strongly monotone variational inequalities. *Journal of Optimization Theory and Applications* 112, no. 1 (2002): 129–143.

[142] Heidelberger, P., X.-R. Cao, M. Zazanis, and R. Suri. Convergence properties of infinitesimal perturbation analysis estimates. *Management Science* 34, no. 11 (1988): 1281–1302.

[143] Heidergott, B. Infinitesimal perturbation analysis for queueing networks with general service time distributions. *Queueing Systems* 31 (1999): 43–58.

[144] Heidergott, B. Analysing sojourn times in queueing networks: A structural approach. *Mathematical Methods of Operations Research* 52, no. 1 (2000): 115–132.

[145] Heidergott, B. *Max-Plus Linear Stochastic Systems and Perturbation Analysis.* Vol. 15. New York: Springer, 2006.

[146] Heidergott, B., and X.-R. Cao. A note on the relation between weak derivatives and perturbation realization. *IEEE Transactions on Automatic Control* 47, no. 7 (2002): 1112–1115.

[147] Heidergott, B., T. Farenhorst-Yuan, and F. Vázquez-Abad. A perturbation analysis approach to phantom estimators for waiting times in the G/G/1 queue. *Discrete Event Dynamic Systems* 20, no. 2 (2010): 249–273.

[148] Heidergott, B., and A. Hordijk. Single-run gradient estimation via measure-valued differentiation. *IEEE Transactions on Automatic Control* 49, no. 10 (2004): 1843–1846.

[149] Heidergott, B., A. Hordijk, and H. Weisshaupt. Measure-valued differentiation for stationary Markov chains. *Mathematics of Operations Research* 31, no. 1 (2006): 154–172.

[150] Heidergott, B., A. Hordijk, and H. Weisshaupt. Derivatives of Markov kernels and their Jordan decomposition. *Journal of Applied Analysis* 14, no. 1 (2008):13–26.

[151] Heidergott, B., and H. Leahu. Weak differentiability of product measures. *Mathematics of Operations Research* 35, no. 1 (2010): 27–51.

[152] Heidergott, B., and Y. Peng. Gradient estimation for smooth stopping criteria. *Advances in Applied Probability* 55, no. 1 (2023): 29–55.

[153] Heidergott, B., and F. Vázquez-Abad. Measure-valued differentiation for Markov chains. *Journal of Optimization and Applications* 136, no. 2 (2008):187–209.

[154] Heidergott, B., and F. Vázquez-Abad. Measure-valued differentiation for random horizon problems. *Markov Processes and Related Fields* 12, no. 3 (2006): 509–536.

[155] Heidergott, B., and F. Vázquez-Abad. Gradient estimation for a class of systems with bulk services: A problem in public transportation. *ACM Transactions on Modeling and Computer Simulation (TOMACS)* 19, no. 3 (2009): 1–27.

[156] Heidergott, B., F. Vázquez-Abad, and W. Volk-Makarewicz. Sensitivity estimation for Gaussian systems. *European Journal of Operational Research* 187, no. 1 (2008): 193–207.

[157] Henderson, S. G. Mathematics for simulation. Chap. 2 in *Handbooks in Operations Research and Management Science*, vol. 13, *Simulation*, edited by S. G. Henderson and B. L. Nelson, 19–53. Amsterdam: North-Holland, 2006.

[158] Hewitt, E., and K. Stromberg. *Real and Abstract Analysis.* New York: Springer, 2013.

[159] Ho, Y., and X.-R. Cao. *Perturbation Analysis of Discrete Event Systems*. Boston: Kluwer, 1991.

[160] Ho, Y., X.-R. Cao, and G. Cassandras. Infinitesimal and finite perturbation analysis for queueing networks. In *21st IEEE Conference on Decision and Control*, vol. 21, 854–855. New York: Institute of Electrical and Electronics Engineers, 1982.

[161] Ho, Y., and C. Cassandras. Computing co-state variables for discrete event systems. In *1980 19th IEEE Conference on Decision and Control including the Symposium on Adaptive Processes*, 697–700. New York: Institute of Electrical and Electronics Engineers, 1980.

[162] Ho, Y., and C. Cassandras. A new approach to the analysis of discrete event dynamic systems. *Automatica* 19, no. 2 (1983): 149–167.

[163] Ho, Y., A. Eyler, and D. Chien. A gradient technique for general buffer storage design in a serial production line. *International Journal of Production Research* 17, no. 6 (1979): 557–580.

[164] Ho, Y., and S. Li. Extensions of infinitesimal perturbation analysis. *IEEE Transactions on Automatic Control* 33, no. 5 (1988): 427–438.

[165] Hu, J., Convexity of sample path performance and strong consistency of infinitesimal perturbation analysis estimates. *IEEE Transactions on Automatic Control* 37, no. 2 (1992): 258–262.

[166] Hu, J., and S. Strickland. Strong consistency of sample path derivative estimates. *Applied Mathematics Letters* 3, no. 4 (1990): 55–58.

[167] Huang, D. On a modified version of the Lindley recursion. *Queueing Systems* 105, no. 3–4 (2023): 271–289.

[168] Hunter, J. Introduction to dynamical systems, Department of Mathematics, University of California at Davis, 2011.

[169] Iusem, A. On the convergence properties of the projected gradient method for convex optimization. *Computational & Applied Mathematics* 22, no. 1 (2003): 37–52.

[170] Jagannathan, R., G. Skoulakis, and Z. Wang. Generalized methods of moments: Applications in finance. *Journal of Business & Economic Statistics* 20, no. 4 (2002): 470–481.

[171] Jais, I., A. Ismail, and Q. Nisa. Adam optimization algorithm for wide and deep neural network. *Knowledge Engineering and Data Science* 2, no. 1 (2019): 41–46.

[172] Jin, B., K. Scheinberg, and M. Xie. High probability complexity bounds for line search based on stochastic oracles. *Advances in Neural Information Processing Systems* 34 (2021).

[173] Johnson, R., and T. Zhang. Accelerating stochastic gradient descent using predictive variance reduction. In *Advances in Neural Information Processing Systems* 26 (2013): 315–323.

[174] Jones, G. On the Markov chain central limit theorem. *Probability Surveys* 1 (2004): 299–320.

[175] Judge, G., W. Griffiths, R. Hill, H. Lütkepohl, and T.-C. Lee. *The Theory and Practice of Econometrics*, Vol. 49. New York: John Wiley & Sons, 1991.

[176] Kaniovski, Y., and H. Young. Learning dynamics in games with stochastic perturbations. *Games and Economic Behavior* 11, no. 2 (1995): 330–363.

[177] Kao, C., W. Song, and S. Chen. A modified quasi-Newton method for optimization in simulation. *International Transactions in Operational Research* 4, no. 3 (1997): 223–233.

[178] Kartashov, N. *Strong Stable Markov Chains*. Berlin: De Gruyter, 2019.

[179] Kass, R., and P. Vos. *Geometrical Foundations of Asymptotic Inference*. Vol. 908. New York: John Wiley & Sons, 2011.

[180] Katkovnik, Y., and O. Kulchits. Convergence of a class of random search algorithms. *Automation and Remote Control* 33 (1972): 1321–1326.

[181] Kesten, H. Accelerated stochastic approximation. *The Annals of Mathematical Statistics* 29, no. 1 (1958): 41–59.

[182] Kiefer, J., and J. Wolfowitz. Stochastic estimation of the maximum of a regression function. *Annals of Mathematical Statistics* 23, no. 3 (1952): 462–466.

[183] Kingma, D., and J. Ba. Adam: A method for stochastic optimization. arXiv preprint, arXiv:1412.6980, 2014.

[184] Kleijnen, J., and R. Rubinstein. Optimization and sensitivity analysis of computer simulation models by the score function method. *European Journal of Operational Research* 88 (1996): 413–427.

[185] Kleywegt, A., A. Shapiro, and T. Homem de Mello. The sample average approximation method for stochastic discrete optimization. *SIAM Journal on optimization* 12, no. 2 (2002): 479–502.

[186] Konečný, J., and P. Richtárik. Semi-stochastic gradient descent methods. *Frontiers in Applied Mathematics and Statistics* 3 (2017): 9.

[187] Kong, X., J. Guo, D. Zheng, J. Zhang, and W. Fu. Quality control for medium voltage insulator via a knowledge-informed SPSA based on historical gradient approximations. *Processes* 8, no. 2 (2020): 146.

[188] König, D., and K. Matthes. Verallgemeinerungen der Erlangschen Formeln. I. *Mathematische Nachrichten* 26, no. 1–4 (1963): 45–56.

[189] König, D., K. Matthes, and K. Nawrotzki. *Verallgemeinerungen der Erlangschen und Engsetschen Formeln: Eine Methode in der Bedienungstheorie*. Berlin: Akademie-Verlag, 1967.

[190] Krishnamurthy, V., and F. Vázquez-Abad. Gradient based policy optimization of constrained Markov decision processes. In *Stochastic Processes, Finance and Control: A Festschrift in Honor of Robert J Elliott*, 503–547. Hackensack, NJ: World Scientific, 2012.

[191] Kroese, D., T. Taimre, and Z. Botev. *Handbook of Monte Carlo Methods*. Vol. 706. Hoboken, NJ: John Wiley & Sons, 2013.

[192] Kumagai, S. An implicit function theorem: Comment. *Journal of Optimization Theory and Applications* 31, no. 2 (1980): 285–288.

[193] Kushner, H., and D. Clark. *Stochastic Approximation Methods for Constrained and Unconstrained Systems.* New York: Springer, 1978.

[194] Kushner, H., and H. Huang. Rates of convergence for stochastic approximation type algorithms. *SIAM Journal on Control and Optimization* 17, no. 5 (1979): 607–617.

[195] Kushner, H., and F. Vázquez-Abad. Stochastic approximation methods for systems over an infinite horizon. *SIAM Journal on Control and Optimization* 34, no. 2 (1996): 712–756.

[196] Kushner, H., and J. Yang. Stochastic approximation with averaging of the iterates: Optimal asymptotic rate of convergence for general processes. *SIAM Journal on Control and Optimization* 31, no. 4 (1993): 1045–1062.

[197] Kushner, H., and G. Yin. *Stochastic Approximation Methods.* New York: Springer, 1997.

[198] Kushner, H., and G. Yin. *Stochastic Approximation and Recursive Algorithms.* New York: Springer, 2003.

[199] Lai, T. Stochastic approximation. *Annals of Statistics* 31, no. 2 (2003): 391–406.

[200] Law, A., and W. Kelton. *Simulation Modeling and Analysis.* Vol. 3. New York: McGraw-Hill, 2007.

[201] L'Ecuyer, P., and G. Yin. Budget dependent convergence rate of stochastic approximation. *SIAM Journal on Optimization* 8, no. 1 (1997): 217–247.

[202] Levitin, E., and B. Polyak. Constrained minimization methods. *USSR Computational Mathematics and Mathematical Physics* 6, no. 5 (1966): 1–50.

[203] Levy, K. Apprentissage par simulation stochastique: Étude de convergence et application à un modèle Markovien de tarification en transport aérien. Mémoire de Maîtrise, Université de Montréal, 2005.

[204] Li, Y., F. Cao, and X.-R. Cao. An improvement of policy gradient estimation algorithm. In *The Proceedings of Workshop on Discrete Event Systems*, edited by B. Lennartson, M. Fabian, K. Akesson, A. Guia, and R. Kumar, 168–172. Goteborg, Sweden, 2008.

[205] Ljung, L. Analysis of recursive stochastic algorithms. *IEEE Transactions on Automatic Control* 22, no. 4 (1997): 551–575.

[206] Löpker, A. and J. van Leeuwaarden. Transient moments of the TCP window size process. *Journal of Applied Probability* 45, no. 1 (2008): 163–175.

[207] Lu, Z., and L. Xiao. On the complexity analysis of randomized block-coordinate descent methods. *Mathematical Programming* 152 (2015): 615–642.

[208] Mangasarian, O., and M. Solodov. Backpropagation convergence via deterministic nonmonotone perturbed minimization. In *Advances in Neural Information Processing Systems* 7: 383–390, 1994.

[209] Marti, K. *Stochastic Optimization Methods*. 3rd ed. Berlin: Springer, 2015.

[210] Marti, K., and E. Fuchs. On solutions of stochastic programming problems by descent procedures with stochastic and deterministic directions. *Methods of Operations Research* 33 (1979): 281–293.

[211] Marti, K., and E. Fuchs. Computation of descent directions and efficient points in stochastic optimization problems without using derivatives. In *Stochastic Programming 84 Part II*, 132–156. Berlin: Springer, 1986.

[212] Marti, K., and E. Fuchs. Rates of convergence of semi-stochastic approximation procedures for solving stochastic optimization problems. *Optimization* 17, no. 2 (1986): 243–265.

[213] Maryak, J., and D. Chin. Global random optimization by simultaneous perturbation stochastic approximation. *Johns Hopkins APL Technical Digest* 25, no. 2 (2004): 91–100.

[214] Matthes, K. Zur Theorie der Bedienungsprozesse (On the theory of service processes). In *Transactions of the Third Prague Conference on Information Theory, Statistical Decision Functions, Random Processes held at Liblice near Prague, from June 5 to 13, 1962*, edited by Jaroslav Kozesnik. N.p.: Academic Press, 1964.

[215] Meyer, C. The role of the generalized inverse in the theory of finite Markov chains. *SIAM Review* 17, no. 3 (1975): 443–464.

[216] Meyn, S. *Control Techniques for Complex Networks*. New York: Cambridge University Press, New York, 2007.

[217] Mikolov, T., A. Deoras, D. Povey, L. Burget, and J. Černocký. RNNLM-recurrent neural network language modeling toolkit. In *Proceedings of the 2011 ASRU Workshop*, 196–201, 2011.

[218] Mikolov, T., A. Deoras, D. Povey, L. Burget, and J. Černocký. Strategies for training large scale neural network language models. In *2011 IEEE Workshop on Automatic Speech Recognition & Understanding*, 196–201. Piscataway, NJ: Institute of Electrical and Electronics Engineers, 2011.

[219] Millar, R. *Maximum Likelihood Estimation and Inference: With Examples in R, SAS and ADMB*, vol. 111. New York: John Wiley & Sons, 2011.

[220] Milnor, J. Analytic proofs of the "hairy ball theorem" and the brouwer fixed point theorem. *The American Mathematical Monthly* 85, no. 7 (1978): 521–524.

[221] Mohamed, S., M. Rosca, M. Figurmov, and A. Mnih. Monte Carlo gradient estimation in machine learning. *Journal of Machine Learning Research* 21, no. 132 (2020): 1–62.

[222] Mokkadem, A., and M. Pelletier. Convergence rate and averaging nonlinear two-time-scale stochastic approximation algorithms. *The Annals of Applied Probability* 16, no. 3 (2006): 1671–1702.

[223] Moulines, E., and F. Bach. Non-asymptotic analysis of stochastic approximation algorithms for machine learning. In *Advances in Neural Information Processing Systems 24* (2011): 451–459.

[224] Murty, K. *Linear Programming*. New York: Wiley, 1983.

[225] Neelakantan, A., L. Vilnis, Q. Le, I. Sutskever, L. Kaiser, K. Kurach, and J. Martens. Adding gradient noise improves learning for very deep networks. arXiv preprint, arXiv:1511.06807, 2015.

[226] Nemirovski, A., A. Juditsky, G. Lan, and A. Shapiro. Robust stochastic approximation approach to stochastic programming. *SIAM Journal on Optimization* 19, no. 4 (2009): 1574–1609.

[227] Nesterov, Y. A method for solving the convex programming problem with convergence rate O $(1/k^2)$. In *Doklady Akademii Nauk SSSR* 269 (1983): 543–547.

[228] Nesterov, Y. Random gradient-free minimization of convex functions. LIDAM Discussion Papers CORE 2011001 Louvain: Université catholique de Louvain, Center for Operations Research and Econometrics (CORE), 2011.

[229] Nesterov, Y., and V. Spokoiny. Random gradient-free minimization of convex functions. *Foundations of Computational Mathematics* 17, no. 2 (2017): 527–566.

[230] Newey, W., and K. West. Hypothesis testing with efficient method of moments estimation. *International Economic Review* 28, no. 3 (1987): 777–787.

[231] Nguyen, N., and G. Yin. Stochastic approximation with discontinuous dynamics, differential inclusions, and applications. *The Annals of Applied Probability* 33, no. 1 (2023): 780–823.

[232] Nocedal, J., and S. Wright. *Numerical Optimization*. 2nd ed. New York: Springer, 2006.

[233] Okamura, A., T. Kirimot, and M. Kondo. A new normalized stochastic approximation algorithm using a time-shift parameter. *Electronics and Communications in Japan, Part 3*, 78 (1995): 41–51.

[234] Ott, T., J. Kemperman, and M. Mathis. The stationary behavior of ideal TCP congestion avoidance. Unpublished manuscript, available at www.teunisott.com, 1996.

[235] Padberg, M. *Linear Programming and Extensions*. 2nd ed. New York: Springer, 1999.

[236] Pasupathy, R., and S. Kim. The stochastic root-finding problem: Overview, solutions, and open questions. *ACM Transactions on Modeling and Computer Simulation (TOMACS)* 21, no. 3 (2011): 1–23.

[237] Patan, K., and T. Parisini. Stochastic learning methods for dynamic neural networks: Simulated and real-data comparisons. In *Proceedings of the 2002 American Control Conference (IEEE Cat. No. CH37301)*, vol. 4, 2577–2582. Piscataway, NJ: Institute of Electrical and Electronics Engineers, 2002.

[238] Peng, Y., M. Fu, B. Heidergott, and H. Lam. Maximum likelihood estimation by Monte Carlo simulation: Toward data-driven stochastic modeling. *Operations Research* 68, no. 6 (2020): 1896–1912.

[239] Peng, Y., M. Fu, J.-Q. Hu, and B. Heidergott. A new unbiased stochastic derivative estimator for discontinuous sample performances with structural parameters. *Operations Research* 66, no. 2 (2018): 487–499.

[240] Perko, L. *Differential Equations and Dynamical Systems.* 3rd ed. New York: Springer, 2000.

[241] Pflug, G. Derivatives of probability measures—Concepts and applications to the optimization of stochastic systems. In *Lecture Notes in Control and Information Science*, no. 103, 252–274. Berlin: Springer, 1988.

[242] Pflug, G. Sampling derivatives of probabilities. *Computing* 42 (1989): 315–328.

[243] Pflug, G. On-line optimization of simulated Markov processes. *Mathematics of Operations Research* 15, no. 3 (1990): 381–395.

[244] Pflug, G. Gradient estimates for the performance of Markov chains and discrete event processes. *Annals of Operations Research* 39, no. 1–4 (1992): 173–194.

[245] Pflug, G. *Optimization of Stochastic Models: The Interface Between Simulation and Optimization.* Boston: Kluwer, 1996.

[246] Pickands, J., and S. Robart. Estimation for an M/G/∞ queue with incomplete information. *Biometrika* 84, no. 2 (1997): 295–308.

[247] Plambeck, E., B.-R. Fu, S. Robinson, and R. Suri. Sample-path optimization of convex stochastic performance functions. *Mathematical Programming* 75, no. 2 (1996): 137–176.

[248] Polyak, B. *Introduction to Optimization.* New York: Optimization Software Inc., 1987.

[249] Polyak, B., and A. Juditsky. Acceleration of stochastic approximation by averaging. *SIAM Journal on Control and Optimization* 30, no. 4 (1992): 838–855.

[250] Polyak, B., and Y. Tsypkin. Pseudogradient adaptation and training algorithms. *Automation and Remote Control* 12 (1973): 83–94.

[251] Prashanth, L., H. Prasad, N. Desai, S. Bhatnagar, and G. Dasgupta. Simultaneous perturbation methods for adaptive labor staffing in service systems. *Simulation* 91, no. 5 (2015): 432–455.

[252] Puterman, M. *Markov Decision Processes: Discrete Stochastic Dynamic Programming.* John Wiley & Sons, 2014.

[253] Ramadge, P., and W. Wonham. The control of discrete event systems. *Proceedings of the IEEE* 77, no. 1 (1990): 81–98.

[254] Ramamurthy, R., C. Bauckhage, R. Sifa, and S. Wrobel. Policy learning using SPSA. In *International Conference on Artificial Neural Networks*, 3–12. Cham, Switzerland: Springer, 2018.

[255] Reiman, M., and A. Weiss. Sensitivity analysis via likelihood ratios. In *Proceedings of the 18th Winter Simulation Conference (WSC)*, 285–289. Washington, DC: Association for Computing Machinery, 1986.

[256] Reiman, M., and A. Weiss. Sensitivity analysis for simulations via likelihood ratios. *Operations Research* 37, no. 5 (1989): 830–844.

[257] Revesz, P. *The Laws of Large Numbers*. New York: Academic Press, 2014.

[258] Rinne, H. *The Weibull Distribution: A Handbook*. Boca Raton, FL: CRC Press, 2008.

[259] Robbins, H., and S. Monro. A stochastic optimization method. *Annals of Mathematical Statistics* 22, no. 3 (1951): 400–407.

[260] Robinson, S. Analysis of sample-path optimization. *Mathematics of Operations Research* 21, no. 3 (1996): 513–528.

[261] Rockefellar, R. *Convex Analysis*. Princeton, NJ: Princeton University Press, 1972.

[262] Ross, J., T. Taimre, and P. Pollett. Estimation for queues from queue length data. *Queueing Systems* 55, no. 2 (2007): 131–138.

[263] Ross, S. *Introduction to Probability Models*. Oxford: Academic Press, 2014.

[264] Ross, S. M. *Simulation*. Amsterdam: Elsevier Science, 2022.

[265] Royden, H. *Real Analysis*. 2nd ed. New York: Macmillan, 1968.

[266] Rubinstein, R. Sensitivity analysis and performance extrapolation for computer simulation models. *Operations Research* 37, no. 1 (1989): 72–81.

[267] Rubinstein, R., and D. Kroese. *Simulation and the Monte Carlo Method*. New York: John Wiley & Sons, 2016.

[268] Rubinstein, R., and B. Melamed. *Moderns Simulation and Modelling*. New York: Wiley, 1998.

[269] Rubinstein, R., and A. Shapiro. Optimization of static simulation models by the score function method. *Mathematics and Computers in Simulation* 32, no. 4 (1990): 373–392.

[270] Rubinstein, R., and A. Shapiro. *Discrete Event Systems: Sensitivity Analysis and Optimization by the Score Function Method*. Chichester, UK: Wiley, 1993.

[271] Ruder, S. An overview of gradient descent optimization algorithms. arXiv preprint, arXiv:1609.04747, 2016.

[272] Rudin, W. *Principles of Mathematical Analysis*. 3rd ed. New York: McGraw-Hill, 1976.

[273] Ruppert, D. A Newton-Raphson version of the multivariate Robbins-Monro procedure. *The Annals of Statistics* 13, no. 1 (1985): 236–245.

[274] Ruppert, D. *Efficient Estimators from a Slowly Convergent Robbins-Monro Procedure*. Technical report 781. Ithaca, NY: Cornell University, 1988.

[275] Ruppert, D. *Stochastic Approximation*. New York: Marcel Dekker, 1991.

[276] Sadegh, P. Constrained optimization via stochastic approximation with a simultaneous perturbation gradient approximation. *Automatica* 33, no. 5 (1997): 889–892.

[277] Saridis, G. Learning applied to successive approximation algorithms. *IEEE Transactions on Systems Science and Cybernetics* 6, no. 2 (1970): 97–103.

[278] Schassberger, R., Insensitivity of steady-state distributions of generalized semi-Markov processes by speeds. *Advances in Applied Probability* 10, no. 4 (1978): 836–851.

[279] Schassberger, R. Insensitivity of steady-state distributions of generalized semi-Markov processes. part ii. *The Annals of Probability* 6, no. 6 (1978): 85–93.

[280] Schmidt, M., N. Le Roux, and F. Bach. Minimizing finite sums with the stochastic average gradient. *Mathematical Programming* 162, no. 1–2 (2017): 83–112.

[281] Shalev-Shwartz, S., and T. Zhang. Accelerated mini-batch stochastic dual coordinate ascent. *Advances in Neural Information Processing Systems* (2013): 378–385.

[282] Shao, J. *Mathematical Statistics: Exercises and Solutions*. New York: Springer Science & Business Media, 2006.

[283] Shapiro, A. Asymptotic analysis of stochastic programs. *Annals of Operations Resarch* 30, no. 1–4 (1991): 169–186.

[284] Shapiro, A. Simulation-based optimization—Convergence analysis and statistical interference. *Communications in Statistics -Stochastic Models* 12 (1996): 425–454.

[285] Sharia, T. Truncated stochastic approximation with moving bounds: Convergence. *Statistical Inference for Stochastic Processes* 17 (2014): 163–179.

[286] Sharia, T., and L. Zhong. Asymptotic behavior of truncated stochastic approximation procedures. *Mathematical Methods of Statistics* 26 (2017): 37–54.

[287] Shi, J., and J. Spall. SQP-based projection SPSA algorithm for stochastic optimization with inequality constraints. In *2021 American Control Conference (ACC)*, 1244–1249. Piscataway, NJ: Institute of Electrical and Electronics Engineers, 2021.

[288] Sigman, K. *Lecture Notes on Stochastic Modelling I*. New York: Columbia University, 2009.

[289] Snijders, T. Models for longitudinal network data. In *Models and Methods in Social Network Analysis*, edited by Peter J. Carrington, John Scott, and Stanley Wasserman, 215–247. Cambridge: Cambridge University Press, 2005.

[290] Snijders, T., and M. Van Duijn. Conditional maximum likelihood estimation under various specifications of exponential random graph models. In *Contributions to Social Network Analysis, Information Theory, and other Topics in Statistics*, edited by J. Hagberg, 117–134. Stockholm: University of Stockholm 2002.

[291] Song, Q., J. Spall, Y. Soh, and J. Ni. Robust neural network tracking controller using simultaneous perturbation stochastic approximation. *IEEE Transactions on Neural Networks* 19, no. 5 (2008): 817–835.

[292] Spall, J. A stochastic approximation algorithm for large-dimensional systems in the Kiefer-Wolfowitz setting. In *Proceedings of the 27th IEEE Conference on Decision and Control*, 1544–1548. Piscataway, NJ: Institute of Electrical and Electronics Engineers, 1988.

[293] Spall, J. Multivariate stochastic approximation using a simultaneous perturbation gradient approximation. *IEEE Transactions on Automatic Control* 37, no. 3 (1992): 332–341.

[294] Spall, J. Adaptive stochastic approximation by the simultaneous perturbation method. *IEEE Transactions on Automatic Control* 45, no. 10 (2000): 1839–1853.

[295] Spall, J. *Introduction to Stochastic Search and Optimization*. New York: Wiley, 2003.

[296] Stein, C. Estimation of the mean of a multivariate normal distribution. *Annals of Statistics* 9, no. 6 (1981):1135–1151.

[297] Styblinski, M., and T. Tang. Experiments in nonconvex optimization: Stochastic approximation with function smoothing and simulated annealing. *Neural Networks* 3, no. 4 (1990): 467–483.

[298] Suri, R., and M. Zazanis. Perturbation analysis gives strongly consistent sensitivity estimates for the M/G/1 queue. *Management Science* 34, no. 1 (1988): 39–64.

[299] Sutton, R., and A. Barto. *Reinforcement Learning: An Introduction*. Cambridge, MA: MIT Press, 2011.

[300] Syski, R. Ergodic potential. *Stochastic Processes and Their Applications* 7, no. 3 (1978): 311–336.

[301] Tsagris, M., C. Beneki, and H. Hassani. On the folded normal distribution. *Mathematics* 2, no. 1 (2014): 12–28.

[302] Tseng, P. An incremental gradient (-projection) method with momentum term and adaptive stepsize rule. *SIAM Journal on Optimization* 8, no. 2 (1998): 506–531.

[303] Tsitsiklis, J., D. Bertsekas, and M. Athans. Distributed asynchronous deterministic and stochastic gradient optimization algorithms. *IEEE Transactions of Automatic Control* 31, no. 9 (1986): 803–812.

[304] Uzawa, H. Iterative methods for concave programming. In *Studies in Linear and Nonlinear Programming*, edited by K. Arrow, L. Hurwicz, and H. Uzawa, 154–165. Redwood City, CA: Stanford University Press, 1958.

[305] Vázquez-Abad, F. Stochastic recursive algorithms for optimal routing in queueing networks. PhD diss., Brown University, 1989.

[306] Vázquez-Abad, F. Strong points of weak convergence: A study using RPA gradient estimation for automatic learning. *Automatica* 35, no. 7 (1999): 1255–1274.

[307] Vázquez-Abad, F. A course on sensitivity analysis for gradient estimation of DES performance measures, part I of III. In *Discrete Event Systems, Analysis and Control*, edited by R. Boel and G. Stremersch, 3–28. Boston: Kluwer, 2000.

[308] Vázquez-Abad, F. RPA pathwise derivative estimation of ruin probabilities. *Insurance: Mathematics and Economics* 26, no. 2–3 (2000): 269–288.

[309] Vázquez-Abad, F., S. Bernabel, and M. Fu. Data-driven adaptive threshold control for bike share systems. In *WSC '17: Proceedings of the 2017 Winter Simulation Conference*, 1–12. Piscataway, NJ: Institute of Electrical and Electronics Engineers, 2017.

[310] Vázquez-Abad, F., C. Cassandras, and V. Julka. Centralized and decentralized asynchronous optimization of stochastic discrete-event systems. *IEEE Transactions on Automatic Control* 43, no. 5 (1998): 631–655.

[311] Vázquez-Abad, F., and M. Cepeda-Jüneman. The phantom SPA method: An inventory problem revisited. In *Proceedings of the 31st Conference on Winter Simulation: Simulation—A bridge to the future*, vol. 2, 1664–1669. New York: Association for Computing Machinery, 1999.

[312] Vázquez-Abad, F., and H. Jacobson. Application of RPA and the harmonic gradient estimators to a priority queueing system. In *Proceedings of Winter Simulation Conference*, 369–376. Washington, DC: IEEE Computer Society, 1994.

[313] Vázquez-Abad, F., and H. Kushner. Estimation of the derivative of a stationary measure with respect to a control parameter. *Journal of Applied Probability* 29, no. 2 (1992): 343–352.

[314] Vázquez-Abad, F. and P. L'Ecuyer. Simulation trees for functional estimation via the phantom method. In *11th International Conference on Analysis and Optimization of Systems Discrete Event Systems*, 449–455. Cham, Switzerland: Springer, 1994.

[315] Vázquez-Abad, F., P. L'Ecuyer, and M. Benoit. On the linear growth of the split-and-merge simulation tree for a multicomponent model. In *Proceedings 1996 WODES*, 57–62. London: Fraunhofer-Institut für Energiewirtschaft und Energiesystemtechnik, 1996.

[316] Vázquez-Abad, F., and P. LeQuoc. Sensitivity analysis for ruin probabilities. *Journal of the Operational Research Society* 52, no. 1 (2001): 71–81.

[317] Venter ,J. An extension of the Robbins-Monro procedure. *Annals of Mathematical Statistics* 38, no. 1 (1967): 181–190.

[318] Wang, H., and A. Banerjee. Randomized block coordinate descent for online and stochastic optimization. arXiv preprint, arXiv:1407.0107, 2014.

[319] Wang, T., J. Ke, K. Wang, and S. Ho. Maximum likelihood estimates and confidence intervals of an M/M/R queue with heterogeneous servers. *Mathematical Methods of Operations Research* 63 (2006): 371.

[320] Ward, M., and S. Ahlquist. *Maximum Likelihood for Social Science: Strategies for Analysis*. Cambridge: Cambridge University Press, 2018.

[321] Wardi, Y., W. Gong, C. Cassandras, and M. Kallmes. Smoothed perturbation analysis for a class of piecewise constant sample performance functions. *Discrete Event Dynamic Systems* 1, no. 4 (1992): 393–414.

[322] Wardi, Y., C. Seatzu, X. Chen, and S. Yalamanchili. Performance regulation of event-driven dynamical systems using infinitesimal perturbation analysis. *Non-Linear Analysis: Hybrid Systems* 22 (2016): 116–136.

[323] Wasan, M. *Stochastic Approximation*. Cambridge: Cambridge University Press, 1969.

[324] Whitney, J., L. Solomon, and S. Hill. Constrained optimization over discrete sets via SPSA with application to non-separable resource allocation. In *Proceedings of the 2001 Winter Simulation Conference*, vol. 1, 313–317. Piscataway, NJ: Institute of Electrical and Electronics Engineers, 2001.

[325] Whitt, W. Continuity of generalized semi-Markov processes. *Mathematics of Operations Research* 5, no. 4 (1980): 494–501.

[326] Williams, R. Simple statistical gradient-following algorithms for connectionist reinforcement learning. *Machine Learning* 8 (1992): 229–256.

[327] Wolf, P. Convergence conditions for ascent methods. *SIAM Review* 11, no. 2 (1969): 226–235.

[328] Wolf, P. Convergence conditions for ascent methods II. *SIAM Review* 13, no. 2 (1971): 185–188.

[329] Wonham, W., and J. Ramadge. Modular supervisory control of discrete-event systems. *Mathematics of Control, Signals and Systems* 1 (1988): 13–30.

[330] Xia, L., and X.-R. Cao. Relationship between perturbation realization factors with queueing models and Markov models. *IEEE Transactions on Automatic Control* 51, no. 10 (2006): 1699–1704.

[331] Xia, L., Q. Xia, and X.-R. Cao. A tutorial on event-based optimization—A new optimization framework. *Discrete Event Dynamic Systems* 24, no. 2 (2014): 103–132.

[332] Xiao, L., and T. Zhang. A proximal stochastic gradient method with progressive variance reduction. *SIAM Journal on Optimization* 24, no. 4 (2014): 2057–2075.

[333] Xu, Z., Y., Li, and X. Zhao. Simulation-based optimization by new stochastic approximation algorithm. *Asia-Pacific Journal of Operational Research* 31, no. 4 (2014): 1450026.

[334] Yakowitz, S., P. L'Ecuyer, and F. Vázquez-Abad. Global stochastic optimization with low-dispersion point sets. *Operations Research* 48, no. 6 (2000): 939–950.

[335] Zazanis, M., and R. Suri. Perturbation analysis of the GI/GI/1 queue. *Queueing Systems* 18 (1994): 199–248.

[336] Zhang, D., and A. Nagurney. On the stability of projected dynamical systems. *Journal of Optimization Theory and Applications* 85, no. 1 (1995): 97–124.

[337] Zhang, J., T. He, S. Sra, and A. Jadbabaie. Why gradient clipping accelerates training: A theoretical justification for adaptivity. arXiv preprint, arXiv:1905.11881, 2019.

[338] Zhao, T., M. Yu, Y. Wang, R. Arora, and H. Liu. Accelerated mini-batch randomized block coordinate descent method. In *Advances in Neural Information Processing Systems* 27 (2014): 3329–3337, 2014.

[339] Zhong, M., and C. Cassandras. Asynchronous distributed optimization with event-driven communication. *IEEE Transactions on Automatic Control* 55, no. 12 (2010): 2735–2750.

[340] Zhou, N., X. Yu, B. Andersson, and C. Cassandras. Optimal event-driven multi-agent persistent monitoring of a finite set of targets. In *Proceedings of 55th IEEE Conference on Decision and Control*, 1814–1819, 2016.

Index

absolutely continuous, 367
absolute measure, 367
accelerated gradient, 346
algorithm
    Adam, 345
    approximate Hessian, 94
    Arrow-Hurwicz, 33, 66
    block-coordinate descent, 347
    Cauchy, 11
    descent, 10
    Frank-Wolfe, 30
    hyperparameter, 34
    Kesten's rule, 94
    momentum method, 346
    Nesterov's accelerated gradient method, 346
    Newton-Raphson method, 10
    penalty, 68
    semi-stochastic gradient, 346
    steepest descent method, 11
    Uzawa, 32
aperiodic chain, 378
Armijo's rules, 11
artillery, French, 83
art of modelling, 34
Arzela-Ascoli theorem, 55, 362
asymptotically stable point, 44
asymptotic convergence rate, 155
asymptotic efficiency, 155
atom, 379

ball constraints, 27
Banach space, 241, 374
batching, 157
    consecutive, 85, 107, 125
    independent, 86, 107
    parallel, 86
    streaming, 85, 125
bifurcation, 44
    diagram, 44

block-coordinate descent, 347
blocked after service, 308
Borel field, 365
box constraints, 27
Brownian motion, 144
budget allocation, 157

cadlag, 373
Cauchy's method, 11
Cauchy term, 363
center of mass, 51
Cesàro sum, 364
change of measure, 260
clipping, 64, 106, 107
C^n, xii, 4
coercivity
    for an NLP, 62
    NLP, 70
    vector field, 62
commuting condition (CC), 307
complementary slackness, 19
computational budget, 154
concave function, 5, 6
constrained optimization, 18
constraint
    active, 19
    hard, 35, 194
    inactive, 19
    qualification, 19
    soft, 35
continuous mapping theorem, 371
contraction mapping, 108
control variates, 178
convergence
    almost surely, 370
    in distribution, 370
    v-norm, 374
    in probability, 370
    total variation, 371
    weak, 370

convex, 5
  function, 6
  non-linear problem (NLP), 19
  problem, 19
  statistical test, 36
coordinate descent, 28, 347
correlated noise model, 122
coupling, 379
cumulative distribution function (cdf), xii, 368
cycles
  of a regenerative process, 377

density
  probability, 368
  Riemann densities, 213–215
derivative-free algorithm, 301
descent direction, 10
deviation matrix, 334
directional monotone, 80
discrete event dynamic system (DEDS), 209
discrete event system (DES), 209
distribution
  Bernoulli, 173
  Beta, 226
  exponential, 168
  Gamma, 181
  Lomax, 171
  Maxwell, 181
  normal, 170, 174, 182, 216
  Pareto type I, 171, 182, 240–241, 328
  Pareto type II, 171, 195, 199, 200
  Poisson, 173
  Weibull, 170, 173
distribution function, 368
dominated convergence theorem, 371
dynamical system, 42

endogenuous noise model, 122
equicontinuity, 55
equilibrium point, 43
ergodic projector, 381
Euclidean norm, 3
Euler method, 40
Euler scheme, 48
exogenous noise model, 100

feasible
  active constraint, 19
  point, 19
  region, 19
fictitious game, 98
finite difference (FD), 17
  convergence, 17
  stochastic, 105
finite horizon problem, 186
first-order optimality condition, 6, 19
Fisher information matrix, 340
fixed-point mapping, 108
Fubini's theorem, 367
functional central limit theorem, 144

gain sequence, 9
  Armijo's rules, 11
  generic decreasing, 94
  non-standard examples, 94
  optimal, 94, 152
  variations, 94
  Wolfe's conditions, 11
gain size, 9
Gamma function, 226
Gaussian smoothed functional
  approximation (GSFA), 300
generalized methods of moments (GMoM), 342
  iterated GMoM, 343
generalized Semi-Markov process (GSMP), 305
global
  maximum, 5
  minimum, 5
globally asymptotically stable point, 44
global optimization, 96, 299
golden rule of optimization, 34
gradient, xii, 4
  descent method, 11
  vanishing, 9

Hadarmard Matrix, 298
Hahn-Jordan decomposition, 180, 240, 242, 367
hard constraint, 35
Harris
  ergodic, 379
  recurrent, 378
hazard rate function, 306
Hessian, xii, 4
  approximate, 94, 298
Hurwitz

condition, 52
matrix, 45
hyperball, 27
hypercube, 27
hyperparameter, 34

importance sampling, 177, 259
induced measure, 367
infinite horizon problem, 188
infinitesimal perturbation analysis (IPA), 167, 303
interpolation process, 53
shifted, 53
inward normals, 48, 49, 50, 51

Karush-Kuhn-Tucker (KKT), 19
point, 19
second-order conditions, 21
Kesten's rule, 94
Kiefer-Wolfowitz procedure, 105

Lagrange
Duality Method, 31
multipliers, 19
term, 363
Landau symbol, xii
lattice distribution, 377
learning
algorithm, 80
Q-learning, 347–348
reinforced, 82
learning rate, 9
least mean square (LMS) algorithm, 82
Lebesgue measure, 366
level set, 4
L'Hôpital's rule, 290, 363
limit process, 41
Lindley recursion, 86, 136, 187
linear regression, 81
Lipschitz
almost surely continuous, 198
constant, 13, 361
continuity, 13, 361
continuity of Markov kernel, 316
continuity of transition kernel, 316
modulus, 198
local
maximum, 5
minimum, 5
location of maximum, 5

location of minimum, 5
logisitic regression, 344
Lyapunov function, 52

MacLaurin series, 363
Markov chain, 378
aperiodicity, 379, 381
atom, 379
d-cycle, 378
ergodic, 381
ergodic projector, 381
Harris ergodic, 378
irreducible, 381
kernel, 316
period, 378
ϕ-irreducible, 378
Poisson equation, 381
positive recurrent, 381
transition kernel, 378
unichain, 382
uniformly ϕ-recurrent, 378
martingale difference noise model, 100
matrix
Hurwitz, 45
positive (negative) definite, 4
semi-definite, 4
maximum
global, 5
local, 5
location, 5
proper, 5
strict, 5
value, 5
maximum likelihood estimation (MLE), 339
mean value theorem, 362
measurable
mapping, 365
space, 365
measure, 366
absolute, 367
absolutely continuous, 367
finite, 366
induced, 367
integral, 225
measurable space, 366
μ-integral, 366
non-negative, 366
probability, 367
Radon-Nikodym, 367
Radon-Nykodym derivative, 225

measure (cont.)
   regular, 368
   σ-finite, 366
   signed, 240, 366
measurable space, 366
measure-valued differentiation (MVD), 179
   randomized, 232
methods of moments (MoM), 342
metric, 360
   complete, 361
   complete metric space, 361
   pseudo, 360
   space, 360
mini-batching, 157
minimum
   global, 5
   local, 5
   location, 5
   proper, 5
   strict, 5
   value, 5
momentum method, 346
μ-integral, 366
multiplier method, 31

Nesterov's accelerated gradient, 346
neural network, 86
Newton-Raphson method, 10
Newton's method, 10
NLP, 3, 18
   convex, 18
noise
   correlated, 104
   independent, 104
   martingale difference noise model, 100
   unpredictable, 104
non-interruption condition, 304
non-lattice distribution, 377
non-linear problem (NLP), 3, 18, 19
   convex, 18
norm, 360
   v-norm, 373
   total variation, 370
normal vector, 357
v-norm, 214, 242, 373
   convergence, 374

observed feedback, 77
optimality condition
   first-order, 6
   second-order, 6
ordinary differential equations (ODEs)
   asymptotically stable point, 44
   autonomous, 42
   bifurcation, 44
   blows up in finite time, 43
   bounded trajectories, 51
   domain of attraction, 44
   equilibrium point, 43
   globally asymptotically stable point, 44
   initial condition, 42
   projected, 50
   projection, 48
   stable point, 44
   stationary point, 43
   surrogate, 74
   target, 41
   trajectory, 42
   unstable point, 44
Ornstein-Uhlenbeck process, 147

parameter
   location, 200–201
   scale, 200–201
penalty method, 23
piecewise differentiable mapping, 201
Poisson equation, 382
Polish space, 361
Polyak-Rupert averaging, 95, 282
probability
   measure, 367
   space, 367
probability density function (pdf), xii, 368
problem
   convex, 18
   linear, 3
   NLP, 3, 18
   non-linear, 3
projected gradient, 38
projection, 26, 158
   approximate, 29
   directional derivative, 49
   Dykstra's method, 29
   Goldstein-Levitin-Polyak, 26
   increasing sequence, 95
   issues with bias, 28
   method, 26
   ODE, 50
proper maximum, 5

proper minimum, 5
pseudo metric, 360

Q-learning, 347–348
quantiles, 113, 127, 200
quantile sensitivity, 200
quantile updating, 113
queuing
  queue length
    network, DES, 308
    single server, 327, 331
    single server, IPA stationary, 270
    single server, SF random horizon, 262
    stationary, 89
    tandem line, DES, 307
    transient, 86
  waiting times
    finite horizon, 86, 187
    infinite horizon, 188
    IPA, finite horizon, 206
    MVD, 318, 325
    MVD, finite horizon, 233
    random horizon, 190
    SF, finite horizon, 223
    stationary, 89, 136, 193

Radon-Nikodym derivative, 225, 367
random horizon problem, 190
randomization, 230
randomized MVD, 232
random variable, 365
regenerative, 377
  classical, 377
  process, 377
  set, 379
regular measure, 368
reinforcement learning, 82, 347, 350
renewal times, 377
Riemann densities, 213–215
risk function, 154
Robbins-Monro theorem, 82
root finding, 79

(s, S) policy, 277
saddle point, 5
saddle point theorem, 32
sample average approach (SAA), 92, 177
score function (SF), 172
  control variates, 178
  product technique for, 223
second-order condition, 6

selection rule, 304
set
  closed, 360
  open, 360
shifted interpolation process, 53
σ-field, 365
simultaneous perturbation stochastic
    approximation (SPSA), 295
  Newton-Raphson, 298
single server queue, 193, 206, 305, 307
Skorohod representation, 372
smoothed perturbation analysis (SPA), 210, 287
Snell's law, 6
soft constraint, 35
space
  Banach, 374
  measurable, 365, 366
  metric, 360
  Polish, 361
  probability, 367
  topological, 360
stable point, 44
static problem, 84
stationary point, 6
  ODE, 43
stationary problem, 121, 192
statistical fitting, 79
statistical learning, 86
steepest descent method, 11
Stein's equation, 300
stepsize, 9
stepsize sequence, 9
stochastic approximation (SA), 76
  accelerated, 95
  global optimization, 96, 299
  robust, 95
stochastic counterpart, 91, 177
strictly convex non-linear problem, 19
strong coupling, 379
supervised learning, 4, 79
surrogate ODE, 74

taboo probability, 378
tangent hyperplane, 358
target ODE, 41
target tracking, 79
Taylor series, 363
  Cauchy term, 363
  Lagrange term, 363

Taylor series (cont.)
  MacLaurin series, 363
  Taylor polynomial, 363
tightness
  of measures, 372
  of random variables, 372
topological space, 360
total variation norm, 370
transition kernel, 316
truncation, 27
  avoiding projection in SA, 66
  principle, 65, 106
two-armed bandit, 96
two-timescale, 25, 31

underlying process, 77, 120
uniformly
  bounded, 362
  continuous families of mappings, 362
  continuous mappings, 362
  ϕ-recurrent Markov chain, 378
unstable point, 44

value
  of maximum, 5
  of minimum, 5
value function, 382
vanishing
  gradient, 9
  update, 34, 70
variance control scheme, 107
vector field, 42
  coercive, 62
  target, 89, 120

Wald's equality, 369
weak continuity, 238
weak convergence, 238
Weierstrass theorem, 359
weighted supremum norm, 214, 242
well-posed optimization ODE, 62
well-posed optimization problem, 61
well-posed problem, 33, 70
Wiener process, 144
Wolfe's conditions, 11